AF691665

CBSE Maths Objective Champion

"From PYQs to Practice:
75+ Series, 180 Sets,
Key Points & 270+ Exercises
with 1100+ MCQs"

CLASS 10
(STANDRAD & BASIC)

notionpress.com

INDIA • SINGAPORE • MALAYSIA

Copyright © Goutam De 2026
All Rights Reserved.

ISBN
Paperback 979-8-90431-697-6
Hardcase 979-8-90431-722-5

This book has been published with all efforts taken to make the material error-free after the consent of the author. However, the author and the publisher do not assume and hereby disclaim any liability to any party for any loss, damage, or disruption caused by errors or omissions, whether such errors or omissions result from negligence, accident, or any other cause.

While every effort has been made to avoid any mistake or omission, this publication is being sold on the condition and understanding that neither the author nor the publishers or printers would be liable in any manner to any person by reason of any mistake or omission in this publication or for any action taken or omitted to be taken or advice rendered or accepted on the basis of this work. For any defect in printing or binding the publishers will be liable only to replace the defective copy by another copy of this work then available.

First Edition: 2026

Designed by:
Scylence PVT LTD
Soro, Balasore, Odisha-756045

Contents

1 Real Numbers

Key Point A chain of factors which is represented in the form of a tree, is called factor tree.

1. Complete the missing entries in the following factor tree: **(2008)**

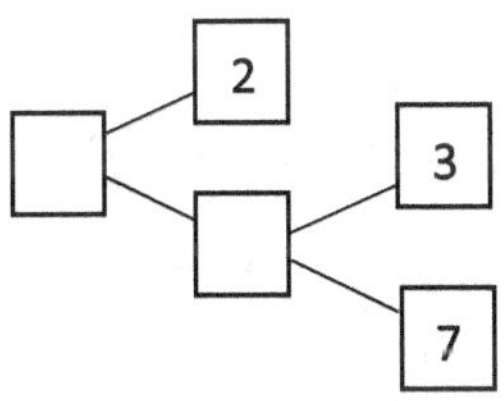

Solution:

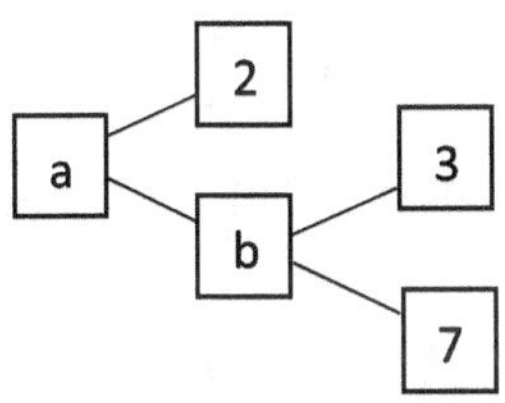

Let the missing entries be a and b

$\therefore b = 3 \times 7 = 21$ and $a = 2 \times b = 2 \times 21 = 42$

2. The value of x in the factor tree is **(2010, 2011)**

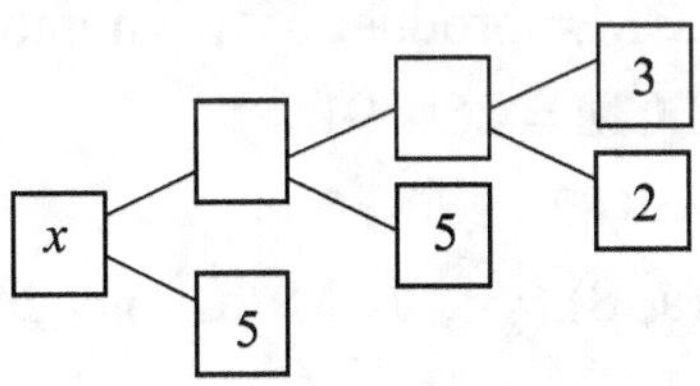

(a) 30 (b) 150
(c) 100 (d) 50

Solution: (b)

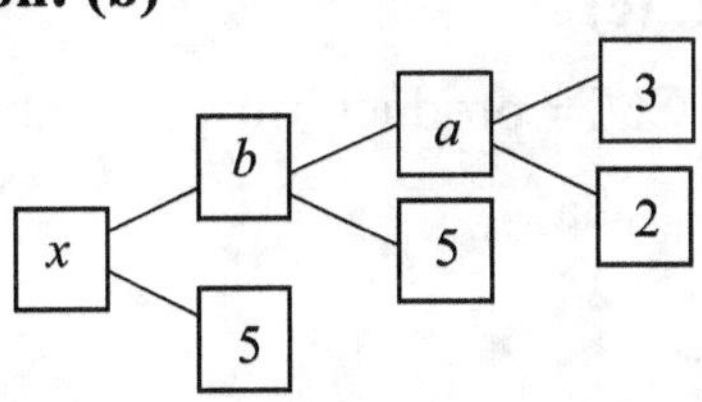

Let the missing entries are a and b

$\therefore a = 2 \times 3 = 6$ and $b = 5 \times a = 5 \times 6 = 30$

$\therefore x = 5 \times b = 5 \times 30 = 150$

Exercise 1.1

(i) Complete the missing entries in the following factor tree:

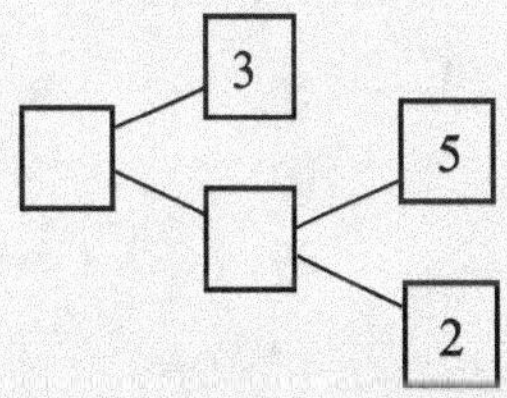

(ii) Complete the missing entries in the following factor tree:

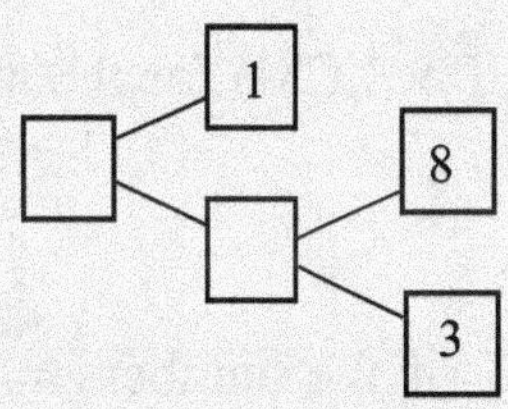

(iii) Complete the missing entries in the following factor tree :

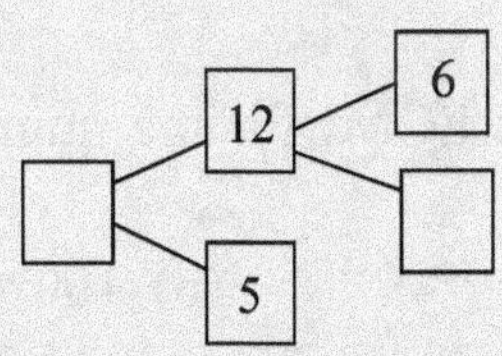

(iv) The value of x in the factor tree is

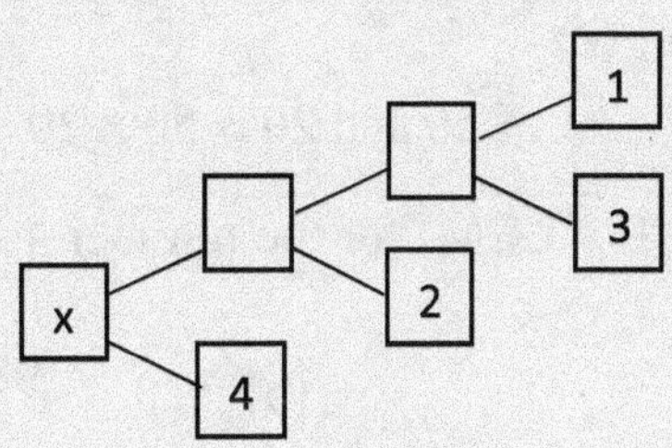

(a) 8 (b) 6
(c) 24 (d) 12

(v) The value of x in the factor tree is

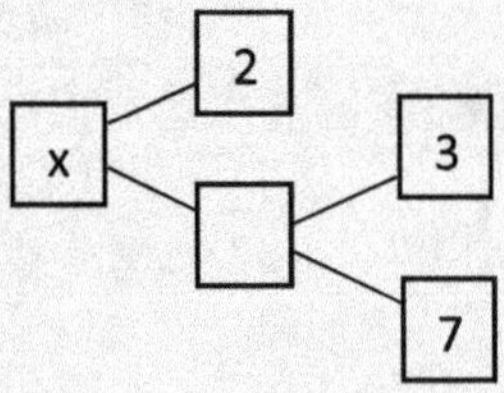

(a) 21 (b) 14

(c) 70 (d) 42

(vi) The value of x in the factor tree is

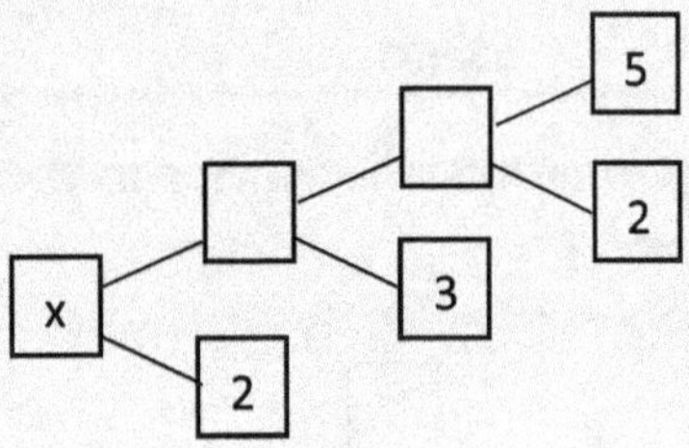

(a) 10 (b) 20

(c) 40 (d) 60

Key Point HCF × LCM of two numbers = Product of the numbers

3. Find the (HCF × LCM) for the number 100 and 190. **(2009)**

Solution:

HCF × LCM of the numbers = Product of the number

∴ HCF × LCM of 100 and 190 = 100 × 190 = 19000

4. The (HCF × LCM) for the number 50 and 20. **(2010)**

(a) 10 (b) 100

(c) 1000 (d) 50

Solution: (c)

HCF × LCM of 50 and 20 = 50 × 20 = 1000

5. The (HCF × LCM) for the number 40 and 30 is. **(2024)**

(a) 12 (b) 120

(c) 1200 (d) 40

Solution: (c)

HCF × LCM of 40 and 30 = 40 × 30 = 1200

Exercise 1.2

(i) Find the (HCF × LCM) for the number 80 and 90.

(ii) Find the (HCF × LCM) for the number 72 and 16.

(iii) Find the (HCF × LCM) for the number a and b.

(iv) The (HCF × LCM) for the number 30 and 18 is.

(a) 30 (b) 118

(c) 540 (d) 1900

(v) The (HCF × LCM) for the number 12 and 18 is.

(a) 216 (b) 180

(c) 120 (d) 300

(vi) The (HCF × LCM) for the number 70 and 80 is.

(a) 560 (b) 5600

(c) 7000 (d) 800

Key Point LCM = product of two numbers ÷ HCF

HCF = product of two numbers ÷ LCM

6. Given that HCF (26,91) = 13, then LCM (26, 91) is. **(2010)**

(a) 2366 (b) 182

(c) 91 (d) 364

Solution: (b)

HCF × LCM = product of two numbers

⇒ 13 × LCM = 26 × 91

∴ LCM = 182

7. If HCF (a, 8) = 4, LCM (a, 8) = 24, then a is. **(2011)**

(a) 8 (b) 10

(c) 12 (d) 14

Solution: (c)

HCF × LCM = product of two numbers

⇒ 4 × 24 = 8a

⇒ 8a = 4 × 24

∴ a = 12

8. If HCF (336,54) = 6, find LCM (336, 54) **(2018)**

Solution : (c)

HCF × LCM = product of two numbers

$\Rightarrow 6 \times \text{LCM} = 336 \times 54$

$\Rightarrow \text{LCM} = 3024$

9. If the HCF of 360 and 64 is 8, then their LCM is **(2023)**

(a) 2480 (b) 2780

(c) 512 (d) 2880

Solution: (d)

LCM × HCF = product of two numbers

$\Rightarrow \text{LCM} \times 8 = 360 \times 64$

$\therefore \text{LCM} = 2880$

10. Given that LCM (91, 26) = 182, then HCF (91, 26) is **(2010)**

(a) 13 (b) 26

(c) 7 (d) 9

Solution: (a)

LCM × HCF = product of two numbers

$\Rightarrow 182 \times \text{HCF} = 91 \times 26$

$\therefore \text{HCF} = 13$

11. If the HCF of two positive integers a and b is 1, then their LCM is: **(2025)**

(a) $a + b$ (b) a

(c) b (d) ab

Solution: (d)

HCF × LCM = product of the numbers

$\Rightarrow 1 \times \text{LCM} = \text{ab}$

$\therefore \text{LCM} = \text{ab}$

12. HCF(x, 20) = 2 and LCM (x, 20) = 60, then value of x is : **(2025)**

(a) 3 (b) 6

(c) 20 (d) 10

Solution: (b)

$\text{HCF}(x, 20) \times \text{LCM}(x, 20) = x \times 20$

$\Rightarrow 2 \times 60 = 20x$

$\Rightarrow x = 6$

Exercise 1.3

(i) Given that HCF (612, 1314) = 18, then LCM (612, 1314) is

(a) 44676 (b) 46766

(c) 40126 (d) 38126

(ii) Given that HCF (135, 225) = 45, find the LCM (135,225) **(2010)**

(iii) Given that HCF (161, 207) = 23, then LCM (161, 207) is

(a) 1229 (b) 13339

(c) 1669 (d) 1449

(iv) Given that HCF (275,308) = 11, then LCM (275, 308) is.

(a) 2750 (b) 7700

(c) 1100 (d) 15260

(v) Given that HCF (65, 117) − 13 then LCM (65, 117) is.

(a) 65 (b) 117

(c) 1585 (d) 605

(vi) Given that HCF (54, 81) = 27, then LCM (54, 81) is

(a) 162 (b) 182

(c) 81 (d) 405

(vii) Given that HCF (156, 78) = 78, then LCM (156, 78) is **(2020)**

(a) 156 (b) 78

(c) 156 × 78 (d) 156 × 12

(viii) If HCF (a, 10) = 5, LCM (a, 10) = 70, then a is

(a) 35 (b) 60

(c) 70 (d) 80

(ix) If HCF (x, 9) = 3, LCM (x, 9) = 45, then x is.

(a) 10 (b) 15

(c) 12 (d) 24

(x) If HCF (a, 12) = 3, LCM (a, 12) = 36, then a is

(a) 36 (b) 18

(c) 26 (d) 24

Key Point • HCF of two or more natural numbers = Product of the smallest power of each common prime factor in the numbers

• LCM of two or more natural numbers = Product of the highest power of all prime factor in the numbers.

13. HCF of 144 and 198 is **(2020)**

(a) 9 (b) 18

(c) 6 (d) 12

Solution: (b)

$144 = 2 \times 2 \times 2 \times 2 \times 3 \times 3$

$198 = 2 \times 3 \times 3 \times 11$

$\therefore$ HCF of 144 and 198 is $2 \times 3 \times 3 = 18$

14. The HCF of 40,110 and 360 is **(2025)**

Solution: (d)

$40 = 2 \times 2 \times 2 \times 5$

$110 = 2 \times 5 \times 11$

$360 = 2 \times 2 \times 2 \times 3 \times 3 \times 5$

$\therefore$ HCF of 40,110 and 360 is $2 \times 5 = 10$

15. HCF of $(3^4 \times 2^2 \times 7^3)$ and $(3^2 \times 5 \times 7)$ is **(2023)**

(a) 630 (b) 63

(c) 729 (d) 567

Solution: (b)

HCF of $(3^4 \times 2^2 \times 7^3)$ and $(3^2 \times 5 \times 7)$ is $3^2 \times 7 = 63$

16. The HCF and the LCM of 12,21,15 respectively **(2020)**

(a) 3,140 (b) 12,420

(c) 3,420 (d) 420,3

Solution: (c)

$12 = 2 \times 2 \times 3$

$15 = 3 \times 5$

$21 = 3 \times 7$

$\therefore$ HCF of 12,15 and 21 is 3

And LCM of 12,15 and 21 is $2 \times 2 \times 3 \times 5 \times 7 = 420$

17. LCM of $(2^3 \times 3 \times 5)$ and $(2^4 \times 5 \times 7)$ is **(2023)**

(a) 40 (b) 560

(c) 1120 (d) 1680

Solution: (d)

LCM of $(2^3 \times 3 \times 5)$ and $(2^4 \times 5 \times 7)$ is $2^4 \times 3 \times 5 \times 7 = 1680$.

18. The LCM of the three number 28, 44 and 132 is **(2024)**

(a) 258 (b) 231

(c) 462 (d) 924

Solution: (d)

$28 = 2 \times 2 \times 7$

$44 = 2 \times 2 \times 11$

$132 = 2 \times 2 \times 3 \times 11$

$\therefore$ LCM is 28,44 and 132 is $2 \times 2 \times 3 \times 7 \times 11 = 924$.

19. If two positive integers A and B can be expressed as $A = ab^2$ and $B = a^3 b$, where a, b are prime numbers, then LCM (A, B) is **(2011, 2019)**

(a) ab (b) a^2b^2

(c) a^3b^2 (d) a^4b^3

Solution: (c)

$A = a \times b \times b$

$B = a \times a \times a \times b$

$\therefore$ LCM (A, B) $= a \times a \times a \times b \times b = a^3b^2$

20. LCM of $2^3 \times 3^2$ and $2^2 \times 3^3$ is **(2012)**

(a) 2^3 (b) 3^3

(c) $2^3 \times 3^3$ (d) $2^2 \times 3^2$

Solution: (c)

LCM of $2^3 \times 3^2$ and $2^2 \times 3^3$ is $2^3 \times 3^3$

21. If two positive integers p and q can be expressed as $p = 18a^2b^4$ and $q = 20a^3b^2$, where a and b are prime numbers, then LCM (p, q) is **(2024)**

(a) $2a^2b^2$ (b) $180a^2b^2$

(c) $12a^2b^2$ (d) $180a^3b^4$

Solution: (d)

$p = 2 \times 3 \times 3 \times a \times a \times b \times b \times b \times b$

$q = 2 \times 2 \times 5 \times a \times a \times a \times b \times b$

$\therefore$ LCM of (p, q) $= 2 \times 2 \times 3 \times 3 \times 5 \times a \times a \times a \times b \times b \times b \times b = 180a^3b^4$

22. The value of (HCF – LCM) for the two numbers 3 and 5 is : **(2025)**

(a) 2 (b) 4
(c) 14 (d) −14

Solution: (d)

HCF of 3 and 5 is 1

LCM of 3 and 5 is 15

∴ their HCF − LCM = 1 − 15 = −14

23. If HCF (98, 28) = m and LCM (98, 28) = n, then the value of n-7m is **(2025)**

(a) 0 (b) 28
(c) 98 (d) 198

Solution: (c)

$m = 14, n = 196$

$\therefore n - 7m$

$= 196 - 7 \times 14$

$= 196 - 98$

$= 98$

24. If $x = ab^3$ and $y = a^3b$ where a and b are prime numbers, then [HCF (x, y)] is equal to: **(2025)**

(a) $1 - a^3b^3$ (b) $ab(1 - ab)$
(c) $ab - a^4b^4$ (d) $ab(1 - ab)(1 + ab)$

Solution: (d)

HCF (x, y) − LCM (x, y)

$= ab - a^3b^3$

$= ab(1 - a^2b^2)$

$= ab(1 + ab)(1 - ab)$

25. Let $x = a^2b^3c^n$ and $y = a^3b^mc^2$, where a, b, c are prime numbers. If LCM of x and y is $a^3b^4c^3$, then the value of $m + n$ is **(2025)**

(a) 10 (b) 7
(c) 6 (d) 5

Solution: (b)

LCM of $x = a^2b^3c^n$ and $y = a^3b^mc^2$ is $a^3b^4c^3$

$\therefore m = 4$ and $n = 3$

$\therefore m + n = 4 + 3 = 7$

26. Let $a = p^2q^3r^n$ and $b = p^3q^mr^2$, where p, q, r are prime numbers. If LCM of a and b is $p^3q^4r^3$, then the value of $3n - 2m$ is **(2025)**

(a) −1 (b) 1
(c) 3 (d) −3

Solution: (b)

LCM of $a = p^2q^3r^n$ and $b = p^3q^mr^2$ is $p^3q^4r^3$

$\therefore m = 4, n = 3$

$\therefore 3n - 2m$

$= 3 \times 3 - 2 \times 4$

$= 9 - 8 = 1$

27. Let $p = x^2y^3z^n$ and $q = x^3y^mz^2$, where x, y, z are prime numbers. If LCM $(p, q) = x^3y^4z^3$, then the value of $(2m + 3n)$ is **(2025)**

(a) 18 (b) 17
(c) 15 (d) 14

Solution: (b)

LCM $(p, q) = x^3y^4z^3$

$\therefore m = 4, n = 3$

$\therefore 2m + 3n$

$= 2 \times 4 + 3 \times 3$

$= 8 + 9 = 17$

28. The HCF and the LCM of 14, 21 and 77 respectively are **(2025)**

(a) 7, 77 (b) 14, 462
(c) 7, 462 (d) 21, 77

Solution: (c)

$14 = 2 \times 7$

$21 = 3 \times 7$

$77 = 11 \times 7$

∴ HCF (14, 21, 77) = 7

LCM (14, 21, 77) = $7 \times 2 \times 3 \times 11 = 462$

29. The LCM of 960 and 240 is : **(2026)**

(a) 960 (b) 240
(c) 60 (d) 15

Solution: (a)

$960 = 2 \times 2 \times 240$

$240 = 1 \times 240$

$\therefore$ The LCM of 960 and 240 is $2 \times 2 \times 240 = 960$

30. The HCF of 960 and 432 is : **(2026)**

(a) 48 (b) 54

(c) 72 (d) 36

Solution: (a)

$960 = 2 \times 2 \times 2 \times 2 \times 2 \times 2 \times 3 \times 5 = 2^6 \times 3 \times 5$

$432 = 2 \times 2 \times 2 \times 2 \times 3 \times 3 \times 3 = 2^4 \times 3^3$

$\therefore$ The HCF of 960 and 432 is $2^4 \times 3 = 48$

31. The HCF of 4^3 and 6^2 is: **(2026)**

(a) 2^2 (b) 2^3

(c) $2^2.3^2$ (d) 1

Solution: (a)

$4^3 = (2^2)^3 = 2^2 \times 2^2 \times 2^2$

$6^2 = (2 \times 3)^2 = 2^2 \times 3^2$

$\therefore$ The HCF of 4^3 and 6^2 is 2^2.

32. The HCF of 3^5 and 5^3 is: **(2026)**

(a) $3^0.5^0$ (b) $3^3.5^3$

(c) $3^1.5^1$ (d) $3^5.5^3$

Solution: (a)

The HCF of 3^5 and 8^3 is $1 = 3^0.5^0$

33. The HCF of 2^3 and 3^2 is: **(2026)**

(a) $2^0.3^0$ (b) $2^1.3^1$

(c) $2^3.3^2$ (d) $2^2.3^3$

Solution: (a)

The HCF of 2^3 and 3^2 is $1 = 2^0.3^0$

34. The HCF of $3^7.7^3$ and $3^3.7^7$ is : **(2026)**

(a) 1 (b) 3.7

(c) $3^3.7^3$ (d) $3^7.7^7$

Solution: (c)

The HCF of $3^7.7^3$ and $3^3.7^7$ is $3^3.7^3$

Exercise 1.4

(i) HCF of 112 and 196 is **(2020)**

(a) 7 (b) 14

(c) 28 (d) 56

(ii) HCF of 117 and 221 is

(a) 11 (b) 13

(c) 17 (d) 19

iii. HCF of 135 and 225 is **(2020)**

(a) 15 (b) 75

(c) 45 (d) 5

iv. HCF of 252 and 539 is

(a) 3 (b) 7

(c) 9 (d) 11

(v) HCF of $(2^3 \times 3^3 \times 7)$ and $(2^2 \times 5^2 \times 7^2)$ is

(a) 28 (b) 520

(c) 36 (d) 144

(vi) HCF of $(3^3 \times 5^2 \times 7)$ and $((3 \times 2^5 \times 7)$ is

(a) 3 (b) 7

(c) 21 (d) 60

(vii) The HCF and LCM of 18,24,36 respectively are

(a) 18, 120 (b) 6, 12

(c) 12, 120 (d) 3, 72

(viii) The HCF and LCM of 15, 18, 21 respectively are

(a) 3, 630 (b) 5, 210

(c) 3, 72 (d) 6, 120

(ix) The HCF and LCM of 18, 24, 42 respectively are

(a) 18, 540 (b) 2, 360

(c) 3, 210 (d) 6, 504

(x) LCM of $(2^2 \times 3^2 \times 5)$ and $(2^3 \times 5 \times 11)$ is

(a) 3960 (b) 3690

(c) 3060 (d) 4800

(xi) LCM of $(2^4 \times 3^2 \times 5^2)$ and $(2^3 \times 3^3 \times 5)$ is

(a) 12800 (b) 10800

(c) 16000 (d) 12300

(xii) If two positive integers P and Q can be expressed as $P = x^3y^2$ and $Q = x^2y^3$, where x, y are prime numbers, then LCM (P, Q) is

(a) xy (b) x^2y^2

(c) x^3y^3 (d) x^3y^4

(xiii) If two positive integers A and B can be expressed as $A = a^4b^2$ and $B = a^3b^4$, where a, b are prime numbers, then LCM (A, B) is

(a) a^4b^4 (b) a^3b^3

(c) a^4b^2 (d) a^3b^4

(xiv) If two positive integers X and Y can be expressed as $X = a^5b^3$ and $Y = a^4b^7$, where a, b are prime number, then LCM (X, Y) is

(a) a^5b^3 (b) a^5b^7

(c) a^4b^7 (d) a^5b^5

(xv) LCM of $3^3 \times 4^2$ and $4^3 \times 3^2$ is

(a) 3×4 (b) $3^3 \times 4^2$

(c) $3^3 \times 4^3$ (d) $3^2 \times 4^3$

(xvi) LCM of a^4b^2 and a^3b^4 is

(a) a^3b^3 (b) a^4b^2

(c) a^3b^4 (d) a^4b^4

(xvii) LCM of x^5y^6 and x^4y^5 is

(a) x^5y^6 (b) x^4y^5

(c) x^5y^5 (d) x^4y^4

(xviii) If two positive integers a and b can be expressed as $a = pq^2$ and $b = p^3q$, p, q being prime numbers, then LCM (a, b) is

(a) pq (b) p^3q^3

(c) p^3q^2 (d) p^2q^2

(xix) If two positive integers a and b can be expressed as $a = x^3y^2$ and $b = xy^3$, x, y are prime numbers, then LCM (a, b) is

(a) x^3y^2 (b) x^3y^3

(c) xy^3 (d) x^3y^2

Key Point If HCF$(a \times b) = c$, LCM $(a \times b) = d \times k$, then $k = (a \times b) \div (c \times d)$

35. Given that HCF $(2520 \times 6600) = 40$, LCM $(2520 \times 6600) = 252 \times K$, then the value of K is **(2010,2011)**

(a) 1650 (b) 1600

(c) 165 (d) 1625

Solution: (a)

HCF × LCM = Product of two numbers

$\Rightarrow 40 \times 252 \times K = 2520 \times 6600$

$\therefore K = 1650$

36. Given that HCF$(253 \times 440) = 11$ and LCM $(253,440) = 253 \times R$, The value of R is. **(2010)**

(a) 400 (b) 40

(c) 440 (d) 253

Solution: (b)

HCF × LCM = Product of two numbers

$\Rightarrow 11 \times 253 \times R = 253 \times 440$

$\therefore R = 40$

Exercise 1.5

(i) Given that HCF $(2230 \times 8000) = 10$, LCM $(2230 \times 8000) = 223 \times a$, then the value of a is

(a) 2230 (b) 8000

(c) 1040 (d) 4200

(ii) Given that HCF $(1600,8400) = 400$, LCM $(1600,8400) = 21 \times a$, then the value of a is

(a) 1600 (b) 8400

(c) 400 (d) 1620

(iii) Given that HCF $(1240,6080) = 40$, LCM $(1240,6080) = 124 \times x$, then the value of x is

(a) 400 (b) 820

(c) 1220 (d) 1520

37. If the HCF of 65 and 117 is expressible in the form 65 m − 117, then the value of m is **(2010, 2013)**

(a) 4 (b) 2

(c) 1 (d) 3

Solution: (b)

HCF of 65 and 117 is 13

$\Rightarrow 65\,m - 117 = 13$

$\Rightarrow 65\,m = 130$

$\therefore m = 2$

38. If the HCF of 85 and 153 is expressible in the form 85 n − 153, then the value of n is. **(2010,2011)**

(a) 3 (b) 2

(c) 4 (d) 1

Solution: (b)

HCF of 85 and 153 is 17

$\Rightarrow 85\,n - 153 = 17$

$\Rightarrow 85\,n = 170$

$\therefore n = 2$

39. HCF of 210 and 55 is expressible in the form of $(210 \times 5 + 55y)$ then the value of y is. **(2013)**

(a) 19 (b) −19
(c) 5 (d) −5

Solution: (b)

HCF of 210 and 55 is 5

$\therefore 210 \times 5 + 55y = 5$

$\Rightarrow 1050 + 55y = 5$

$\Rightarrow 55y = -1045$

$\Rightarrow y = -19$

Exercise 1.6

(i) If the HCF of 57 and 152 is expressible in the form 8a – 13, then the value of a is

(a) 2 (b) 4
(c) 1 (d) 8

(ii) If HCF of 85 and 153 is expressible in the form 14a – 123, then the value of a is

(a) 5 (b) 8
(c) 10 (d) 12

(iii) If the HCF of 42 and 147 is expressible in the form 9p – 132, then the value of p is

(a) 17 (b) 19
(c) 11 (d) 13

(iv) If the HCF of 35 and 55 is expressible in the form 2x – 3, then the value of x is

(a) 3 (b) 4
(c) 8 (d) 12

(v) If the HCF of 72 and 136 is expressible in the form 11a – 14, then the value is a is

(a) 1 (b) 2
(c) 4 (d) 23

(vi) HCF of 180 and 200 is expressible in the form of $(140 \times 3 + 40y)$, then the value of y is

(a) – 10 (b) 10
(c) 40 (d) – 40

(vii) HCF of 84 and 60 is expressible in the form of $(18 \times 2 + 4y)$, then the value of y is

(a) 6 (b) – 6
(c) 18 (d) – 18

(viii) HCF of 42 and 91 is expressible in the form of $(13 \times 3 + 4a)$, then the value of a is

(a) 6 (b) – 6
(c) 8 (d) – 8

40. The HCF of two numbers a and b is 5 and their LCM is 200. Find the product of ab. **(2019)**

Solution:

HCF × LCM = Product of two numbers

$\Rightarrow 5 \times 200 = ab$

$\therefore ab = 1000$

Exercise 1.7

(i) The HCF of two numbers a and b is 6 and their LCM is 378. Find the product is ab.

(ii) The HCF of two numbers x and y is 4 and their LCM is 280. Find the product of xy.

(iii) The HCF a two number p and q is a and their LCM is b. Find the product of pq.

41. HCF of two numbers is 27 and their LCM is 162. If one of the number is 54, then other number is. **(2020)**

(a) 36 (b) 35
(c) 9 (d) 81

Solution: (d)

Number 1 × Number 2 = their LCM × HCF

$\Rightarrow 54 \times \text{Number } 2 = 162 \times 27$

$\therefore \text{Number } 2 = 81$

42. The LCM of two numbers is 182 and their HCF is 13. If one of the numbers is 26, find the other. **(2020)**

Solution:

Number 1 × Number 2 = their LCM × HCF

$\Rightarrow 26 \times \text{Number } 2 = 13 \times 182$

$\therefore \text{Number } 2 = 91$

Exercise 1.8

(i) HCF of two number is 12 and their LCM is 336. If one number be 48, the other number is

(a) 84 (b) 80
(c) 48 (d) 108

(ii) HCF of two numbers is 13 and their LCM is 819. If one number be 91, the other number is

(a) 91 (b) 89

(c) 117 (d) 221

(iii) HCF of two number is 16 and their LCM is 160. If one number be 32, the other number is

(a) 48 (b) 96

(c) 80 (d) 112

43. If p and q are two co-prime numbers, HCF (p,q) is **(2010)**

(a) p (b) q

(c) pq (d) 1

Solution: (d) 1

44. HCF of two consecutive even numbers is **(2011)**

(a) 0 (b) 1

(c) 4 (d) 2

Solution : (d) 2

45. If p and q are natural numbers and p is the multiple of q, then what is the HCF of p and q ? **(2023)**

(a) pq (b) p

(c) q (d) p + q

Solution: (c)

Let p = kq

∴ HCF of p and q is q

46. If the product of two co-prime numbers is 553, then their HCF is **(2024)**

(a) 1 (b) 553

(c) 7 (d) 70

Solution: (a) 1

47. For any prime number *p*, if *p* divides a^2, where a is any real number, then *p* also divides **(2025)**

(a) a (b) $a^{\frac{1}{2}}$

(c) $a^{\frac{3}{2}}$ (d) $a^{\frac{1}{8}}$

Solution: (a)

p divides $a^2 \Rightarrow$ p divides a.

48. The LCM of two consecutive natural numbers p and $p+1$ is **(2026)**

(a) p (b) p^2+p

(c) 1 (d) $2p+1$

Solution: (b)

LCM two consecutive natural numbers p and $p+1$ is $p\,(p+1) = p^2 + p$

Exercise 1.9

(i) If a^2, b^2 are two co-prime numbers, HCF (a^2, b^2) is.

(a) a^2 (b) b^2

(c) a^2b^2 (d) 1

(ii) If (a + b) and (c + d) are two co-prime numbers, HCF (a + b, c + d) is

(a) a + b

(b) (c + d)

(c) (a + b), (c + d)

(d) 1

49. The product of the HCF and LCM of the smallest prime number and smallest composite number is **(2010, 2018)**

(a) 2 (b) 4

(c) 6 (d) 8

Solution: (d)

Smallest prime number is 2

Smallest composite number is 4

∴ Their HCF is 2 and LCM is 4

and their HCF × LCM = 2 × 4 = 8

Exercise 1.10

(i) The product of the HCF and LCM of the greatest prime numbers of one digit and smallest prime number of one digit is

(a) 1 (b) 9

(c) 7 (d) 4

(ii) The product of the HCF and LCM of the smallest even number and smallest composite number is.

(a) 1 (b) 2

(c) 4 (d) 8

50. The HCF of the smallest number of two digit and the largest multiple of 5 which is less then 40 is. **(2020)**

Solution: (c)

Smallest number of two digits is 10

Largest multiple of 5 which is less than 40 is 35

HCF of 10 and 35 is 5

Exercise 1.11

(i) The HCF of the smallest even number of two digits and the largest multiple of 6 which is less then 50 is.

(a) 2 (b) 3
(c) 6 (d) 8

(ii) The HCF of the smallest multiple of 6 of two digits and the largest multiple of 6 which is less than 60 is

(a) 8 (b) 10
(c) 6 (d) 18

(iii) The HCF of the LCM of 3 and 5 and the largest multiple of 6 which is less than 62 is.

(a) 3 (b) 5
(c) 15 (d) 60

(iv) The HCF of the smallest number of two digits and the largest multiple of 8 which is less then 80 is

(a) 8 (b) 4
(c) 2 (d) 5

51. The LCM of smallest two digit number and smallest composite number is **(2023)**

(a) 12 (b) 4
(c) 20 (d) 40

Solution: (c)

Smallest 2 digit number is 10

Smallest composite number is 4

$\therefore$ required LCM is 20

52. The LCM of the smallest prime number and the smallest odd composite number is. **(2024)**

(a) 10 (b) 6
(c) 9 (d) 18

Solution: (d)

Smallest prime number is 2

Smallest odd composite number is 9

$\therefore$ their LCM is 18

Key Point The least positive integer divisible a and b is LCM of a and b.

53. The least positive integer divisible 20 and 24 is **(2013)**

(a) 240 (b) 480
(c) 120 (d) 960

Solution: (c)

LCM of 20 and 24 is 120.

54. The least number which is a perfect square and is divisible by each of 16, 20 and 50 is **(2025)**

(a) 1200 (b) 100
(c) 3600 (d) 2400

Solution: (c)

LCM of 16, 20 and 50 is 400

$\therefore$ The required least perfect square number is $400 \times 9 = 3600$

Exercise 1.12

(i) The least positive integer divisible by 8 and 15 is **(2013)**

(a) 120 (b) 140
(c) 108 (d) 150

(ii) The least positive integer divisible by 12 and 18 is.

(a) 18 (b) 12
(c) 48 (d) 36

(iii) The least positive integer divisible by 20 and 25 is

(a) 80 (b) 160
(c) 100 (d) 120

(iv) The LCM of smallest composite number and smallest odd composite number is.

(a) 4 (b) 1
(c) 36 (d) 9

(v) The LCM of smallest prime number and smallest odd composite number is

(a) 2 (b) 9
(c) 18 (d) 1

Key Point Smallest number which is divisible by both a and b is LCM of a and b.

55. Write the smallest number which is divisible by both 306 and 657. **(2018)**

Solution: LCM of 306 and 657 is 22,338

Exercise 1.13

(i) Write the smallest number which is divisible by both 228 and 665.

(ii) Write the smallest number which is divisible by both 273 and 355.

(iii) Write the smallest number which is divisible by both 391 and 493.

Key Point Greatest number which is divides a and b, leaving remainder c is HCF of (a – c) and (b – c).

56. What is the largest number that divides 245 and 1029, leaving remainder 5 is each ? **(2020)**

(a) 15 (b) 16

(c) 9 (d) 5

Solution: (b)

$245 - 5 = 240$ and $1029 - 5 = 1024$

Now HCF of 240 and 1024 is 16

Exercise 1.14

(i) What is the largest number that divides 5191 and 5854, leaving remainder 4 is each?

(a) 39 (b) 41

(c) 47 (d) 67

(ii) What is the largest number that divider 211 and 939 leaving remainder 3 in each.

(a) 88 (b) 104

(c) 112 (d) 208

(iii) What is the largest number that divider 615 and 963, leaving remainder 6 in each?

(a) 73 (b) 67

(c) 47 (d) 87

(iv) What is the largest number that divider 972 and 2016, leaving remainder 15 in each ?

(a) 87 (b) 107

(c) 93 (d) 113

Key Point Greatest number which is divides a and b ,leaving remainder c and d is HCF of (a – c) and (b – d).

57. The greatest number which divides 281 and 1249, leaving remainder 5 and 7 respectively. **(2024)**

(a) 23 (b) 276

(c) 138 (d) 69

Solution: (c)

$281 - 5 = 276$

$1249 - 7 = 1242$

Now, HCF of 276 and 1242 is 138

58. The greatest number which divides 70 and 125, leaving remainders 5 and 8 respectively, is: **(2025)**

(a) 13 (b) 65

(c) 875 (d) 1750

Solution: (a)

$70 - 5 = 65$, $125 - 8 = 117$

Now, HCF of 65 and 117 is 13

Exercise 1.15

(i) The greatest number which divides 70 and 125 leaving remainder 5 and 8 respectively.

(a) 13 (b) 65

(c) 875 (d) 1750

(ii) The greatest number which divides 83 and 138, leaving remainder 5 and 8 respectively.

(a) 13 (b) 26

(c) 39 (d) 65

(iii) The greatest number which divides 280 and 1245 leaving remainder 4 and 3 respectively.

(a) 138 (b) 142

(c) 152 (d) 160

59. How many prime factors are there in prime factorization of 5005. **(2010)**

(a) 2 (b) 4
(c) 6 (d) 7

Solution: (b)

$5005 = 5 \times 7 \times 11 \times 13$

60. Express 429 as a product of its prime factor. **(2019)**

Solution: $429 = 3 \times 11 \times 13$

Exercise 1.16

(i) How many prime factors are there in prime factorization of 3003.
(a) 2 (b) 4
(c) 6 (d) 7

(ii) How many prime factors are there in prime factorization of 100947.
(a) 2 (b) 4
(c) 5 (d) 6

(iii) How many prime factors are there in prime factorization of 3289
(a) 2 (b) 3
(c) 4 (d) 6

(iv) Express 1463 as a product of its prime factors.

(v) Express 399 as a product of its prime factors.

(vi) Express 741 as a product of its prime factors.

(vii) Prime factorization of 424 is. **(2024)**
(a) $2 \times 53 \times 4$ (b) $2 \times 53 \times 2$
(c) $2^3 \times 53$ (d) $2^3 \times 53$

61. 225 can be expressed as. **(2020)**
(a) 5×3^2 (b) $5^2 \times 3$
(c) $3^2 \times 5^2$ (d) $5^3 \times 3$

Solution: (c)

$225 = 3 \times 3 \times 5 \times 5 = 3^2 \times 5^2$

62. 1260 can be expressed as a product of prime numbers as. **(2020)**
(a) $3^2 \times 4 \times 5 \times 4$ (b) $3^2 \times 2^2 \times 5 \times 7^2$
(c) $3^2 \times 2^2 \times 5 \times 7$ (d) $2^2 \times 3^2 \times 5^2 \times 7$

Solution: (c)

$1260 = 2 \times 2 \times 3 \times 3 \times 5 \times 7 = 3^2 \times 2^2 \times 5 \times 7$

63. The prime factorization of natural number 288 is. **(2023)**
(a) $2^4 \times 3^3$ (b) $2^4 \times 3^2$
(c) $2^5 \times 3^2$ (d) $2^5 \times 3^1$

Solution: (c)

$288 = 2 \times 2 \times 2 \times 2 \times 2 \times 3 \times 3 = 2^5 \times 3^2$

64. The prime factorization of 1728 is. **(2023)**
(a) $2^5 \times 3^3$ (b) $2^5 \times 3^4$
(c) $2^6 \times 3^3$ (d) $2^6 \times 3^2$

Solution: (c)

$$1728 = 2 \times 2 \times 2 \times 2 \times 2 \times 2 \times 3 \times 3 \times 3 = 2^6 \times 3^3$$

Exercise 1.17

(i) Express 288 as the product of its prime factors. **(2021)**

(ii) 180 can be expressed as
(a) $2^2 \times 3^2 \times 5$ (b) $2^3 \times 3 \times 5^2$
(c) $2^2 \times 3^3 \times 5^2$ (d) $2^3 \times 3^3 \times 5^3$

(iii) 504 can be expressed as
(a) $2^2 \times 3^3 \times 7^2$ (b) $2^3 \times 3^2 \times 7$
(c) $2 \times 3^3 \times 7^2$ (d) $2^4 \times 3^3 \times 7^2$

65. Write the number of zero's in the end of a number whose prime factorization is $2^2 \times 5^3 \times 3^2 \times 17$. **(2019)**

Solution:

$2^2 \times 5^3 \times 3^2 \times 17 = 4 \times 125 \times 9 \times 17 = 76500$

∴ Number of zero's in the end of the number 76500 is 2.

Exercise 1.18

(i) Write the number zero's in the end of a number whose prime factorization is $2^3 \times 3^2 \times 5^2 \times 13$.

(ii) Write the number of zero's in the end of a number whose prime factorization is $2^4 \times 3 \times 5^3 \times 11$.

(iii) Write the number of zero's in the end of a number whose prime factorization is $2^5 \times 3^2 \times 5^2 \times 17$.

66. The sum of the exponents of prime-factorization of 196 is. **(2020)**
(a) 3 (b) 4
(c) 5 (d) 2

Solution: (b)

$196 = 2^2 \times 7^2$

$\therefore$ the sum of exponents is $2 + 2 = 4$

67. The sum of the exponents of prime-factorization of 4004 is. **(2025)**

(a) 5 (b) 4
(c) 3 (d) 2

Solution: (a)

$4004 = 2^2 \times 7 \times 11 \times 13$

$\therefore$ the sum of exponents is $2 + 1 + 1 + 1 = 5$

68. The exponents of 2 in the prime factorization of 144 is. **(2020)**

(a) 2 (b) 4
(c) 1 (d) 6

Solution: (b)

$144 = 2^4 \times 3^2$

$\therefore$ exponent of 2 in the prime factorization of 144 is 4.

Exercise 1.19

(i) The sum of exponents of prime-factorization of 125 is.

(a) 5 (b) 4
(c) 3 (d) 2

(ii) The sum of exponents of prime- factorization of 216 is.

(a) 4 (b) 6
(c) 8 (d) 10

(iii) The sum of exponents of prime- factorization of 648 is.

(a) 6 (b) 7
(c) 8 (d) 10

(iv) The exponent of 3 in the prime factorization of 216 is.

(a) 2 (b) 3
(c) 4 (d) 6

(v) The exponent of 5 in the prime factorization of 500 is.

(a) 3 (b) 4
(c) 5 (d) 0

(vi) The exponent of 7 in the prime factorization of 392 is.

(a) 1 (b) 3
(c) 4 (d) 2

(vii) Find the exponent of 2 in the prime-factorization of 288. **(2021)**

69. $119^2 - 111^2$ is : **(2010, 2011)**

(a) Prime number
(b) Composite number
(c) An odd Prime number
(d) an odd composite number

Solution: (b)

$119^2 - 111^2 = (119 + 111)(119 - 111)$

$= 230 \times 8$

$= 1840$

= composite number.

70. $(3 \times 11 \times 13 + 3)$ is: **(2026)**

(a) a prime number
(b) divisible by 13
(c) a composite number
(d) an odd number

Solution: (c)

$3 \times 11 \times 13 + 3$

$= 3(11 \times 13 + 1)$

71. The natural number 1 is: **(2026)**

(a) a prime number
(b) a composite number
(c) prime as well as composite
(d) neither prime nor composite

Solution: (d)

neigther prime nor composite.

72. $7 \times 11 \times 13 + 5$ is **(2026)**

(a) a prime number
(b) an odd number
(c) a composite number
(d) a multiple of 5

Solution: (c)

$7 \times 11 \times 13 + 5 = 1001 + 5 = 1006$ which is a composite number.

73. The natural number 2 is: **(2026)**

(a) a prime number
(b) a composite number
(c) prime as well as composite
(d) neither prime nor composite

Solution: (a)

a prime number.

Exercise 1.20

(i) $112^2 - 109^2$ is :
(a) Prime number
(b) Composite number
(c) An odd prime number
(d) An odd composite number

(ii) $117^2 - 103^2$ is :
(a) Prime number
(b) Composite number
(c) An odd prime number
(d) An odd composite number

74. $(2 + \sqrt{5})(2 + \sqrt{5})$ expression is. **(2010)**

(a) a rational number (b) a whole number
(c) an irrational number (d) a natural number

Solution: (c)

$(2 + \sqrt{5})(2 + \sqrt{5})$
$= (2 + \sqrt{5})^2$
$= 2^2 + 2.2.\sqrt{5} + (\sqrt{5})^2$
$= 4 + 4\sqrt{5} + 5$
$= 9 + 4\sqrt{5}$

Exercise 1.21

(i) $(3 - \sqrt{2})(3 - \sqrt{2})$ expansion is
(a) a rational number
(b) a whole number
(c) an irrational number
(d) a natural number

(ii) $(4 + \sqrt{3})(4 + \sqrt{3})$ expansion is
(a) a rational number
(b) a whole number
(c) an irrational number
(d) a natural number

(iii) $(3 - \sqrt{4})(3 + \sqrt{4})$ expansion is
(a) a rational number
(b) a whole number
(c) an irrational number
(d) a natural number

75. Which is not an irrational number ? **(2010)**

(a) $5 - \sqrt{3}$ (b) $\sqrt{2} + \sqrt{5}$
(c) $4 + \sqrt{2}$ (d) $6 + \sqrt{9}$

Solution: (d)

$6 + \sqrt{9} = 6 + 3 = 9$

Exercise 1.22

(i) Which is not an irrational number ?
(a) $2 - \sqrt{3}$ (b) $\sqrt{4} + \sqrt{6}$
(c) $4 - \sqrt{9}$ (d) $1 + \sqrt{3}$

(ii) Which is not an irrational number ?
(a) $\sqrt{3} - \sqrt{2}$ (b) $\sqrt{4} - \sqrt{2}$
(c) $10 - \sqrt{2}$ (d) $\sqrt{9} - \sqrt{4}$

(iii) Which is not an irrational number ?
(a) $\sqrt{16} - \sqrt{9}$ (b) $\sqrt{25} - \sqrt{3}$
(c) $\sqrt{5} - \sqrt{3}$ (d) $\sqrt{9} - \sqrt{2}$

(iv) Which is not an irrational number ?
(a) $(\sqrt{2} - 1)(\sqrt{2} + 1)$
(b) $(\sqrt{2} + 1)(\sqrt{2} + 1)$
(c) $(\sqrt{2} - 1)(\sqrt{2} - 1)$
(d) $(\sqrt{3} - \sqrt{2})$

76. Which of the following is not an irrational number ? **(2010)**

(a) $5 - \sqrt{3}$ (b) $\sqrt{5} + \sqrt{3}$
(c) $4 + \sqrt{2}$ (d) $5 + \sqrt{9}$

Solution: (d)

$5 + \sqrt{9} = 5 + 3 = 8$

77. A pair of irrational numbers whose product is a rational number is : **(2011)**

(a) $\sqrt{16}, \sqrt{4}$ (b) $\sqrt{5}, \sqrt{2}$
(c) $\sqrt{3}, \sqrt{27}$ (d) $\sqrt{36}, \sqrt{2}$

Solution: (c)

$\sqrt{3}\times\sqrt{27}=\sqrt{3\times 27}=\sqrt{81}=9$

Exercise 1.23

(i) A pair of irrational numbers whose product is a rational number is :

(a) $\sqrt{2},\sqrt{4}$ (b) $\sqrt{5},\sqrt{125}$

(c) $\sqrt{6},\sqrt{36}$ (d) $\sqrt{3},\sqrt{6}$

(ii) A pair of irrational numbers whose product is a rational number is:

(a) $\sqrt{5},\sqrt{10}$ (b) $\sqrt{6},\sqrt{12}$

(c) $\sqrt{2},\sqrt{8}$ (d) $\sqrt{7},\sqrt{8}$

(iii) A pair of irrational numbers whose product is a rational number is :

(a) $\sqrt{7},\sqrt{21}$ (b) $\sqrt{20},\sqrt{5}$

(c) $\sqrt{12},\sqrt{7}$ (d) $\sqrt{3},\sqrt{17}$

78. Write whether $\frac{2\sqrt{45}+3\sqrt{20}}{2\sqrt{5}}$ on simplification gives an irrational or a rational number. **(2018)**

Solution: $\frac{2\sqrt{45}+3\sqrt{20}}{2\sqrt{5}}$

$=\frac{2.3\sqrt{5}+3.2\sqrt{5}}{2\sqrt{5}}$

$=\frac{6\sqrt{5}+6\sqrt{5}}{2\sqrt{5}}$

$=\frac{12\sqrt{5}}{\sqrt{5}}$

$=12$

Exercise 1.24

(i) Write whether $\frac{3\sqrt{50}+2\sqrt{18}}{7\sqrt{2}}$ on simplification gives an irrational or a rational number.

(ii) Write whether $\frac{12\sqrt{3}+35\sqrt{3}}{2\sqrt{3}}$ on simplification gives an irrational or a rational number.

(iii) Write whether $\frac{4\sqrt{50}-2\sqrt{98}}{13\sqrt{2}}$ on simplification gives an irrational or a rational number.

79. Find a rational number between $\sqrt{2}$ and $\sqrt{3}$ **(2019)**

Solution: 2

80. Find a rational number between $\sqrt{2}$ and $\sqrt{7}$ **(2019)**

Solution: 2

81. Which of the following is a rational number between $\sqrt{3}$ and $\sqrt{5}$? **(2025)**

(a) 1.4142387954012.......

(b) $2.32\overline{6}$

(c) π

(d) 1.857142

Solution: (d)

$\sqrt{3}\approx 1.732$

$\sqrt{5}\approx 2.236$

1.857142 is a rational number between $\sqrt{3}$ and $\sqrt{5}$

Exercise 1.25

(i) Find a rational number between $\sqrt{3}$ and 3.

(ii) Find a rational number between $\sqrt{3}$ and 4.

(iii) Find a rational number between $\sqrt{3}$ and 6.

82. $2\sqrt{2}$ is **(2020)**

(a) an integer

(b) a rational number

(c) an irrational number

(d) a whole number

Exercise 1.26

(i) $3\sqrt{2}$ is

(a) an integer

(b) a rational number

(c) an irrational number

(d) a whole number

(ii) $2+\sqrt{3}$ is

(a) an integer
(b) a rational number
(c) an irrational number
(d) a whole number

(iii) $\sqrt{5} - 2$ is
(a) an integer
(b) a rational number
(c) an irrational number
(d) a whole number

83. $\left(1+\sqrt{3}\right)^2 - \left(1-\sqrt{3}\right)^2$ is: **(2025)**

(a) a positive rational number
(b) a negative integer
(c) a positive irrational number
(d) a negative irrational number.

Solution: (c)

$\left(1+\sqrt{3}\right)^2 - \left(1-\sqrt{3}\right)^2$
$= 4 \times 1 \times \sqrt{3} = 4\sqrt{3}$

84. $\left(\sqrt{3}+\sqrt{2}\right)^2 + \left(\sqrt{3}-\sqrt{2}\right)^2$ in a/an **(2025)**

(a) positive rational number
(b) negative rational number
(c) positive irrational number
(d) negative irrational number.

Solution: (a)

$\left(\sqrt{3}+\sqrt{2}\right)^2 + \left(\sqrt{3}-\sqrt{2}\right)^2$
$= 2\left\{\left(\sqrt{3}\right)^2 + \left(\sqrt{2}\right)^2\right\}$
$= 2(3 + 2) = 10$

85. $\sqrt{0.4}$ is a/an **(2025)**

(a) natural number
(b) integer
(c) rational number
(d) irrational number

Solution: (d)

$\sqrt{0.4} = \sqrt{\frac{4}{10}} = \frac{2}{\sqrt{10}}$

86. The number $3+\sqrt{2}$ is **(2025)**

(a) a rational number
(b) an irrational number
(c) an integer
(d) a natural number

Solution: (b)

an irrational number

87. $\frac{\sqrt{3}-3}{\sqrt{3}}$ is: **(2025)**

(a) a rational number
(b) an irrational number
(c) an integer
(d) a natural number.

Solution: (b)

$\frac{\sqrt{3}-3}{\sqrt{3}}$

$= \frac{\sqrt{3}\left(1-\sqrt{3}\right)}{\sqrt{3}} = 1-\sqrt{3}$

88. $\left(2+\sqrt{2}\right)^2$ is: **(2025)**

(a) a rational number
(b) an irrational number
(c) an integer
(d) a natural number

Solution: (b)

$\left(2+\sqrt{2}\right)^2$
$= 2^2 + 2.2.\sqrt{2} + \left(\sqrt{2}\right)^2$
$= 4 + 4\sqrt{2} + 2$
$= 6 + 4\sqrt{2}$

89. $\left(2-5\sqrt{3}\right)^2$ is **(2025)**

(a) a negative integer
(b) an irrational number
(c) a rational number
(d) a positive integer

Solution: (b)

$\left(2-5\sqrt{3}\right)^2$

$= 2^2 - 2.2.5\sqrt{3} + \left(5\sqrt{3}\right)^2$

$= 4 - 20\sqrt{3} + 75$

$= 79 - 20\sqrt{3}$ which is an irrational number.

90. $\sqrt{2}\left(\sqrt{2}-1\right)$ is **(2025)**

(a) an integer

(b) a rational number

(c) an irrational number

(d) equal to 1

Solution: (c)

$\sqrt{2}\left(\sqrt{2}-1\right)$

$= 2-\sqrt{2}$ which is an irrational number.

91. The simplest form of $\frac{1095}{1168}$ is **(2020)**

(a) $\frac{17}{26}$ (b) $\frac{25}{26}$

(c) $\frac{13}{16}$ (d) $\frac{15}{16}$

Solution: (d)

$\frac{1095}{1168} = \frac{73\times15}{73\times16} = \frac{15}{16}$

Exercise 1.27

(i) The simplest form of $\frac{1106}{1343}$ is.

(a) $\frac{14}{17}$ (b) $\frac{17}{19}$

(c) $\frac{16}{17}$ (d) $\frac{19}{23}$

(ii) The simplest form of $\frac{534}{801}$ is.

(a) $\frac{7}{9}$ (b) $\frac{11}{13}$

(c) $\frac{2}{3}$ (d) $\frac{17}{19}$

(iii) The simplest form of $\frac{258}{473}$ is.

(a) $\frac{6}{11}$ b) $\frac{11}{13}$

(c) $\frac{13}{17}$ d) $\frac{17}{19}$

92. If $P^2 = \frac{32}{50}$, then P is a/an. **(2023)**

(a) Whole number (b) Integer

(c) Rational number (d) Irrational number.

Solution: (c)

$P^2 = \frac{32}{50}$,

$\Rightarrow P = \pm\sqrt{\frac{32}{50}} = \pm\frac{4}{5}$

Exercise 1.28

(i) If $P^2 = \frac{125}{80}$, then P is a/an.

(a) Whole number

(b) Integer

(c) Rational number

(d) Irrational number

(ii) If $X^2 = \frac{343}{28}$, then X is a/an.

(a) Whole number

(b) Integer

(c) Rational number

(d) Irrational number

93. Let N be the product of the first 11 positive integers i.e $N = 1\times2\times3\times4\times5\times6\times7\times8\times9\times10\times11$. What is the smallest positive integer which is not a factor of N ? **(2011)**

(a) 12 (b) 13

(c) 14 (d) 15

Solution: (b) 13

94. $2.\overline{35}$ is **(2020)**

(a) an integer

(b) a rational number

(c) an irrational number

(d) a whole number

Solution: (b)

$2.\overline{35} = \frac{235-2}{99} = \frac{233}{99} = \frac{7}{3}$

Exercise 1.29

(i) $2.\overline{16}$ is
(a) an integer
(b) a rational number
(c) an irrational number
(d) a whole number

(ii) $1.6\bar{1}$ is
(a) an integer
(b) a rational number
(c) an irrational number
(d) a whole number

(iii) $32.\overline{23}$ is
(a) an integer
(b) a rational number
(c) an irrational number
(d) a whole number

95. The total number of factors of a prime number is. **(2020)**

(a) 1 (b) 0
(c) 2 (d) 3

Solution: (c) 2

96. If x is a whole number, then 8^x ends with an even digit, except for which value of x ? **(2024)**

(a) 6 (b) 4
(c) 2 (d) 0

Solution: (d) 0

97. The number 2^n, where n is a natural number, cannot end with the digit : **(2025)**

(a) 4 (b) 6
(c) 2 (d) 0

Solution: (d) 0

98. The number 3^n, where n is a natural number, cannot end with the digit: **(2025)**

(a) 1 (b) 3
(c) 5 (d) 7

Solution: (c) 5

99. If the number a^n, where n is a natural number, always ends with digit a, then the possible value of a is: **(2025)**

(a) 2 (b) 4
(c) 6 (d) 8

Solution: (c)

$6^1 = 6$, $6^2 = 36$, $6^3 = 216$ and so on.

100. If the prime factorization of 2520 is $2^3 \times 3^a \times b \times 7$, then the value of a + 2b is. **(2024)**

(a) 12 (b) 10
(c) 9 (d) 7

Solution: (a)

$2520 = 2 \times 2 \times 2 \times 3 \times 3 \times 5 \times 7 = 2^3 \times 3^2 \times 5 \times 7$

$\Rightarrow 2^3 \times 3^a \times b \times 7 = 2^3 \times 3^2 \times 5 \times 7$

$\therefore$ a = 2, b = 5

$\therefore$ a + 2b = 2 + 2 × 5 = 12

101. If $1080 = 2^p \times 3^q \times 5$, then (p – q) is equal to: **(2025)**

(a) 6 (b) –1
(c) 1 (d) 0

Solution: (d)

$1080 = 2^3 \times 3^3 \times 5$

$\Rightarrow 2^p \times 3^q \times 5 = 2^3 \times 3^3 \times 5$

$\therefore p = 3, q = 3$

$\therefore p - q = 3 - 3 = 0$

102. If $a^b = 32$, where 'a' and 'b' are positive integers, then the value of b^{ab} is **(2025)**

(a) 72 (b) 5^{10}
(c) 2^{10} (d) 5^{12}

Solution: (c)

$a^b = 2^5$

$\therefore a = 2$ and $b = 5$

$\therefore b^{ab} = 2^{2\times 5} = 2^{10}$

103. If $(-1)^n + (-1)^8 = 0$, then *n* is: **(2025)**

(a) any positive integer
(b) any negative integer
(c) any odd number
(d) any even number

Solution:

$(-1)^n + (-1)^8 = 0$

$\Rightarrow (-1)^n + 1 = 0$

$\Rightarrow (-1)^n = -1$

$\therefore$ *n* is any odd number.

104. Which of the following cannot be the unit digit of 8^n, where n is a natural number? **(2025)**

(a) 4 (b) 2
(c) 0 (d) 6

Solution: (c) 0

105. The LCM of two numbers is 3600. Which of the following can not be their HCF? **(2025)**

(a) 600 (b) 400
(c) 500 (d) 150

Solution: (c)

3600 is not divisible by 500.

106. The sum of first five prime numbers is **(2025)**

(a) 18 (b) 26
(c) 28 (d) 39

Solution: (c)

$2 + 3 + 5 + 7 + 11$

$= 28$

107. For any natural number n, 6^n ends with the digit : **(2026)**

(a) 0 (b) 6
(c) 3 (d) 2

Solution: (b) 6

108. Which of the following numbers will not end with 0 for any natural number **(2026)**

(a) $4n$ (b) 4^n
(c) $3^n + 1$ (d) 10^{n+1}

Solution: (b)

for $n = 10$ $4n = 40$

For $n = 3$ $3^n + 1 = 3^3 + 1 = 10$

For any value of n 10^{n+1} always end with zero

109. For any natural number n,5^nends with the digit : **(2026)**

(a) 0 (b) 5
(c) 3 (d) 2

Solution: (b) 5

Exercise 1.30

(i) If the prime factorization of 6300 is. $2^2 \times 3^a \times b \times 7$, then the value of $2a + b$ is.

(a) 25 (b) 29
(c) 31 (d) 36

(ii) If the prime factorization of 363825 is. $2^3 \times 5^a \times b \times 11$, then the value of $b - 4a$ is.

(a) 40 (b) 36
(c) 39 (d) 41

(iii) If the prime factorization of 18900 is. $2^2 \times 3^a \times b \times 7$, then the value of $b - a$ is.

(a) 20 (b) 24
(c) 32 (d) 22

Question on 110 to 124 consist of statements, namely, **Assertion (A)** and **Reason (R)**. For selecting the correct answer, use the following code :

(a) Both Assertion (A) and Reason (R) are true and Reason (R) is a correct explanation of Assertion (A).
(b) Both Assertion (A) and Reason (R) are true but Reason (R)is not a correct explanation of Assertion (A).
(c) Assertion (A) is true and Reason (R) is false.
(d) Assertion (A) is false and Reason (R) is true.

110. Assertion (A) : The number 5^n cannot end with the digit 0, where n is a natural number

Reason (R) : Prime factorization of 5 has only two factors, 1 and 5. **(2023)**

Solution: (b)

Both A and R are true but R is not the correct explanation of A.

111. Assertion(A) : The prime number which divides 36 also divide 6

Reason(R) : Any number which divides p^2 also divides p. **(2025)**

Solution: (c)

A is true but R is false.

112. Assertion (A): 4^n end with digit 0 for some natural number n.

Reason (R): For a number 'x' having 2 and 5 an its prime factors, x^n always ends with digit 0 for every natural number n. **(2025)**

Solution: (d)

Assertion (A) is false and Reason (R) is true.

113. Assertion (A): Unit digit of 3^n cannot be an even number for any natural number n.

Reason (R): 2 is not a prime factor of 3^n for any natural number *n*. **(2025)**

Solution: (a)

$3^1 = 3$

$3^2 = 9$

$3^3 = 27$

$3^4 = 81$

$3^5 = 243$

$\therefore$ The unit digits are 3, 9, 7, 1 and this cycle repeats.

None of these unit digits are even number.

114. Assertion (A): For any two prime numbers *p* and *q*, their HCF is 1 and LCM is $p + q$

Reason (R): For any two natural numbers, HCF $\times$ LCM = product of numbers. **(2025)**

Solution: (d)

For any two prime numbers *p* and *q*, their HCF is 1 and LCM is *pq*.

115. Assertion (A): For two prime numbers *x* and *y* $(x < y)$, HCF $(x, y) = x$ and LCM $(x, y) = y$

Reason (R): HCF $(x, y) \leq$ LCM (x, y), where *x*, *y* are any two natural numbers. **(2025)**

Solution: (d)

For two prime numbers *x* and *y* their HCF is 1 and LCM is *xy*

116. Assertion (A): For two odd prime numbers *x* and *y* $(x \neq y)$, LCM $(2x, 4y) = 4xy$

Reason (R): LCM (x, y) is a multiple of HCF (x, y) **(2025)**

Solution:

Let $x = 3$ and $y = 9$

$\therefore$ LCM $(2 \times 3, 4 \times 9) = 4 \times 9 = 4y$

$\therefore$ Assertion (A) is incorrect.

Now, LCM $(x, y) \times$ HCF $(x, y) = xy$

$$\Rightarrow \text{LCM}(x, y) = \frac{xy}{\text{HCF}(x, y)}$$

Since HCF (x, y) is a factor of *xy*

$\therefore$ LCM (x, y) will always be a multiple of HCF (x, y).

117. Assertion (A): For any two natural numbers *a* and *b*, the HCF of *a* and *b* is a factor of the LCM of *a* and *b*.

Reason (R): HCF of any two natural numbers divides both the numbers. **(2025)**

Solution: (a)

HCF of *a* and *b* is a factor of LCM of *a* and *b*.

HCF of any two natural numbers divides both the numbers.

118. Assertion (A): The prime numbers which divide 36 also divide 6.

Reason (R): Any number which divides p^2 also divides p. **(2025)**

Solution: (c)

The prime numbers which divide 36 are 2, 3

2 and 3 are divide 6.

$\therefore$ Assertion (A) is true

36 is divided by 9 but $\sqrt{36} = 6$ is not divided by 9.

119. Assertion (A): $\left(2-\sqrt{5}\right)^2$ is a rational number

Reason (R): $\left(\sqrt{5}\right)^2$ is a rational number **(2026)**

Solution: (D)

$\left(2-\sqrt{5}\right)^2 = 4 - 4\sqrt{5} + 5 = 9 - 4\sqrt{5}$ which is a irrational number.

$\left(\sqrt{5}\right)^2 = 5$ which is a rational number.

120. Assertion (A): $\left(3+2\sqrt{2}\right)^2$ is a rational number.

Reason (R): $\sqrt{2}$ is an irrational number. **(2026)**

Solution: (D)

$\left(3+2\sqrt{2}\right)^2$

$= 3^2 + 2.3.2\sqrt{2} + \left(2\sqrt{2}\right)^2$

$= 9 + 12\sqrt{2} + 8$

$= 17 + 12\sqrt{2}$ which is a irrational number.

121. Assertion (A): $\left(\sqrt{3}+1\right)^2$ is a rational number

Reason (R): $\left(\sqrt{3}\right)^2 - 3$ is a rational number **(2026)**

Solution: (d)

$\left(\sqrt{3}+1\right)^2$

$= \left(\sqrt{3}\right)^2 + 2.\sqrt{3}.1 + (1)^2$

$= 3 + 2\sqrt{3} + 1$

$= 4 + 2\sqrt{3}$ which is a irrational number

$\left(\sqrt{3}\right)^2 - 3 = 3 - 3 = 0$ which is a rational number

122. Assertion (A): H.C.F (36 m^2, 18 m) = 18 m, where m is a prime number.

Reason (R): H.C.F of two numbers is always less than or equal to the smaller number **(2026)**

Solution: (a)

H.C.F of 36 m^2 and 18 m is 18 m

123. Assertion (A): $7 \times 2 + 3$ is a composite number.

Reason(R): A composite number has more than two factors. **(2026)**

Solution: (d)

$7 \times 2 + 3 = 17$ which is a prime number

124. Assertion (A): $\left(\sqrt{3}+\sqrt{2}\right)$ is an irrational number

Reason(R): sum of the any two irrational number is always irrational **(2026)**

Solution: (c)

Assertion (A) is true ,but Reason (R) is false.

Direction: Question number **i** to **x** are Assertion and Reason based questions. Two statements are given, one labelled as Assertion (A) and the other labelled as Reason (R). Select the correct answer to these questions from the codes (A), (B), (C) and (D) as given below.

(a) Both Assertion (A) and Reason (R) are true and Reason (R)

(b) Both Assertion (A) and Reason (R) are true, but Reason (R) is not the correct explanation of Assertion (A).

(c) Assertion (A) is true, but Reason (R) is false.

(d) Assertion (A) is false, but Reason (R) is true.

Exercise 1.31

(i) Assertion (A): $\sqrt{a}$ is an irrational number, where a is a prime number.

Reason (R): Square root of any prime number is an irrational number.

(ii) Assertion (A): The HCF of two number is 12. And their product is 18000. Then their LCM is 1500.

Reason (R): If a, b are two positive. Integers, then HCF $\times$ LCM = a $\times$ 3

(iii) Assertion (A): If two positive integers a and b are written as $a = p^3q^2$, and $b = p^2q^3$, where p and q are prime numbers, then their LCM is p^3q^3.

Reason (R): LCM of two numbers is the product of the highest powers of all prime present in the numbers.

(iv) Assertion (A): The number $2+\sqrt{3}$ is irrational.

Reason (R): The sum of a rational and an irrational number is irrational.

(v) Assertion (A): For any two positive integers a and b, HCF (a, b) $\times$ LCM (a, b) $= x \times y$.

Reason (R): If the HCF of two numbers is 5 and their product is 150, then their LCM is 40.

(vi) Assertion (A): $2 \times 3 \times 4 \times 5 \times 6 + 6$ is a composite number.

Reason (R): Every composite number can be expressed as product of primes.

(vii) Assertion (A): HCF of (7, 11) is 1

Reason (R): If a and b are primers, then HCF (a,b) = 1

(viii) Assertion (A): 8 is a rational number.

Reason (R): The square roots of all positive integers are irrationals.

(ix) Assertion (A): If the HCF of two numbers is 1, they are said to be co-prime.

Reason (R): The LCM of two co-prime numbers is always equal to their product.

(x) Assertion (A): The HCF of two numbers is 8 and their product is 120, then their LCM is 15.

Reason (R): For any two positive integers a and b, HCF(a,b) LCM (a, b) = a × b.

ANSWER KEYS

Exercise – 1.1

(i) 30, 10 **(ii)** 24, 24 **(iii)** 60, 2
(iv) (d) 24 **(v)** (d) 42 **(vi)** (d) 60

Exercise – 1.2

(i) 7200 **(ii)** 1152 **(iii)** ab
(iv) (c) 540 **(v)** (a) 216 **(vi)** (b) 5600

Exercise – 1.3

(i) (a) 44676 **(ii)** (b) 675 **(iii)** (c) 1449
(iv) (b) 7700 **(v)** (c) 585 **(vi)** (a) 162
(vii) (a) 156 **(viii)** (a) 35 **(ix)** (b) 18
(x) (b) 15

Exercise – 1.4

(i) (c) 28 **(ii)** (b) 13 **(iii)** (c) 45
(iv) (b)7 **(v)** (a) 28 **(vi)** (c) 21
(vii) (b) 6,72 **(viii)** (a) 3, 630
(ix) (d)6, 504 **(x)** (a) 3960
(xi) (b) 10800 **(xii)** (c) x^3y^3
(xiii) a^4b^4 **(xiv)** (b)a^5b^7
(xv) (c) 33×43 **(xvi)** (d) a^4b^4
(xvii) (a) x5y6 **(xviii)** (c) p^3q^2
(xix) (b) x^3y^3

Exercise – 1.5

(i) (b) 8000 **(ii)** (a) 1600 **(iii)** (d) 1520

Exercise – 1.6

(i) (b) 4 **(ii)** (c) 10 **(iii)** (a) 17 **(iv)** (b) 4
(v) (b) 4 **(vi)** (a) –10 **(vii)** (b) –6
(viii) (d) –8

Exercise – 1.7

(i) 2268 **(ii)** 1120 **(iii)** ab

Exercise – 1.8

(i) (a) 84 **(ii)** (c) 117 **(iii)** (c) 80

Exercise – 1.9

(i) (d) 1 **(ii)** (d) 1

Exercise – 1.10

(i) (d) 14 **(ii)** (d) 8

Exercise – 1.11

(i) (a) 2 **(ii)** (c) 6 **(iii)** (c) 15 **(iv)** (c) 2

Exercise – 1.12

(i) (a) 120 **(ii)** (d) 36 **(iii)** (c) 100
(iv) (c) 36 **(v)** (c) 18

Exercise – 1.13

(i) 7980 **(ii)** 15015 **(iii)** 11339

Exercise – 1.14

(i) (a) 39 **(ii)** (b) 104 **(iii)** (d) 87
(iv) (a) 87

Exercise – 1.15

(i) (a) 13 **(ii)** (b) 26 **(iii)** (a) 138

Exercise – 1.16

(i) (b) 4 **(ii)** (c) 5 **(iii)** (b) 3
(iv) $7 \times 11 \times 19$ **(v)** $3 \times 7 \times 19$
(vi) $3 \times 13 \times 19$ **(vii)** (c) $2^3 \times 53$

Exercise – 1.17

(i) $2^5 \times 3^2$ **(ii)** $2^2 \times 3^2 \times 5$ **(iii)** $2^3 \times 3^2 \times 7$

Exercise – 1.18

(i) two **(ii)** three **(iii)** two

Exercise – 1.19

(i) (c) 3 **(ii)** (b) 6 **(iii)** (b) 7 **(iv)** (b) 3
(v) (a) 3 **(vi)** (d) 2 **(vii)** 5

Exercise – 1.20

(i) (d) An odd composite number
(ii) (b) composite number

Exercise – 1.21

(i) (c) an irrational number
(ii) (c) an irrational number
(iii) (d) a natural number

EXERCISE – 1.22

(i) (c) $\sqrt{5}-\sqrt{9}$ **(ii)** (c) $\sqrt{2}, \sqrt{8}$

(iii) (a) $\sqrt{16}-\sqrt{19}$ **(iv)** (a) $(\sqrt{2}-1)(\sqrt{2}+1)$

EXERCISE – 1.23

(i) (b) $\sqrt{5}, \sqrt{125}$ **(ii)** (c) $\sqrt{2}, \sqrt{8}$

(iii) (b) $\sqrt{20}, \sqrt{5}$

EXERCISE – 1.24

(i) 3 **(ii)** $\frac{47}{2}$ **(iii)** $\frac{6}{13}$

EXERCISE – 1.25

(i) 2 **(ii)** 2 **(iii)** 3

EXERCISE – 1.26

(i) (c) an irrational number
(ii) (c) an irrational number
(iii) (c) an irrational number

EXERCISE – 1.27

(i) (a) $\frac{14}{17}$ **(ii)** (c) $\frac{2}{3}$ **(iii)** (a) $\frac{6}{11}$

EXERCISE – 1.28

(i) (c) rational number
(ii) (c) rational number

EXERCISE – 1.29

(i) (b) a rational number
(ii) (b) a rational number
(iii) (b) a rational number

EXERCISE – 1.30

(i) (b) 29 **(ii)** (d) 41 **(iii)** (d) 22

EXERCISE – 1.31

(i) a **(ii)** a **(iii)** a **(iv)** a
(v) c **(vi)** a **(vii)** a **(viii)** c
(ix) a **(x)** c.

2 Polynomials

Key Point Total number of zeroes in any polynomial equation = total number of times the curve intersects x-axis.

1. In the figure 1, the graph of the Polynomial p (x) is given. The number of zeroes of the polynomial is. **(2020)**

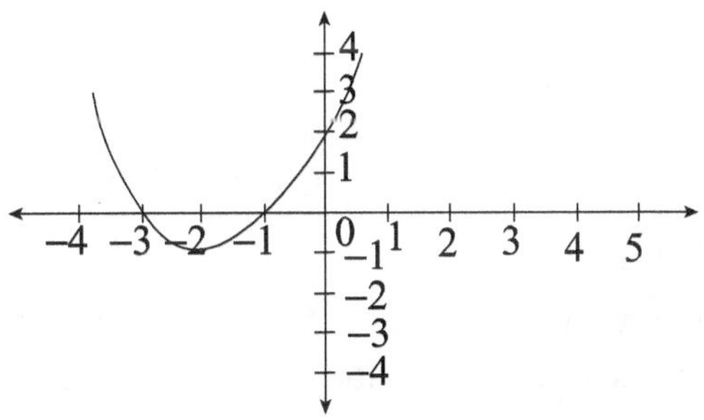

(a) 1 (b) 2

(c) 3 (d) 0

Solution: (b)

The number of zeroes is 2 as the graph p(x) intersects the x-axis at two points.

2. The number of zeroes for a Polynomial p(x) where graph of y = p(x) is given in figure 1 is. **(2020)**

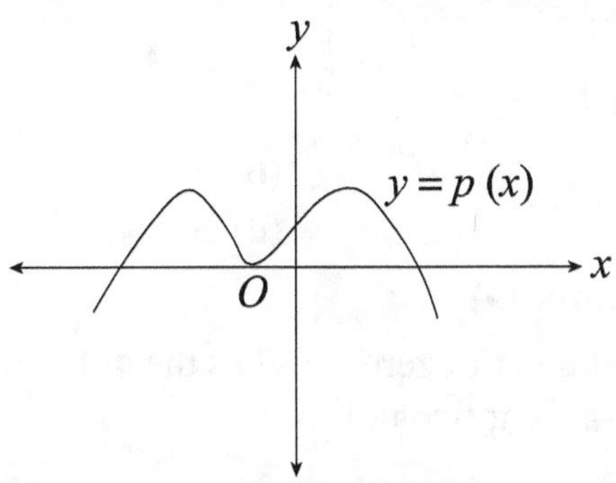

(a) 3 (b) 4

(c) 0 (d) 5

Solution: (a)

The number of zeroes is 3 as the graph y = p(x) intersects the x-axis at three points.

3. The graph of a Polynomial is shown in figure,2, the number of its zeroes is. **(2020)**

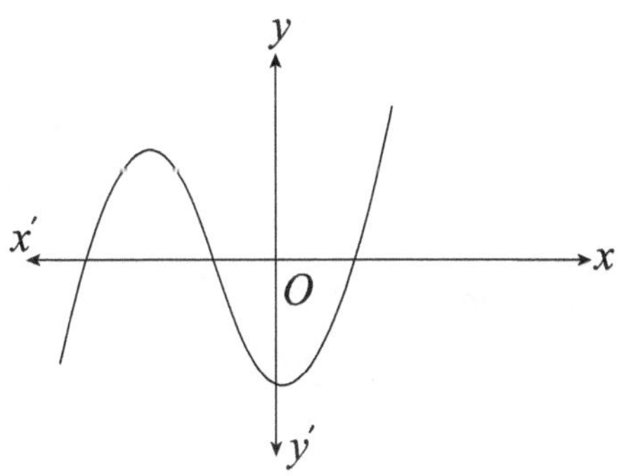

(a) 3 (b) 1

(c) 2 (d) 4

Solution: (a)

The number of zeroes is 3 as the graph intersects the x-axis at three points.

4. Number of zeroes of the Polynomial p(x) shown in Figure – 1, are **(2020)**

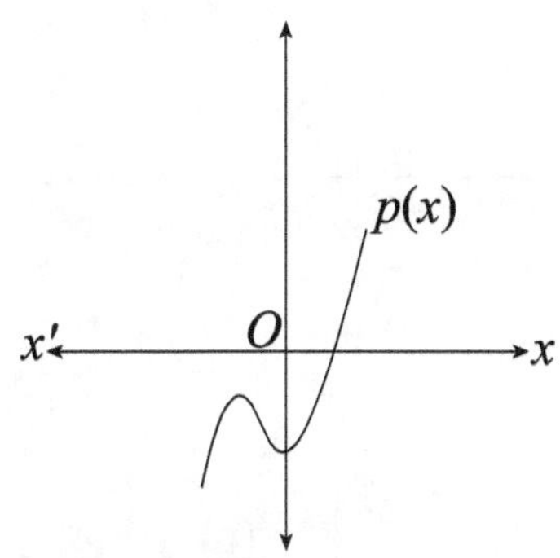

Figure -1

(a) 3 (b) 2

(c) 1 (d) 0

Solution: (c)

The number of zeroes is 1 as the graph intersects the x-axis at one point.

5. In Figure-1, number of zeroes of the polynomial p(x) shown in the graph are. **(2020)**

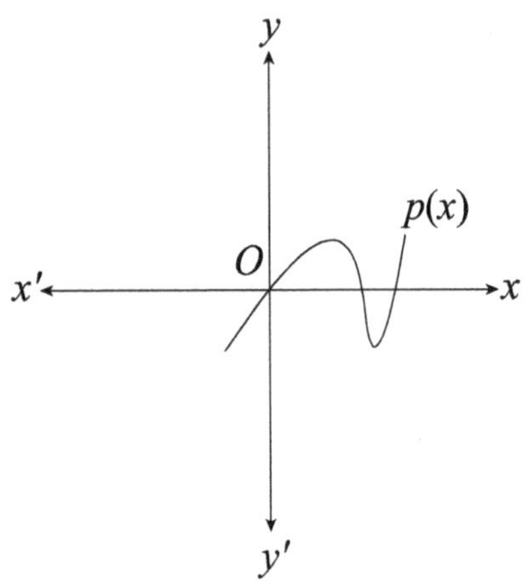

Figure -1

(a) 3 (b) 2

(c) 1 (d) 4

Solution: (a)

The number of zeroes is 3 as the graph p(x) intersect the x-axis at three points.

6. The graph of y = p(x) is given for a polynomial p(x). The number of zeroes of p(x) from the graph is. **(2023)**

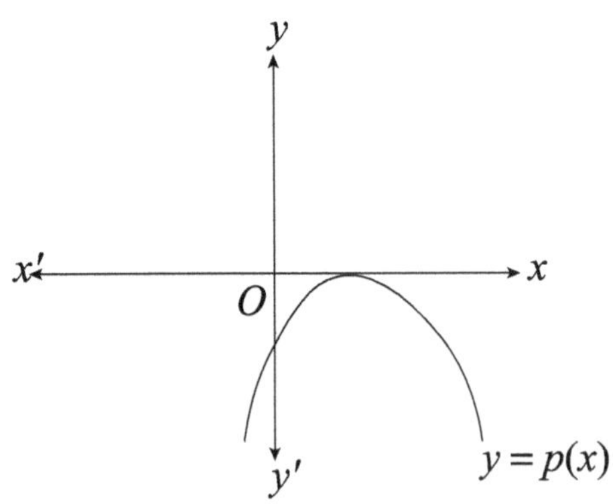

(a) 3 (b) 1

(c) 2 (d) 0

Solution: (b)

The number of zeroes is 1 as the graph y = p(x) intersect the x-axis at one point.

7. The graph of y = f(x) is shown in the figure for the some polynomial f(x). **(2023)**

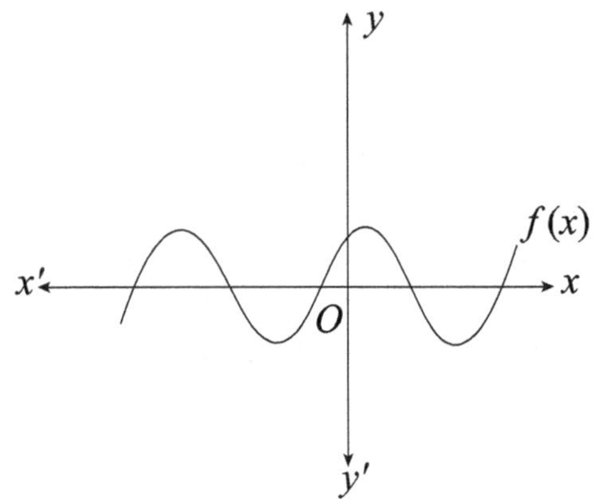

Number of zeroes of f(x) is

(a) 5 (b) 6

(c) 4 (d) 8

Solution: (a)

The number of zeroes is 5 as the graph intersects the x-axis at five points.

8. Graph of a polynomial p(x) is given in the figure. The number of zeroes of p(x) is. **(2023)**

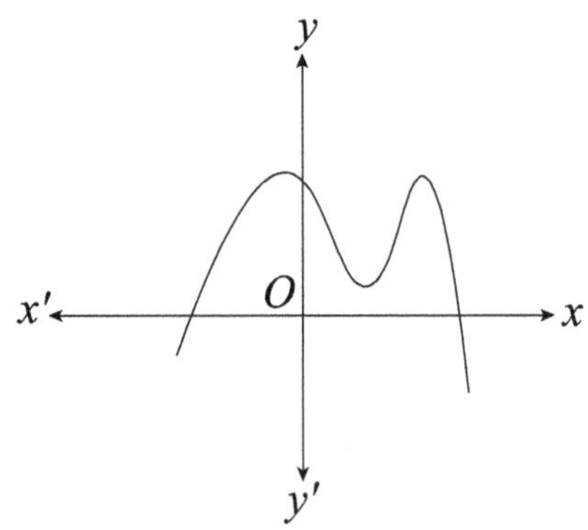

(a) 2 (b) 3

(c) 4 (d) 5

Solution: (a)

The number of zeroes is 2 as the graph intersects the x-axis at two points.

9. The graph of y = p(x) is shown in the figure for some polynomial p(x). The number of zeroes of p(x) is/are. **(2023)**

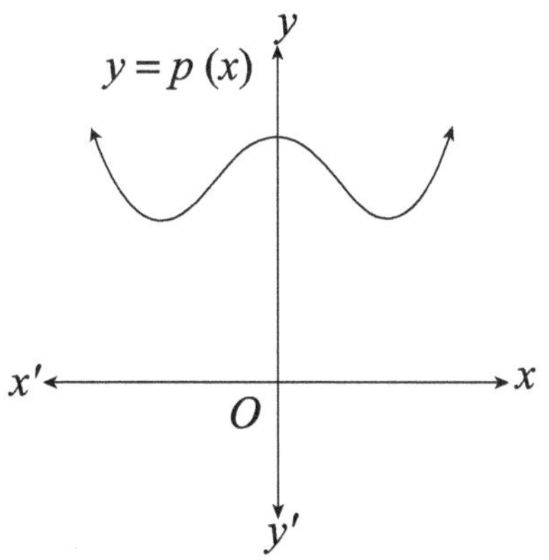

(a) 0 (b) 1

(c) 2 (d) 3

Solution: (a)

The number of zeroes is 0 as the graph y = p(x) does not intersect any points on x-axis.

10. The graph of y = f(x) is shown in the figure for some polynomial f(x). **(2023)**

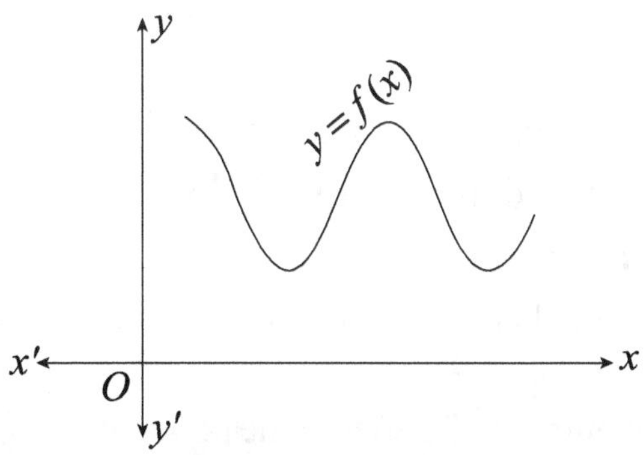

The number of zeroes of f(x) is

(a) 0 (b) 2

(c) 3 (d) 4

Solution: (a)

The number of zero is 0 as the graph y = f(x) does not intersect any point on x-axis.

11. The graph of y = f(x) is shown in the figure for some polynomial f(x). The number of zeroes of f(x) are. **(2023)**

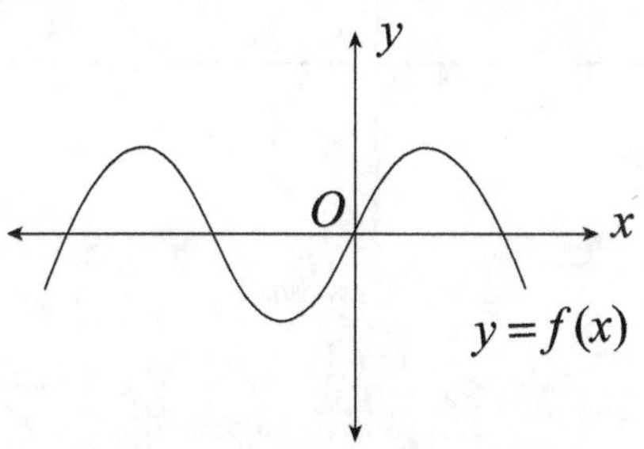

(a) 4 (b) 3

(c) 2 (d) 1

Solution: (a)

The number of zeroes are 4 as the graph or y = f(x) intersect the x-axis at 4 points.

12. The graph of y = p(x) is given in the adjoining figure. Zeroes of the polynomial p(x) are. **(2023)**

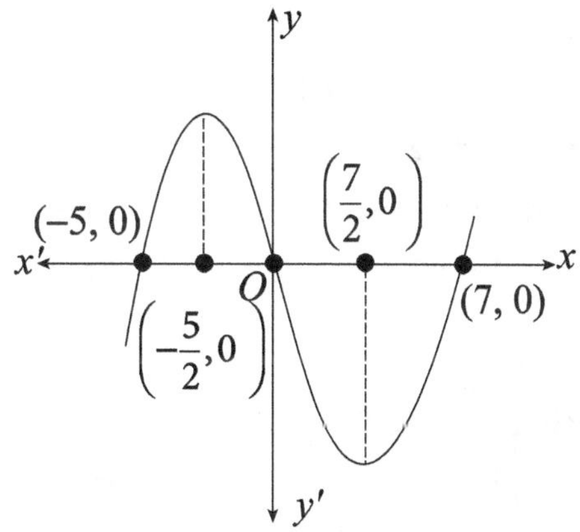

(a) –5,7 (b) $\frac{-5}{2}, \frac{-7}{2}$

(c) –5,0,7 (d) $-5, \frac{-5}{2}, \frac{7}{2}, 7$

Solution: (c)

The graph y = p(x) intersect the x-axis at (-5,0), (0,0) and (7,0) points

∴ zeroes of p(x) are = –5,0,7

13. Two polynomial are shown in the graph below the number of distinct zeroes of both the polynomial is : **(2025)**

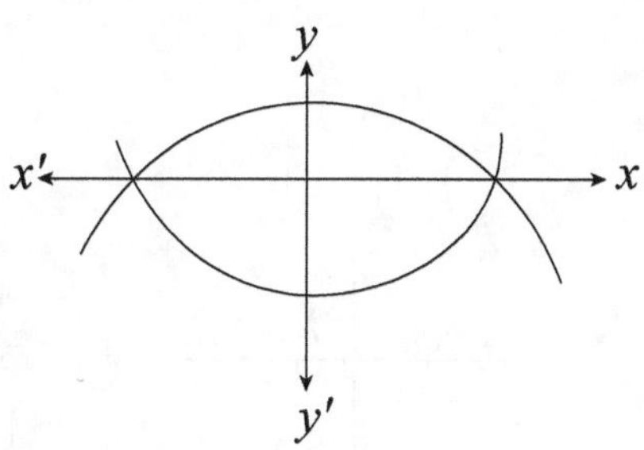

(a) 3 (b) 5

(c) 2 (d) 4

Solution: (c) 2

14. Observe the given graph of polynomial p(x), the number of zeroes of $p(x)$ is **(2025)**

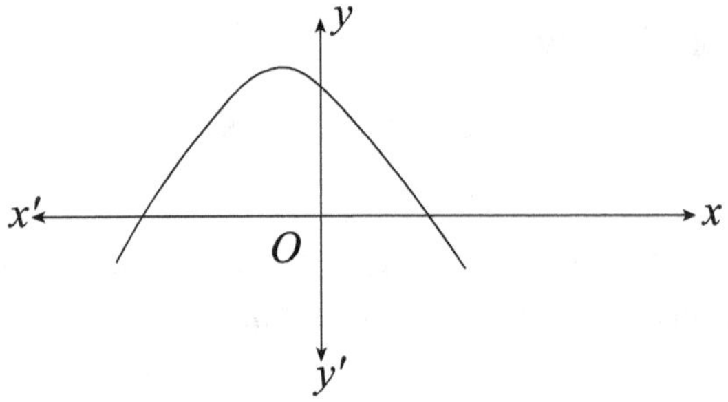

(a) 0 (b) 1
(c) 3 (d) 2

Solution: (d) 2

15. The graph of the polynomial $ax^2 + bx + c$ is a down ward parabola, is

(a) $a > 0$ (b) $a < 0$
(c) $a = 0$ (d) $a > 1$ **(2025)**

Solution:(b) $a < 0$

16. In the given figure, graph of $p(x)$ is shown. Number of distinct zeroes of $p(x)$ is **(2025)**

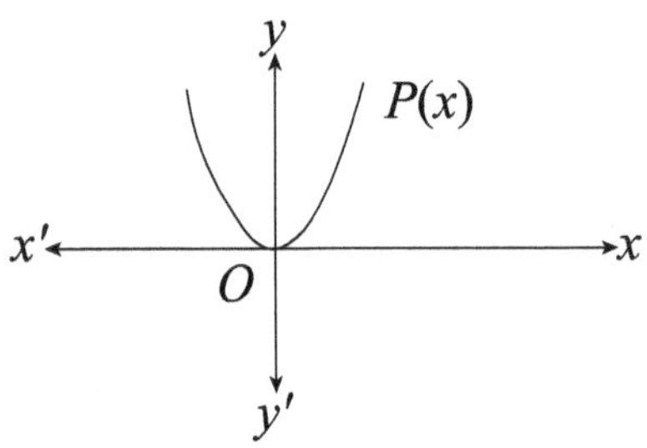

(a) 0 (b) 1
(c) 2 (d) many

Solution:(b) 1

17. In which of the following graphs of the polynomial is one of the zeroes twice the other ? **(2026)**

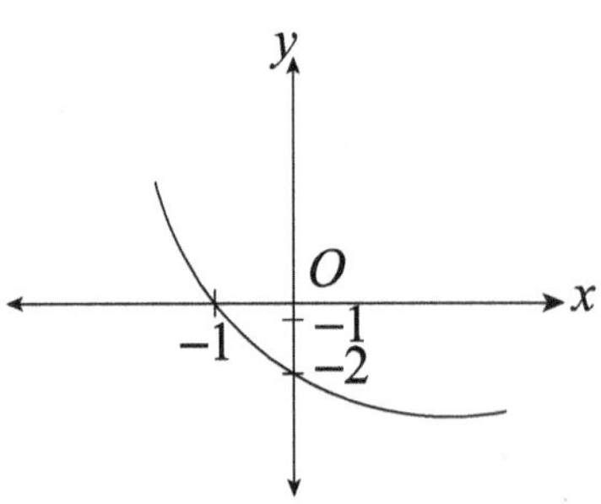

(i)

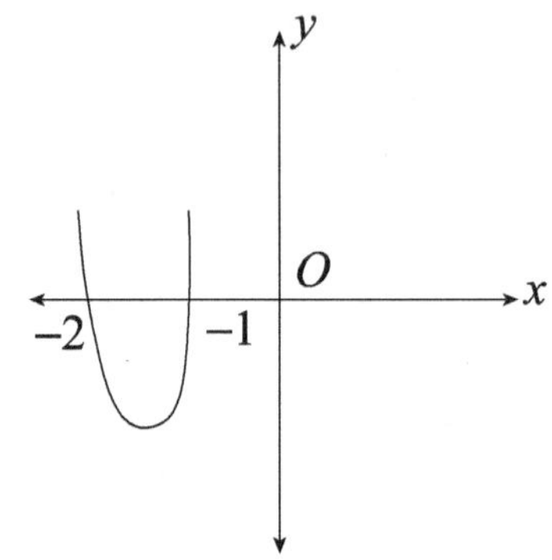

(ii)

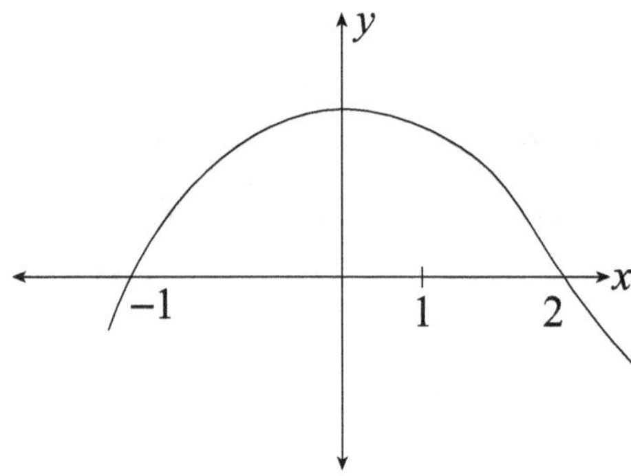

(iii)

(a) (i) only (b) (i) and (iii)
(c) (ii) and (iii) (d) (ii) only

Solution: (b)

The number of zero of polynomial of graph (i) is 1.

The number of zreoes of the polynomial of graph (ii) is 2

The number of zeroes of the polynomial of graph (iii) is 2

18. The graph of $y = f(x)$ in given The number of distinct zeroes of $y = f(x)$ is : **(2026)**

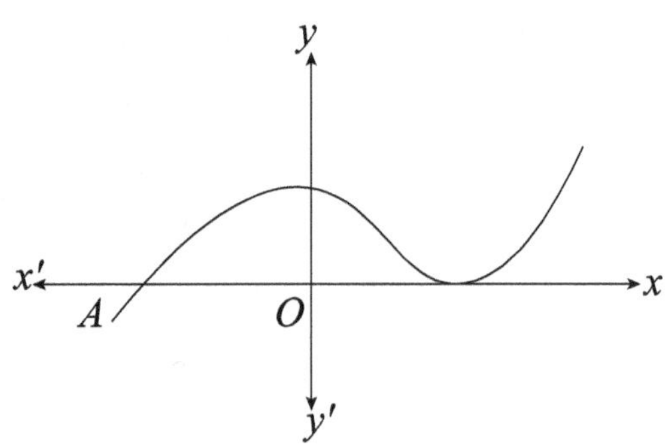

(a) 0 (b) 1
(c) 2 (d) 3

Solution: (c)

Since the graph tauches x – axis is at two distinct points ,the zeroes of the graph $y = f(x)$ is 2

19. The graph of a polynomial $p(x)$ is shown here The number of zeroes of the polynimal $p(x)$ is **(2026)**

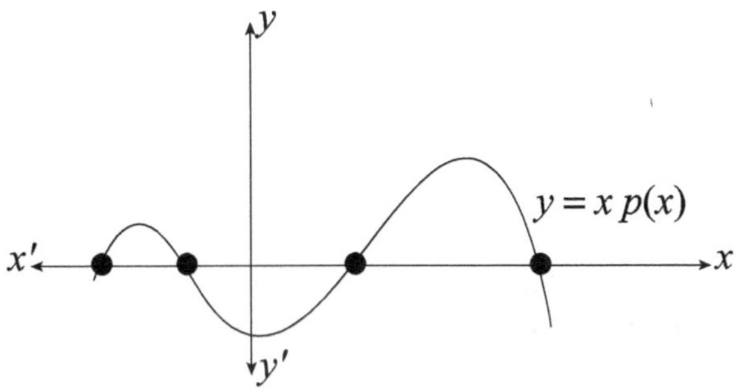

(a) 5 (b) 1

(c) 0 (d) 4

Solution: (d)

The given graph cuts the $x - axis$ at 4 points

$\therefore$ the number of zeroes of the Polynomial $p(x)$ is 4

20. The graph of $y = f(x)$ is given The number of zeroes of $f(x)$ is **(2026)**

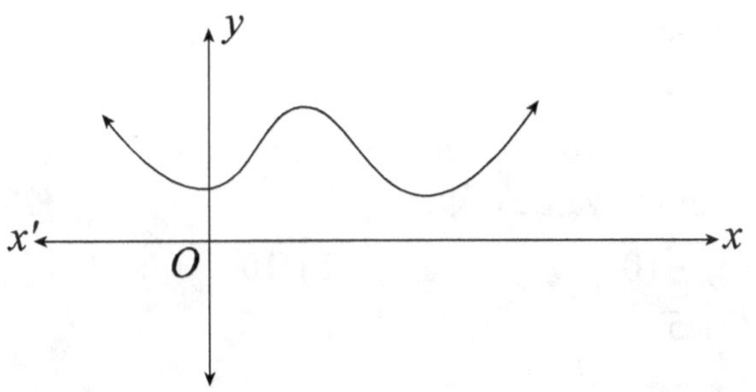

(a) 0 (b) 1

(c) 2 (d) 4

Solution: (a)

The given graph does not cut any point on x – axis

$\therefore$ The number of zeroes of the graph $y = f(x)$ is 0

21. Which of the following graphs repesents a polynomial with both zeroes being positive ? **(2026)**

(a)

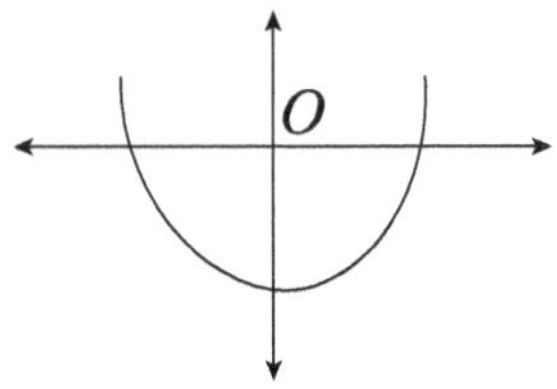

(b)

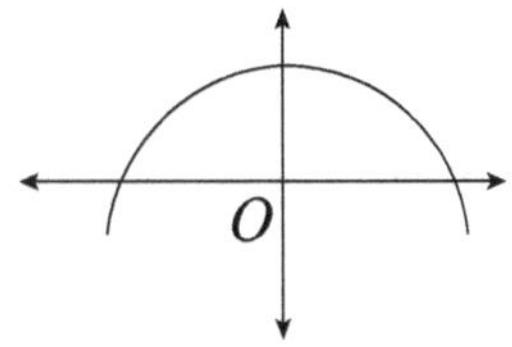

(c)

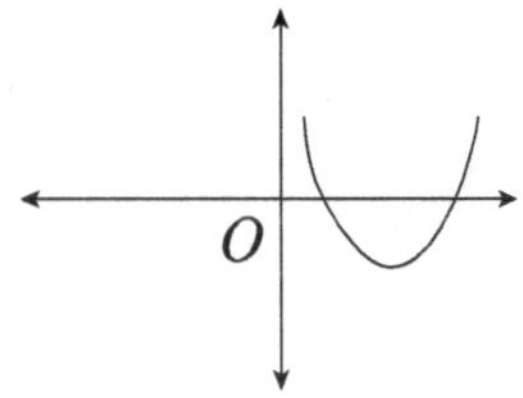

(d)

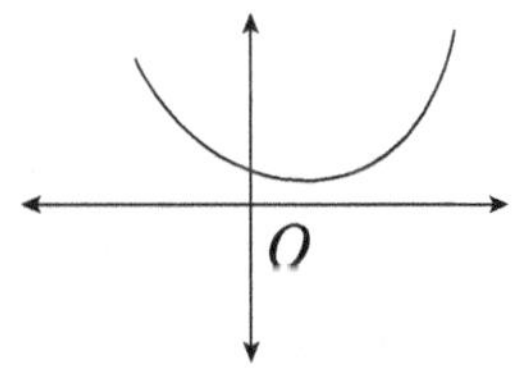

Solution: (c)

22. Observe the graph of polynomial $p(x)$ number of zeroes of $p(x)$ is : **(2026)**

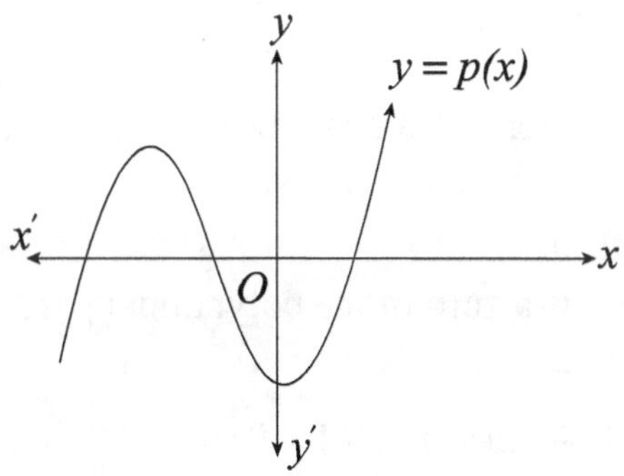

(a) 5 (b) 4

(c) 6 (d) 3

Solution: (d) 3

Exercise 2.1

(i) In the figure-1, the graph of the polynomial p(x) is given. The number of zeroes of the polynomial is

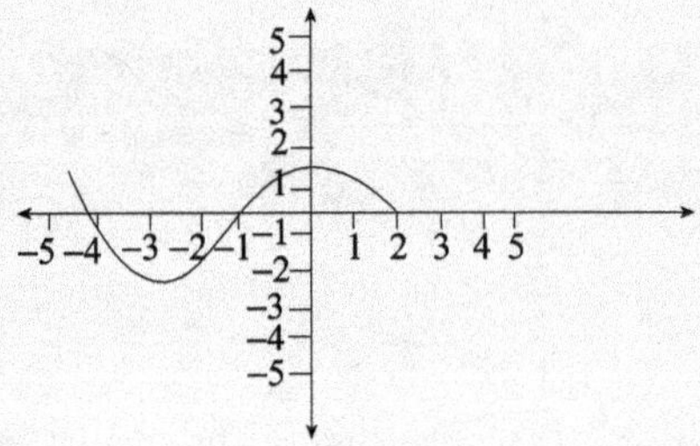

(a) 2 (b) 4

(c) 6 (d) 1

(ii) In the figure-1, the graph of the polynomial p(x) is given. The number of zeroes of the polynomial is

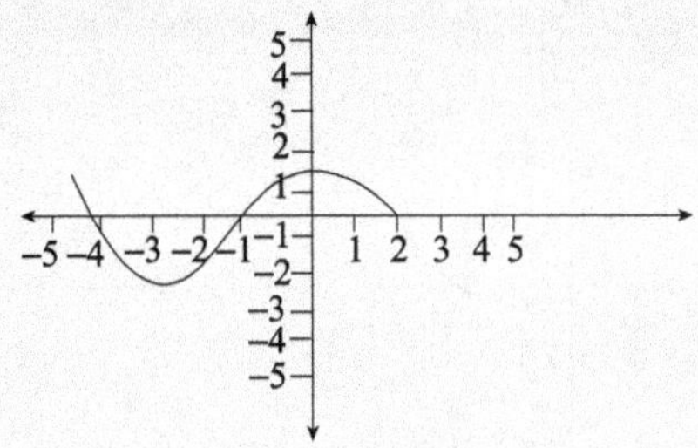

(a) 0 (b) 1

(c) 2 (d) 3

Key Point If α be a zero of a polynomial $ax^2 + bx + c$, then $a\alpha^2 + b\alpha + c = 0$

23. If 1 is a zero of the polynomial $p(x) = ax^2 - 3(a - 1)x - 1$, then find the value of a. **(2009)**

Solution:

since 1 is a zero of the polynomial p(x).

$\therefore p(1) = 0$

$\Rightarrow a.1^2 - 3(a - 1).1 - 1 = 0$

$\Rightarrow a - 3(a - 1) - 1 = 0$

$\Rightarrow a - 3a + 3 - 1 = 0$

$\Rightarrow -2a + 2 = 0$

$\Rightarrow -2a = -2$

$\therefore a = 1$

24. If one zero of the quadratic polynomial $x^2 + 3x + k$ is 2, then the value of k is. **(2020)**

(a) 10 (b) −10

(c) −7 (d) −2

Solution: (b)

$2^2 + 3.2 + k = 0$

$\Rightarrow 10 + k = 0$

$\therefore k = -10$

25. If one zero of a quadratic polynomial (kx^2+3x+k) is 2, then the value of k is. **(2020)**

(a) $\frac{5}{6}$ (b) $\frac{-5}{6}$

(c) $\frac{6}{5}$ (d) $-\frac{6}{5}$

Solution: (d)

$k.2^2 + 3.2 + k = 0$

$\Rightarrow 4k + 6 + k = 0$

$\Rightarrow 5k + 6 = 0$

$\Rightarrow k = -\frac{6}{5}$

26. If 2 is a zero of the polynomial $ax^2 - 2x$, then the value of 'a'.is_____ **(2020)**

Solution:

$a.2^2 - 2.2 = 0$

$\Rightarrow 4a - 4 = 0$

$\Rightarrow 4a = 4$

$\therefore a = 1$

27. If one zero of the polynomial $x^2 + 3x + k$ is 2, then the value of k. **(2023)**

(a) −10 (b) 10

(c) 5 (d) −5

Solution: (a)

$2^2 + 3.\ 2 + k = 0$

$\Rightarrow 4 + 6 + k = 0$

$\Rightarrow k = -10$

28. If one of the zeroes of the quadratic polynomial $(\alpha - 1)x^2 + \alpha x + 1$ is −3, then the value of α is. **(2024)**

(a) $\frac{-2}{3}$ (b) $\frac{2}{3}$

(c) $\frac{4}{3}$ (d) $\frac{3}{4}$

Solution: (c)

$(\alpha - 1)(-3)^2 + \alpha.(-3) + 1 = 0$

$\Rightarrow 9(\alpha - 1) - 3\alpha + 1 = 0$

$\Rightarrow 9\alpha - 9 - 3\alpha + 1 = 0$

$\Rightarrow 6\alpha = 8$

$\Rightarrow \alpha = \frac{4}{3}$

29. If –4 is a Zero of the polynomial $p(x) = x^2 - x - (2 + 2k)$, then the value of k is **(2025)**

(a) 3 (b) 9

(c) 6 (d) –9

Solution: (b)

$(-4)^2 - (-4) - (2 + 2k) = 0$

$\Rightarrow 16 + 4 - 2 - 2k = 0$

$\Rightarrow -2k = -18$

$\therefore k - 9$

30. One of the zeroes of the polynomial $p(x) = kx^2 - 9x + 3$ is $\left(\frac{-3}{2}\right)$. The value of k is **(2025)**

(a) $\frac{22}{3}$ (b) $\frac{-14}{3}$

(c) $\frac{14}{3}$ (d) $-\frac{22}{3}$

Solution: (d)

$p\left(\frac{-3}{2}\right) = 0$

$\Rightarrow k\left(\frac{-3}{2}\right)^2 - 9\left(\frac{-3}{2}\right) + 3 = 0$

$\Rightarrow \frac{9k}{4} + \frac{27}{2} + 3 = 0$

$\Rightarrow \frac{9k}{4} = -\frac{33}{2}$

$\Rightarrow k = -\frac{22}{3}$

Exercise 2.2

(i) If –1 is a zero of the Polynomial $f(x) = x^2 - 2(a - 3)x + 3$, then find the value of a.

(ii) If one zero of the quadratic Polynomial $x^2 + 3x + k$ is 2, then find the value of k.

(iii) If one zero of the quadratic polynomial $x^2 - 2x + k$ is 1, then the value of k is. **(2020)**

(a) 1 (b) –1

(c) 2 (d) –2

(iv) If one zero of the quadratic polynomial $2x^2 - 5x + k$ is –2, then the value of k is

(a) –8 (b) –10

(c) –18 (d) 18

(v) If one zero of the quadratic polynomial $3x^2 + 2x + a$ is –1, then the value of a is

(a) 1 (b) –1

(c) 0 (d) –2

(vi) If one zero of the quadratic polynomial $(k - 1)x^2 + kx + 1$ is –4, then the value of k is **(2020)**

(a) $\frac{-5}{4}$ (b) $\frac{5}{4}$

(c) $\frac{-4}{3}$ (d) $\frac{4}{3}$

(vii) If one zero of the quadratic polynomial $(k + 1)x^2 - kx + 3$ is 2, then the value of k is

(a) $\frac{-7}{2}$ (b) $\frac{7}{2}$

(c) $\frac{5}{2}$ (d) $\frac{-3}{2}$

(viii) If one zero of the quadratic polynomial $(k + 1)x^2 + 2kx + 2$ is –1, then the value of k is

(a) 3 (b) –3

(c) 2 (d) 4

Key Point If α, β are the zeroes of a polynomial,then the polynomial will be $k[x^2 - (\alpha + \beta)x + \alpha\beta]$ i.e $k[x^2$ – sum of the zeroes × x + product of the zeroes]

31. If α, β are the zeroes of a polynomial such that $\alpha + \beta = 6$ and $\alpha\beta = 4$, then write the polynomial. **(2010)**

Solution:

$k(x^2 - 6x + 4)$

32. The quadratic polynomial, the sum of whose zeroes is –5 and their product is 6, is. **(2020)**

(a) $x^2 + 5x + 6$ (b) $x^2 - 5x + 6$
(c) $x^2 - 5x - 6$ (d) $-x^2 + 5x + 6$

Solution: (a)

$x^2 - (-5)x + 6 = x^2 + 5x + 6$

33. Form a quadratic polynomial, the sum and product of whose zeroes are (–3) and 2 respectively. **(2020)**

Solution:

$K\{x^2 - (-3)x + 2\} = K(x^2 + 3x + 2)$

34. The sum and product of the zeroes of a quadratic polynomial are 3,–10 respectively the quadratic polynomial is. **(2020)**

(a) $x^2 - 3x + 10$ (b) $x^2 + 3x - 10$
(c) $x^2 - 3x - 10$ (d) $x^2 + 3x + 10$

Solution: (c)

$x^2 - 3x + (-10) = x^2 - 3x - 10$

35. A quadratic polynomial the sum and product of whose zeroes are –3 and 2 respectively is. **(2023)**

(a) $x^2 + 3x + 2$ (b) $x^2 - 3x + 2$
(c) $x^2 - 3x - 2$ (d) $x^2 + 3x - 2$

Solution: (a)

$x^2 - (-3)x + 2 = x^2 + 3x + 2$

36. A quadratic polynomial the sum and product of whose zeroes are 2 and –1 respectively is. **(2023)**

(a) $x^2 + 2x + 1$ (b) $x^2 - 2x - 1$
(c) $x^2 + 2x - 1$ (d) $x^2 - 2x + 1$

Solution: (b)

$x^2 - 2x + (-1) = x^2 - 2x - 1$

37. If the sum and the product of a quadratic polynomial are $2\sqrt{3}$ and 3 respectively, then a quadratic polynomial is. **(2024)**

(a) $x^2 + 2\sqrt{3}x - 3$ (b) $(x - \sqrt{3})^2$
(c) $x^2 - 2\sqrt{3}x - 3$ (d) $x^2 + 2\sqrt{3}x + 3$

Solution: (b)

$x^2 - 2\sqrt{3}x + 3 = x^2 - 2\sqrt{3}x + \left(\sqrt{3}\right)^2$
$= (x - \sqrt{3})^2$

38. A quadratic polynomial, the sum and product of whose zeroes are –1 and –2 respectively, is **(2025)**

(a) $x^2 - x - 2$ (b) $2x^2 + x - 1$
(c) $x^2 + x - 2$ (d) $\frac{1}{2}x^2 + x - 4$

Solution: (c)

The required polynomial is
$x^2 - (-1)x - 2$
$= x^2 + x - 2$

39. The sum and product of zeroes of a quadratic polynomial p(x) are $-\frac{1}{3}$ and 2 respectively. The polynomial p(x) is: **(2026)**

(a) $3x^2 - x + 6$ (b) $x^2 + \frac{1}{3}x - 2$
(c) $3x^2 - x + 2$ (d) $-3x^2 - x - 6$

Solution: (d)

The required polynomial will be

$k\left(x^2 - \left(-\frac{1}{3}\right)x + 2\right)$
$= \frac{k}{3}\left(3x^2 + x + 6\right)$
$= \frac{-3}{3}\left(3x^2 + x + 6\right)[\text{let } k = -3]$
$= -\left(3x^2 + x + 6\right)$
$= -3x^2 - x - 6$

40. If the zeroes of a polynomial p(x) are – 3 and 8, then p(x) equals **(2026)**

(a) $x^2 + 5x - 4$ (b) $(x+3)(-x+8)$
(c) $a\left(x^2 + 5x - 24\right)$ (d) $x^2 - 24$

Solution: (b)

sum of the zeroes = – 3 + 8 = 5
product of the zeroes $= (-3)8 = -24$

$\therefore$ the required polynomial is

$k(x^2 - 5x - 24)$

$= -(x^2 - 5x - 24)$ [let $k = -1$]

$= -(x-8)(x+3)$

$= (-x+8)(x+3)$

Exercise 2.3

(i) If α, β are the zeroes of a polynomial such that α + 3 = −2 and α.β = 3, then write the polynomial.

(ii) If α, β are the zeroes of a polynomial such that α + β = −1 and αβ = −3, then write the polynomial.

(iii) If α,β are the zeroes of a polynomial such that $\alpha + \beta = -\frac{3}{4}$ and $\alpha,\beta = -\frac{3}{2}$, then write the polynomial.

(iv) The quadratic polynomial, the sum of whose zeroes is −3 and their product is −2, is

(a) $x^2 + 3x + 2$ (b) $x^2 - 3x + 2$

(c) $x^2 + 3x - 2$ (d) $x^2 - 3x - 2$

(v) The quadratic polynomial, the sum of whose zeroes is 5 and their product is 7, is

(a) $x^2 + 5x + 7$ (b) $x^2 - 5x - 7$

(c) $x^2 - 5x + 7$ (d) $-x^2 - 5x + 7$

(vi) The quadratic polynomial, the sum of whose zeroes is −6 and their product is 3, is

(a) $x^2 - 6x + 3$ (b) $x^2 + 6x + 3$

(c) $x^2 - 6x - 3$ (d) $-x^2 - 6x + 3$

Key Point If two zeroes of a polynomial $ax^2 + bx + c$ are reciprocal to each other, then a = c.

41. If one zero of the polynomial $(3x^2 + 8x + k)$ is the reciprocal of the other, then the value of k is. **(2020)**

(a) 3 (b) −3

(c) $\frac{1}{3}$ (d) $-\frac{1}{3}$

Solution :

$\alpha.\frac{1}{\alpha} = \frac{k}{3}$

$\Rightarrow 1 = \frac{k}{3}$

$\therefore k = 3$

42. If one zero of the polynomial $6x^2 + 37x - (k-2)$ is reciprocal of the other, then what is the value of k ? **(2023)**

(a) −4 (b) −6

(c) 6 (d) 4

Solution: (a)

$\alpha.\frac{1}{\alpha} = \frac{-(k-2)}{6}$

$\Rightarrow 1 = -\frac{(k-2)}{6}$

$\Rightarrow 6 = -k + 2$

$\Rightarrow -k + 2 = 6$

$\Rightarrow -k = 4$

$\therefore k = -4$

43. If one zero the polynomial $q(x) = (p^2 + 4)x^2 + 65x + 4p$ is reciprocal of the other, then the value of 'p' is **(2025)**

(a) −1 (b) 1

(c) −2 (d) 2

Solution: (d)

Let the zeros of the polynomial q (x) are α and $\frac{1}{\alpha}$

$\therefore \alpha.\frac{1}{\alpha} = \frac{4p}{p^2+4}$

$\Rightarrow \frac{4p}{p^2+4} = 1$

$\Rightarrow p^2 + 4 = 4p$

$\Rightarrow p^2 - 4p + 4 = 0$

$\Rightarrow (p-2)^2 = 0$

$\Rightarrow p - 2 = 0$

$\therefore p = 2$

44. If the zeroes of the polynomial $ax^2 + bx + \frac{2a}{b}$ are reciprocal of each other, then the value of b is **(2025)**

(a) 2 (b) $\frac{1}{2}$

(c) −2 (d) $-\frac{1}{2}$

Solution: (a)

Let the zeroes of the polynomial be

α and $\frac{1}{\alpha}$

$\therefore \alpha \frac{1}{\alpha} = \frac{\frac{2a}{b}}{a}$

$\Rightarrow 1 = \frac{2a}{b} \times \frac{1}{a}$

$\Rightarrow b = 2$

Exercise 2.4

(i) Find the value of k for which the zeroes of the polynomial $2x^2 + 5x + k$ are reciprocal of each other.

(ii) Find the value of a for which the zeroes of the polynomial $ax^2 + 3x + 2$ are reciprocal of each other.

(iii) Find the value of k for which the zeroes of the polynomial $5x^2 - 2x + (k - 1)$ are reciprocal of each other.

(iv) If one zero of the polynomial $(a^2+9)x^2+13x+6a$ is the reciprocal of the other, find the value of a.

(v) If one zero of the polynomial $(k^2+4)x^2+9x+4k$ is the reciprocal of the other, find the value of a.

Key Point If α and β are the zeroes of a polynomial $ax^2 + bx + c$, then $\alpha + \beta = -\frac{b}{a}$

45. If α and β are the zeroes of a polynomial $2x^2 - 13x + 6$, then $\alpha + \beta$ is equal to. **(2020)**

(a) −3 (b) 3

(c) $\frac{13}{2}$ (d) $\frac{-13}{2}$

Solution: (c)

$$\alpha+\beta = -\left(\frac{-13}{2}\right) = \frac{13}{2}$$

46. In the quadratic polynomial $t^2 - 16$, sum of the zeroes is. **(2020)**

Solution:

$t^2 - 16 = t^2 + 0.t - 16$

$\therefore$ Sum of the zeroes $= -\frac{0}{1} = 0$

47. The sum of zeroes of the polynomial $\sqrt{2}x^2 - 17$ are given as **(2023)**

(a) $\frac{17\sqrt{2}}{2}$ (b) $-\frac{17\sqrt{2}}{2}$

(c) 0 (d) 1

Solution: (c)

$\sqrt{2}x^2 - 17 = \sqrt{2}\,x^2 + 0.x - 17$

$\therefore$ Sum of the zeroes $= -\frac{0}{\sqrt{2}} = 0$

48. If α,β are zeroes of the polynomial $x^2 - 1$, then value of $(\alpha + \beta)$ is. **(2023)**

(a) 2 (b) 1

(c) −1 (d) 0

Solution :(d)

$x^2 + 0.x - 1.$

$\therefore \alpha+\beta = \frac{0}{1} = 0$

49. The sum of the zeroes of the polynomial;

$p(x) = 5x - 7x^2 + 3$ is **(2025)**

(a) $\frac{-7}{5}$ (b) $\frac{7}{5}$

(c) $\frac{5}{7}$ (d) $\frac{-5}{7}$

Solution: (c)

$p(x) = -7x^2 + 5x + 3$

$\therefore$ sum of the zeros $= -\left(\frac{5}{-7}\right)$

$= \frac{5}{7}$

50. The value of k for which sum of the zeroes of the polynomial $p(x) = 3x^2 - kx + 6$ is 2, is **(2026)**

(a) 2 (b) −6
(c) −2 (d) 6

Solution: (d)

sum of zeroes $= -\left(-\frac{k}{3}\right)$

$2 = \frac{k}{3}$

$\therefore k = 6$

Exercise 2.5

(i) If α and β are the zeroes of a polynomial $x^2 - 3x + 2$, then $\alpha + \beta$ is equal to.

(a) 3 (b) −3
(c) 2 (d) −2

(ii) If α and β are the zeroes of a polynomial $3x^2 + 4x - 5$, then $\alpha + \beta$ is equal to.

(a) $\frac{4}{3}$ (b) $-\frac{4}{3}$
(c) 3 (d) −5

(iii) If α and β are the zeroes of a polynomial $2x^2 - 4x + 5$, then $\alpha + \beta$ is equal to.

(a) −2 (b) $\frac{5}{2}$
(c) 5 (d) 2

Key Point If α and β are the zeroes of a polynomial $ax^2 + bx + c$, then $\alpha\beta = \frac{c}{a}$

51. If α and β are the zeroes of $(2x^2 + 5x - 9)$, then the value of αβ is. **(2020)**

(a) $\frac{-5}{2}$ (b) $\frac{5}{2}$
(c) $\frac{-9}{2}$ (d) $\frac{9}{2}$

Solution: (c)

$\alpha\beta = \frac{-9}{2}$

52. The product of the two zeroes of the polynomial $3x^2 - 7x - 27$ is. **(2020)**

(a) 27 (b) 9
(c) −9 (d) $\frac{7}{3}$

Solution: (c)

Product of zeroes $= \frac{-27}{3} = -9$

53. If α and β are the zeroes of $(3x^2 + 4x - 3)$, then the value of αβ is. **(2020)**

(a) 1 (b) $\frac{4}{3}$
(c) $\frac{-4}{3}$ (d) −1

Solution: (d)

$\alpha\beta = \frac{-3}{3} = -1$

54. In the quadratic polynomial $x^2 - 6x$, product of the zeroes is. **(2020)**

Solution :

$x^2 - 6x = x^2 - 6x + 0$

$\therefore$ Product of the zeroes $= \frac{0}{1} = 0$

Exercise 2.6

(i) If α and β are the zeroes of $x^2 - 5x + 3$, then the value of αβ is.

(a) 3 (b) −3
(c) 5 (d) −5

(ii) If α and β are the zeroes of $2x^2 + 3x + 5$, then the value of αβ is.

(a) $\frac{3}{2}$ (b) $\frac{-3}{2}$
(c) $\frac{5}{2}$ (d) $\frac{2}{5}$

(iii) If α and β are the zeroes of $3x^2 - 2x + 4$, then the value of αβ is.

(a) 4 (b) 3
(c) $\frac{4}{3}$ (d) $\frac{3}{4}$

(iv) The product of the two zeroes of the polynomial $4x^2 - 3x + 8$ is.

(a) 2 (b) −2
(c) $\frac{-3}{4}$ (d) $\frac{3}{4}$

(v) The product of the two zeroes of the polynomial $3x^2 + 2x + 8$ is.

(a) $\frac{-8}{3}$ (b) $\frac{8}{3}$

(c) $\frac{2}{3}$ (d) $\frac{-2}{3}$

(vi) The product of the two zeroes of the polynomial $x^2 - 3x + 10$ is.

(a) 3 (b) −3

(c) 10 (d) −10

55. The sum and product of zeroes of the polynomial $p(x) = x^2 + 5x + 6$ are respectively. **(2023)**

(a) 5, −6 (b) −5, 6

(c) 2, 3 (d) −2, −3

Solution: (b)

Sum of zeroes $= \frac{-5}{1} = -5$

Product of zeroes $= \frac{6}{1} = 6$

56. The sum and product of zeroes of the polynomial $p(x) = 3x^2 - 5x + 2$ are respectively. **(2023)**

(a) $\frac{5}{3}, \frac{2}{3}$ (b) $\frac{-5}{3}, \frac{2}{3}$

(c) $1, \frac{2}{3}$ (d) $\frac{-5}{3}, \frac{-2}{3}$

Solution: (a)

Sum of zeroes $= -\left(\frac{-5}{3}\right) = \frac{5}{3}$

Product of zeroes $= \frac{2}{3}$

57. If α and β are the zeroes of the polynomial $x^2 + 2x + 1$, then $\frac{1}{\alpha} + \frac{1}{\beta}$ is equal to. **(2020)**

(a) −2 (b) 2

(c) 0 (d) 1

Solution: (a)

$\alpha + \beta = -2$ and $\alpha.\beta = 1$

Now, $\frac{1}{\alpha} + \frac{1}{\beta}$

$= \frac{\alpha + \beta}{\alpha\beta}$

$= \frac{-2}{1}$

$= -2$

58. If α and β are the zeros of the polynomial $p(x)) = x^2 + x - 1$, then $+\frac{1}{\alpha} + \frac{1}{\beta}$ is equal to. **(2010)**

(a) 1 (b) 2

(c) −1 (d) $-\frac{1}{2}$

Solution: (a)

$\alpha + \beta = -1 \text{and } \alpha.\beta = -1$

Now, $\frac{1}{\alpha} + \frac{1}{\beta}$

$= \frac{\alpha + \beta}{\alpha\beta}$

$= \frac{-1}{-1}$

$= 1$

59. If α and β are the zeroes of the polynomial $p(x) = 4x^2 - 3x - 7$, then $\frac{1}{\alpha} + \frac{1}{\beta}$ is equal to. **(2023)**

Solution: (d)

$\alpha + \beta = -\left(\frac{-3}{4}\right) = \frac{3}{4}$

$\alpha.\beta = \frac{-7}{4}$

Now, $\frac{1}{\alpha} + \frac{1}{\beta}$

$= \frac{\alpha + \beta}{\alpha\beta}$

$= \frac{\frac{3}{4}}{\frac{-7}{4}}$

$= \frac{-3}{7}$

60. If α and β are the zeroes of the polynomial p (x) $= 6x^2 - 5x - 4$, then $\frac{1}{\alpha}+\frac{1}{\beta}$ is equal to. **(2023)**

(a) $\frac{5}{4}$ (b $\frac{-5}{4}$

(c) $\frac{4}{5}$ (d) $\frac{5}{24}$

Solution: (b)

$\alpha+\beta=-\left(\frac{-5}{6}\right)=\frac{5}{6}\cdot\alpha\beta=\frac{-4}{6}$

Now, $\frac{1}{\alpha}+\frac{1}{\beta}=\frac{\alpha+\beta}{\alpha\beta}=\frac{\frac{5}{6}}{\frac{-4}{6}}=\frac{-5}{4}$

61. α, β are zero of the polynomial $2x^2 + 5x + 1$:

The value of $\left(\frac{1}{\alpha}+\frac{1}{\beta}\right)$ is. **(2025)**

(a) $\frac{-5}{4}$ (b) 5

(c) $\frac{5}{4}$ (d) -5

Solution: (d)

$\alpha+\beta=\frac{-5}{2}$

$\alpha\beta=\frac{1}{2}$

Now, $\frac{1}{\alpha}+\frac{1}{\beta}$

$=\frac{\alpha+\beta}{\alpha\beta}$

$=\frac{\frac{-5}{2}}{\frac{1}{2}}$

$=-5$

62. If α,β are zeroes of the polynomial $3x^2 + 14x - 5$, then the value of $3\left(\frac{\alpha+\beta}{\alpha\beta}\right)$ is **(2025)**

(a) $\frac{14}{5}$ (b) $\frac{42}{5}$

(c) $-\frac{14}{5}$ (d) $-\frac{42}{5}$

Solution: (b)

$\alpha+\beta=\frac{-14}{3}$

$\alpha\beta=\frac{-5}{3}$

Now, $3\left(\frac{\alpha+\beta}{\alpha\beta}\right)$

$=3\times\frac{\frac{-14}{3}}{\frac{-5}{3}}=\frac{42}{5}$

63. If α and β are zeroes of the polynomial $x^2 - x - 4$, then the value of $\left(\frac{1}{\alpha}+\frac{1}{\beta}\right)$ is **(2025)**

(a) -4 (b) $-\frac{1}{4}$

(c) $\frac{1}{4}$ (d) 4

Solution:(b)

$\alpha+\beta=-(-1)=1$

$\alpha\beta=-4$

$\frac{1}{\alpha}+\frac{1}{\beta}$

$=\frac{\alpha+\beta}{\alpha\beta}$

$=\frac{1}{-4}$

$=-\frac{1}{4}$

64. If the square of the difference of the zeroes of the quadratic polynomial $y^2 + py + 36$ is equal to 81, then the value of p are. **(2025)**

(a) ±5 (b) ±15
(c) ±18 (d) ±12

Solution: (b)

let the zeroes of the polynomial be α and β

$\therefore \alpha + \beta = -p$ ...(i)

$\alpha\beta = 36$ ...(ii)

$(\alpha - \beta)^2 = 81$

$\Rightarrow (\alpha + \beta)^2 - 4\alpha\beta = 81$

$\Rightarrow (-p)^2 - 4 \times 36 = 81$

$\Rightarrow p^2 = 81 + 144$

$\Rightarrow p^2 = 225$

$\therefore p = \pm 15$

65. If the zeroes of the polynomial $p(x) = 2x^2 - 7x + 6$ are α and β, then the value of $\frac{1}{\alpha} + \frac{1}{\beta}$ is **(2026)**

(a) $\frac{7}{2}$ (b) $\frac{6}{7}$
(c) $-\frac{7}{6}$ (d) $\frac{7}{6}$

Solution: (d)

$\alpha + \beta = -\left(\frac{-7}{2}\right) = \frac{7}{2}$

$\alpha\beta = \frac{6}{2} = 3$

Now, $\frac{1}{\alpha} + \frac{1}{\beta}$

$= \frac{\beta + \alpha}{\alpha\beta}$

$= \frac{\alpha + \beta}{\alpha\beta}$

$= \frac{\frac{7}{2}}{3}$

$= \frac{7}{6}$

Exercise 2.7

(i) If α and β are the zeroes of the polynomial $2x^2 - 3x + 1$, then $\frac{1}{\alpha} + \frac{1}{\beta}$ is equal to.

(a) 3 (b) –3
(c) –2 (d) 2

(ii) If α and β are the zeroes of the polynomial $x^2 + 3x + 2$, then $\frac{1}{\alpha} + \frac{1}{\beta}$ is equal to.

(a) 3 (b) 2
(c) $\frac{3}{2}$ (d) $-\frac{3}{2}$

(iii) If α and β are the zeroes of the polynomial $x^2 + 3x + 8$, then $\alpha + \beta + \alpha\beta$ is.

(a) –3 (b) 8
(c) 0 (d) 5

(iv) If α and β are the zeroes of the polynomial $2x^2 + 5x - 9$, then $\alpha + \beta + \alpha\beta$ is.

(a) 0 (b) 7
(c) –7 (d) $-\frac{9}{2}$

(v) If α and β are the zeroes of the polynomial $x^2 - p(x + 1) - c$, then $(\alpha + 1)(\beta + 1)$ is.

(a) 1 + c (b) c
(c) 1 – c (d) 1

66. If the sum of the zeroes of the quadratic polynomial $kx^2 + 2x + 3k$ is equal to their product, then k equals. **(2020)**

(a) $\frac{1}{3}$ (b) $-\frac{1}{3}$
(c) $\frac{2}{3}$ (d) $-\frac{2}{3}$

Solution: (d)

$\alpha + \beta = \alpha\beta$

$\Rightarrow \frac{-2}{k} = \frac{3k}{k}$

$\Rightarrow k = -\frac{2}{3}$

67. If α and β are the zeroes of the polynomial $p(x) = kx^2 - 30x + 45k$ and $\alpha + \beta = \alpha\beta$, then the value of k is. **(2024)**

(a) $-\frac{2}{3}$ (b) $-\frac{3}{2}$
(c) $\frac{3}{2}$ (d) $\frac{2}{3}$

Solution: (d)

$\alpha+\beta=\alpha\beta$

$\Rightarrow -\left(\frac{-30}{k}\right)=\frac{45k}{k}$

$\Rightarrow \frac{30}{k}=45$

$\Rightarrow k=\frac{30}{45}=\frac{2}{3}$

68. If 'α' and 'β' are the zeroes of the polynomial $ax^2 - 5x + c$ and $\alpha + \beta = \alpha\beta = 10$,then. **(2023)**

(a) $a = 5, c = \frac{1}{2}$ (b) $a = 1, c = \frac{5}{2}$

(c) $a = \frac{5}{2}, c = 1$ (d) $a = \frac{1}{2}, c = 5$

Solution: (d)

$\alpha+\beta=-\left(\frac{-5}{a}\right)=\frac{5}{a}$

$\alpha\beta=\frac{c}{a}$

$\therefore \alpha+\beta=10$

$\Rightarrow \frac{5}{a}=10$

$\Rightarrow a=\frac{1}{2}$

and $\alpha\beta=\frac{c}{a}$

$\Rightarrow 10=\frac{c}{\frac{1}{2}}$

$\Rightarrow 2c = 10$

$\therefore c = 5$

69. If α and β are the zeroes of polynomial $3x^2 + 6x + k$ such that $\alpha+\beta+\alpha\beta=-\frac{2}{3}$ then the value of k is: **(2025)**

(a) −8 (b) 8
(c) −4 (d) 4

Solution: (d)

$\alpha+\beta=\frac{-6}{3}=-2$

$\alpha\beta=\frac{k}{3}$

Now, $\alpha+\beta+\alpha\beta=-\frac{2}{3}$

$\Rightarrow -2+\frac{k}{3}=-\frac{2}{3}$

$\Rightarrow \frac{k}{3}=2-\frac{2}{3}$

$\Rightarrow \frac{k}{3}=\frac{4}{3}$

$\therefore k=4$

70. If α and β are two zeroes of a polynomial $f(x) = px^2 - 2x + 3p$ and $\alpha + \beta = \alpha\beta$,then value of p is **(2026)**

(a) $-\frac{2}{3}$ (b) $\frac{2}{3}$

(c) $\frac{1}{3}$ (d) $-\frac{1}{3}$

Solution: (b)

$\alpha+\beta=-\left(\frac{-2}{p}\right)=\frac{2}{p}$

$\alpha\beta=\frac{3p}{p}=3$

Now, $\alpha+\beta=\alpha\beta$

$\Rightarrow \frac{2}{p}=3$

$\therefore p=\frac{2}{3}$

Exercise 2.8

(i) If the sum of the zeroes of the quadratic polynomial $(k + 1)x^2 - 3x + 2(k + 1)$ is equal to their product, then k equals.

(a) $\frac{1}{2}$ (b) $-\frac{1}{2}$

(c) 1 (d) $\frac{5}{3}$

(ii) If the sum of the zeroes of the quadratic polynomial $ax^2 - 3x + 4a$ is equal to their product, then k equals.

(a) $\frac{3}{4}$ (b) $\frac{4}{3}$

(c) 4 (d) 13

(iii) If the sum of the zeroes of the quadratic polynomial $(2a + 1) x^2 - 3x + 5(2a + 1)$ is equal to their product, then k equals.

(a) $\frac{1}{3}$ (b) $-\frac{2}{3}$

(c) $\frac{3}{2}$ (d) $\frac{-1}{5}$

(iv) If 'α' and 'β' are the zeroes of the polynomial $f(x) = kx^2 - 12x + 6k$ and $\alpha + \beta = \alpha\beta$, then, the value of k is.

(a) 2 (b) – 2

(c) $\frac{1}{2}$ (d) $\frac{-1}{2}$

(v) If 'α' and 'β' are the zeroes of the polynomial $f(x) = kx^2 - 6x + 18k$ and $\alpha + \beta = \alpha\beta$, then, the value of k is.

(a) 3 (b) $\frac{1}{3}$

(c) – 3 (d) $\frac{-1}{3}$

(vi) If 'α' and 'β' are the zeroes of the polynomial $p(x) = ax^2 - 7x + 3a$ and $\alpha + \beta = \alpha\beta$, then, the value of a is.

(a) $\frac{7}{3}$ (b) $\frac{-1}{3}$

(c) 3 (d) 7

(vii) If 'α' and 'β' are the zeroes of the polynomial $ax^2 - 16x + c$ and $\alpha + \beta = \alpha\beta = 8$, then.

(a) $a = 2, c = 16$ (b) $a = 16, b = 2$

(c) $a = \frac{1}{2}, c = 4$ (d) $a = 16, b = \frac{1}{2}$

(viii) If 'α' and'β' are the zeroes of the polynomial $ax^2 - 2x + c$ and $\alpha + \beta = \alpha\beta = 4$, then.

(a) $a = \frac{1}{2}, c = 2$ (b) $a = 2, b = \frac{1}{2}$

(c) $a = 4, c = \frac{1}{2}$ (d) $a = 4, b = \frac{1}{4} + 1$

(ix) If 'α' and 'β' are the zeroes of the polynomial $ax^2 - 3x + c$ and $\alpha + \beta = \alpha\beta = 12$, then.

(a) $a = \frac{1}{4} + 1, c = 4$

(b) $a = \frac{1}{4} + 1, c = 3$

(c) $a = 3, c = \frac{1}{4} + 1$

(d) $a = 6, c = \frac{1}{2}$

71. If the sum of zeroes of the polynomial $p(x) = 2x^2 - k\sqrt{2}x + 1$ is $\sqrt{2}$, then value of k is. **(2024)**

(a) $\sqrt{2}$ (b) 2

(c) $2\sqrt{2}$ (d) $\frac{1}{2}$

Solution: (b)

$$\alpha + \beta = -\frac{(-k\sqrt{2})}{2}$$

$$\Rightarrow \sqrt{2} = \frac{k\sqrt{2}}{2}$$

$$\therefore k = 2$$

72. For what value of k, the product of zeroes of the polynomial $kx^2 - 4x - 7$ is 2. **(2024)**

(a) $-\frac{1}{14}$ (b) $-\frac{7}{2}$

(c) $\frac{7}{2}$ (d) $\frac{-2}{7}$

Solution: (b)

$$\frac{-7}{k} = 2$$

$$\Rightarrow 2k = -7$$

$$\therefore k = -\frac{7}{2}$$

73. If 'α' and 'β' are the zeroes of the polynomial $p(x) = 2x^2 - x - 1$ then $\alpha^2 + \beta^2$ is equal to. **(2023)**

(a) $-\frac{3}{4}$ (b) $\frac{5}{4}$

(c) $\frac{1}{4}+1$ (d) $\frac{3}{4}$

Solution: (b)

$\alpha+\beta=-\left(\frac{1}{2}\right)$

$\Rightarrow \alpha+\beta=\frac{1}{2}$

Now, $\alpha^2+\beta^2=(\alpha+\beta)^2-2\,\alpha\beta$

$=\left(\frac{1}{2}\right)^2-2\times\left(-\frac{1}{2}\right)$

$=\frac{1}{4}+1$

$=\frac{5}{4}$

74. If 'α' and 'β' are the zeroes of the quadratic polynomial $p(x) = x^2 - ax - b$, then $\alpha^2 + \beta^2$ is. **(2023)**

(a) $a^2 - 2b$ (b) $a^2 + 2b$

(c) $b^2 - 2a$ (d) $b^2 + 2a$

Solution: (b)

$\alpha+\beta=-\left(\frac{-a}{1}\right)=a$

$\alpha\beta=-b$

Now, $\alpha^2+\beta^2$

$=(\alpha+\beta)^2-2\,\alpha\beta$

$=a^2-2(-b)$

$=a^2+2b$

Exercise 2.9

(i) If 'α' and 'β' are the zeroes of the polynomial $f(x) = 2x^2 + 3x + 5$ then $\alpha^2 + \beta^2$ is equal to.

(a) $\frac{11}{2}$ (b) $\frac{-11}{2}$

(c) $\frac{11}{4}$ (d) $\frac{-11}{4}$

(ii) If 'α' and 'β' are the zeroes of the polynomial $f(x) = x^2 - 3x + 2$ then $\alpha^2 + \beta^2$ is equal to.

(a) 9 (b) 0

(c) 5 (d) 4

(iii) If 'α' and 'β' are the zeros of the polynomial $f(x) = x^2 + 5x + 8$ then $\alpha^2 + \beta^2$ is equal to.

(a) 9 (b) 7

(c) 0 (d) 40

(iv) If 'α' and 'β' are the zeroes of the polynomial $p(x) = 2x^2 - 3x - 9$ then $\alpha^2 + \beta^2$ is equal to.

(a) $\frac{23}{4}$ (b) $\frac{-23}{4}$

(c) $\frac{45}{4}$ (d) $\frac{-27}{4}$

(v) If 'α' and 'β' are the zeroes of the polynomial $p(x) = ax^2 + bx + c$, then $\alpha^2 + \beta^2$ is equal to.

(a) $\frac{a^2-2bc}{b^2}$ (b) $\frac{b^2-2ac}{a^2}$

(c) $\frac{a^2+2bc}{b^2}$ (d) $\frac{b^2+2ac}{a^2}$

75. The zeroes of the polynomial $p(x) = x^2 + 4x + 3$ are given by. **(2023)**

(a) 1, 3 (b) – 1, 3

(c) 1, – 3 (d) – 1, – 3

Solution: (d)

$x^2 + 4x + 3 = 0$

$\Rightarrow (x + 1)(x + 3) = 0$

$\Rightarrow x + 1 = 0$ or $x + 3 = 0$

$\Rightarrow x = -1$ or $x = -3$

76. The zeroes of the polynomial $p(x) = 3x^2 + 11x - 4$ are given by. **(2023)**

(a) $\frac{1}{3}, -4$ (b) $-\frac{1}{3}, 4$

(c) $\frac{1}{3}, 4$ (d) $-\frac{1}{3}, -4$

Solution; (a)

$3x^2 + 11x - 4 = 0$

$\Rightarrow 3x^2 + 12x - x - 4 = 0$

$\Rightarrow 3x(x + 4) - 1\ (x + 4) = 0$

$\Rightarrow (x + 4)\ (3x - 1) = 0$

$\Rightarrow x + 4 = 0$ or $3x - 1 = 0$

$x = -4$ or $x = \frac{1}{3}$

77. The zeroes of the polynomial $p(x) = x^2 + 3x + 2$ are given by. **(2023)**

(a) 1,2 (b) 2, – 1

(c) – 2,1 (d) –2, – 1

Solution: (d)

$\Rightarrow x^2 + 3x + 2 = 0$

$\Rightarrow (x + 1)\ (x + 2) = 0$

$\Rightarrow x + 1 = 0$ or $x + 2 = 0$

$\Rightarrow x = -1$ or $x = -2$

78. The zeroes of the quadratic polynomial $16x^2 - 9$ are. **(2023)**

(a) $\frac{3}{4}, \frac{3}{4}$ (b) $-\frac{3}{4}, \frac{3}{4}$

(c) $\frac{9}{16}, \frac{9}{16}$ (d) $-\frac{3}{4}, -\frac{3}{4}$

Solution: (b)

$\Rightarrow 16x^2 - 9 = 0$

$\Rightarrow 16x^2 = 9$

$\Rightarrow x^2 = \frac{9}{16}$

$\Rightarrow x = \pm\frac{3}{4}$

79. The zeroes of the quadratic polynomial $25x^2 - 49$ are. **(2023)**

(a) $\frac{49}{25}, \frac{49}{25}$ (b) $\frac{-49}{25}, \frac{+49}{25}$

(c) $\frac{7}{5}, \frac{-7}{5}$ (d) $\frac{7}{5}, \frac{7}{5}$

Solution: (c)

$\Rightarrow 25x^2 - 49 = 0$

$\Rightarrow 25x^2 = 49$

$\Rightarrow x^2 = \frac{49}{25}$

$\therefore x = \pm\frac{7}{5}$

80. The zeroes of the quadratic polynomial $p(x) = 2x^2 - x - 3$ are. **(2023)**

(a) $\frac{-3}{2}, 1$ (b) $\frac{3}{2}, 1$

(c) $\frac{-3}{2}, -1$ (d) $\frac{3}{2}, -1$

Solution: (d)

$2x^2 - x - 3 = 0$

$\Rightarrow (2x - 3)\ (x + 1) = 0$

$\Rightarrow 2x - 3 = 0$ or $x + 1 = 0$

$\Rightarrow x = \frac{3}{2}$ or $x = -1$

81. The zeroes of the polynomial $x^2 - 3x - m\ (m+3)$ are. **(2020)**

(a) m,m + 3 (b) – m, m + 3

(c) m, – (m + 3) (d) – m, – (m + 3)

Solution: (b)

$\Rightarrow x^2 - 3x - m\ (m + 3) = 0$

$\Rightarrow x^2 - \{(m + 3) - m\}x - m(m + 3) = 0$

$\Rightarrow x^2 - (m + 3)x + mx - m(m + 3) = 0$

$\Rightarrow x\ (x - m - 3) + m\ (x - m - 3) = 0$

$\Rightarrow (x - m - 3)\ (x + m) = 0$

$\Rightarrow x - m - 3 = 0$ or $x + m = 0$

$\Rightarrow x = m + 3$ or $x = -m$

Exercise 2.10

(i) The zeroes of the quadratic polynomial $x^2 + 12x + 35$ are.

(a) 5,7 (b) –5,7

(c) 5,–7 (d) –5,–7

(ii) The zeroes of the quadratic polynomial $6x^2 + 11x + 3$ are.

(a) $\frac{-3}{2}, -\frac{1}{3}$ (b) $\frac{3}{2}, -\frac{1}{3}$

(c) $\frac{3}{2}, \frac{1}{3}$ (d) $\frac{-3}{2}, \frac{1}{3}$

(iii) The zeroes of the quadratic polynomial $x^2 - 3x - 10$ are.

(a) –5, 2 (b) –5, –2

(c) 5, –2 (d) 5, 2

82. Which of the following is a quadratic polynomial having zeroes $\frac{-2}{3}$ and $\frac{2}{3}$? **(2023)**

(a) $4x^2 - 9$ (b) $\frac{4}{9}(9x^2 + 4)$

(c) $x^2 + \frac{9}{4}$ (d) $5(9x^2 - 4)$

Solution: (d)

Sum of zeroes $= \frac{-2}{3} + \frac{2}{3} = 0$

Product of zero $= \left(\frac{-2}{3}\right)\left(\frac{2}{3}\right) = \frac{-4}{9}$

$\therefore$ the polynomial is

$k(x^2 - 0.x - \frac{4}{9})$

$= k(x^2 - \frac{4}{9})$

$= \frac{k}{9}(9x^2 - 4)$

$= \frac{45}{9}(9x^2 - 4)$ [let k = 45]

$= 5(9x^2 - 4)$

83. Which of the following is a quadratic polynomial with zeroes $\frac{5}{3}$ and 0 ? **(2023)**

(a) $3x(3x - 5)$ (b) $3x(x - 5)$

(c) $x^2 - \frac{5}{3}$ (d) $\frac{5}{3}x^2$

Solution: (a)

Sum of zeroes $= \frac{5}{3} + 0 = \frac{5}{3}$

Product of zeroes $= \frac{5}{3} \times 0 = 0$

The polynomial will be

$k(x^2 - \frac{5}{3}x + 0)$

$= \frac{kx}{3}(3x - 5)$

$= \frac{9x}{3}(3x - 5)$ [Let k = 9]

$= 3x(3x - 5)$

84. A, polynomial $p(x)$, which has sum of its zeroes equal to their product, is: **(2026)**

(a) $3x^2 + 2x + 2$ (b) $3x^2 - 2x - 3$

(c) $3x^2 - 2x + 2$ (d) $x^2 - 3x + 2$

Solution: (c)

Sum of the zeroes of the polynomial $3x^2 - 2x + 2$ is $-\left(\frac{-2}{3}\right) = \frac{2}{3}$

Product of the zeroes is $\frac{2}{3}$

Exercise 2.11

(i) Which of the following is a quadratic polynomial having zeroes 2 and –3 ?

(a) $x^2 + x - 6$ (b) $x^2 - x - 6$

(c) $x^2 + x + 6$ (d) $x^2 + 2x + 6$

(ii) Which of the following is a quadratic polynomial having zeroes $\frac{3}{2}$ and $-\frac{1}{2}$?

(a) $4x^2 + 4x + 3$ (b) $4x^2 - 4x - 3$

(c) $4x^2 + 4x - 3$ (d) $4x^2 - 4x + 3$

(iii) Which of the following is a quadratic polynomial having zeroes $\frac{3}{2}$ and $-\frac{3}{2}$?

(a) $9(4x^2 + 9)$ (b) $4(x^2 - 9)$

(c) $9(4x^2 - 9)$ (d) $4x^2 - x + 3$

85. If one zeroes of the polynomial $x^2 - 3kx + 4k$ be twice the other, then the value of k is. **(2023)**

(a) –2 (b) 2

(c) $\frac{1}{2}$ (d) $-\frac{1}{2}$

Solution: (b)

$\alpha + 2\alpha = -\left(\frac{-3k}{1}\right)$

$\Rightarrow 3 = 3k$

$\Rightarrow \alpha = k$ (i)

and $\alpha . 2\alpha = 4k$

$\Rightarrow 2\alpha^2 = 4k$

$\Rightarrow \alpha^2 = 2k$

$\Rightarrow k^2 = 2k$ [form (i)]

$\Rightarrow k^2 - 2k = 0$

$\Rightarrow k(k - 2) = 0$

$\Rightarrow (k - 2) = 0$ $[k \neq 0]$

$\therefore k = 2$

Exercise 2.12

(i) If one zeroes of the polynomial $x^2 - 4kx + 3k$ be twice the other, then the value of k is.

(a) $\frac{27}{32}$ (b) $\frac{32}{27}$

(c) 8 (d) $\frac{3}{2}$

(ii) If one zeroes of the polynomial $x^2 - 6kx + 8k$ be twice the other, then the value of k is.

(a) –1 (b) 0

(c) –2 (d) 1

(iii) If one zero of the polynomial $x^2 + 9kx + 16k$ be twice the other, then the value of k is.

(a) $\frac{9}{8}$ (b) $\frac{8}{9}$

(c) 8 (d) 9

86. If the zeroes of the quadratic polynomial $x^2 + (a+1)x + b$ are 2 and –3, then. **(2023)**

(a) $a = -7, b = -1$ (b) $a = 5, b = -1$

(c) $a = 2, b = -6$ (d) $a = 0, b = -6$

Solution: (d)

$2^2 + (a+1).2 + b = 0$

$\Rightarrow 4 + 2a + 2 + b = 0$

$\Rightarrow 2a + b = -6$ (i)

and $(-3)^2 + (a+1)(-3) + b = 0$

$\Rightarrow 9 - 3a - 3 + b = 0$

$\Rightarrow -3a + b = -6$

$\Rightarrow 3a - b = 6$ (ii)

Solving (i) and (ii) we get

$a = 0, b = -6$

Exercise 2.13

(i) If the zeroes of the quadratic polynomial $x^2 + (a+1)x + b$ are 1 and –1, then.

(a) $a = 1, b = -1$

(b) $a = -1, b = 1$

(c) $a = -1, b = -1$

(d) $a = 0, b = 1$

(ii) If –2 and 3 are the zeroes of the quadratic polynomial $x^2 + (a + 1)x + b$, then.

(a) $a = -2, b = 6$

(b) $a = 2, b = -6$

(c) $a = -2, b = -6$

(d) $a = 2, b = 6$

(iii) If 4 and –3 are the zeroes of the quadratic polynomial $x^2 + (a + 1)x + b$, then.

(a) $a = -2, b = -12$

(b) $a = -1, b = -10$

(c) $a = 2, b = 12$

(d) $a = -2, b = 12$

87. The zeroes of a polynomial $x^2 + px + q$ are twice the zeroes of the polynomial $4x^2 - 5x - 6$. The value of p is. **(2024)**

(a) $-\frac{5}{2}$ (b) $\frac{5}{2}$

(c) –5 (d) 10

Solution: (a)

$4x^2 - 5x - 6 = 0$

$\Rightarrow 4x^2 - 8x + 3x - 6 = 0$

$\Rightarrow 4x(x - 2) + 3(x - 2) = 0$

$\Rightarrow (x - 2)(4x + 3) = 0$

$\Rightarrow x = 2$ or $x = -\frac{3}{4}$

$\therefore$ Sum of zeroes of the polynomial $x^2 + px + q =$

$$4+\left(\frac{-3}{2}\right)=\frac{5}{2}$$

$$\therefore -p=\frac{5}{2}$$

$$\Rightarrow p=\frac{-5}{2}$$

Exercise 2.14

(i) The zeroes of a polynomial $x^2 - px + q$ are twice the zeroes of the polynomial $2x^2 - 5x - 3$. The value of p is.

(a) -2 (b) 6
(c) 5 (d) -4

(ii) The zeroes of a polynomial $x^2 + x - q$ are twice the zeroes of the polynomial $2x^2 - 5x - 3$. The value of q is.

(a) 8 (b) 6
(c) 10 (d) -4

(iii) The zeroes of a polynomial $x^2 + px + q$ are twice the zeroes of the polynomial $x^2 + 5x + 6$. The value of p is.

(a) 8 (b) 6
(c) 10 (d) -4

(iv) The zeroes of a polynomial $x^2 - px + q$ are twice the zeroes of the polynomial $x^2 + 2x - 15$. The value of p is.

(a) -10 (b) -6
(c) -16 (d) 16

(v) The zeroes of a polynomial $x^2 + px + q$ are twice the zeros of the polynomial $6x^2 + x - 12$. The value of p is

(a) $\frac{1}{3}$ (b) $-\frac{1}{3}$
(c) $\frac{4}{3}$ (d) $-\frac{3}{2}$

88. The degree of polynomial having zeroes -3 and 4 only is. **(2020)**

(a) 2 (b) 1
(c) more than 3 (d) 3

Solution: (a) 2

89. The maximum number of zeroes of a cubic polynomial can have is. **(2020)**

(a) 1 (b) 4
(c) 2 (d) 3

Solution: (d) 3

90. The number of polynomial having zeroes -3 and 5 is. **(2023)**

(a) only one (b) infinite
(c) exactly two (d) at most two

Solution: (b) infinite

91. The number of polynomial having zeroes -1 and 2 is. **(2023)**

(a) exactly 2 (b) only 1
(c) at most 2 (d) infinite

Solution: (d) infinite

92. The number of quadratic polynomial having zeroes -5 and -3 only is. **(2023)**

(a) 1 (b) 2
(c) 3 (d) more than 3

Solution: (d) more than 3

93. The number of polynomial having zeroes -3 and 4 is. **(2023)**

(a) 1 (b) 2
(c) 3 (d) more than 3

Solution: (d) more than 3

94. If $p(x) = x^2 + 5x + 6$, then $p(-2)$ is. **(2023)**

(a) 20 (b) 0
(c) -8 (d) 8

Solution: (b)

$p(-2) = (-2)^2 + 5(-2) + 6$
$= 4 - 10 + 6$
$= 0$

95. Which of the following statements is true for a polynomial p(x) of degree 3? **(2025)**

(a) p(x) has at most two distnct zeroes.
(b) p(x) has at teast two distinct zeroes.
(c) p(x) has exactly three distinct zeroes.
(d) p(x) has at most three distinct zeroes.

Solution: (d)

p(x) has at most three distinct zeroes.

96. The value of x for which the polynomials $9 - x^2$ and $6x + x^2 + 9$ vanish simultaneously, is **(2025)**

(a) 3 (b) 2
(c) –2 (d) –3

Solution:(d)

$9 - x^2 = 0$

$\Rightarrow x^2 = 9$

$\Rightarrow x = \pm 3$

∴ The zeroes of the poynomial are 3, –3

$6x + x^2 + 9 = 0$

$\Rightarrow x^2 + 6x + 9 = 0$

$\Rightarrow (x+3)^2 = 0$

$\Rightarrow x = -3, -3$

∴ The zeroes of the polynomial $6x + x^3 + 9$ are –3, –3

∴ The required value of x is –3

97. A quadratic polynomial having only zero (–2) is **(2025)**

(a) $(x-2)^2$ (b) $x^2 - 2$
(c) $x^2 + 2x$ (d) $(x+2)^2$

Solution: (d)

$(x+2)^2$

98. How many zeroes does p (x) = $(x-2)(x+3)$ have ? **(2026)**

(a) zero (b) one
(c) two (d) three

Solution: (c)

Two

Question on **99** to **103** consist of statements, namely, **Assertion (A)** and **Reason (R)**. For selecting the correct answer, use the following code :

(a) Both Assertion (A) and Reason (R) are true and Reason (R) is a correct explanation of Assertion (A).
(b) Both Assertion (A) and Reason (R) are true but Reason (R)is not a correct explanation of Assertion (A).
(c) Assertion (A) is true and Reason (R) is false.
(d) Assertion (A) is false and Reason (R) is true.

99. Assertion (A): Polynomial $x^2 + 4x$ has two real zeroes.

Reason (R): Zeroes of the polynomial $x^2 + ax$ $(a \neq 0)$ are 0 and a. **(2023)**

Solution: (c)

Assertion (A) is true, but Reason (R) is false.

100. Assertion (A): Degree of a zero polynomial is not defined.

Reason (R): Degree of a non – zero constant polynomial is 0. **(2024)**

Solution: (b)

Both Assertion (A) and Reason (R) are true but Reason (R)is not a correct explanation of Assertion (A).

101. Both **Assertion (A):** Zeroes of a polynomial p (x) = $x^2 - 2x - 3$ are – 1 and 3.

Reason (R): The graph of polynomial p (x) = $x^2 - 2x - 3$ intersects x – axis at (– 1, 0) and (3, 0). **(2024)**

Solution: (a)

Both Assertion (A) and Reason (R) are true and Reason (R) is a correct explanation of Assertion (A).

102. Assertion (A): if the graph of a polynomial touches x – axis at only one point, then the polynomial cannot be a quadratic polynomial.

Reason (R): A polynomial of degree n $(n > 1)$ can have at most n zeros. **(2024)**

Solution: (b)

Both Assertion (A) and Reason (R) are true but Reason (R)is not a correct explanation of Assertion (A).

103. Assertion (A): The polynomial $p\ (y) = y^2 + 4y + 3$ has two zeroes.

Reason(R): A quadratic polynomial can have at most two zeroes **(2026)**

Solution: (b)

$p\ (y) = y^2 + 4y + 3$

$= (y+3)(y+1)$

∴ zeroes are –3, –1

Direction: Question number **i to x** are Assertion and Reason based questions. Two statements are given, one labelled as Assertion (A) and the other labelled as Reason (R). Select the correct answer to these questions from the codes (A), (B), (C) and (D) as given below.

(a) Both Assertion (A) and Reason (R) are true and Reason (R)

(b) Both Assertion (A) and Reason (R) are true, but Reason (R) is not the correct explanation of Assertion (A).

(c) Assertion (A) is true, but Reason (R) is false.

(d) Assertion (A) is false, but Reason (R) is true.

Exercise 2.15

(i) Assertion (A): If the sum of the zeroes of the quadratic polynomial $x^2 + 2kx + 8$ is 2, then the value of k is 1

Reason (R): Sum of zeroes of the quadratic polynomial $ax^2 + bx + c$ is $-\frac{b}{a}$.

(ii) Assertion (A): If one zero of polynomial $f(x) = (k^2 + 9)x^2 + 11x + 6k$ is reciprocal of the other, then $k = 3$

Reason (R): Product of the zeroes of the polynomial $ax^2 + bx + c$ is $\frac{c}{a}$

(iii) Assertion (A): The graph of a polynomial intersect x- axis at 4 points and y axis at 1 point, the polynomial has 4 zeroes.

Reason (R): The number of zeroes that a polynomial f(x) can have is the number of times polynomial intersect x and y axis.

(iv) Assertion (A): The sum and product of the zeroes of a quadratic polynomial are $-\frac{1}{3}$ and $\frac{1}{3}$ respectively, then the quadratic polynomial is $3x^2 + x + 1$

Reason (R): The quadratic polynomial whose sum and product of zeroes are given is x^2 – (sum of zeroes)x + product of zeroes .

(v) Assertion (A): Both zeroes of quadratic polynomial $x^2 - 3(k + 1)x + 2$ are equal in magnitude but oppsite in sign then value of k is $\frac{1}{2}$.

Reason (R): Sum of zeroes of a quadratic polynomial $ax^2 + bx + c$ is $-\frac{b}{a}$.

(vi) Assertion (A): If the product of the zeroes of the quadratic polynomial $x^2 + 2x + 3k$ *is* – 6 then value of k is –2.

Reason (R): Sum of zeroes of a quadritc polynomial $ax^2 + bx + c$ is $\frac{c}{a}$

(vii) Assertion (A): If the zeroes of the quadratc polynomial $(k - 1)x^2 + kx + 1$ *is* –3, then the value of k is $\frac{4}{3}$.

Reason (R): If α is a zero of the polynomial $p(x) = ax^2 + bx + c$, then $a\alpha^2 + b\alpha + c = 0$

(viii) Assertion (A): Degree of a zero polynomial is not defined.

Reason (R): Degree of a non - zero constant polynomial 0.

(ix) Assertion (A): zeroes of $f(x) = x^2 - 4x - 5$ are 5,–1.

Reason (R): The polynomial whose zeroes are $2 + \sqrt{3}, 2 - \sqrt{3}$ is $x^2 - 4x + 7$.

(x) Assertion (A): The polynomial $f(x) = x^2 + 2$ has no real zeroes.

Reason (R): A quadratic polynomial always has real zeroes.

ANSWER KEYS

Exercise – 2.1

(i) (c) 6 **(ii)** (a) 0

Exercise – 2.2

(i) 1 **(ii)** – 10

(iii) (a) 1 **(iv)** (c) –18

(v) (b) – 1

Exercise – 2.3

(i) k $(x^2 + 2x + 3)$ **(ii)** k $(x^2 + x - 3)$
(iii) k $(4x^2 + 3x - 6)$ **(iv)** $x^2 + 3x - 2$
(v) $x^2 - 5x + 7$ **(vi)** (b) $x^2 + 6x + 3$

Exercise – 2.4

(i) 2 **(ii)** 2 **(iii)** 6
(iv) 3 **(v)** 2

Exercise – 2.5

(i) (a) 3 **(ii)** (b) $-\frac{4}{3}$ **(iii)** (d) 2

Exercise – 2.6

(i) (a) 3 **(ii)** (c) $\frac{5}{2}$ **(iii)** **(c)** $\frac{4}{3}$

(iv) (a) 2 **(v)** (b) $\frac{8}{3}$ **(vi)** (c) 10

Exercise – 2.7

(i) (a) 3 **(ii)** (d) $\frac{-3}{2}$ **(iii)** (d) 5

(iv) (c) – 7 **(v)** (c) 1 – c

Exercise – 2.8

(i) (a) $\frac{1}{2}$ **(ii)** (a) $\frac{3}{4}$ **(iii)** (d) $-\frac{1}{5}$

(iv) (a) 2 **(v)** (b) $\frac{1}{3}$ **(vi)** (a) $\frac{7}{3}$

(vii) (c) $a = \frac{1}{2}, c = 4$ **(viii)** $a = \frac{1}{2}, c = 2$

(ix) $a = \frac{1}{4}, c = 3$

Exercise – 2.9

(i) (d) $-\frac{11}{4}$ **(ii)** (c) 5 **(iii)** (a) – 7

(iv) (c) $\frac{27}{4}$ **(v)** (b) $\frac{b^2 - 2ac}{a^2}$

Exercise – 2.10

(i) (d) –5, –7 **(ii)** (a) $-\frac{3}{2}, -\frac{1}{3}$
(iii) (c) 5, –2

Exercise – 2.11

(i) (a) $x^2 + x - 6$ **(ii)** (b) $4x^2 - 4x - 3$
(iii) (c) $9(4x^2 - 9)$

Exercise – 2.12

(i) (a) $\frac{27}{32}$ **(ii)** (d) 1 **(iii)** (b) $\frac{8}{9}$

Exercise – 2.13

(i) (c) $a = -1, b = -1$ **(ii)** (c) $a = -2, b = -6$
(iii) $a = -2, b = -12$

Exercise – 2.14

(i) (c) 5 **(ii)** (a) 6 **(iii)** (c) 10

(iv) (c) – 16 **(v)** (a) $\frac{1}{3}$

3 Pair of Linear Equations in Two Variables

Key Point The System of equations $a_1x + b_1y + c_1 = 0$ and $a_2x + b_2y + c_2 = 0$ has infinite number of solutions, if $\frac{a_1}{a_2} = \frac{b_1}{b_2} = \frac{c_1}{c_2}$

1. Find the number of solution of the following pair of linear equation. **(2009)**

$x + 2y - 8 = 0$
$2x + 4y = 16$

Solution:

$x + 2y - 8 = 0$
$2x + 4y - 16 = 0$

Now, $\frac{a_1}{a_2} = \frac{1}{2}, \frac{b_1}{b_2} = \frac{2}{4} = \frac{1}{2}, \frac{c_1}{c_2} = \frac{-8}{-16} = \frac{1}{2}$

$\therefore \frac{a_1}{a_2} = \frac{b_1}{b_2} = \frac{c_1}{c_2}$

Hence the given system of equations has infinitely many solutions.

2. Equation of *a* line coincide with $2.5x - 2y = 3$ is: **(2026)**

(a) $5x - 4y = 3$ (b) $5x - 4y + 6 = 0$
(c) $15x - 12y - 3 = 0$ (d) $5x - 4y - 6 = 0$

Solution: (d)

condition for coincident lines is

$\frac{a_1}{a_2} = \frac{b_1}{b_2} = \frac{c_1}{c_2}$

now, $2.5x - 2y - 3 = 0$
and $5x - 4y - 6 = 0$

$\frac{a_1}{a_2} = \frac{2.5}{5} = \frac{1}{2}$

$\frac{b_1}{b_2} = \frac{-2}{-4} = \frac{1}{2}$

$\frac{c_1}{c_2} = \frac{-3}{-6} = \frac{1}{2}$

3. If the pair of linear equations: $a_1x + b_1y + c_1 = 0$ and $a_2x + b_2y + c_2 = 0$ is consistent and dependent then **(2026)**

(a) $\frac{a_1}{a_2} \neq \frac{b_1}{b_2}$ (b) $\frac{a_1}{a_2} \neq \frac{b_1}{b_2} = \frac{c_1}{c_2}$
(c) $\frac{a_1}{a_2} = \frac{b_1}{b_2} \neq \frac{c_1}{c_2}$ (d) $\frac{a_1}{a_2} = \frac{b_1}{b_2} = \frac{c_1}{c_2}$

Solution: (d)

$\frac{a_1}{a_2} = \frac{b_1}{b_2} = \frac{c_1}{c_2}$

4. If a pair of linear equations in two variables is represented by two coincident lines, then the pair of equations has: **(2026)**

(a) a unique solution
(b) two solutions
(c) no solution
(d) an infinite number of solutions

Solution: (d)

an infinite number of solutions

Exercise 3.1

(i) Find the number of solutions of the following pair of linear equations.

$2x - 3y - 6 = 0$
$12x - 18y = 36$

(ii) Find the number of solutions of the following pair of linear equations.

$3x + 2y - 6 = 0$
$9x + 6y = 18$

(iii) Find the number of solutions of the following pair of linear equations.
$2x - 3y + 6 = 0$
$4x - 6y + 12 = 0$

5. The value of k, for which the pair of linear equations $kx + y = k^2$ and $x + ky = 1$ have infinitely many solutions is. **(2020)**

(a) ± 1 (b) 1
(c) -1 (d) 2

Solution:

For infinitely many solutions, we must have

$$\frac{a_1}{a_2} = \frac{b_1}{b_2} = \frac{c_1}{c_2}$$

i, e $\frac{k}{1} = \frac{1}{k} = \frac{-k^2}{-1}$

Now, the following cases arise

Case – I	Case – II	Case - III
$\frac{k}{1} = \frac{1}{k}$	$\frac{1}{k} = \frac{-k^2}{-1}$	$\frac{k}{1} = \frac{-k^2}{-1}$
$\Rightarrow k^2 = 1$	$\Rightarrow k^3 = 1$	$\Rightarrow k^2 - k = 0$
$\Rightarrow k = \pm 1$	$\Rightarrow k = 1$	$\Rightarrow k(k-1) = 0$
		$\Rightarrow k - 1 = 0$ [$\because k \neq 0$]
		$\Rightarrow k = 1$

$\therefore k = 1$

6. The value of 'p' for which the equations $px + 3y = p - 3$, $12x + py = p$ has infinitely many solution is : **(2025)**

(a) -6 only
(b) 6 only
(c) ± 6
(d) Any real number except ± 6

Solution: (b)

$px + 3y - (p - 3) = 0$
$12x + py - p = 0$

The given pair of linear equations has infinitely many solutions, if

$$\frac{p}{12} = \frac{3}{p} = \frac{p-3}{p}$$

$\Rightarrow \frac{p}{12} = \frac{3}{p}$ and $\frac{3}{p} = \frac{p-3}{p}$

$\Rightarrow p^2 = 36$ and $\Rightarrow 3 = p - 3$ [$\because p \neq 0$]

$\Rightarrow p = \pm 6$ and $p = 6$

$\therefore p = 6$

7. If the system of equations $3x + 2y = 4$, $4ax + (a + b)y = 16$ has infinitely many solutions, then **(2025)**

(a) $5a = 3b$ (b) $3a = 5a$
(c) $a + b = 15$ (d) $a - b = 2$

Solution: (a)

The given system of equations has infinitely many solutions.

$$\therefore \frac{3}{4a} = \frac{2}{a+b} = \frac{-4}{-16}$$

$\Rightarrow \frac{3}{4a} = \frac{2}{a+b}$ and $\frac{2}{a+b} = \frac{-4}{-16}$

$\Rightarrow 3a + 3b = 8a$ and $8 = a + b$

$\Rightarrow 5a = 3b$ and $a + b = 8$

8. For what value of p, pair of equations $2x + 3y = 4$, $(p + 2)x + 6y = (3p + 2)$ will have infinitely many solutions? **(2025)**

(a) 1 (b) 4
(c) 5 (d) 2

Solution: (d)

The given system of equations will have infinitely many solutions, if

$$\frac{2}{p+2} = \frac{3}{6} = \frac{-4}{-(3p+2)}$$

$\Rightarrow \frac{2}{p+2} = \frac{3}{6}$ and $\frac{3}{6} = \frac{4}{3P+2}$

$\Rightarrow p + 2 = 4$ and $3p + 2 = 8$

$\Rightarrow p = 2$ and $p = 2$

9. The system of equations $2x + ky = 4$ and $3x - 6y = 6$ has infinite many solutions, if **(2026)**

(a) $k \neq 4$ (b) $k = 4$
(c) $k = -4$ (d) No value of k

Solution: (c)

$\frac{a_1}{a_2} = \frac{b_1}{b_2} = \frac{c_1}{c_2}$

$\Rightarrow \frac{2}{3} = \frac{k}{-6} = \frac{-4}{-6}$

$\Rightarrow \frac{2}{3} = \frac{k}{-6}$

$\therefore k = -4$

Exercise 3.2

(i) The value of k, for which the pair of linear equations $5x+2y=2k$ and $2(k+1)x+ky=(3k+4)$ have infinitely many solutions

(a) 5 (b) 4
(c) 2 (d) –2

(ii) The value of k for which the pair of linear equations $2x + 3y = 6$ and $kx + 6y = 12$ have infinitely many solutions

(a) 2 (b) –2
(c) 4 (d) –4

(iii) The value of k for which the pair of linear equations $2x - 3y + 10 = 0$ and $3x + ky + 15 = 0$ have infinitely many solutions

(a) $\frac{9}{2}$ (b) $-\frac{9}{2}$
(c) 2 (d) 4

10. The value of k for which the equations $3x-y+8=0$ and $6x + ky = -16$ represent coincident lines is. **(2020)**

(a) $-\frac{1}{2}$ (b) $\frac{1}{2}$
(c) 2 (d) –2

Solution: (d)

condition for coincident lines is

$\frac{a_1}{a_2} = \frac{b_1}{b_2} = \frac{c_1}{c_2}$

$\Rightarrow \frac{3}{6} = \frac{-1}{k} = \frac{8}{16}$

i. e $\frac{-1}{k} = \frac{3}{6}$

$\Rightarrow -k = 2$

$\therefore k = -2$

11. If the pair of equations $3x - y + 8 = 0$ and $6x - ry + 16 = 0$ represent coincident lines, then the value of 'r' is. **(2023)**

(a) $-\frac{1}{2}$ (b) $\frac{1}{2}$
(c) –2 (d) 2

Solution: (d)

Condition for coincident lines is

$\frac{a_1}{a_2} = \frac{b_1}{b_2} = \frac{c_1}{c_2}$

$\Rightarrow \frac{3}{6} = \frac{-1}{-r} = \frac{8}{16}$

i. e = $\frac{3}{6} = \frac{-1}{-r}$

$\Rightarrow r = 2$

Exercise 3.3

(i) The value of k for which the equations $2x-y=4$ and $3x + ky - 6 = 0$ represent coincident lines is.

(a) $\frac{-2}{3}$ (b) $\frac{-3}{2}$
(c) –3 (d) –2

(ii) The value of k for which the equations $3x - ky = 6$ and $2x - 3y = 4$ represent coincident lines is.

(a) 3 (b) 2
(c) $\frac{9}{2}$ (d) $\frac{2}{9}$

(iii) The value of k for which the equations $3x - 4y = 6$ and $x + 3ky = 2$ represent coincident lines is.

(a) $-\frac{4}{9}$ (b) $-\frac{9}{4}$
(c) –4 (d) 9

(iv) The value of k for which the equations $3x - y + 8 = 0$ and $6x - ky + 16 = 0$ represent coincident lines is. **(2023)**

(a) $\frac{1}{2}$ (b) $-\frac{1}{2}$

(c) 2 (d) –2

12. The pair of linear equations $2x = 5y + 6$ and $15y = 6x - 18$ represents two lines which are. **(2023)**

(a) intersecting
(b) parallel
(c) coincident
(d) either intersecting or parallel

Solution: (c)

$2x - 5y - 6 = 0$

$6x - 15y - 18 = 0$

Now, $\frac{a_1}{a_2} = \frac{2}{6} = \frac{1}{3}$

$\frac{b_1}{b_2} = \frac{-5}{-15} = \frac{1}{3}$

$\frac{c_1}{c_2} = \frac{-6}{-18} = \frac{1}{3}$

$\therefore \frac{a_1}{a_2} = \frac{b_1}{b_2} = \frac{c_1}{c_2}$

Hence, the given system of equations is coincident.

Exercise 3.4

(i) The pair of linear equations $x + y = 8$ and $4x + 4y = 32$ represent two lines which are.

(a) intersecting
(b) parallel
(c) coincident
(d) either intersecting or parallel

(ii) The pair of linear equations $3y = 2x - 5$ and $4x = 6y + 10$ represent two lines which are.

(a) intersecting
(b) parallel
(c) coincident
(d) either intersecting or parallel

(iii) The pair of linear equations $x = 4y + 5$ and $2x = 8y + 10$ represent two lines which are.

(a) intersecting
(b) parallel
(c) coincident
(d) either intersecting or parallel

Key Point The System of equations $a_1x + b_1y + c_1 = 0$ and $a_2x + b_2y + c_2 = 0$ has no solutions(i. e inconsistent), if $\frac{a_1}{a_2} = \frac{b_1}{b_2} \neq \frac{c_1}{c_2}$

13. The value of k for which the system of equations $x + y = 4$ and $2x + ky = 3$ for no solution is. **(2020)**

(a) –2 (b) $\neq 2$

(c) 3 (d) 2

Solution: (d)

In order that the given system has no solution, we must have

$\frac{a_1}{a_2} = \frac{b_1}{b_2} \neq \frac{c_1}{c_2}$

i, e $\frac{1}{2} = \frac{1}{k} \neq \frac{4}{3}$

$\frac{1}{2} = \frac{1}{k}$

$\Rightarrow k = 2$

14. The value of k for which the system of equations $x + 2y = 3$ and $5x + ky = 7$ has no solution is. **(2020)**

Solution:

In order that the given system has no solution, we must have

$\frac{a_1}{a_2} = \frac{b_1}{b_2} \neq \frac{c_1}{c_2}$

i, e $\frac{1}{5} = \frac{2}{k} \neq \frac{-3}{-7}$

i, e $\frac{1}{5} = \frac{2}{k}$

$\Rightarrow k = 10$

15. The value of k for which the system of equations $kx + 2y = 5$ and $3x + 4y = 1$ have no solution is. **(2023)**

(a) $k = \frac{3}{2}$ (b) $k \neq \frac{3}{2}$

(c) $k \neq \frac{2}{3}$ (d) $k = 15$

Solution: (a)

The given system of equations have no solution if

$$\frac{k}{3} = \frac{2}{4} \neq \frac{-5}{-1}$$

i, e $\frac{k}{3} = \frac{2}{4}$

i, e $k = \frac{3}{2}$

16. The pair of equations $7x - 14y = -7$ and $3x - 6y = 2$ has **(2025)**

(a) Unique solution
(b) Two solutions
(c) No solution
(d) Infinitely many solutions

Solution: (c)

$$\frac{a_1}{a_2} = \frac{7}{3}$$

$$\frac{b_1}{b_2} = \frac{-14}{-6} = \frac{7}{3}$$

$$\frac{c_1}{c_2} = \frac{7}{-2} = \frac{-7}{2}$$

$$\therefore \frac{a_1}{a_2} = \frac{b_1}{b_2} \neq \frac{c_1}{c_2}$$

$\therefore$ The given system of equations has no solution.

17. The system of equations $x + 5 = 0$ and $2x - 1 = 0$, has **(2025)**

(a) No solution (b) Unique solution

(c) Two solutions (d) Infinite solutions

Solution: (a)

No solutions.

Exercise 3.5

(i) The value of k for which the system of equations $kx + 2y = 0$ and $3x - 4y = 10$ has no solution is.

(a) $\frac{3}{2}$ (b) $\frac{-3}{2}$

(c) 1 (d) $\frac{2}{3}$

(ii) The value of k for which the system of equations $x + 2y = 3$ and $3x + ky + 15 = 0$ has no solution is.

(a) $k = -6$ (b) $k \neq 6$

(c) $k = 6$ (d) $k = -2$

(iii) The value of k for which the system of equations $x + 2y = 3$ and $5x + ky + 7 = 0$ has no solution is.

(a) 10 (b) -10

(c) $\neq 10$ (d) 5

(iv) The value of k for which the system of equations $kx + 3y = 4$ and $12x + ky = 6$ for no solution is.

(a) 6 (b) -6

(c) ± 6 (d) $\neq 6$

(v) The value of k for which the system of equations $8x + 5y = 9$, $kx + 10y = 12$ for no solution is.

(a) 16 (b) -12

(c) ± 12 (d) $\neq 12$

18. A pair of linear equations $\frac{3x}{2} + \frac{5y}{3} = 7$ and $9x + 10y = 14$ is. **(2020)**

(a) consistent
(b) inconsistent
(c) consistent with one solutions
(d) consistent with many solution

Solution: (b)

$$\frac{3x}{2} + \frac{5y}{3} = 7$$

$$\Rightarrow \frac{9x+10y}{6} = 7$$

$$\Rightarrow 9x + 10y = 42$$

$$\Rightarrow 9x + 10y - 42 = 0$$

$9x + 10y = 14$

$\Rightarrow 9x + 10y - 14 = 0$

Now, $\frac{a_1}{a_2} = \frac{9}{9} = 1$

$\frac{b_1}{b_2} = \frac{10}{10} = 1$

$\frac{c_1}{c_2} = \frac{-42}{-14} = 3$

$\therefore \frac{a_1}{a_2} = \frac{b_1}{b_2} \neq \frac{c_1}{c_2}$

The given system of equation is inconsistent.

Exercise 3.6

(i) The pair of linear equation $3x - 5y = 11$, $6x - 10y = 7$ is.

(a) consistent

(b) inconsistent

(c) consistent with one solutions

(d) consistent with many solution

(ii) The pair of linear equation $5x - 15y = 8$ and $3x - 9y = \frac{3}{4}$ is.

(a) consistent

(b) inconsistent

(c) consistent with one solutions

(d) consistent with many solution

(iii) The pair of linear equation $3x + 8y = 5$ and $9x + 24y = 16$ is.

(a) consistent

(b) inconsistent

(c) consistent with one solutions

(d) consistent with many solution

19. The value of k for which the system of linear equations $x + 2y = 3$, $5x + ky + 7 = 0$ is inconsistent. **(2020)**

(a) $-\frac{14}{3}$ (b) $\frac{2}{5}$

(c) 5 (d) 10

Solution: (d)

In order that the given system of equations is inconsistent.

$\frac{a_1}{a_2} = \frac{b_1}{b_2} \neq \frac{c_1}{c_2}$

i, e $\frac{1}{5} = \frac{2}{k} \neq \frac{-3}{7}$

i, e $\frac{1}{5} = \frac{2}{k}$

i, e $k = 10$

20. The pair of linear equation $x + 2y + 5 = 0$ and $-3x - 6y + 1 = 0$ has. **(2023)**

(a) a unique solution

(b) exactly two solution

(c) infinitely many solution

(d) no solution

Solution: (d)

$\frac{a_1}{a_2} = \frac{1}{-3} = -\frac{1}{3}$

$\frac{b_1}{b_2} = \frac{2}{-6} = -\frac{1}{3}$

$\frac{c_1}{c_2} = \frac{1}{5}$

$\therefore \frac{a_1}{a_2} = \frac{b_1}{b_2} \neq \frac{c_1}{c_2}$

The pair of linear equation has no solution.

21. If the lines represented by equations $3x + 2my = 2$ and $2x + 5y + 1 = 0$ are parallel, then the value of m is. **(2023)**

(a) $\frac{2}{5}$ (b) $-\frac{5}{4}$

(c) $\frac{3}{2}$ (d) $\frac{15}{4}$

Solution:

The given system of linear equations represent parallel lines,

$\therefore \frac{3}{2} = \frac{2m}{5} \neq \frac{-2}{1}$

i, e $\frac{3}{2} = \frac{2m}{5}$

i, e $m = \frac{15}{4}$

22. The value of k for which the system of linear equations $\frac{x}{2} + \frac{y}{3} = 5$ and $2x + ky = 7$ in inconsistent is **(2026)**

(a) $\frac{3}{4}$ (b) $\frac{4}{3}$

(c) $\frac{1}{3}$ (d) 3

Solution: (b)

$\frac{x}{2} + \frac{y}{3} = 5$

$\Rightarrow 3x + 2y = 30$

$\Rightarrow 3x + 2y - 30 = 0$

$2x + ky - 7 = 0$

The given system of linear equations is inconsistent, if

$\frac{a_1}{a_2} = \frac{b_1}{b_2} \neq \frac{c_1}{c_2}$

$\Rightarrow \frac{3}{2} = \frac{2}{k} \neq \frac{-30}{-7}$

$\Rightarrow \frac{3}{2} = \frac{2}{k}$

$\Rightarrow k = \frac{4}{3}$

Exercise 3.7

(i) If the lines represented by equation $rx + 2y = 5$ and $(r - 1)x + 5y = 2$ are parallel, then the value of r is.

(a) $\frac{2}{3}$ (b) $\frac{-2}{3}$

(c) $\frac{3}{2}$ (d) $\frac{-3}{2}$

(ii) If the lines represented by equation $(k + 1)x + y = 5$ and $3x + (k - 1)y = 10$ are parallel, then the value of k is.

(a) 2 or –2 (b) 10 or 2

(c) –2 or 3 (d) –1 or 1

(iii) If the lines represented by equation $3x + ky = 5$ and $4x + (k + 1)y = 6$ are parallel, then the value of k is.

(a) 1 (b) 2

(c) 3 (d) 0

(iv) If the lines represented by equation $2x + ky = 8$ and $(k - 3)x + 14y = 32$ are parallel, then the value of k is.

(a) –4 or 3 (b) 4 or –3

(c) –4 or 7 (d) 3 or –5

(v) If the lines represented by equation $kx - 3y = 1$ and $(4 - k)x + y = 3$ are parallel, then the value of k is.

(a) 6 (b) 2

(c) 4 (d) 8

23. The pair of equations $ax + 2y = 9$ and $3x + by = 18$ represent parallel lines, where a, b are integers if. **(2023)**

(a) $a = b$ (b) $3a = 2b$

(c) $2a = 3b$ (d) $ab = 6$

Solution: (d)

The given system of equations are parallel if

$\frac{a_1}{a_2} = \frac{b_1}{b_2} \neq \frac{c_1}{c_2}$

$\Rightarrow \frac{a}{3} = \frac{2}{b} \neq \frac{-9}{-18}$

$\Rightarrow \frac{a}{3} = \frac{2}{b}$

$\Rightarrow ab = 6$

24. Which out of the following type of straight lines will be represent by the system of equations $3x + 4y = 5$ and $6x + 8y = 7$? **(2024)**

(a) parallel

(b) intersecting

(c) coincident

(d) perpendicular to each other.

Solution: (a)

$\frac{a_1}{a_2} = \frac{3}{6} = \frac{1}{2}$

$\frac{b_1}{b_2} = \frac{4}{8} = \frac{1}{2}$

$\frac{c_1}{c_2} = \frac{-5}{-7}$

$\therefore \frac{a_1}{a_2} = \frac{b_1}{b_2} \neq \frac{c_1}{c_2}$

The given pair of equations are parallel to each other.

25. Equation of another line parallel to the line represented by $2x - 6y = 7$ is **(2026)**

(a) $y = 3x - 7$ (b) $2x = 9 - 6y$

(c) $x - 3y = 7$ (d) $x = \frac{7}{2} - 3y$

Solution: (c)

The equation of the first line $2x - 6y - 7 = 0$

and $x - 3y = 7$

$\Rightarrow x - 3y - 7 = 0$

Now $\frac{a_1}{a_1} = \frac{2}{1} = 2$

$\frac{b_1}{b_2} = \frac{-6}{-3} = 2$

$\frac{c_1}{c_2} = \frac{-7}{-7} = 1$

$\therefore \frac{a_1}{a_2} = \frac{b_1}{b_2} \neq \frac{c_1}{c_2}$

$\therefore$ The two lines $2x - 6y = 7$ and $x - 3y = 7$ are parallel to each other.

Exercise 3.8

(i) The pair of equations $ax + 2y = 2$ and $3x + by = 6$ represent parallel lines, where a, b are integers if.

(a) $ab = 2$ (b) $a = 2b$

(c) $b = 3a$ (d) $ab = 6$

(ii) The pair of equations $ax - 4y = 3$ and $3x + by = 6$ represent parallel lines, where a, b are integers if.

(a) $a = b$ (b) $a = 3b$

(c) $ab = 6$ (d) $ab = -12$

(iii) The pair of equations $x - ax = 3$ and $bx + 4y = 4$ represent parallel lines, where a, b are integers if.

(a) $a - b = 0$ (b) $a + b = 0$

(c) $ab = -4$ (d) $a = 3b$

Key Point The System of equations $a_1x + b_1y + c_1 = 0$ and $a_2x + b_2y + c_2 = 0$ has unique solution, if $\frac{a_1}{a_2} \neq \frac{b_1}{b_2}$

26. If the equation $kx - 2y = 3$ and $3x + y = 5$ represent two interacting lines at unique point, then the value of k is. **(2020)**

Solution:

For unique solution

$\frac{a_1}{a_2} \neq \frac{b_1}{b_2}$

$\Rightarrow \frac{k}{3} \neq \frac{-2}{1}$

$\Rightarrow k \neq -6$

27. The system of linear equations $px + qy = r$ and $p_1x + q_1y = r$, has a unique solutions if: **(2026)**

(a) $pq \neq p_1q_1$ (b) $pp_1 \neq qq_1$

(c) $pq_1 \neq qp_1$ (d) $pqr \neq p_1q_1r_1$

Solution: (c)

The given system of linear equations has a unique solutions ,if $\frac{p}{p_1} \neq \frac{q}{q_1} \Rightarrow pq_1 \neq p_1q$

Exercise 3.9

(i) If the equations $kx + 3y = 4$ and $2x - 3y = 2$ represent two interacting lines at unique point, then the value of k is __________.

(ii) If the equation $kx - 4y = 5$ and $3x - 2y = 7$ represent two interacting lines at unique point, then the value of k is __________.

(iii) If the equation $kx - y = 7$ and $3x + 7y = 3$ represent two interacting lines at unique point, then the value of k is _________.

28. The pair of linear equation $y = 0$ and $y = -6$ has. **(2020)**

(a) a unique solution
(b) no solution
(c) infinite many solution
(d) only solution (0,0)

Solution: (b)

$y = 0$ and $y = -6$ are two parallel lines and having no solution.

29. The pair of linear equation $x = 0$ and $x = -4$ has. **(2020)**

(a) a unique solution
(b) no solution
(c) infinite many solution
(d) only solution (0,0)

Solution: (b)

The equations $x = 0$ and $x = -4$ are two parallel lines and having no solution.

30. The number of solutions of the system of equations given by $x = -2$ and $y = -2$ is **(2026)**

(a) 0 (b) 1
(c) 2 (d) infinite

Solution: (b)

1

31. The system of equations $x = 2$ and $x = 3$ has: **(2026)**

(a) Unique solution (2, 3)
(b) two solutions (2, 0) and (3, 0)
(c) no solution
(d) infinitely many solutions

Solution: (b)

two solutions (2, 0) and (3, 0)

Exercise 3.10

(i) The pair of linear equation $y = 2$ and $y = 5$ has.
(a) a unique solution
(b) no solution
(c) infinite many solution
(d) only solution (0,0)

(ii) The pair of linear equation $x = 0$ and $x = -4$ has.
(a) a unique solution
(b) no solution
(c) infinite many solution
(d) only solution (0,0)

32. In a pair of linear equations in consident, then the lines represented by them are. **(2020)**

(a) parallel
(b) intersecting or coincident
(c) always coincident
(d) always intersecting

Solution: (b)

intersecting or coincident

33. The pair of equation $x = a$ and $y = b$ graphically represents linear which are. **(2023)**

(a) parallel
(b) intersecting at (b, a)
(c) coincident
(d) intersecting at (a, b)

Solution: (d)

intersecting at (a, b)

34. The value of k for which the pair of equations $kx = y + 2$ and $6x = 2y + 3$ has infinite many solutions. **(2023)**

(a) is $k = 3$ (b) does not exist
(c) is $k = -3$ (d) is $k = 4$

Solution: (b)

$kx - y - 2 = 0$

$6x - 2y - 3 = 0$

For infinite many solutions

$$\frac{a_1}{a_2} = \frac{b_1}{b_2} = \frac{c_1}{c_2}$$

Here $\frac{b_1}{b_2} = \frac{-1}{-2} = \frac{1}{2}$

and $\frac{c_1}{c_2} = \frac{-2}{-3} = \frac{2}{3}$

$\therefore \frac{b_1}{b_2} \neq \frac{c_1}{c_2}$

35. If the pair of linear equations $x - y = 1, x + ky = 5$ has a unique solution $x = 2, y = 1$, then the value of k is. **(2023)**

(a) -2 (b) -3

(c) 3 (d) 4

Solution: (c)

$2 + k.\ 1 = 5$

$\Rightarrow k = 3$

Exercise 3.11

(i) If the pair of linear equations $2x - 3y = 5$, $x - ky = 3$ has a unique solution $x = 7, y = 3$, then the value of k is.

(a) $\frac{-3}{4}$ (b) $\frac{4}{3}$

(c) 3 (d) -4

(ii) If the pair of linear equations $2x - y = 3$, $3x + ky = 17$ has a unique solution $x = 4, y = 5$, then the value of k is.

(a) 0 (b) -1

(c) 1 (d) 5

(iii) If the pair of linear equations $x - y = 3$, $2x + ky = 7$ has a unique solution $x = 4, y = 1$, then the value of k is.

(a) 1 (b) -1

(c) 0 (d) 2

36. In the given figure, graph of two linear equation are shown. The pair of these linear equation is. **(2024)**

(a) consistent with unique solution,

(b) consistent with infinitely many solution.

(c) inconsistent

(d) inconsistent but can be made consistent by extending these lines.

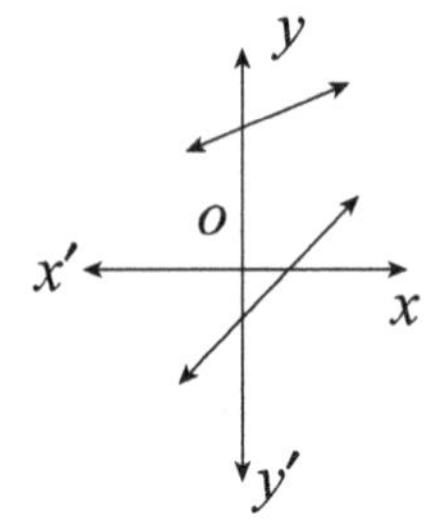

Solution: (d)

inconsistent list can be made consistent by extending these lines.

37. The pair of linear equations $2kx + 5y = 7$, $6x + 5y = 11$ have a unique solution, if. **(2024)**

(a) $k \neq 3$ (b) $k \neq -3$

(c) $k \neq \frac{1}{3}$ (d) $k \neq -\frac{1}{3}$

Solution: (a)

The pair of linear equations have a unique solution, if

$$\frac{a_1}{a_2} \neq \frac{b_1}{b_2}$$

$$\Rightarrow \frac{2k}{6} \neq \frac{5}{5}$$

$$\Rightarrow \frac{k}{3} \neq 1$$

$\Rightarrow k \neq 3$

38. If $2x + 3y = 15$ and $3x + 2y = 25$, then the value of $x - y$ is. **(2023)**

(a) -10 (b) 8

(c) 10 (d) -8

Solution: (c)

$2x + 3y = 15$ ....(i)

$3x + 2y = 25$ ....(ii)

Multiplying (i) by 3 and (ii) by 2 we get

$6x + 9y = 45$ ....(iii)

$6x + 4y = 50$ ....(iv)

Subtracting (iv) from (iii) we get

$5y = -5$

$\Rightarrow y = -1$

Putting $y = -1$ in (i) we get

$2x + 3(-1) = 15$

$\Rightarrow 2x = 18$

$\Rightarrow x = 9$

$\therefore x - y = 9 - (-1) = 9 + 1 = 10$

Exercise 3.12

(i) If $2x + y = 7$ and $4x - 3y + 1 = 0$, then the value of $x + y$ is.

(a) 2 (b) 3

(c) –5 (d) 5

(ii) If $x + y = 8$ and $2x - 3y = 1$, then the value of $x - y$ is.

(a) 3 (b) 2

(c) 5 (d) 10

(iii) If $x + y = 8$ and $4x - 3y = 26$, then the value of $x - y$ is.

(a) $\frac{44}{7}$ b) –7

(c) –2 (d) 3

39. The solution of the pair of equations $x + y = a + b$ and $ax - by = a^2 - b^2$ is. **(2023)**

(a) $x = b, y = a$ (b) $x = -a, y = b$

(c) $x = a, y = b$ (d) $x = a, y = -b$

Solution: (c)

$x + y = a + b$ ....(i)

$ax - by = a^2 - b^2$ ....(ii)

Multiplying equation (i) by a and equation (ii) by 1 we get

$ax + ay = a^2 + ab$ ....(iii)

$ax - by = a^2 - b^2$ ....(iv)

subtracting (iv) from (iii), we get $ay + by = ab + b^2$

$\Rightarrow (a + b)y = (a + b)\, b$

$\Rightarrow y = b$

Putting $y = b$ in equation (i) we get

$x + b = a + b$

$\Rightarrow x = a$

Exercise 3.13

(i) The solution of the pair of equations $x + y = 8$ and $2x - 3y = 1$ is

(a) $x = 3, y = -5$ (b) $x = 5, y = 3$

(c) $x = -2, y = 0$ (d) $x = 1, y = 2$

(ii) The solution of the pair of equations $2x + 3y = 18$ and $x - 2y = 2$ is.

(a) $x = 6, y = 2$ (b) $x = 1, y = 4$

(c) $x = 3, y = 4$ (d) $x = -1, y = 5$

(iii) The solution of the pair of equations $3x + 4y = -7$ and $5x - 7y = 2$ is

(a) $x = 1, y = 1$ (b) $x = -1, y = -1$

(c) $x = 0, y = 4$ (d) $x = 5, y = 3$

40. 3 chairs and 1 table cost ₹ 900; where 5 chairs and 3 tables cost ₹ 2,100, if the cost of 1 chair is ₹ x and the cost of 1 table is ₹ y, then the situation can be represented algebraically as. **(2023)**

(a) $3x + y = 900,\ 3x + 5y = 2100$

(b) $x + 3y = 900,\ 3x + 5y = 2100$

(c) $3x + y = 900,\ 5x + 3y = 2100$

(d) $x + 3y = 900,\ 5x + 3y = 2100$

Solution: (c)

$3x + y = 900$

$5x + 3y = 2100$

Exercise 3.14

(i) 7 chairs and 4 table cost ₹ 7000; where 5 chairs and 3 tables cost ₹ 5080. If the cost of 1 chair is ₹ x and the cost of 1 table is ₹ y, then the situation can be represented algebraically as. **(2023)**

(a) $7x + 4y = 5080,\ 5x + 3y = 7000$

(b) $7x + 4y = 7000,\ 5x + 3y = 5080$

(c) $5x + 4y = 7000,\ 5x + 3y = 5080$

(d) $7x + 5y = 7080,\ 4x + 3y = 7000$

(ii) 36 pens and 24 pencils cost ₹ 702, whereas 24 pens and 36 pencils cost ₹ 558. If the cost of 1 pen ₹ x and the cost of 1 pencil in ₹ y, then the situation can be represented algebraically as

(a) $36x + 24y = 702$
$24x + 36y = 558$

(b) $36x + y = 702$
$24x + y = 558$

(c) $24x + 36y = 702$
$x + y = 558$

(d) $36x + 24y = 558$
$24x + 36y = 702$

41. The pair of linear equations $x + 2y - 5 = 0$ and $2x - 4y + 6 = 0$. **(2023)**

(a) is inconsistent
(b) is consistent
(c) is consistent with a unique solution
(d) is consistent with two solutions.

Solution: (c)

$$\frac{a_1}{a_2} = \frac{1}{2}$$

$$\frac{b_1}{b_2} = \frac{2}{-4} = -\frac{1}{2}$$

$$\therefore \frac{a_1}{a_2} \neq \frac{b_1}{b_2}$$

∴ The pair of linear equations is consistent with a unique solution.

Direction for questions 42 to 45 : in question number 33 to 35, a statement of Assertion (A) is followed by a statement of Reason (R). choose the correct option as :

(a) Both assertion (A) and Reason (R) are true and Reason (R) is the correct explanation of Assertion (A).
(b) Both assertion (A) and Reason (R) are true and Reason (R) is not the correct explanation of Assertion (A).
(c) Assertion (A) is true, but Reason (R) is false.
(d) Assertion (A) is false, but Reason (R) is true.

42. Assertion (A) : The system of linear equations $3x + 5y - 4 = 0$ and $15x + 25y - 25 = 0$ is inconsistent.

Reason (R) : The pair of linear equation $a_1x + b_1y + c_1 = 0$ and $a_2x + b_2y + c_2 = 0$ is inconsistent if $\frac{a_1}{a_2} = \frac{b_1}{b_2} \neq \frac{c_1}{c_2}$ **(2023)**

Solution:

$$\frac{a_1}{a_2} = \frac{3}{15} = \frac{1}{5}$$

$$\frac{b_1}{b_2} = \frac{5}{25} = \frac{1}{5}$$

$$\frac{c_1}{c_2} = \frac{-4}{-25} = \frac{4}{25}$$

$$\therefore \frac{a_1}{a_2} = \frac{b_1}{b_2} \neq \frac{c_1}{c_2}$$

Hence the pair of linear equation is inconsistent. Hence both assertion and reason are true and reason is the correct explanation of assertion.

43. Assertion (A): The pair of linear equations $px + 3y + 59 = 0$, $2x + 6y + 118 = 0$ will have infinite many solution if $p = 1$

Reason (R): If the pair of linear equations $px + 3y + 19 = 0$ and $2x + 6y + 157 = 0$ has a unique solution, then $p \neq 1$ **(2025)**

Solution: (b)

The pair of linear equations $px + 3y + 59 = 0$ and $2x + 6y + 118 = 0$ will have infinite many solutions, if

$$\frac{a_1}{a_2} = \frac{b_1}{b_2} = \frac{c_1}{c_2}$$

$$\Rightarrow \frac{p}{2} = \frac{3}{6} = \frac{59}{118}$$

$$\Rightarrow \frac{p}{2} = \frac{3}{6}$$

$$\Rightarrow p = 1$$

∴ Assertion(A) is true

Now, the pair of linear equations $px + 3y + 19 = 0$ and $2x + 6y + 157 = 0$ has a unique solution

$\therefore \frac{a_1}{a_2} \neq \frac{b_1}{b_2}$

$\Rightarrow \frac{p}{2} \neq \frac{3}{6}$

$\Rightarrow p \neq 1$

$\therefore$ Reason (R) is true.

44. Assertion (A): The value of p for which the system of equations $4x + py + 8 = 0$ and $2x + 2y + 2 = 0$ is consistent is 4.

Reason(R): The system of equations, $a_1x + b_1y = c_1$ and $a_2x + b_2y = c_2$ is consistent with infitely many solution, if $\frac{a_1}{a_2} = \frac{b_1}{b_2} = \frac{c_1}{c_2}$ **(2025)**

Solution: (d)

$\frac{a_1}{a_2} = \frac{4}{2} = 2$

$\frac{b_1}{b_2} = \frac{p}{2}$

$\frac{c_1}{c_2} = \frac{8}{2} = 4$

if $p = 4$

$\frac{a_1}{a_2} = \frac{b_1}{b_2} \neq \frac{c_1}{c_2}$

$\therefore$ The system of equations is inconsistent.

45. Assertion (A) : The system of linear equations $3x - 5y + 7 = 0$ and $-6x + 10y + 14 = 0$ is in consistent

Reason (R) : When two linear equations don't have unique solution, they always represent parallel lines.

Solution: (c)

$\frac{a_1}{a_2} = \frac{3}{-6} = -\frac{1}{2}$

$\frac{b_1}{b_2} = \frac{-5}{10} = -\frac{1}{2}$

$\frac{c_1}{c_2} = \frac{7}{14} = \frac{1}{2}$

$\therefore \therefore \frac{a_1}{a_2} = \frac{b_1}{b_2} \neq \frac{c_1}{c_2}$

$\therefore$ The system of given linear equations is inconsistent.

Direction for question (i) to (v) : in question number (i) to (v), a statement of Assertion (A) is followed by a statement of Reason (R). choose the correct option as :

(a) both assertion (A) and reason (R) are true and reason (R) is the correct explanation of assertion (A).

(b) both assertion (A) and reason (R) are but reason (R) is not the correct explanation of assertion (A).

(c) Assertion (A) is true but reason (R) is false.

(d) Assertion (A) is false but reason (R) is true.

Exercise 3.15

(i) Assertion (A) : The system of linear equations $2x + 3y + 5 = 0$ and $4x + 6y + 7 = 0$ is inconsistent.

Reason (R) : The pair of linear equations $a_1x + b_1y + c_1 = 0$ and $a_2x + b_2y + c_2 = 0$ is inconsistent if $\frac{a_1}{a_2} = \frac{b_1}{b_2} \neq \frac{c_1}{c_2}$

(ii) Assertion (A) : The system of linear equations $4x - 2y + 1 = 0$ and $8x - 4y + 3 = 0$ is inconsistent.

Reason (R) : The pair of linear equations $a_1x + b_1y + c_1 = 0$ and $a_2x + b_2y + c_2 = 0$ is inconsistent if $\frac{a_1}{a_2} = \frac{b_1}{b_2} \neq \frac{c_1}{c_2}$

(iii) Assertion (A) : The system of linear equations $3x + 5y + 2 = 0$ and $6x + 10y + 3 = 0$ is inconsistent.

Reason (R) : The pair of linear equations $a_1x + b_1y + c_1 = 0$ and $a_2x + b_2y + c_2 = 0$ is inconsistent if $\frac{a_1}{a_2} \neq \frac{b_1}{b_2}$

(iv) Assertion (A) : The system of linear equations $3x + 4y + 8 = 0$ and $6x + 8y + 3 = 0$ is inconsistent.

Reason (R) : The pair of linear equations $a_1x + b_1y + c_1 = 0$ and $a_2x + b_2y + c_2 = 0$ is inconsistent if $\frac{a_1}{a_2} = \frac{b_1}{b_2} = \frac{c_1}{c_2}$.

(v) **Assertion (A) :** The system of linear equations $2x + 3y + 4 = 0$ and $3x + 2y + 5 = 0$ has a unique solutions.

Reason (R) : The system of equations $a_1x + b_1y + c_1 = 0$ and $a_2x + b_2y + c_2 = 0$ has a unique solution when $\frac{a_1}{a_2} \neq \frac{b_1}{b_2}$

46. If $x = 1$ and $y = 2$ is a solution of the pair of linear equations $2x - 3y + a = 0$ and $2x + 3y - b = 0$, then **(2025)**

(a) $a = 2b$ (b) $2a = b$

(c) $a + 2b = 0$ (c) $2a + b = 0$

Solution: (b)

$2 \times 1 - 3 \times 2 + a = 0$

$\Rightarrow 2 - 6 + a = 0$

$\therefore a = 4$

$2 \times 1 + 3 \times 2 - b = 0$

$\Rightarrow b = 8$

$\therefore 2a = b$

47. The line represented by the equation $x - y = 0$ is **(2025)**

(a) parallel to x axis

(b) parallel to y axis

(b) passing through the origin

(d) passing through the pair (3,2)

Solution: (c)

Passing through the origin.

48. The equation of a line parallel to x - axis and at the distance of 3 units below x - axis is **(2025)**

(a) $x = 3$ (b) $x = -3$

(c) $y = -3$ (d) $y = 3$

Solution: (c)

$y = -3$

49. The equation of a line parallel to y axis and at a distance of 5 units to the rights of y - axis is **(2025)**

(a) $x = 5$ (b) $x = -5$

(c) $y = 5$ (d) $y = -5$

Solution: (a)

$x = 5$

50. The distance of a point A from x - axis is 3 units. Which of the following can not be coordinates of the point A? **(2025)**

(a) (1,3) (b) (−3, −3)

(c) (−3, 3) (d) (3, 1)

Solution: (d) (3, 1)

51. The system of equations $y + a = 0$ and $2x = b$ has **(2025)**

(a) No solutions

(b) $\left(-a, \frac{b}{2}\right)$ as its soluton

(c) $\left(\frac{b}{2}, -a\right)$ as its solutions

(d) Infinite many solution

Solution: (c)

$\left(\frac{b}{2}, -a\right)$ an its solutions.

52. If (0, 0) is the solution of equation $x + y = c - 1$, then the value of c is **(2025)**

(a) 0 (b) 1

(c) − 1 (d) Any real number

Solution: (b)

Since (0,0) is a solution of the equation $x + y = c - 1$

$\therefore 0 + 0 = c - 1$

$\Rightarrow c = 1$

53. The number of solution of the system of equations $x = 3, y = -1$ is: **(2025)**

(a) 0 (b) 1

(c) 2 (d) Infinite

Solution: (b) 1

54. The number of solutions of the system of equations $x = a, x = b$ $(a \neq b)$ is: **(2025)**

(a) 0 (b) 1

(c) 2 (d) infinite

Solution: (a) 0

ANSWER KEYS

Exercise – 3.1

(i) infinitely many solutions
(ii) infinitely many solutions
(iii) infinitely many solutions

Exercise – 3.2

(i) (b) 4 **(ii)** (c) 4 **(iii)** (b) $\frac{-9}{2}$

Exercise – 3.3

(i) (b) $\frac{-3}{2}$ **(ii)** (c) $\frac{9}{2}$ **(iii)** (a) $\frac{-4}{9}$,

(iv) (c) $k = 2$

Exercise – 3.4

(i) (c) coincident **(ii)** (c) coincident
(iii) (c) coincident

Exercise – 3.5

(i) (b) $\frac{-3}{2}$ **(ii)** (c) $k = 6$ **(iii)** (a) 10

(iv) (c) ±6 **(v)** (a) 16

Exercise – 3.6

(i) (b) inconsistent **(ii)** (b) inconsistent
(iii) (b) inconsistent

Exercise – 3.7

(i) (b) $\frac{-2}{3}$ **(ii)** (a) 2 or –2 **(iii)** (c) 3

(iv) (c) –4 or 7 **(v)** (a) 6

Exercise – 3.8

(i) (d) $ab = 6$ **(ii)** (d) $ab = -12$ **(iii)** (c) $ab = -4$

Exercise – 3.9

(i) $k \neq -2$ **(ii)** $k \neq 6$ **(iii)** $k \neq \frac{-3}{7}$

Exercise – 3.10

(i) (b) no solution **(ii)** (b) no solution

Exercise – 3.11

(i) (b) $\frac{4}{3}$ **(ii)** (c) 1 **(iii)** (b) –1

Exercise – 3.12

(i) (d) 5 **(ii)** (b) 2 **(iii)** (a) $\frac{44}{7}$

Exercise – 3.13

(i) (b) $x = 5, y = 3$ **(ii)** (a) $x = 6, y = 2$
(iii) (b) $x = -1, y = -1$

Exercise – 3.14

(i) (b) **(ii)** (a)

Exercise – 3.15

(i) (a) **(ii)** (a) **(iii)** (c) **(iv)** (c) **(v)** (a)

4 Quadratic Equations

Key Point The discriminant of quadratic equation $ax^2 + bx + c = 0$ is $b^2 - 4ac$

1. Find the discriminant of the quadratic equation $3\sqrt{3}x^2 + 10x + \sqrt{3} = 0$ **(2009)**

Solution :

Given equation is $3\sqrt{3}x^2 + 10x + \sqrt{3} = 0$

On comparing it with $ax^2 + bx + c = 0$, we get $a = 3\sqrt{3}$, $b = 10$ and $c = \sqrt{3}$

Then, discriminant,

$D = b^2 - 4ac$

$= (10)^2 - 4.\ 3\ \sqrt{3}.\sqrt{3}$

$= 100 - 36$

$= 64$

2. Write the discriminant of the quadratic equation $(x + 5)^2 = 2(5x - 3)$. **(2019)**

Solution :

Given equation is

$(x + 5)^2 = 2(5x - 3)$

$\Rightarrow x^2 + 10x + 25 = 10x - 6$

$\Rightarrow x^2 + 31 = 0$

On comparing it with $ax^2 + bx + c = 0$, we get $a = 1$, $b = 0$, $c = 31$

Then, discriminant,

$D = b^2 - 4ac$

$= 0^2 - 4 \times 1 \times 31$

$= -124$

3. The discriminant of the quadratic equation $2x^2 - 4x + 3 = 0$ is. **(2020)**

(a) -8 (b) 10

(c) 8 (d) $2\sqrt{2}$

Solution: (a)

Given equation is $2x^2 - 4x + 3 = 0$

On comparing it with $ax^2 + bx + c = 0$, we get $a = 2$, $b = -4$ and $c = 3$

Then, discriminant,

$D = b^2 - 4ac$

$= (-4)^2 - 4.\ 2.\ 3$

$= 16 - 24$

$= -8$

4. The discriminant of the quadratic equation $4x^2 - 6x + 3 = 0$ is.

(a) 12 (b) 84

(c) $2\sqrt{3}$ (d) -12

Solution : (d)

$D = b^2 - 4ac$

$= (-6)^2 - 4.\ 4.\ 3$

$= 36 - 48$

$= -12$

5. The discriminant of the quadratic equation $2x^2 - 5x - 3 = 0$ is. **(2023)**

(a) 1 (b) 49

(c) 7 (d) 19

Solution : (b)

$D = b^2 - 4ac$

$= (-5)^2 - 4.2.\ (-3)$

$= 25 + 24$

$= 49$

6. The discriminant of the quadratic equation $2x^2 + x - 1 = 0$ is. **(2023)**

(a) -9 (b) -7

(c) 9 (d) 7

Solution : (c)

$D = b^2 - 4ac$

$= 1^2 - 4.2.(-1)$

$= 1 + 8$

$= 9$

7. The discriminant of the quadratic equation $x^2 - 3x - 2 = 0$ is: **(2025)**

(a) 1 (b) 17

(c) $\sqrt{17}$ (d) $-\sqrt{17}$

Solution: (b)

$D = b^2 - 4ac$

$= (-3)^2 - 4 \times 1 \times (-2)$

$= 9 + 8$

$= 17$

8. The discriminant of the quadratic equation $-x^2 - 5x + 6 = 0$ is: **(2025)**

(a) 1 (b) -1

(c) 49 (d) 7

Solution: (c)

$D = b^2 - 4ac$

$= (-5)^2 - 4 \times (-1) \times 6$

$= 25 + 24$

$= 49$

9. The discriminant of the quadratic equation $2x^2 - 3x - 5 = 0$ is **(2025)**

(a) -31 (b) 49

(c) 7 (d) $\sqrt{-31}$

Solution: (b)

$D = b^2 - 4ac$

$= (-3)^2 - 4 \times 2 \times (-5)$

$= 9 + 40$

$= 49$

10. The discriminant of the quadratic equation $ax^2 + x + a = 0$ is: **(2026)**

(a) $\sqrt{1-4a^2}$ (b) $1-4a^2$

(c) $4a^2 - 1$ (d) $\sqrt{4a^2-1}$

Solution: (b)

$D = 1^2 - 4.a.a = 1 - 4a^2$

Exercise 4.1

(i) Find the discriminant of the quadratic equation $3x^2 - 4\sqrt{3}x + 4 = 0$

(ii) Find the discriminant of the quadratic equation $12x^2 - 4\sqrt{15}x + 5 = 0$

(iii) Find the discriminant of the quadratic equation $3x^2 - 2\sqrt{6}x + 2 = 0$

(iv) Find the discriminant of the quadratic equation $4\sqrt{3}x^2 + 5x - 2\sqrt{3} = 0$

(v) Write the discriminant of the quadratic equation $(x + 5)^2 = 2(5x - 3)$

(vi) Write the discriminant of the quadratic equation $2x^2 + \sqrt{5}\,x - 5 = 0$

(vii) Write the discriminant of the quadratic equation $3x^2 - 11ax + 9a^2 = 0$

Key Point If α be a root of a quadratic equation $ax^2 + bx + c = 0$, then $a\alpha^2 + b\alpha + c = 0$

11. Show that $x = -3$ is a solution of $x^2 + 6x + 9 = 0$ **(2009)**

Solution :

LHS $= (-3)^2 + 6(-3) + 9$

$= 9 - 18 + 9.$

$= 0$

$\therefore x = -3$ is a solution of $x^2 + 6x + 9 = 0$

Exercise 4.2

(i) Show that $x = -4$ is a solution of $x^2 + 8x + 16 = 0$

(ii) Show that $x = \frac{2}{3}$ is a solution of $9x^2 - 6x + 4 = 0$

(iii) Show that $x = 3$ is a solution of $3x^2 - 10x + 3 = 0$

(iv) Show that $x = 5$ is a solution of $x^2 - 12x + 35 = 0$

12. If $x = \frac{-1}{2}$ is a solution of the quadratic equation $3x^2 + 2kx - 3 = 0$, find the value of k. **(2015)**

Solution :

Since $x = \frac{-1}{2}$ is a solution of the quadratic equation $3x^2 + 2kx - 3 = 0$

$$\therefore 3\left(-\frac{1}{2}\right)^2 + 2k\left(-\frac{1}{2}\right) - 3 = 0$$

$\Rightarrow 3.\ \frac{1}{4} - k - 3 = 0$

$\Rightarrow -k = 3 - \frac{3}{4}$

$\Rightarrow -k = \frac{9}{4}$

$\therefore\ k = -\frac{9}{4}$

13. If one root of the quadratic equation $6x^2 - x - k = 0$ is $\frac{2}{3}$, Then find the value of k. **(2007)**

Solution :

Since one root of the quadratic equation $6x^2 - x - k = 0$ is $\frac{2}{3}$,

$\therefore\ 6.\left(\frac{2}{3}\right)^2 - \frac{2}{3} - k = 0$

$\Rightarrow 6.\ \frac{4}{9} - \frac{2}{3} - k = 0$

$\Rightarrow \frac{8}{3} - \frac{2}{3} - k = 0$

$\Rightarrow \frac{6}{3} - k = 0$

$\Rightarrow 2 - k = 0$

$\therefore\ k = 2$

14. If $x = 3$ is a one root of the quadratic equation $x^2 - 2kx - 6 = 0$, then find the value of k. **(2018)**

Solution :

Since $x = 3$ is a one root of the quadratic equation $x^2 - 2kx - 6 = 0$

$\therefore\ 3^2 - 2k.\ 3 - 6 = 0$

$= 9 - 6\text{k} - 6 = 0$

$= 3 - 6k = 0$

$= -6\text{k} = -3$

$\therefore\ k = \frac{1}{2}$

15. Find the value of k for which $x = 2$ is a solution of the equation $kx^2 + 2x - 3 = 0$ **(2019)**

Solution :

Since $x = 2$ is a solution of the equation $kx^2 + 2x - 3 = 0$

$\therefore\ k.\ 2^2 + 2.2 - 3 = 0$

$\Rightarrow 4k + 1 = 0$

$\Rightarrow 4k = -1$

$\Rightarrow k = -\frac{1}{4}$

16. If $x = 0.3$ is a root of the equation $x^2 - 0.9k = 0$, then k is equal to

(a) 1 (b) 10

(c) 0.1 (d) 100

Solution : (c)

Since $x = 0.3$ is a root of the equation $x^2 - 0.9k = 0$, then

$(0.3)^2 - 0.9k = 0$

$\rightarrow 0.09 - 0.9k = 0$

$\Rightarrow -0.9k = -0.09$

$\Rightarrow k = 0.1$

17. If 'p' is a root of the quadratic equation $x^2 - (p + q)\ x + k = 0$, then the value of 'k' is **(2023)**

(a) p (b) q

(c) $p + q$ (d) pq

Solution :(d)

Since p is a root of the quadratic equation $x^2 - (p + q)\ x + k = 0$

$\therefore\ p^2 - (p + q).\ p + k = 0$

$\Rightarrow p^2 - p^2 - pq + k = 0$

$\Rightarrow k = pq$

Exercise 4.3

(i) If $x = 3$, is a solution of the quadratic equation $2x^2 - 7x + k = 0$, find the value of k.

(a) −3 (b) 2

(c) 3 (d) −2

(ii) If $x = 3$, is a solution of the quadratic equation $x^2 - 5x + b = 0$, find the value of k.

(a) 6 (b) 0

(c) 2 (d) −3

(iii) If $x = 3$, is a solution of the quadratic equation $2x^2 - 8x + k = 0$, find the value of k.

(a) 6 (b) −6
(c) 3 (d) −3

(iv) If $x = 3$, is a solution of the quadratic equation $x^2 - 5x + p^2 - p = 0$, find the value of $p = ?$.

(a) −2, −3 (b) 2, −3
(c) −2, 3 (d) 2, 3

(v) If $x = \frac{5}{2}$ is a solution of the quadratic equation $px^2 - 3x - 5 = 0$, then $p = ?$

(a) −2 (b) 2
(c) 1 (d) −1

(vi) If $x = 0.2$, is a solution of the quadratic equation $x^2 - 0.4k = 0$, find the value of k.

(a) 1 (b) 10
(c) 0.1 (d) 100

(vii) If $x = 0.4$, is a solution of the equation $x^2 - 0.16k = 0$, then k is equal to

(a) 1 (b) 10
(c) 0.1 (d) 0.01

18. If 1 is a root of the equations $ay^2 + ay + 3 = 0$ and $y^2 + y + b = 0$, then ab equals. **(2012)**

(a) 3 (b) $-\frac{7}{2}$
(c) 6 (d) −3

Solution : (a)

Since 1 is a root of the equations $ay^2 + ay + 3 = 0$ and $y^2 + y + b = 0$

$\therefore\ a.1^2 + a.1 + 3 = 0$

$= a + a + 3 = 0$

$= 2a = -3$

$= a = -\frac{3}{2}$

and $1^2 + 1 + b = 0$

$\Rightarrow 1 + 1 + b = 0$

$\Rightarrow b = -2$

$\therefore\ ab = \left(-\frac{3}{2}\right)(-2) = 3$

Exercise 4.4

(i) If −1 is a root of the equations $x^2 + ax - 2 = 0$ and $x^2 + 2bx + 3 = 0$, then $a + b$ equals.

(a) 1 (b) 2
(c) 0 (d) 3

(ii) If 2 is a root of the equations $x^2 + 2x + b = 0$ and $x^2 - 3ax + 2 = 0$, then ab equals.

(a) 8 (b) −8
(c) 2 (d) 6

(iii) If 1 is a root of the equations $y^2 - ay + 3 = 0$ and $y^2 + y + b = 0$, then ab equals.

(a) 4 (b) 2
(c) 8 (d) −8

Key Point If the quadratic equation $ax^2 + bx + c = 0$ has two equal roots, then $b^2 - 4ac = 0$.

19. If the quadratic equation $px^2 - 2\sqrt{15}\,px + 15 = 0$ has two equal roots, then find the value of p. **(2015)**

Solution :

Since the roots of the quadratic equation $px^2 - 2\sqrt{15}\,px + 15 = 0$ are equal

$D = 0$

$\Rightarrow \left(-2\sqrt{15}\right)^2 - 4p.15 = 0$

$\Rightarrow 60 - 60p = 0$

$\Rightarrow -60p = -60$

$\therefore\ p = 1$

20. The value of k for which the quadratic equation $2x^2 + kx + 2 = 0$ has equal roots, is **(2020)**

(a) 4 (b) ±4
(c) −4 (d) 0

Solution: (b)

The given equation is $2x^2 + kx + 2 = 0$

$D = k^2 - 4.2.2 = k^2 - 16$

For equal roots, we must have $D = 0$

Now, $D = 0$

$\Rightarrow k^2 - 16 = 0$

$\Rightarrow k^2 = 16$

$k = \pm 4$

21. If quadratic equation $3x^2 - 4x + k = 0$ has equal roots, then the value of k is ________. **(2020)**

Solution :

The given equation is $3x^2 - 4x + k = 0$

$\therefore D = (-4)^2 - 4.3.k = 16 - 12k$

For equal roots, we must have $D = 0$

Now, $D = 0$

$\Rightarrow 16 - 12k = 0$

$\Rightarrow -12k = -16$

$\Rightarrow k = \dfrac{16}{12} = \dfrac{4}{3}$

22. If the quadratic equation $9x^2 + bx + \dfrac{1}{4} = 0$ has equal roots, then the value of 'b' is **(2023)**

(a) 0 (b) -3 only

(c) 3 only (d) ± 3

Solution :

The given equation is $9x^2 + bx + \dfrac{1}{4} = 0$

$\therefore D = b^2 - 4.9.\dfrac{1}{4} = b^2 - 9$

For equal roots, we must have $D = 0$

Now, $D = 0$

$\Rightarrow b^2 - 9 = 0$

$\Rightarrow b^2 = 9$

$\Rightarrow b = \pm 3$

23. If the roots of equation $ax^2 + bx + c = 0, a \neq 0$ are real and equal, then which or the following relation is true ? **(2024)**

(a) $a = \dfrac{b^2}{c}$ (b) $b^2 = ac$

(c) $ac = \dfrac{b^2}{4}$ (d) $c = \dfrac{b^2}{a}$

Solution :(c)

Since the roots of the equation $ax^2 + bx + c = 0$ are real and equal

$\therefore D = 0$

$\Rightarrow b^2 = 4ac$

$\Rightarrow 4ac = b^2$

$\Rightarrow ac = \dfrac{b^2}{4}$

24. If the roots of quadratic equation $4x^2 - 5x + k = 0$ are real and equal then value of k is. **(2024)**

(a) $\dfrac{5}{4}$ (b) $\dfrac{25}{16}$

(c) $-\dfrac{5}{4}$ (d) $-\dfrac{25}{16}$

Solution : (b)

Since the roots of the quadratic equation are real and equal

$\therefore D = 0$

$\Rightarrow (-5)^2 - 4.4.k = 0$

$\Rightarrow 25 - 16k = 0$

$\Rightarrow -16k = -25$

$\Rightarrow k = \dfrac{25}{16}$

25. If the roots of quadratic equation $ax^2 + bx + c = 0$ are real and equal then value of k is. **(2023)**

(a) $-\dfrac{b}{2a}$ (b) $\dfrac{b}{2a}$

(c) $\dfrac{-b^2}{4a}$ (d) $\dfrac{b^2}{4a}$

Solution : (d)

The given equation is $ax^2 + bx + c = 0$

$\therefore D = b^2 - 4ac$

For real and equal roots, we must have $D = 0$

Now $D = 0$

$\Rightarrow b^2 - 4ac = 0$

$\Rightarrow -4ac = -b^2$

$\Rightarrow c = \dfrac{b^2}{4a}$

26. Find the value (s) of k for which the quadratic equation $3x^2 + kx + 3 = 0$ has real and equal roots. **(2019)**

Solution :

The given equation is $3x^2 + kx + 3 = 0$

$\therefore D = k^2 - 4.3.3 = k^2 - 36$

For real and equal roots, we must have

$D = 0$

$\therefore k^2 - 36 = 0$

$\Rightarrow k^2 = 36$

$\Rightarrow k = \pm 6$

27. The Roots of the equation $x^2 + bx + c = 0$ are equal, if ________. **(2020)**

Solution :

$b^2 - 4c = 0$

28. The two roots of the equation $3x^2 - 2\sqrt{6}x + 2 = 0$ are **(2023)**

(a) real and district (b) not real

(c) real and equal (d) rational

Solution : (c)

The given equation is $3x^2 - 2\sqrt{6}x + 2 = 0$

This is of the form $ax^2 + bx + c = 0$,

Where $a = 3$, $b = -2\sqrt{6}$ and $c = 2$

$\therefore D = b^2 - 4ac = \left(2\sqrt{6}\right)^2 - 4.3.2$

$= 4 \times 6 - 24$

$= 24 - 24$

$= 0$

Hence the roots of the equation are real and equal.

29. The value of 'a' for which $ax^2 + 3x + 1 = 0$ has real and equal roots is: **(2024)**

(a) $\frac{4}{9}$ (b) $\frac{9}{4}$

(c) $\frac{3}{2}$ (d) $\frac{2}{3}$

Solution: (b)

The given quadratic equation has real and equal roots, if

$D = 0$

$\Rightarrow 3^2 - 4 . a . 1 = 0$

$\Rightarrow 4a = 9$

$\therefore a = \frac{9}{4}$

30. One of the values of 'p' for which $px^2 + 4x + p = 0$ has real and equal roots is: **(2025)**

(a) 4 (b) –4

(c) 2 (d) 0

Solution:

The given equation has real and equal roots, if

$D = 0$

i e $4^2 - 4p . p = 0$

$\Rightarrow 16 - 4p^2 = 0$

$\Rightarrow 4p^2 = 16$

$\Rightarrow p^2 = 4$

$\Rightarrow p = 2$

31. The value of k for which the roots of the quadratic equation $6x^2 + 4kx + k = 0$ are real and equal, is **(2025)**

(a) 0 (b) $\frac{3}{4}$

(c) $-\frac{3}{2}$ (d) $\frac{2}{3}$

Solution: (a)

The roots of the given quadratic equation are real and equal, if

$D = 0$

$\Rightarrow (4k)^2 - 4 . 6 . k = 0$

$\Rightarrow 16k^2 - 24k = 0$

$\Rightarrow 8k\,(2k - 3) = 0$

$\Rightarrow 8k = 0$ or $2k - 3 = 0$

$\Rightarrow k = 0$ or $k = \frac{3}{2}$

32. Which of the following quadratic equation has real and equal roots ? **(2025)**

(a) $(x + 1)^2 = 2x + 1$

(b) $x^2 + x = 0$

(c) $x^2 - 4 = 0$

(d) $x^2 + x + 1 = 0$

Slloution: (a)

$(x + 1)^2 = 2x + 1$

$\Rightarrow x^2 + 2x + 1 = 2x + 1$

$\Rightarrow x^2 = 0$

$\therefore D = b^2 - 4ac = 0^2 - 4 . 1 . 0 = 0$

33. For what value of k, the roots of the equation $x^2 + 4x + k = 0$ are real ? **(2019)**

Solution:

The give n equation is $x^2 + 4x + k = 0$

This is of the the form $ax^2 + bx + c = 0$

Where $a = 1$, $b = 4$ and $c = k$

$\therefore D = b^2 - 4ac = 4^2 - 4 \cdot 1 \cdot k = 16 - 4k$

The roots of the quadratic equation are real if $D \geq 0$

$\Rightarrow 16 - 4k \geq 0$

$\Rightarrow -4k \geq -16$

$\Rightarrow k \leq 4$

34. If the roots of the quadratic equation $\sqrt{3}x^2 - kx + 2\sqrt{3} = 0$ are real and equal, Then the value (*s*) of *k* is / are : **(2026)**

(a) $\pm\sqrt{24}$ (b) 0

(c) 4 (d) –5

Solution: (a)

$b^2 - 4ac = 0$

$\Rightarrow (-k)^2 - 4\sqrt{3}.2\sqrt{3} = 0$

$\Rightarrow k^2 - 24 = 0$

$\Rightarrow k^2 = 24$

$\Rightarrow k = \pm\sqrt{24}$

35. The value of '*m*' for which the quadratic equation $x^2 - 2x + m = 0$ has real and equal roots is: **(2026)**

(a) –1 (b) 0

(c) 4 (d) 1

Solution:

$b^2 - 4ac = 0$

$\Rightarrow (-2)^2 - 4.1.m = 0$

$\Rightarrow 4 - 4m = 0$

$\Rightarrow -4m = -4$

$\Rightarrow m = 1$

36. The value of '*p*' for which the equation $x^2 - 2x - p = 0$ has real and equal roots is: **(2026)**

(a) –1 (b) 0

(c) 1 (d) 4

Solution: (a)

$b^2 - 4ac = 0$

$\Rightarrow (-2)^2 - 4(-p) = 0$

$\Rightarrow 4 + 4p = 0$

$\Rightarrow 4p = -4$

$\Rightarrow p = -1$

37. The value of '*m*' for which the equation $x^2 - mx - m = 0$ has real and equal roots is: **(2026)**

(a) 0 only (b) 4 only

(c) 0 or 4 (d) 0 or –4

Solution: (d)

$b^2 - 4ac = 0$

$\Rightarrow (-m)^2 - 4.1.(-m) = 0$

$\Rightarrow m^2 + 4m = 0$

$\Rightarrow m(m+4) = 0$

$\Rightarrow m = 0$ or $m + 4 = 0$

$\Rightarrow m = 0$ or $m = -4$

38. The value of *m* for which the quadratic equation $3x^2 - 7x + m = 0$ has real and equal roots, is **(2026)**

(a) 7 (b) $\frac{49}{12}$

(c) $\frac{49}{3}$ (d) 4

Solution: (b)

$b^2 - 4ac = 0$

$\Rightarrow (-7)^2 - 4 \times 3 \times m = 0$

$\Rightarrow 49 - 12m = 0$

$\Rightarrow m = \frac{49}{12}$

39. The value of *k* for which the equation $kx^2 - 6x - 4 = 0$ has real and equal roots is **(2026)**

(a) $\frac{9}{4}$ (b) –4

(c) $-\frac{9}{4}$ (d) –2

Solution: (c)

$D = 0$

$\Rightarrow (-6)^2 - 4.k.(-4) = 0$

$\Rightarrow 36 + 16k = 0$

$\Rightarrow k = -\frac{9}{4}$

40. The value of *k* for which the equation $2kx^2 - 6x + 3 = 0$ has real and equal roots, is: **(2026)**

(a) $\frac{3}{2}$ (b) $\frac{1}{2}$

(c) $-\frac{3}{2}$ (d) 2

Solution: (a)

$D=0$

$\Rightarrow (-6)^2 - 4\times 2k\times 3 = 0$

$\Rightarrow 36-24k=0$

$\Rightarrow 24k=36$

$\Rightarrow k=\frac{3}{2}$

41. If roots of the quadratic equation $x^2-k\sqrt{3}\,x+2=0$ are real and equal, then value of k is: **(2026)**

(a) −2 (b) $\sqrt{\frac{8}{3}}$

(c) 1 (d) 2

Solution: (b)

$D=0$

$\Rightarrow \left(-k\sqrt{3}\right)^2 - 4.1.2 = 0$

$\Rightarrow 3k^2=8$

$\Rightarrow k^2=\frac{8}{3}$

$\Rightarrow k=\sqrt{\frac{8}{3}}$

Exercise 4.5

(i) If the quadratic equation $9x^2 + 6px + 4 = 0$ has equal roots, then p = ?

(a) 2 or −1 (b) −2 or 1

(c) 2 or −2 (d) −1 or 1

(ii) If the quadratic equation $4x^2 - kx + 2 = 0$ has equal roots, then k = ?

(a) 2 or −2

(b) $2\sqrt{3}$ or $-2\sqrt{3}$

(c) $2\sqrt{2}$ or $-2\sqrt{2}$

(d) $\sqrt{2}$ or $-\sqrt{2}$

(iii) If the quadratic equation $x^2+2(k+2)x+9k=0$ has equal roots, then k = ?

(a) 1 or 4 (b) −1 or 2

(c) −1 or 1 (d) 0 or 2

(iv) If the quadratic equation $4x^2 - 3x + 1 = 0$ has equal roots, then k = ?

(a) $\frac{4}{3}$ or $-\frac{4}{3}$ (b) $\frac{1}{3}$ or $\frac{2}{3}$

(c) 2 or $\frac{1}{3}$ (d) $\frac{2}{3}$ or $\frac{-2}{3}$

(v) If the quadratic equation $x^2 - 2x(1 + 3k) + 7(3 + 2k) = 0$ has equal roots, then k = ?

(a) 1 or $\frac{3}{7}$ (b) −2 or $\frac{10}{3}$

(c) 2 or $-\frac{10}{9}$ (d) 4 or −5

(vi) If the quadratic equation $x^2 + (m - 3)x + m = 0$ has equal roots, then m = ?

(a) 1 or 9 (b) −1 or 3

(c) −1 or −3 (d) 0 or 4

(vii) If the quadratic equation $kx(x - 2) + 6 = 0$ has equal roots, then k = ? **(2019)**

(a) 3 (b) −2

(c) 6 (d) 1

(viii) If the quadratic equation $(3k + 1)x^2 + 2(k + 1)x + k = 0$ has two equal roots, then k = ?

(a) 1 or $\frac{1}{2}$ (b) 1 or $-\frac{1}{2}$

(c) −1 or $-\frac{1}{2}$ (d) −1 or $\frac{1}{2}$

(ix) If the quadratic equation $9x^2 + kx + 16 = 0$ has equal roots, then k = ?

(a) 4 or 24 (b) 24 or −8

(c) 24 or −24 (d) 8 or −8

(x) If the quadratic equation $x^2 + (7 - 2k)x = 12$ has equal roots, then k = ?

(a) 3 or 4 (b) –3 or 4

(c) –3 or –4 (d) 0 or 2

(xi) If the roots of the equation $x^2 - 4ax + b = 0$ are real and equal, then which of the following relations is true ?

(a) $a^2 = 4b$ (b) $b^2 = 4a$

(c) $4a^2 = b$ (d) $ab = 4$

(xii) If the roots of the equation $x^2 + ax + b = 0$ are real and equal, then which of the following relations is true ?

(a) $a = b$ (b) $a^2 = b$

(c) $a = b^2$ (d) $4a^2 = b$

(xiii) If the roots of the equation $ax^2 + cx + b = 0$ are real and equal, then which of the following relations is true ?

(a) $ab = \frac{c^2}{4}$ (b) $4b = c^2$

(c) $4\,ab = c$ (d) $a^2b = c$

Key Point **(i)** The roots of the quadratic equation $ax^2 + bx + c = 0$ to be real and distinct if $b^2 - 4ac > 0$

(ii) The roots of the quadratic equation $ax^2 + bx + c = 0$ to be imaginary if $b^2 - 4ac < 0$

(iii) The roots of the quadratic equation $ax^2 + bx + c = 0$ to be rational, if $b^2 - 4ac$ is a perfect square number.

42. Roots of the quadratic equation $2x^2 - 4x + 3 = 0$ are **(2020)**

(a) Real and equal

(b) Real and distinct

(b) Not real

(d) Real

Solution:

The given equation is $2x^2 - 4x + 3 = 0$

This is of the form $ax^2 + bx + c = 0$ where $a = 2, b = -4$ and $c = 3$

$\therefore D = b^2 - 4ac = (-4)^2 - 4{\cdot}2{\cdot}3 = 16 - 24 = -8 < 0$

$\therefore$ The roots of the quadratic equation are not real

43. Find the nature of roots of the quadratic equations $2x^2 - 4x + 3 = 0$ **(2019)**

Solution:

$b^2 - 4ac = (-4)^2 - 4 \times 2 \times 3$

$= 16 - 24$

$= -8 < 0$

$\therefore$ The roots of the equation are not real.

44. The roots of the equation $x^2 - 8 = 0$ are **(2025)**

(a) Rational and distinct

(b) Irrational and distinct

(c) Real and equal

(d) Not real

Solution: (b)

$D = b^2 - 4ac$

$= 0^2 - 4 \times 1 \times (-8)$

$= 32 > 0$

$\therefore$ The roots of the equation are irrational and distinct.

45. Find the nature of the roots of the quadratic equation $4x^2 + 4\sqrt{3}\,x + 3 = 0$ **(2019)**

Solution:

The given equation is $4x^2 + 4\sqrt{3}\,x + 3 = 0$

This is of the from $ax^2 + bx + c = 0$, where

$a = 4, b = 4\sqrt{3}$ and $c = 3$

$\therefore D = b^2 - 4ac = \left(4\sqrt{3}\right)^2 - 4{\cdot}4{\cdot}3$

$= 48 - 48$

$= 0$

$\therefore$ The roots of the equation are equal.

46. For what values of k does the quadratic equation $4x^2 - 12x - k = 0$ have no real roots ? **(2019)**

Solution:

The given equation is $4x^2 - 12x - k = 0$

This is of the from $ax^2 + bx + c = 0$ where

$a = 4, b = -12$ and $c = -k$

$\therefore D = b^2 - 4ac$

$= (-12)^2 - 4{\cdot}4{\cdot}(-k)$

$= 144 + 16k$

For non real roots, we must have $D < 0$

Now, $D < 0$

$\Rightarrow 144 + 16k < 0$

$\Rightarrow 16k < -144$

$\Rightarrow k < -9$

47. The quadratic equations $x^2 - 4x + k = 0$ has distinct real roots if **(2020)**

(a) $k = 4$ (b) $k > 4$

(c) $k = 16$ (d) $k < 4$

Solution: (d)

The given equation is $x^2 - 4x + k = 0$

This is of the from $ax^2 + bx + c = 0$

Where $a = 1, b = -4$ and $c = k$

$\therefore D = (-4)^2 - 4 \cdot 1 \cdot k$

$= 16 - 4k$

For distinct real roots, we must have $D > 0$

Now, $D > 0$

$\Rightarrow 16 - 4k > 0$

$\Rightarrow -4k > -16$

$\Rightarrow k < 4$

48. The equation $x^2 - 8x + k = 0$ has real and distinct roots if **(2020)**

(a) $k = 16$ (b) $k > 16$

(c) $k = 8$ (d) $k < 16$

Solution: (d)

The given equation is $x^2 - 8x + k = 0$

This is of the from $ax^2 + bx + c = 0$

Where $a = 1, b = -8$ and $c = k$

$\therefore D = b^2 - 4ac$

$= (-8)^2 - 4 \cdot 1 \cdot k$

$= 64 - 4k$

For real and distinct roots, we must have $D > 0$

Now, $D > 0$

$\Rightarrow 64 - 4k > 0$

$\Rightarrow -4k > -64$

$\Rightarrow k < 16$

49. Which of the following quadratic equations has real and distinct roots ? **(2025)**

(a) $x^2 + 2x = 0$ (b) $x^2 + x + 1 = 0$

(c) $(x - 1)^2 = 1 - 2x$ (d) $2x^2 + x + 1 = 0$

Solution: (a)

Here $a = 1, b = 2, c = 0$

$\therefore D = b^2 - 4ac$

$= 2^2 - 4 \times 1 \times 0$

$= 4 > 0$

$\therefore$ The quadratic equation $x^2 + 2x = 0$ has real and distinct roots.

50. The least positive value of k, for which the quadratic equation $2x^2 + kx - 4 = 0$ has rational roots, is **(2023)**

(a) $\pm 2\sqrt{2}$ (b) 2

(c) ± 2 (d) $-\frac{1}{2}$

Solution: (b)

Here $D = b^2 - 4ac$

$= k^2 - 4 \cdot 2 \cdot (-4)$

$= k^2 + 32$

Now, the roots of the equation are rational if $D > 0$ and D is a square number.

Clearly the least positive value of k such that the quadratic equation $2x^2 + kx - 4 = 0$ has rational roots is 2.

51. Which of the following equations does not have a real root ? **(2025)**

(a) $x^2 = 0$ (b) $2x - 1 = 3$

(c) $x^2 + 1 = 0$ (d) $x^3 + x^2 = 0$

Solution: (c)

$x^2 + 1 = 0$

Here $a = 1, b = 0, c = 1$

$\therefore D = b^2 - 4ac = 0^2 - 4 \times 1 \times 1 = -4 < 0$

52. The value of p for which roots of the quadratic equation $x^2 - px + 6 = 0$ are rational, is : **(2026)**

(a) 1 (b) -5

(c) 25 (d) $\sqrt{5}$

Solution: (b)

$D = b^2 - 4ac$

$\Rightarrow D = (-p)^2 - 4.1.6$

$\Rightarrow D = p^2 - 24$

if $p = 1, D = 1^2 - 24 = -23$ (Not a perfect square)

if $p = -5, D = (-5)^2 - 24 = 1$ (perfect square)

if $p = 25, D = (25)^2 - 24 = 601$ (Not a perfect square)

if $p = \sqrt{5}$, $D = \left(\sqrt{5}\right)^2 - 24 = -19$ (Not a perfect square)

53. The roots of the quadratic equation $x^2 + 9 = 0$ are **(2026)**

(a) real and equal
(b) not real
(c) real and negative of each other
(d) rational numbers

Solution: (b)

$D = 0^2 - 4 \times 1 \times 9 = -36 < 0$

Exercise 4.6

(i) For what value of k, the roots of the equation $x^2 - 3x + 2k = 0$ are real.

(ii) For what value of k, the roots of the equation $x^2 - x + 2k = 0$ are real.

(iii) For what value of k, the roots of the equation $x^2 - 4x + k = 0$ are real.

(iv) Find the nature of roots of the quadratic equation $4x^2 - 5x + 3 = 0$.

(v) Find the nature of roots of the quadratic equation $3x^2 - 4x + 8 = 0$

(vi) Find the nature of roots of the quadratic equation $x^2 - 2x + 4 = 0$

(vii) Find the nature of roots of the quadratic equation $4x^2 + 3x + 2 = 0$

(viii) Find the nature of the roots of the quadratic equation $3x^2 - 4\sqrt{3}\,x + 4 = 0$.

(ix) Find the nature of the roots of the quadratic equation $x^2 + 8x + 16 = 0$

(x) Find the nature of the roots of the quadratic equation $9x^2 - 12x + 4 = 0$

(xi) Find the nature of the roots of the quadratic equation $12x^2 - 4\sqrt{15}\,x + 5 = 0$

(xii) For what values of k does the quadratic eqution $2x^2 - 3x - k = 0$ have no real roots ?

(a) $\frac{-9}{8}$ (b) $\frac{-8}{9}$

(c) $\frac{9}{8}$ (d) $\frac{8}{9}$

(xiii) For what values of k does the quadratic equation $3x^2 + 2x - 2k = 0$ have no real roots ?

(a) $\frac{1}{6}$ (b) $-\frac{1}{6}$

(c) $\frac{3}{8}$ (d) $\frac{4}{5}$

(xiv) For what values of k does the quadratic equation $x^2 - 3x - k = 0$ have no real roots ?

(a) $-\frac{9}{4}$ (b) $\frac{9}{4}$

(c) $\frac{2}{3}$ (d) $\frac{3}{2}$

(xv) The quadratic equation $x^2 - 3x + k = 0$ has distinct real roots if

(a) $\frac{9}{4}$ (b) $\frac{4}{9}$

(c) 4 (d) 9

(xvi) The quadratic equation $x^2 - 2x + k = 0$ has distinct real roots, if

(a) $k = 1$ (b) $k > 1$

(c) $k < 1$ (d) $k = -1$

(xvii) The quadratic equation $x^2 - 5x + k = 0$ has distinct real roots if

(a) $k = \frac{25}{4}$ (b.) $k > \frac{25}{4}$

(c) $k < \frac{25}{4}$ (d.) $k = 4$

(xviii) The least positive value of k, for which the quadratic equation $3x^2 + kx - 1 = 0$ has rational roots, is

(a) $\pm 2\sqrt{2}$ (b) 2

(c) ± 2 (d) $-\frac{1}{2}$

(xix) The least positive value of k, for which the quadratic equation $x^2 - 2kx - 3 = 0$ has rational roots, is

(a) 1 (b) ± 1

(c) 2 (d) ± 4

(xx) The least positive value of k, for which the quadratic equation $4x^2 - kx - 1 = 0$ has rational roots, is

(a) 1 (b) 2
(c) ± 3 (d) ± 4

54. The value of λ for which ($x^2 + 4x + \lambda$) is a perfect square, is **(2020)**

(a) 16 (b) 9
(c) 1 (d) 4

Solution:

The given equation is $x^2 + 4x + \lambda = 0$. This is of the form $ax^2 + bx + c = 0$. where $a = 1$, $b = 4$ and $c = \lambda$

$D = 4^2 - 4 \cdot 1 \cdot \lambda$

$= 16 - 4\lambda$

The given equation is perfect square if $D = 0$

Now, $D = 0$

$\Rightarrow 16 - 4\lambda = 0$

$\Rightarrow -4\lambda = -16$

$\therefore \lambda = 4$

Exercise 4.7

(i) The value of λ for which $x^2 + 2x + \lambda = 0$ is a perfect square, is

(a) 1 (b) –1
(c) 2 (d) 4

(ii) The value of λ for which $x^2 + 6x + \lambda = 0$ is a perfect square,is

(a) 6 (b) 4
(c) 9 (d) –1

(iii) The value of λ for which $x^2 - 4x + \lambda = 0$ is a perfect square, is

(a) –4 (b) 4
(c) 2 (d) 1

55. The roots of the quadratic equation $x^2 - 0.04 = 0$ are. **(2020)**

(a) ± 0·2 (b) ± 0·02
(c) 0·4 (d) 2

Solution: (a)

$x^2 - 0{\cdot}04 = 0$

$\Rightarrow x^2 = 0{\cdot}04$

$\Rightarrow x = \pm 0{\cdot}2$

56. The roots of the quadratic equation $(x-1)^2 = 16$ are: **(2026)**

(a) 5, 3 (b) 4, –4
(c) 5, –3 (d) –53

Solution: (c)

$(x-1)^2 = 16$

$\Rightarrow x - 1 = \pm 4$

$\Rightarrow x - 1 = 4$ or $x - 1 = -4$

$\Rightarrow x = 5$ or $x = -3$

57. The roots of the quadratic equation $4x^2 - (a-1)^2 = 0$ are: **(2026)**

(a) $a-1, a+1$ (b) $\frac{a-1}{2}, \frac{-a+1}{2}$
(c) $\frac{a-1}{2}, \frac{-a-1}{2}$ (d) $\pm(a-1)$

Solution: (b)

$4x^2 - (a-1)^2 = 0$

$\Rightarrow 4x^2 = (a-1)^2$

$\Rightarrow x^2 = \left(\frac{a-1}{2}\right)^2$

$\Rightarrow x = \pm\frac{a-1}{2}$

$\Rightarrow x = \frac{a-1}{2}$ or $x = -\frac{a-1}{2} = \frac{-a+1}{2}$

Exercise 4.8

(i) The roots of the quadratic equation $x^2 - 0.09 = 0$ are

(a) ± 0.03 (b) ± 0.3
(c) 3 (d) ± 3

(ii) The roots of the quadratic equation $x^2 - 0.16 = 0$ are

(a) ± 0.4 (b) 0.4
(c) – 0.4 (d) ± 2

(iii) The roots of the quadratic equation $x^2 - 0.81 = 0$ are

(a) ± 0.3 (b) ± 0.9
(c) ± 0.03 (d) ± 0.09

58. A quadratic equation whose roots are $(2 + \sqrt{3})$ and $(2 - \sqrt{3})$ is **(2023)**

(a) $x^2 - 4x + 1 = 0$
(b) $x^2 + 4x + 1 = 0$
(c) $4x^2 - 3 = 0$
(d) $x^2 - 1 = 0$

Solution: (a)

Sum of the roots $= (2+\sqrt{3})+(2-\sqrt{3}) = 4$

Product of the roots $= (2+\sqrt{3})(2-\sqrt{3})$
$= 4 - 3 = 1$

$\therefore$ The quadratic equation is $x^2 - 4x + 1 = 0$

59. A quadratic equation whose roots are $(3-\sqrt{2})$ and $(3+\sqrt{3})$ is **(2023)**

(a) $x^2 - 6x + 7 = 0$
(b) $x^2 + 6x + 7 = 0$
(c) $9x^2 - 2 = 0$
(d) $x^2 - 7 = 0$

Solution: (a)

Sum of the roots $= (3-\sqrt{2})+(3+\sqrt{2}) = 6$

Product of the roots $= (3-\sqrt{2})(3+\sqrt{2})$
$= 9 - 2 = 7$

$\therefore$ The quadratic equation is
$x^2 - 6x + 7 = 0$

Exercise 4.9

(i) A quadratic equation whose roots are $(3+\sqrt{3})$ and $(3-\sqrt{3})$ is

(a) $x^2 - 9 = 0$
(b) $x^2 - 3 = 0$
(c) $x^2 - 6x + 6 = 0$
(d) $x^2 - 6x - 6 = 0$

(ii) A quadratic equation whose roots are $(4+\sqrt{3})$ and $(4-\sqrt{3})$ is

(a) $x^2 - 16x + 3 = 0$
(b) $x^2 - 3x + 4 = 0$
(c) $x^2 - 8x - 3 = 0$
(d) $x^2 - 8x + 13 = 0$

(iii) A quadratic equation whose roots are $(4+2\sqrt{3})$ and $(4-2\sqrt{3})$ is

(a) $x^2 - 8x + 4 = 0$ (b) $x^2 - 4x + 8 = 0$
(c) $x^2 - 8x - 4 = 0$ (d) $x^2 - 4x - 8 = 0$

60. The roots of the equation $x^2 + 3x - 10 = 0$ are **(2023)**

(a) 2, –5 (b) –2, 5
(c) 2, 5 (d) –2, –5

Solution: (a)

$x^2 + 3x - 10 = 0$
$\Rightarrow x^2 + 5x - 2x - 10 = 0$
$\Rightarrow x(x + 5) - 2(x + 5) = 0$
$\Rightarrow (x + 5)(x - 2) = 0$
$\Rightarrow x + 5 = 0$ or $x - 2 = 0$
$\Rightarrow x = -5$ or $x = 2$

61. The roots of the equation $x^2 + x - p(p + 1) = 0$, where p is a constant, are **(2011)**

(a) $p, p + 1$ (b) $-p, p + 1$
(c) $p, -(p + 1)$ (d) $-p, -(p + 1)$

Solution: (c)

$x^2 + x - p(p + 1) = 0$
$\Rightarrow x^2 + (p + 1)x - px - p(p + 1) = 0$
$\Rightarrow x(x + p + 1) - p(x + p + 1) = 0$
$\Rightarrow x + p + 1 = 0$ or $x - p = 0$
$\Rightarrow x = -(p + 1)$ or $x = p$

62. Value of the roots of the quadratic equation $x^2 - x - 6 = 0$ are **(2020)**

Solution:

$x^2 - x - 6 = 0$
$\Rightarrow x^2 - 3x + 2x - 6 = 0$
$\Rightarrow x(x - 3) + 2(x - 3) = 0$
$\Rightarrow (x - 3)(x - 2) = 0$
$\Rightarrow x - 3 = 0$ or $x + 2 = 0$
$\Rightarrow x = 3$ or $x = -2$

Exercise 4.10

(i) The roots of the equation $x^2 - 5x - (p - 3)(p + 2) = 0$ are

(a) $p - 3, p + 2$
(b) $p - 3, -(p - 2)$

(c) $p+3, -(p+2)$
(d) $(p-3), -(p+2)$

(ii) The roots of the equation $abx^2+(b^2-ac)x-bc=0$ are

(a) $\frac{c}{b}, -\frac{b}{a}$ (b) $\frac{-c}{b}, \frac{b}{a}$

(c) $\frac{c}{b}, \frac{b}{a}$ (d) $\frac{-c}{b}, \frac{-b}{a}$

(ii) The roots of the equation $4x^2-4ax+(a^2-b^2)=0$ are

(a) $a+b, a-b$
(b) $2(a+b), 2(a-b)$
(c) $\frac{a+b}{2}, \frac{a-b}{2}$
(d) $\frac{a+b}{4}, \frac{a-b}{4}$

63. Which of the following quadratic equations has sum of its roots as 4 ? **(2023)**

(a) $2x^2-4x+8=0$
(b) $-x^2+4x+4=0$
(c) $\sqrt{2}x^2-\frac{4}{\sqrt{2}}x+1=0$
(d) $4x^2+4x+4=0$

Solution: (b)

$-x^2+4x+4=0$

$\Rightarrow -(x^2-4x-4)=0$

$\Rightarrow x^2-4x-4=0$

Sum of the roots $= \frac{-(-4)}{1} = 4$

Exercise 4.11

(i) Which of the following quadratic equation has sum of its roots as 2 ?

(a) $3x^2-3x+6=0$
(b) $6x^2+12x+5=0$
(c) $4x^2-8x+5=0$
(d) $9x^2-14x+2=0$

(ii) Which of the following quadratic equation has sum of its root as –1 ?

(a) $x^2+x+1=0$
(b) $2x^2-2x+3=0$
(c) $6x^2-7x+6=0$
(d) $x^2+3x+3=0$

(iii) Which of the following quadratic equations has sum of its roots as $\frac{3}{4}$

(a) $3x^2-4x+2=0$
(b) $8x^2+12x+3=0$
(c) $5x^2-15x+2=0$
(d) $x^2-3x+4=0$

64. A quadratic equation whose one root is 2 and the sum of whose root in zero one is **(2023)**

(a) $x^2+4=0$ (b) $x^2-2=2$
(c) $4x^2-1=0$ (d) $x^2-4=0$

Solution: (d)

Let the another root of the quadratic equation is α

$\therefore \alpha+2=0$

$\Rightarrow \alpha=-2$

$\therefore$ Product of the root is $2 . (-2) = -4$

$\therefore$ The quadratic eqution is

$(x^2-0 . x-4)=0$

$\Rightarrow (x^2-4)=0$

Exercise 4.12

(i) A quadrtic equation whose one root is 3 and the sum of whose root is zero is

(a) $x^2+9=0$ (b) $x^3-3=0$
(c) $x^2-9=0$ (d) $3x^2-1=0$

(ii) A quadratic equation whose one root is –2 and the sum of whose root in zero is

(a) $x^2-2=0$ (b) $x^2+2=0$
(c) $x^2+4=0$ (d) $x^2-4=0$

(iii) A quadratic equation whose one root is 1 and the sum of whose root is zero is

(a) $x^2+1=0$ (b) $x^2-1=0$
(c) $x^2-2=0$ (d) $x^2+2=0$

65. Which of the following is not a quadratic equation ? **(2023)**

(a) $2(x-1)^2=4x^2-2x+1$
(b) $2x-x^2=x^2+5$
(c) $(\sqrt{2}x+\sqrt{3})^2+x^2=3x^2-5x$
(d) $(x^2+2x)^2=x^4+3+4x^3$

Solution: (c)

$(\sqrt{2}\,x + \sqrt{3}\,)^2 + x^2 = 3x^2 - 5x$

$\Rightarrow (\sqrt{2}\,x\,)^2 + 2\sqrt{2}\,x \,.\, \sqrt{3} + (\sqrt{3}\,)^2 + x^2 = 3x^2 - 5x$

$\Rightarrow 3x^2 + 2\sqrt{6}\,x + 3 = 3x^2 - 5x$

$\Rightarrow 2\sqrt{6}\,x + 5x + 3 = 0$

$\Rightarrow (2\sqrt{6} + 5\,)\,x + 3 = 0$

which is a linear equation

66. Which of the following equations is/are not quadratic ? **(2026)**

$q_1: (x-1)^2 = x^2$ $\quad$ $q_2: (x-1)^3 = x^3$

$q_3: (x+1)^3 = 3x^2$ $\quad$ $q_4: (\sqrt{x}+1)^2 = 2\sqrt{x}$

(a) q_1, q_3 and q_4 (b) q_2 and q_3

(c) q_2, q_3 and q_4 (d) q_1 and q_4

Solution: (a)

$q_1: x^2 - 2x + 1 = x^2$

$\Rightarrow -2x + 1 = 0$

$\Rightarrow 2x - 1 = 0$

$q_2: (x-1)^3 = x^3$

$\Rightarrow x^3 - 3x^2 + 3x - 1 = x^3$

$\Rightarrow -3x^2 + 3x - 1 = 0$

$\Rightarrow 3x^2 - 3x + 1 = 0$

$q_3: (x+1)^3 = 3x^2$

$\Rightarrow x^3 + 3x^2 + 3x + 1 = 3x^2$

$\Rightarrow x^3 + 3x + 1 = 0$

$q_4: (\sqrt{x}+1)^2 = 2\sqrt{x}$

$\Rightarrow x + 2\sqrt{x} + 1 = 2\sqrt{x}$

$\Rightarrow x + 1 = 0$

Exercise 4.13

(i) Which of the following is not a quadratic equation

(a) $x^2 + 5x - 6 = 0$

(b) $2x + \frac{3}{x} = 3$

(c) $x + \frac{1}{x} = x^2$

(d) $x^2 - 6 = 0$

(ii) Which of the following is not a quadratic equation

(a) $x^3 - x^2 = (\,x - 1\,)^3$

(b) $x^2 - 6x + 7 = 0$

(c) $(\,x^2 + 1\,) = (\,2 - x\,)^2 + 4$

(d) $(\,x^2 + 1\,)^2 = x^4 + 7$

(iii) Which the following is not a quadratic equation

(a) $2x - x^2 = x^2 + 1$

(b) $(\,x + 1\,)^2 = 2(\,x^2 + 2)$

(c) $(\sqrt{2}\,x + 1\,)^2 = 2x^2 + 7$

(d) $x\,(\,x - 3\,) + 2 = 0$

67. Which of the following equations is a quadratic equation ? **(2025)**

(a) $x^2 = (x + 1\,)^2$

(b) $(x - 1\,)\,(x + 2\,) = 2x + 1$

(c) $(x + 2\,)^3 = 2x\,(\,x^2 - 1\,)$

(d) $\sqrt{x} - x^2$

Solution: (b)

$(x - 1\,)\,(x + 2\,) = 2x + 1$

$\Rightarrow x^2 + x - 2 = 2x + 1$

$\Rightarrow x^2 - x - 3 = 0$

68. Which of the following equations is a quadratic equation ? **(2025)**

(a) $x^3 = (x - 1\,)^3 + 3x^2$

(b) $x^3 = (x + 1\,)^3$

(c) $x^2 = \sqrt{x}$

(d) $x^2 + 1 = \frac{1}{x}$

Solution: (b)

$x^3 = (x + 1\,)^3$

$\Rightarrow x^3 = x^3 + 3x^2 + 3x + 1$

$\Rightarrow 3x^2 + 3x + 1 = 0$

69. Which of the following equation is a quadratic equation ? **(2025)**

(a) $\left(x + \frac{1}{x}\right)^2 = 2$

(b) $\left(x - \sqrt{x}\right)^2 + 2x\sqrt{x} = 0$

(c) $(x + 1\,)^3 = (1 - x\,)^3$

(d) $\left(\sqrt{x} + 1\right)^2 = x^2$

Solution: (b)

$\left(x-\sqrt{x}\right)^2 + 2x\sqrt{x} = 0$

$\Rightarrow x^2 - 2x\sqrt{x} + x + 2x\sqrt{x} = 0$

$\Rightarrow x^2 + x = 0$

70. The equation $x + \frac{1}{x} = 3\ (x \neq 0)$ is expressed as a quadratic equation is the form of $ax^2 + bx + c = 0$. The value of $a - b + c$ is: **(2025)**

(a) 5 (b) 2

(c) 1 (d) −1

Solution: (a)

$x + \frac{1}{x} = 3$

$\Rightarrow \frac{x^2+1}{x} = 3$

$\Rightarrow x^2 + 1 = 3x$

$\Rightarrow x^2 - 3x + 1 = 0$

Here $a = 1, b = -3, c = 1$

$\therefore a - b + c = 1 - (-3) + 1 = 1 + 3 + 1 = 5$

71. The number of possible solutions of a quadratic equation are **(2025)**

(a) Exactly two (b) Atmost two

(c) Atleast two (d) Atleast one

Solution: (b)

Atmost two

Direction for questions (72) and (73) : In question number (58) and (59), a statement of Assertion (A) is followed by a statement of Reason (R) choose the correct option on:

(a)Both assertion (A) and reason (R) are true and reason (R) is the correct explanation of assertion (A).

(b)Both assertion (A) and reason (R) are true but reason (R) is not the correct explanation of assertion (A).

(c)Assertion (A) is true but reason (R) is false.

(d)Assertion (A) is false but reason (R) is false.

72. Statement A (Assertion): If $5 + \sqrt{7}$ is a root of a quadratic equation with rational co-efficients then its other root is $5 - \sqrt{7}$.

Statement R (Reason): Surds roots of a quadratic equation with rational co-efficients occur in conjugate pairs. **(2023)**

Solution: (a)

Both assertion (A) and reason (R) are true and reason (R) is the correct explanation of assertion (A).

73. Assertion (A): If one root of the quadratic equation $4x^2 - 10x + (k - 4) = 0$ is reciprocal of the other, then the value of k is 8.

Reason (R): Roots of the quadratic equation $x^2 - x + 1 = 0$ are real **(2023)**

Solution: (c)

Let the roots of the quadratic equation are α and $\frac{1}{\alpha}$

$\therefore \alpha \frac{1}{\alpha} = \frac{k-4}{4}$

$\Rightarrow 1 = \frac{k-4}{4}$

$\Rightarrow k - 4 = 4$

$\therefore k = 8$

Now, the discriminant of the quadratic equation $x^2 - x + 1 = 0$ is $D = (-1)^2 - 4 . 1 . 1 = 1 - 4 = -3 < 0$

$\therefore$ Roots of the quadratic equation $x^2 - x + 1 = 0$ are not real.

$\therefore$ Assertion (A) is true but reason (R) is false.

Exercise 4.14

(i) Assertion (A): If one root of the quadratic equation $6x^2 - x - k = 0$ is $\frac{2}{3}$, then the value of k is 2.

Reason (R): The quadratic equation $ax^2 + bx + c = 0, a \neq 0$ has almost two roots.

(ii) Assertion (A): $9x^2 - 12x + 16 = 0$ has repeated roots.

Reason (R): The quadratic equation $ax^2 + bx + c = 0$ have repeated roots, if discriminant $D > 0$.

(iii) Assertion (A): The roots of the quadratic equation $x^2 + x + 1 = 0$ are imaginary.

Reason (R): If discriminant $D = b^2 - 4ac < 0$ then the roots of quadratic equation $ax^2 + bx + c = 0$ are imaginary.

(iv) Assertion (A): The quadratic equation $2x^2 + 5kx + 2 = 0$ has equal roots for $k = \pm\frac{4}{5}$.

Reason (R): If discriminant D of a quadratic equation is equal to zero then the roots of equation are real and equal.

(v) Assertion (A): Sum and product of roots of $3x^2 - 4x + 5 = 0$ are $\frac{4}{3}$ and $\frac{5}{3}$ respectively.

Reason (R): If α and β are the roots of $ax^2 + bx + c = 0$, $a \neq 0$, then sum of are roots $\alpha + \beta = -\frac{b}{a}$ and product of roots $\alpha\,\beta = \frac{c}{a}$

(vi) Assertion (A): $(3x + 1)^2 - 9x^2 + 5 = 0$ is not a quadratic equation.

Reason (R): An equation of form $ax^2 + bx + c = 0$, $a \neq 0$, where $a, b, c \in R$ is called a quadratic equation.

(vii) Assertion (A): The quadratic equation $2x^2 + 3x + 1 = 0$ has real roots.

Reason (R): The discriminant is greater than zero.

(viii) Assertion (A): The roots of the quadratic equation $ax^2 + bx + c = 0$ given by the quadratic formula $x = \frac{-b \pm \sqrt{b^2 - 4ac}}{2a}$

Reason (R): The formula is derived using the method of completing the square.

(ix) Assertion (A): The equation $x^2 - 6x + 9 = 0$ has only one root.

Reason (R): A quadratic equation always has two roots.

(x) Assertion (A): The equation $3x^2 + 4kx + 12 = 0$ has equal roots then the value of k is ± 3.

Reason (R): The equation $ax^2 + bx + c = 0$ has equal roots if $D = b^2 - 4ac = 0$.

74. The quadratic eqution whose roots are 7 and $\frac{1}{7}$ is **(2025)**

(a) $7x^2 - 50x + 7 = 0$ (b) $7x^2 - 50x + 1 = 0$
(c) $7x^2 + 50x - 7 = 0$ (d) $7x^2 + 50x - 1 = 0$

Solution: (a)

Sum of the roots $= 7 + \frac{1}{7} = \frac{50}{7}$

Product of the roots $= 7 \times \frac{1}{7} = 1$

$\therefore$ The quadratic equation will be

$x^2 - \frac{50}{7}x + 1 = 0$

$\Rightarrow 7x^2 - 50x + 7 = 0$

75. The quadratic equation whose sum and product of roots are 'a' and '$\frac{1}{a}$' respectively is: **(2025)**

(a) $ax^2 - ax + 1 = 0$ (b) $ax^2 - a^2x + 1 = 0$
(c) $ax^2 + ax + 1 = 0$ (d) $ax^2 + a^2x - 1 = 0$

Solution: (b)

Sum of the roots $= a$

Product of the roots $= \frac{1}{a}$

$\therefore$ The required quadratic equation is

$x^2 - ax + \frac{1}{a} = 0$

$\Rightarrow ax^2 - a^2x + 1 = 0$

76. Which of the following statements is true for the quadratic equation $ax^2 + x + a = 0\ (a \neq 0)$? **(2026)**

(a) The roots of the given equation are reciprocal of each other.
(b) The roots of the given equation are always real.
(c) The roots of the given equation are always positive
(d) The sum of roots of the given equation is 0.

Solution: (a)

$\alpha\beta = \frac{a}{a} = 1$

ANSWER KEYS

Exercise – 4.1

(i) 0 **(ii)** 0 **(iii)** 0

(iv) −71 **(v)** 14 **(vi)** $\sqrt{45}$ **(vii)** $13a^2$

Exercise – 4.3

(i) (c) 3 **(ii)** (a) 6 **(iii)** (a) 6

(iv) (c) −2, 3 **(v)** (b) 2 **(vi)** (c) 0.1

(vii) (c) 0.1

Exercise – 4.4

(i) (d) 3 **(ii)** (b) −8 **(iii)** (c) 8

Exercise – 4.5

(i) (c) 2 or −2 **(ii)** (c) $2\sqrt{2}$ or $-2\sqrt{2}$

(iii) (a) 1 or 4 **(iv)** (a) $\frac{4}{3}$ or $-\frac{4}{3}$

(v) (c) $2 \text{ or } \frac{-10}{9}$ **(vi)** (a) 1 or 9 **(vii)** (c) 6

(viii) $1 \text{ or} -\frac{1}{2}$ **(ix)** (c) 24 or −24 **(x)** (a) 3 or 4

(xi) (c) $4a^2 = b$ **(xii)** (d) $4a^2 = b$ **(xiii)** $ab = \frac{c^2}{4}$

Exercise – 4.6

(i) $k \le \frac{9}{8}$ **(ii)** $k \le \frac{1}{8}$ **(iii)** $k \le 4$

(iv) not real **(v)** not real **(vi)** not real

(vii) not real **(viii)** equal **(ix)** equal

(x) equal **(xi)** equal **(xii)** (a) $-\frac{9}{8}$

(xiii) (b) $-\frac{1}{6}$ **(xiv)** (a) $-\frac{9}{4}$ **(xv)** (a) $\frac{9}{4}$

(xvi) (c) $k < 1$ **(xvii)** $k < \frac{25}{4}$ **(xviii)** (c) ±1

(xix) (b) ±1 **(xx)** (c) ±3

Exercise – 4.7

(i) (a) 1 **(ii)** (c) 9 **(iii)** (b) 4

Exercise – 4.8

(i) (b) ± 0.3 **(ii)** (a) ± 0.4 **(iii)** ± 0.9

Exercise – 4.9

(i) (c) $x^2 - 6x + 6 = 0$ **(ii)** (d) $x^2 - 8x + 13 = 0$

(iii) (a) $x^2 - 8x + 4 = 0$

Exercise – 4.10

(i) (a) $(p-3), -(p+2)$ **(ii)** (a) $\frac{c}{b}, \frac{-b}{a}$

(iii) (c) $\frac{a+b}{2}, \frac{a-b}{2}$

Exercise – 4.11

(i) (c) $4x^2 - 8x + 5 = 0$ **(ii)** (a) $x^2 + x + 1 = 0$

(iii) (a) $3x^2 - 4x + 2 = 0$

Exercise – 4.12

(i) (c) $x^2 - 9 = 0$ **(ii)** (d) $x^2 - 4 = 0$

(iii) (b) $x^2 - 1 = 0$

Exercise – 4.13

(i) (c) $x + \frac{1}{x} = x^2$ **(ii)** (c) $(x^2 + 1) = (2 - x)^2 + 4$

(iii) (c) $\left(\sqrt{2}\,x + 1\right)^2 = 2x^2 + 7$

Exercise – 4.14

(i) a **(ii)** d **(iii)** a

(iv) a **(v)** a **(vi)** b

(vii) c **(viii)** a **(ix)** d **(x)** a

5 Arithmetic Progressions

Key Point The common difference of a A. P a_1, a_2, a_3, ⋯ ⋯ ⋯ ⋯ a_n is $d = a_2 - a_1$

1. The common difference of the A. P. $\frac{1}{p}$, $\frac{1-p}{p}$, $\frac{1-2p}{p}$, ...is. **(2013, 2020)**

(a) p (b) $-p$

(c) -1 (d) 1

Solution: (c)

common difference = difference of any two consecutive terms

$= \frac{1-p}{p} - \frac{1}{p}$

$= \frac{1-p-1}{p}$

$= \frac{-p}{p}$

$= -1$

2. Find the common difference of the A. P. $\frac{1}{a}$, $\frac{3-a}{3a}$, $\frac{3-2a}{3a}$, ... $(a \neq 0)$ is. **(2019)**

Solution: (c)

$= \frac{3-a}{3a} - \frac{1}{a}$

$= \frac{3-a-3}{3a}$

$= \frac{-a}{3a}$

$= \frac{-1}{3}$

3. The common difference of the A.P: $\sqrt{2}, 2\sqrt{2}, 3\sqrt{2}, 4\sqrt{2}$, is: **(2026)**

(a) $\sqrt{2}$ (b) 1

(c) $2\sqrt{2}$ (d) $-\sqrt{2}$

Solution: (a)

$d = 2\sqrt{2} - \sqrt{2} = \sqrt{2}$

4. If the numbers $1 - a, 3 - a$ and $5 - a$ are in A P, then the common difference is: **(2026)**

(a) $1 - a$ (b) $-2 + a$

(c) 2 (d) $2 - a$

Solution: (c)

$d = (3-a) - (1-a)$

$= 3 - a - 1 + a$

$= 2$

Exercise 5.1

(i) The common difference of the A. P. 4, $4\frac{1}{2}$, 5, $5\frac{1}{2}$, ...

(a) 4 (b) $\frac{3}{2}$

(c) $\frac{1}{2}$ (d) 0

(ii) The common difference of the A. P. $\sqrt{2}$, $\sqrt{8}$, $\sqrt{18}$, $\sqrt{32}$, ...is.

(a) $\sqrt{2}$ (b) $2\sqrt{2}$

(c) 2 (d) $\sqrt{3}$

(iii) The common difference of the A. P. $\frac{1}{a}, \frac{2}{a}, \frac{3}{a}$ .... is.

(a) a (b) $\frac{1}{a}$

(c) $\frac{2}{a}$ (d) 1

5. Write the common difference of the A.P. $\sqrt{3}$, $\sqrt{12}$, $\sqrt{27}$, $\sqrt{48}$, ... **(2019)**

Solution: (c)

$= \sqrt{27} - \sqrt{3}$

$= 3\sqrt{3} - \sqrt{3}$

$= 2\sqrt{3}$

Exercise 5.2

(i) Write the common difference of the A.P. $\sqrt{2}$, $\sqrt{18}$, $\sqrt{50}$, $\sqrt{98}$, ...

(ii) Write the common difference of the A.P. $\sqrt{3}$, $\sqrt{27}$, $\sqrt{50}$, $\sqrt{147}$, ...

(iii) Write the common difference of the A.P. $-\sqrt{20}$, $-\sqrt{45}$, $-\sqrt{320}$, ...

Key Point Let S_p be the sum of first p terms of the given A. P, then S_1 = First term of the A.P

6. If the sum of first p terms of an A. P. is $ap^2 + bp$, find its common difference **(2010)**

Solution: (a)

Let S_p be the sum of first p terms of the given A.P.

$\therefore S_p = ap^2 + bp$

$\therefore S_1 = a.1^2 + b.1$

$\Rightarrow t_1 = a + b$

$S_2 = a.2^2 + b.2$

$\Rightarrow t_1 + t_2 = 4a + 2b$

$t_2 = S_2 - S_1$

$= (4a + 2b) - (a + b)$

$= 3a + b$

$\therefore d = t_2 - t_1$

$= (3a + b) - (a + b)$

$= 2a$

Exercise 5.3

(i) If the sum of first n^{th} term of an A. P. is $(2n + 1)$, find its common difference.

(ii) If the sum of first n^{th} term of an A. P. is $(3n^2 - 4n)$, find its common difference.

(iii) If the sum of first n^{th} term of an A. P. is $\left(\frac{3n^2}{2} + \frac{5n}{2}\right)$, find its common difference.

Key Point If the first term and common difference of the A.P are a and d respectively, then the nth term of the A.P is $a_n = a + (n - 1)d$

7. What is the common difference of an A. P. in which $a_{21} - a_7 = 84$? **(2017)**

Solution:

Let a be the first term and d be the common difference of the given A. P.

$a_{21} - a_7 = 84$

$\Rightarrow (a + 20\,d) - (a + 6d) = 84$

$\Rightarrow 14d = 84$

$\therefore d = 6$

8. In an AP, $a_n - a_{n-4} = 32$. Its common difference is: **(2025)**

(a) −8 (b) 8

(c) $4n$ (d) 4

Solution: (b)

$a_n - a_{n-4} = 32$

$\Rightarrow \{a + (n - 1)\,d\} - [a + \{(n - 4) - 1\}\,d] = 32$

$\Rightarrow a + (n - 1)\,d - a - (n - 5)\,d = 32$

$\Rightarrow (n-1)\,d - (n-5)\,d = 32$

$\Rightarrow (n - 1 - n + 5)\, d = 32$

$\Rightarrow 4d = 32$

$\therefore d = 8$

Exercise 5.4

(i) What is the common difference of an A. P. in which $a_{18} - a_{12} = 30$?

(ii) What is the common difference of an A. P. in which $a_{16} - a_9 = 35$?

(iii) What is the common difference of an A. P. in which $a_{32} - a_{26} = 24$?

9. The common difference of an A.P, whose n^{th} term is $a_n = (3n + 7)$ is. **(2020)**

(a) 3 (b) 7

(c) 10 (d) 6

10. The common difference of an A.P, whose n^{th} term is $a_n = (3n + 7)$ is. **(2023)**

(a) 7 (b) 3

(c) $3n$ (d) 1

Solution:

$a_n = 3n + 7$

$\therefore a_1 = 3.1 + 7 = 10$

and $a_2 = 3.2 + 7 = 13$

$\therefore d = a_2 - a_1$

$= 13 - 10$

$= 3$

11. The nth term of an A. P. is $(7 - 4n)$, then what is its common difference? **(2020)**

Solution:

$a_n = 7 - 4n$

$\therefore a_1 = 7 - 4.1$

$= 3$

and $a_2 = 7 - 4.2$

$= -1$

$\therefore d = a_2 - a_1$

$= -1 - 3$

$= -4$

12. The common difference of an A.P, whose n^{th} term is $a_n = 5n - 7$ is. **(2023)**

(a) -7 (b) 7

(c) 5 (d) -2

Solution : (c)

$a_n = 5n - 7$

$\therefore a_1 = 5.1 - 7 = -2$

and $a_2 = 5.2 - 7 = 3$

$\therefore d = a_2 - a_1$

$= 3 - (-2)$

$= 3 + 2$

$= 5$

13. n^{th} term of an A.P. is $7n + 4$, the common difference is : **(2024)**

(a) $7n$ (b) 4

(c) 7 (d) 1

Solution : (c)

$a_n = 7n + 4$

$\therefore a_1 = 7.1 + 4 = 11$

and $a_2 = 7.2 + 4 = 18$

$\therefore d = a_2 - a_1$

$= 18 - 11$

$= 7$

Exercise 5.5

(i) The common difference of an A. P. whose n^{th} term is $a_n = (4n - 10)$.

(a) -6 (b) 4

(c) 5 (d) 2

(ii) The common difference of an A. P. whose n^{th} term is $a_n = (7 - 3n)$.

(a) 0 (b) 2

(c) -3 (d) 3

(iii) The common difference of an A. P. whose n^{th} term is $a_n = (4n + 1)$.

(a) 4 (b) -4

(c) 0 (d) 8

14. The 8th term of an A.P. is 17 and its 14th term is 29. The common difference of this A.P. is. **(2023)**

(a) 3 (b) 2

(c) 5 (d) –2

Solution :

Let a and d be the first term and common difference of the given A.P.

$\therefore a_8 = a + 7d$

$\Rightarrow 17 = a + 7d$

$\Rightarrow a + 7d = 17$..................(i)

and $a_{14} = a + 13d$

$\Rightarrow 29 = a + 13d$

$\Rightarrow a + 13d = 29$............(ii)

Solving (i) and (ii) we get

$d = 2$

15. The 11th and 13th term of an AP are 39 and 45, respectively. What is the common difference of the AP? **(2025)**

(a) 42 (b) 21

(c) 6 (d) 3

Solution: (d)

$a_{11} = 39$

$\Rightarrow a + 10d = 39$ ...(i)

$a_{13} = 45$

$\Rightarrow a + 12d = 45$ ...(ii)

Subtracting (i) from (ii), we get

$2d = 6$

$\Rightarrow d = 3$

Exercise 5.6

(i) The 13th term of an A.P. is 37 and its 20th term is 72. The common difference of this A.P. is.

(a) 10 (b) 2

(c) 18 (d) 8

(ii) The 4th term of an A.P. is 14 and its 12th term is 70. The common difference of this A.P. is.

(a) 7 (b) 2

(c) –8 (d) 10

(iii) The 3rd term of an A.P. is 4 and its 9th term is –8. The common difference of this A.P. is.

(a) 5 (b) 4

(c) –2 (d) –3

16. In an A.P., if $a = 8$ and $a_{10} = -19$, then the value of d is. **(2024)**

(a) 3 (b) $\frac{-11}{9}$

(c) $\frac{-27}{10}$ (d) –3

Solution: (d)

$a_{10} = -19$

$\Rightarrow a + 9d = -19$

$\Rightarrow 8 + 9d = -19$

$\Rightarrow 9d = -27$

$\therefore d = -3$

Exercise 5.7

(i) In an A.P., if $a = 12$ and $a_{16} = 72$, then the value of d is.

(a) 4 (b) 2

(c) 3 (d) 5

(ii) In an A.P., if $a = 10$ and $a_{21} = 110$, then the value of d is.

(a) 5 (b) 4

(c) 3 (d) 2

(iii) In an A.P., if $a = -3$ and $a_8 = 63$, then the value of d is.

(a) 2 (b) 10

(c) –8 (d) 9

17. The first term of an A.P. is p and its common difference is q. Find its 10th term. **(2008)**

18. The first term of an A.P. is p and its common difference is q, then the 10th term, is **(2020)**

(a) $q + 9p$ (b) $p - 9q$

(c) $p + 9q$ (d) $2p + 9q$

Solution: (c)

$a_n = a + (n - 1)d$

$\therefore a_{10} = p + (10 - 1)q$

$= p + 9q$

19. The 7th term of an A.P. whose first term is 28 and common difference (–4), is **(2023)**

(a) 0 (b) 4
(c) 52 (d) 56

Solution: (b)

$a_n = a + (n-1)d$

$\Rightarrow a_7 = 28 + (7-1)(-4)$

$= 28 - 24$

$= 4$

20. The 20th term of an A.P. whose first term is –2 and the common difference is 4, is **(2023)**

(a) 78 (b) 74
(c) –36 (d) –34

Solution: (b)

$a_n = a + (n-1)d$

$\therefore a_{20} = -2 + (20-1)4$

$= -2 + 76$

$= 74$

21. Find the 21st term of the A.P. $-4\frac{1}{2}, -3, -1\frac{1}{2}$ .... **(2019)**

Solution:

First term $(a) = -4\frac{1}{2} = -\frac{9}{2}$

Common difference $(d) = -3 - \left(-\frac{9}{2}\right)$

$= -3 + \frac{9}{2}$

$= \frac{3}{2}$

$n = 21$

$a_n = a + (n-1)d$

$\therefore a_{21} = -\frac{9}{2} + (21-1)\frac{3}{2}$

$= -\frac{9}{2} + 30$

$= \frac{51}{2}$

$= 25\frac{1}{2}$

22. The nth term of the A.P. $a, 3a, 5a$...........is **(2020)**

(a) na (b) $(2n-1)a$
(c) $(2n+1)a$ (d) $2na$

Solution: (b)

$d = 3a - a = 2a$

$\therefore a_n = a + (n-1)d$

$= a + (n-1)2a$

$= 2na - a$

$= (2n-1)a$

23. The 9th tern of the A.P. –15, –11, –7, ...49 is **(2020)**

(a) 32 (b) 0
(c) 17 (d) 18

Solution: (c)

$d = -11 - (-15)$

$= -11 + 15$

$= 4$

$a_n = a + (n-1)d$

$\therefore a_9 = -15 + (9-1)4$

$= -15 + 32$

$= 17$

24. Write the 26th term of the A.P. 7, 4, 1, –2, ... **(2020)**

Solution:

Here $a = 7$

$d = 4 - 7 = -3$

$n = 26$

$a_n = a + (n-1)d$

$\therefore a_{26} = 7 + (26-1)(-3)$

$= 7 - 75$

$= -68$

25. Write the 31st term of the A.P. – 50, –47, –44, ... **(2020)**

Solution:

Here $a = -50$

$d = -47 - (-50)$

$= -47 + 50$

$= 3$

$n = 31$

$a_n = a + (n-1)d$

$= -50 + (31-1)3$

$= -50 + 90$

$= 40$

26. Find the 11th term of the A.P. –27, –22, –17, –12, ... **(2020)**

Solution:

Here $a = -27$

$d = -22 - (-27)$

$= -22 + 27$

$= 5$

$n = 11$

Now, $a_n = a + (n-1)d$

$\therefore a_{11} = -27 + (11-1)5$

$= -27 + 50$

$= 23$

27. Find the 25th term of the A.P. $-5, \frac{-5}{2}, 0, \frac{5}{2}$, ... **(2015)**

Solution:

$a = -5$

$d = \frac{-5}{2} - (-5)$

$= \frac{-5}{2} + 5$

$= \frac{5}{2}$

$n = 25$

$a_n = a + (n-1)d$

$\therefore a_{25} = -5 + (25-1)\frac{5}{2}$

$= -5 + 60$

$= 55$

28. The 10th term of the *AP* $5, \frac{19}{4}, \frac{9}{2}, \frac{17}{4}$, ... is: **(2025)**

(a) $\frac{11}{4}$ (b) $\frac{4}{11}$

(c) $\frac{13}{4}$ (d) $\frac{4}{13}$

Solution: (a)

$d = \frac{19}{4} - 5 = -\frac{1}{4}$

$a_n = a + (n-1)d$

$\therefore a_{10} = 5 + (10-1) \times \left(-\frac{1}{4}\right)$

$= 5 - \frac{9}{4}$

$= \frac{11}{4}$

29. The 20th term of the *A.P*: $10\sqrt{2}, 6\sqrt{2}, 2\sqrt{2}$, ... is **(2025)**

(a) $-76+10\sqrt{2}$ (b) $-62\sqrt{2}$

(c) $-66\sqrt{2}$ (d) $86\sqrt{2}$

Solution: (c)

$d = 6\sqrt{2} - 10\sqrt{2}$

$= -4\sqrt{2}$

$a_n = a + (n-1)d$

$a_{20} = 10\sqrt{2} + (20-1) \times (-4\sqrt{2})$

$= 10\sqrt{2} + 19(-4\sqrt{2})$

$= 10\sqrt{2} - 76\sqrt{2}$

$= -66\sqrt{2}$

30. The 16th term of the *A.P*: $5\sqrt{3}, 2\sqrt{3}, -\sqrt{3}$, ... is **(2025)**

(a) $-25\sqrt{3}$ (b) $-40\sqrt{3}$

(c) $50\sqrt{3}$ (d) $-45+5\sqrt{3}$

Solution: (b)

$d = 2\sqrt{3} - 5\sqrt{3} = -3\sqrt{3}$

$a_n = a + (n - 1)\, d$

$\Rightarrow a_{16} = 5\sqrt{3} + (16-1)\times\left(-3\sqrt{3}\right)$

$= 5\sqrt{3} - 45\sqrt{3}$

$= -40\sqrt{3}$

31. 10th term of the *AP*: −12, −19, −26, ... is **(2025)**

(a) −75 (b) −65

(c) 51 (d) −82

Solution: (a)

Here $a = -12$

$d = -19 - (-12)$

$= -19 + 12$

$= -7$

$n = 10$

$a_{10} = a + 9d$

$= -12 + 9 \times (-7)$

$= -12 - 63$

$= -75$

32. The nth term of the *A.P.* $\frac{-1}{3}, \frac{2}{3}, \frac{5}{3}, \frac{8}{3}$, is **(2026)**

(a) $3n - 4$ (b) $n - \frac{4}{3}$

(c) $\frac{n-2}{3}$ (d) $\frac{n-4}{3}$

Solution: (b)

$a = -\frac{1}{3}$

$d = \frac{2}{3} - \left(-\frac{1}{3}\right)$

$= \frac{2}{3} + \frac{1}{3}$

$= \frac{3}{3}$

$= 1$

$\therefore a_n = a + (n-1)d$

$= -\frac{1}{3} + (n-1)1$

$= n - \frac{4}{3}$

33. If a_n represents n^{th} term of the A.P $-\frac{15}{4}, -\frac{10}{4}, -\frac{5}{4}$, then value of $a_{16} - a_{12}$ **(2026)**

(a) 4 (b) $\frac{5}{4}$

(c) 5 (d) $\frac{25}{4}$

Solution: (c)

$d = -\frac{10}{4} - \left(-\frac{15}{4}\right)$

$= -\frac{10}{4} + \frac{15}{4}$

$= \frac{5}{4}$

$a_{16} - a_{12}$

$(a_1 + 15d) - (a_1 + 11d)$

$= 4d$

$= 4 \times \frac{5}{4}$

$= 5$

34. If $-\frac{20}{9}, \frac{-2}{9}, \frac{16}{9}$ ... are in A.P, then next term of the sequence is : **(2026)**

(a) $\frac{31}{9}$ (b) $\frac{46}{9}$

(c) $\frac{2}{9}$ (d) $\frac{34}{9}$

Solution: (d)

$a_1 = \frac{-20}{9}$

$d = \frac{-2}{9} - \left(\frac{-20}{9}\right)$

$= \frac{-2}{9} + \frac{20}{9}$

$= \frac{18}{9}$

$= 2$

$a_4 = a_1 + 3d$

$$= \frac{-20}{9} + 3 \times 2$$
$$= \frac{-20}{9} + 6$$
$$= \frac{-20+54}{9}$$
$$= \frac{34}{9}$$

35. The nth term of the A.P :$-\frac{1}{3}, \frac{4}{3}, 3$ is : **(2026)**

(a) $\frac{5n-9}{3}$ (b) $\frac{5n-6}{3}$

(c) $\frac{3n-4}{3}$ (d) $\frac{3n+2}{3}$

Solution: (b)

$$a = \frac{-1}{3}$$
$$d = \frac{4}{3} - \left(-\frac{1}{3}\right)$$
$$= \frac{4}{3} + \frac{1}{3}$$
$$= \frac{5}{3}$$
$$\therefore a_n = a + (n-1)d$$
$$= -\frac{1}{3} + (n-1)\frac{5}{3}$$
$$= -\frac{1}{3} + \frac{5n}{3} - \frac{5}{3}$$
$$= \frac{5n}{3} - \frac{6}{3}$$
$$= \frac{5n-6}{3}$$

Exercise 5.8

(i) The first term of an A.P. is 2 and its common difference is 3. Find its 12th term.

(ii) The first term of an A.P. is 3 and its common difference is a. Find its 27th term.

(iii) The first term of an A.P. is –2 and its common difference is –3. Find its 26th term.

(iv) Find the 25th term of the A.P. $-3, -\frac{1}{2}, 2,$ ………………… .

(v) Find the 20th term of the A.P. $4, 4\frac{1}{2}, 5, 5\frac{1}{2}, 6,$ ………………… .

(vi) Find the 29th term of the A.P. $-1, \frac{3}{2}, -2, \frac{-5}{2},$ ………………… .

36. In an A.P., if $d = -2$, $n = 5$ and $a_n = 0$, then the value of a is **(2011)**

(a) 10 (b) 5

(c) –8 (d) 8

Solution: (d)

$$a_n = a + (n-1)d$$
$$\Rightarrow 0 = a + (5-1)(-2)$$
$$\Rightarrow 0 = a - 8$$
$$\therefore a = 8$$

37. In an A.P, if the common difference $(d) = -4$, and the seventh term (a_7) is 4, then find the first term. **(2018)**

Solution:

Let a be the first term of the given A.P.

$$\therefore a_7 = a + (7-1)(-4)$$
$$\Rightarrow -4 = a - 24$$
$$\therefore a = 20$$

38. In an A.P., if $d = -4$, $n = 7$ and $a_n = 4$, then the value of a is. **(2023)**

(a) 6 (b) 7

(c) 20 (d) 28

Solution: (d)

$$a_n = a + (n-1)d$$
$$\Rightarrow 4 = a + (7-1)(-4)$$
$$\Rightarrow 4 = a - 24$$
$$\therefore a = 28$$

Exercise 5.9

(i) In an A.P., if $d = 3$, $n = 6$ and $a_n = 21$, then the value of a is.

(a) –6 (b) 6

(c) 7 (d) 8

(ii) In an A.P., if $d = -3$, $n = 11$ and $a_n = -30$, then the value of a is.

(a) 0 (b) −2

(c) 6 (d) 2

(iii) In an A.P., if $d = -4$, $n = 21$ and $a_n = 20$, then the value of a is.

(a) 0 (b) 20

(c) 80 (d) 100

(iv) In an A.P., if the common difference $(d) = 5$, and the 10th term (a_{10}) is 48, then find the first term.

(v) In an A.P., if the common difference $(d) = -3$, and the 12th term (a_{12}) is 7, then find the first term.

(vi) In an A.P., if the common difference $(d) = 3$, and the 9th term (a_9) is 12, then find the first term.

39. The next term of the A.P. $\sqrt{6}$, $\sqrt{24}$, $\sqrt{54}$ ……is **(2023)**

(a) $\sqrt{60}$ (b) $\sqrt{96}$

(c) $\sqrt{72}$ (d) $\sqrt{216}$

Solution: (b)

$a = \sqrt{6}$

$d = \sqrt{24} - \sqrt{6}$

$= 2\sqrt{6} - \sqrt{6}$

$= \sqrt{6}$

$n = 4$

$\therefore\ a_4 = \sqrt{6} + (4-1).\sqrt{6}$

$= \sqrt{6} + 3\sqrt{6}$

$= 4\sqrt{6}$

$= \sqrt{96}$

40. The next term of the A.P. $\sqrt{7}$, $\sqrt{28}$, $\sqrt{63}$ .... is **(2023)**

(a) $\sqrt{70}$ (b) $\sqrt{80}$

(c) $\sqrt{97}$ (d) $\sqrt{112}$

41. The next (4th) term of the A.P. $\sqrt{7}$, $\sqrt{28}$, $\sqrt{63}$ ……. is **(2024)**

(a) $\sqrt{70}$ (b) $\sqrt{80}$

(c) $\sqrt{97}$ (d) $\sqrt{112}$

Solution: (d)

$a = \sqrt{7}$

$n = 4$

$d = \sqrt{28} - \sqrt{7}$

$= 2\sqrt{7} - \sqrt{7}$

$= \sqrt{7}$

$\therefore a_4 = \sqrt{7} + (4-1).\sqrt{7}$

$= \sqrt{7} + 3\sqrt{7}$

$= 4\sqrt{7}$

$= \sqrt{112}$

42. The next (4th) term of the A.P. $\sqrt{18}$, $\sqrt{50}$, $\sqrt{98}$ ....is **(2024)**

(a) $\sqrt{128}$ (b) $\sqrt{140}$

(c) $\sqrt{162}$ (d) $\sqrt{200}$

Solution: (c)

$a = \sqrt{18} = 3\sqrt{2}$

$n = 4$

$d = \sqrt{50} - \sqrt{18}$

$= 5\sqrt{2} - 3\sqrt{2}$

$= 2\sqrt{2}$

$\therefore a_4 = 3\sqrt{2} + (4-1).\sqrt{2}$

$= 3\sqrt{2} + 6\sqrt{2}$

$= 9\sqrt{2}$

$= \sqrt{162}$

Exercise 5.10

(i) Find the next term of the A.P. $\sqrt{3}$, $\sqrt{12}$, $\sqrt{27}$

(ii) Find the next term of the A.P. $\sqrt{2}$, $\sqrt{18}$, $\sqrt{50}$

(iii) Find the next term of the A.P. $\sqrt{2}$, $\sqrt{27}$, $\sqrt{50}$

43. Find the number of terms in the A.P. 18, $15\frac{1}{2}$, 13, ..., –47 **(2019)**

Solution:

Let n be the number of terms of the given A.P.

Common difference $(d) = 15\frac{1}{2} - 18$

$= \frac{31}{2} - 18$

$= \frac{31-36}{2}$

$= \frac{-5}{2}$

Now, $a_n = a + (n - 1)d$

$\Rightarrow -47 = 18 + (n-1)\left(\frac{-5}{2}\right)$

$\Rightarrow \frac{-5}{2}(n-1) = -65$

$\Rightarrow n - 1 = 26$

$\therefore n = 27$

44. The number of terms of an A. P. 5, 9, 13,......., 185 is. **(2020)**

(a) 31 (b) 51

(c) 41 (d) 40

Solution:

Let n be the number of terms of the given A.P.

$d = 9 - 5 = 4$

$a_n = a + (n - 1)d$

$\Rightarrow 185 = 5 + (n - 1)\,4$

$\Rightarrow 4\,(n - 1) = 180$

$\Rightarrow n - 1 = 45$

$\therefore n = 46$

45. How many terms are there in the A.P. given below ?

14, 19, 24, 29, ..., 119. **(2023)**

(a) 18 (b) 14

(c) 22 (d) 21

Solution: (c)

Let n be the number of terms of the given A.P.

$d = 19 - 14 = 5$

$a_n = a + (n - 1)d$

$\Rightarrow 119 = 14 + (n - 1)\,5$

$\Rightarrow 5\,(n - 1) = 105$

$\Rightarrow n - 1 = 21$

$\therefore n = 22$

46. The number of terms in the A.P. 3, 6, 9, 12, ..., 111 is. **(2024)**

(a) 36 (b) 40

(c) 37 (d) 30

Solution: (c)

Let n be the number of term of the given A.P.

$d = 6 - 3 = 3$

$a_n = a + (n - 1)d$

$\Rightarrow 111 = 3 + (n - 1)\,3$

$\Rightarrow 3(n - 1) = 108$

$\Rightarrow n - 1 = 36$

$\therefore n = 37$

Exercise 5.11

(i) Find the number of terms in the A.P. 4, $4\frac{1}{2}$, 5, $5\frac{1}{2}$, 6, $6\frac{1}{2}$, ..., 56.

(ii) Find the number of terms in the A.P. $\frac{5}{6}$, 1, $1\frac{1}{6}$, $1\frac{1}{3}$, ..., 3.

(iii) Find the number of terms in the A.P. –3, $-\frac{1}{2}$, 2, ..., 22

47. If in an A.P., $a = 15$, $d = -3$ and $a_n = 0$, then find the value of n. **(2019)**

Solution:

$a_n = a + (n - 1)d$

$\Rightarrow 0 = 15 + (n - 1)(-3)$

$\Rightarrow -3(n-1) = -15$

$\Rightarrow n - 1 = 5$

$\therefore n = 6$

Exercise 5.12

(i) If in an A.P., $a = 7$, $d = 4$ and $a_n = 139$, then find the value of n.

(ii) If in an A.P., $a = 5$, $d = 4$ and $a_n = 81$, then find the value of n.

(iii) If in an A.P., $a = 3$, $d = 5$ and $a_n = 88$, then find the value of n.

48. Which term of the A.P. –4, –1, 2,........is 101 ? **(2019)**

Solution:

lent n^{th} term of the given A.P. be 101

$d = -1 - (-4) = -1 + 4 = 3$

$a_n = a + (n-1)d$

$\Rightarrow 101 = -4 + (n-1)3$

$\Rightarrow 3(n-1) = 105$

$\Rightarrow n - 1 = 35$

$\therefore n = 36$

Exercise 5.13

(i) Which term of the A.P. 21, 42, 63, 84,.....is 210 ?

(ii) Which term of the A.P. 10, 7, 4, ...is –77 ?

(iii) Which term of the A.P. 5, 9, 13, 17,.....is 81 ?

Key Point The nth term from end of an A.P. is $\ell - (n-1)d$, where ℓ is the last term, d is common difference and n is the number of terms.

49. Find the 9th term from the end (towards the first term) of the A.P. 5, 9, 13,........., 185 **(2016)**

Solution:

The nth term from end of an A.P. is $1 - (n-1)d$ there

L = last term = 185

d = common difference = 9 – 5 = 4

n = n number of term = 9

$\therefore$ 9th term from the end of the given A.P. is

$185 - (9-1) \times 4$

$= 185 - 32$

$= 153$

50. The second term from the end of the A.P. 5,8,11, ..., 47 is **(2020)**

(a) 50 (b) 45

(c) 44 (d) 41

Solution: (c)

The nth term from end of an A.P. is $1 - (n-1)d$ where

1 = last term = 47

d = common difference = 8 – 5 = 3

n = number of term = 2

$\therefore$ second term from end of the A.P. is $47 - (2-1)3$

$= 47 - 3$

$= 44$

51. The 11th term from the end of the A.P. 10,7,4, ..., –62 is **(2023)**

(a) 25 (b) 16

(c) –32 (d) 20

Solution: (c)

n^{th} term from the end of an A.P. is $1 - (n-1)d$

$1 = -62$

$d = 7 - 10 = -3$

$n = 11$

$\therefore$ 11th term from end of the given A.P. is

$-62 - (11-1) \times (-3)$

$= -62 + 30$

$= -32$

52. The 13th term from the end of the A.P. 20, 13, 6, –1, ..., –148 is **(2023)**

(a) 57 (b) –57

(c) 64 (d) –64

Solution: (d)

n^{th} term from the end of an A.P. is

$a_n = 1 - (n-1)d$

$1 = -148$

$d = 13 - 20 = -7$

$n = 13$

$\therefore$ 13th term from end of the given A.P. is

$-148 - (13 - 1) \times (-7)$

$= -148 + 84$

$= -64$

53. The 14th term from the end of the A.P. –11,–8,–5, ..., 49 is **(2023)**

(a) 7 (b) 10

(c) 13 (d) 28

Solution: (b)

Here

$l = 49$

$d = -8 - (-11) = -8 + 11 = 3$

$n = 14$

$\therefore$ 14th term from end of the given A.P. is $l - (n - 1)d$

$= 49 - (14 - 1) \times 3$

$= 49 - 39$

$= 10$

54. The 8th term from the end of the A.P 3, 7, 11, 15, ..., 143 is. **(2025)**

(a) 135 (b) 125

(c) 115 (d) 111

Solution: (c)

The 8th term from the end of the given A.P $= \ell - (n - 1)\, d$

$= 143 - (8 - 1) \times 4$

$= 143 - 28$

$= 115$

55. The 9th term from the end (towards first term) of the AP 7, 11, 15, 19, ..., 147 is **(2025)**

(a) 135 (b) 125

(c) 115 (d) 39

Solution: (c)

n^{th} term of a AP from the end is $a_n = \ell - (n - 1)d$

Where, ℓ = last term = 147

d = Common difference = $11 - 7 = 4$

n = Number of term = 9

$a_n = n^{th}$ term from the end.

$\therefore a_9 = 147 - (9 - 1) \times 4$

$= 147 - 32$

$= 115$

Exercise 5.14

(i) Find the 17th term from the end (towards the first term) of the A.P. 3, 8, 13, 18,........., 88

(ii) Find the 11th term from the end (towards the first term) of the A.P. 5, 13, 21, ..., 181

(iii) Find the 15th term from the end (towards the first term) of the A.P. 3, 8, 13, ..., 253

Key Point If a, b, c are in A.P then $2b = a + c$.

56. if $\frac{4}{5}$, a, 2 are three consecutive terms of an A.P., then find the value of a. **(2009)**

Solution:

If x, y, z are in A. P., then

$2y = x + z$

$\therefore 2a = \frac{4}{5} + 2$

$\Rightarrow 2a = \frac{14}{5}$

$\therefore a = \frac{7}{5}$

57. If k, $2k - 1$ and $2k + 1$ are three consecutive terms of an A.P., the value of k is **(2014)**

(a) 2 (b) 3

(c) –3 (d) 5

Solution: (b)

$2(2k - 1) = k + (2k + 1)$

$\Rightarrow 4k - 2 = 3k + 1$

$\Rightarrow k = 3$

58. For what value of k will $k + 9$, $2k - 1$ and $2k + 7$ are the consecutive terms of an A.P.? **(2016)**

Solution:

$2(2k - 1) = (k + 9) + (2k + 7)$

$\Rightarrow 4k - 2 = 3k + 16$

$\Rightarrow k = 18$

59. For what value of k will the consecutive terms $2k + 1$, $3k + 3$ and $5k - 1$ form an A.P.? **(2016)**

Solution:

$2(3k + 3) = (2k + 1) + (5k - 1)$

$\Rightarrow 6k + 6 = 7k$

$\Rightarrow -k = -6$

$\therefore k = 6$

60. The value of x for which $2x$, $(x + 10)$ and $(3x + 2)$ are the three consecutive terms of an A.P. is. **(2020)**

(a) 6 (b) −6

(c) 18 (d) −18

Solution: (a)

$2(x + 10) = 2x + (3x + 2)$

$\Rightarrow 2x + 20 = 5x + 2$

$\Rightarrow -3x = -18$

$\therefore x = 6$

61. The value of p for which $(2p + 1)$, 10 and $(5p + 5)$ are the three consecutive terms of an A.P. is. **(2020)**

(a) −1 (b) −2

(c) 1 (d) 2

Solution: (d)

$2 \times 10 = (2p + 1) + (5p + 5)$

$\Rightarrow 20 = 7p + 6$

$\Rightarrow 7p = 14$

$\therefore p = 2$

62. If $3k - 2$, $4k - 6$ and $k + 2$ are three consecutive terms of A.P., then find the value of k. **(2020)**

Solution:

$2(4k - 6) = (3k - 2) + (k + 2)$

$\Rightarrow 8k - 12 = 4k$

$\Rightarrow 4k = 12$

$\therefore k = 3$

63. If $3y - 1$, $3y + 5$ and $5y + 1$ are three consecutive terms of A.P., then find the value of y. **(2020)**

Solution:

$2(3y + 5) = (3y - 1) + (5y + 1)$

$\Rightarrow 6y + 10 = 8y$

$\Rightarrow -2y = -10$

$\therefore y = 5$

64. Find the value of x so that -6, x, 8 are in A.P. **(2020)**

Solution:

$2x = -6 + 8$

$\Rightarrow 2x = 2$

$\therefore x = 1$

65. If $p - 1$, $p + 1$ and $2p + 3$ are in A.P., then the value of p is **(2023)**

(a) −2 (b) 4

(c) 0 (d) 2

Solution: (c)

$2(p + 1) = (p - 1) + (2p + 3)$

$\Rightarrow 2p + 2 = 3p + 2$

$\Rightarrow p = 0$

66. If $k + 2$, $4k - 6$ and $3k - 2$ are three consecutive terms of an A.P., then the value of k is **(2023)**

(a) 3 (b) −3

(c) 4 (d) −4

Solution: (a)

$2(4k - 6) = (k + 2) + (3k - 2)$

$\Rightarrow 8k - 12 = 4k$

$\Rightarrow 4k = 12$

$\therefore k = 3$

67. If -5, x, 3 are three consecutive terms of an A.P., then the value of x is **(2023)**

(a) −2 (b) 2

(c) 1 (d) −1

Solution: (d)

$2x = -5 + 3$

$\Rightarrow 2x = -2$

$\therefore x = -1$

68. If a b and c are in A.P., then : **(2026)**

(a) $2a$, $3b$ and $4c$ are also in A.P.

(b) $a - 2$, $b - 2$ and $c - 2$ are also in A.P.

(c) $\frac{a}{2}, \frac{b}{3}$ and $\frac{c}{4}$ are in A.P.

(d) $a+2, b+3$ and $c+4$ are also in A.P.

Solution: (b)

$2b = a + c$

$\Rightarrow 2b - 4 = a + c - 4$

$\Rightarrow 2(b-2) = (a-2) + (c-2)$

$\therefore a-2, b-2$ and $c-2$ are in A.P.

69. If $p-1, q-1$ and $r-1$ are in A.P., then: **(2026)**

(a) $p-2$, q, $r-2$ are also in A P

(b) $2p, 3q, 4r$ are also in A p

(c) $\frac{p}{2}, \frac{q}{3}, \frac{r}{4}$ are also in A P

(d) $2p, 2q, 2r$ are also in A P

Solution: (d)

$2(q-1) = (p-1) + (r-1)$

$\Rightarrow 2q - 2 = p + r - 2$

$\Rightarrow 2q = p + r$

$\Rightarrow 2(2q) = 2p + 2r$[Mulyiplying both sides by 2]

$2p, 2q, 2r$ are in A.P

70. If $x+1, y+1$ and $z+1$ are in A.P., then which of the following three numbers are in A.P.? **(2026)**

(a) $2x.3y, 4z$ (b) $x+1, 2y+1, 3z+1$

(c) $x-2, y-3, z-4$ (d) $x-3, y-3, z-3$

Solution: (d)

$2(y+1) = (x+1) + (z+1)$

$\Rightarrow 2y + 2 = x + z + 2$

$\Rightarrow 2y = x + z$

$\Rightarrow 2y - 6 = x + z - 6$

$\Rightarrow 2(y-3) = (x-3) + (z-3)$

$\therefore (x-3), (y-3)$ and $(z-3)$ are in A P

71. If -26, x, 2 are in A.P., then the value of x is **(2026)**

(a) 14 (b) -13

(c) -12 (d) -14

Solution: (c)

$2x = -26 + 2$

$\Rightarrow 2x = -24$

$\therefore x = -12$

Exercise 5.15

(i) If 8, a, 12 are three consecutive terms of an A.P., then find value of a.

(ii) if $\frac{8}{3}$, a, -2 are three consecutive terms of an A.P., then find the value of a.

(iii) If $\frac{-3}{5}$, a, $\frac{2}{3}$ are three consecutive terms of an A.P., then find the value of a.

(iv) If $(a+2)$, $2a$, $(2a+3)$ are three consecutive terms of an A.P., then the value of a is

(a) 2 (b) 5

(c) 0 (d) -3

(v) If $(2a-3)$, 4, $(a+8)$ are three consecutive terms of an A.P., then the value of a is

(a) 0 (b) -1

(c) 1 (d) 3

(vi) If $(x+1)$, $3x$, $(4x+7)$ are three consecutive terms of an A.P., then the value of x is

(a) 7 (b) 3

(c) 1 (d) 8

72. If a, b, c form an A.P. with common difference d, then the value of $a - 2b - c$ is equal to.**(2023)**

(a) $2a + 4d$ (b) 0

(c) $-2a - 4d$ (d) $-2a - 3d$

Solution: (c)

Since a, b, c are in A.P.

$\therefore 2b = a + c$

and $c = a + 2d$

Now, $a - 2b - c$

$= a - (a + c) - c$

$= -2c$

$= -2(a + 2d)$

$= -2a - 4d$

Exercise 5.16

(i) If a,b,c form an A.P. with common difference d, then the value of $a - 2b + c$ is equal to.

(a) 0 (b) $a + d$

(c) $a + 2d$ (d) $-a - 2d$

(ii) If a, b, c form an A.P. with common difference d, then the value of $-a - 2b - c$ is equal to.

(a) $-a - d$ (b) $a - 4d$

(c) $-3a - 4d$ (d) $-4a - 4d$

(iii) If a, b, c form an A.P. with common difference d, then the value of $a - 4b + c$ is equal to.

(a) $-2a - d$ (b) $a - d$

(c) $-2a - 2d$ (d) $4d$

73. How many two digits numbers are divisible by 3 ? **(2019)**

Solution:

The two digit numbers divisible by 3 are 3,6,9,12,........, 99

Let n be the number of terms of the given A. P.

Here $a = 3$

$d = 6 - 3 = 3$

$a_n = 99$

$\Rightarrow a + (n - 1)d = 99$

$\Rightarrow 3 + (n - 1)3 = 99$

$\Rightarrow n - 1 = 32$

$\therefore n = 33$

74. The number of multiples of 6 lying between 25 and 363 is: **(2026)**

(a) 56 (b) 56.5

(c) 57 (d) 58

Solution: (a)

$a = 30$

$d = 6$

$a_n = 360$

Now, $a_n = a + (n-1)d$

$\Rightarrow 360 = 30 + (n-1)6$

$\Rightarrow 6(n-1) = 330$

$\Rightarrow n - 1 = 55$

$\therefore n = 56$

75. The number of multiples of 4 lying between 12 and 250 is: **(2026)**

(a) 59 (b) 59.5

(c) 60 (d) 61

Solution:

$a = 16$

$d = 4$

$a_n = 248$

Now, $a_n = a + (n-1)d$

$\Rightarrow 248 = 16 + (n-1)4$

$\Rightarrow 4(n-1) = 232$

$\Rightarrow n - 1 = 58$

$\therefore n = 59$

Exercise 5.17

(i) How many two digits numbers are divisible by 2 ?

(ii) How many two digits numbers are divisible by 8 ?

(iii) How many two digits numbers are divisible by 5 ?

76. Which of the following is not an A.P. ? **(2020)**

(a) $-1.2, 0.8, 2.8,$

(b) $3, 3 + \sqrt{2}, 3 + 2\sqrt{2}, 3 + 3\sqrt{2}, \ldots\ldots$

(c) $\frac{4}{3}, \frac{7}{3}, \frac{9}{3}, \frac{12}{3}, \ldots\ldots$

(d) $-\frac{1}{5}, -\frac{2}{5}, \frac{-3}{5}, \ldots\ldots$

Solution: (c)

In the series $\frac{4}{3}, \frac{7}{3}, \frac{9}{3}, \frac{12}{3},$

second term – first term $= \frac{7}{3} - \frac{4}{3} = \frac{3}{3} = 1$

third term – second term $= \frac{9}{3} - \frac{7}{3} = \frac{2}{3}$

therefore difference between two consecutive terms are not equal

$\therefore$ The given series is not an A.P.

77. Which of the following sequence in not an A.P.? **(2026)**

(a) $2, \frac{5}{2}, 3, \frac{7}{2},$

(b) $-1.2, -3.2, -5.2, -7.2,$

(c) $\sqrt{2}, \sqrt{3}, \sqrt{18},$

(d) $1^2, 3^2, 5^2, 7^2,$

Solution: (d)

$1^2 = 1, 3^2 = 9, 5^2 = 25, 7^2 = 49,$

Now, $9 - 1 = 8$

$25 - 9 = 16$

$49 - 25 = 24$

Exercise 5.18

(i) Which of the following is not an A.P. ?

(a) 4, 4.5, 5, 5.5, 6,.........

(b) 14, 9, 4, −1, −6,..........

(c) $\sqrt{2}$, $\sqrt{8}$, $\sqrt{18}$,..............

(d) $-3, -\frac{1}{2}, 3$...................

(ii) Which of the following is not an A.P. ?

(a) 3, 7, 9, 11, 15, 17,.............

(b) 3, −1, −5, −9,..............

(c) 2.1, 2.6, 3.1, 3.6,................

(d) $\frac{1}{2}, \frac{1}{3}, \frac{1}{4}, \frac{1}{5}$,...................

(iii) Which of the following is not an A.P. ?

(a) −11, −8, −5,..........

(b) 1.2, 1.4, 1.7, 1.9,............

(c) $2, 2\frac{1}{2}, , 3, 3\frac{1}{2}$,...............

(d) 3, 7, 11, 15,...................

Key Point sum of first n natural numbers = $\frac{n(n+1)}{2}$

78. Find the sum of first 100 natural number. **(2020)**

Solution:

sum of first n natural numbers

$= \frac{n(n+1)}{2}$

∴ sum of first 100 natural numbers

$= \frac{100(100+1)}{2}$

$= 50 \times 101$

$= 5050$

Exercise 5.19

(i) Find the sum of first 50 natural numbers.

(ii) Find the sum of first 20 natural numbers.

(iii) Find the sum of first 30 natural numbers.

Key Point If the first term, common difference and last term of a A.P are respectively a, d and ℓ, then its nth term is

(i) $S_n = \frac{n}{2}\{2a + (n-1)d\}$

(ii) $S_n = \frac{n}{2}\{a + \ell\}$

79. The sum of first 20 odd natural numbers is. **(2012)**

(a) 100 (b) 210

(c) 400 (d) 420

Solution: (c)

Here $a = 1$

$d = 3 - 1 = 2$

$n = 20$

$a_n = a + (n-1)d$

$\therefore a_{20} = 1 + (20 - 1)\,2$

$= 1 + 38$

$= 39$

∴ sum of first 20 odd natural numbers

$= \frac{20}{2}(1 + 39)$

$= 10 \times 40$

$= 400$

80. The sum of first 50 odd natural numbers is. **(2023)**

(a) 5000 (b) 2500

(c) 2550 (d) 5050

Solution:

Here $a = 1$

$d = 3 - 1 = 2$

$n = 50$

$S_n = \frac{n}{2}\{2a + (n-1)d\}$

$\therefore S_{50} = \frac{50}{2}\{2 \times 1 + (50-1)\,2\}$

$= 25 \times 100$

$= 2500$

Exercise 5.20

(i) The sum of first 30 odd natural numbers is

(a) 400 (b) 450

(c) 380 (d) 280

(ii) The sum of first 26 odd natural numbers is

(a) 604 (b) 676

(c) 582 (d) 724

(iii) The sum of first 100 odd natural numbers is

(a) 10000 (b) 9998

(c) 8024 (d) 10240

81. The sum of the first 100 even natural numbers is. **(2023)**

(a) 10100 (b) 2550

(c) 5050 (d) 10010

Solution:

Here $a = 2$

$d = 4 - 2 = 2$

$n = 100$

$S_n = \frac{n}{2}\{2a + (n-1)d\}$

$\therefore S_{100} = \frac{100}{2}\{2 \times 2 + (100-1)\,2\}$

$= 50 \times 202$

$= 10100$

Exercise 5.21

(i) The sum of the first 50 even natural numbers is.

(a) 2550 (b) 2500

(c) 2020 (d) 2628

(ii) The sum of the first 30 even natural numbers is.

(a) 900 (b) 884

(c) 930 (d) 1010

(iii) The sum of the first 200 even natural numbers is.

(a) 40000 (b) 42000

(c) 40200 (d) 50200

82. The sum of the first 21 terms of an A.P : 16, 12, 8, 4, ... is: **(2023)**

(a) –480 (b) –504

(c) 1176 (d) –484

Solution: (b)

$S_n = \frac{n}{2}\{2a + (n-1)d\}$

Here $a = 16$

$d = 12 - 16 = -4$

$n = 21$

$\therefore S_{21} = \frac{21}{2}\{2 \times 16 + (21-1)(-4)\}$

$= \frac{21}{2}(32 - 80)$

$= \frac{21}{2} \times (-48)$

$21 \times (-24)$

-504

Exercise 5.22

(i) Find the sum of the first 20th terms of an A.P. 5, 9, 13, 17,.................

(a) 804 (b) 860

(c) 784 (d) 1025

(ii) Find the sum of the first 8th terms of an A.P. 15, 11, 7,................., –13

(a) 8 (b) 12

(c) –10 (d) 16

(iii) Find the sum of the first 14th terms of an A.P. 3, 7, 11, 15,................

83. Find the sum of first 10 multiple of 6. **(2019)**

Solution:

Here $a = 6$

$d = 6$

$n = 10$

$a_{10} = 60$

$S_n = \frac{n}{2}(a + a_n)$

$\therefore S_{10} = \frac{10}{2}(6 + 60)$

$= 5 \times 66$

$= 330$

84. Find the sum of first 10 multiple of 3. **(2019)**

Solution:

Here $a = 3$

$d = 3$

$n = 10$

$a_{10} = 30$

$S_n = \frac{n}{2}(a + a_n)$

$\therefore S_{10} = \frac{10}{2}(3 + 30)$

$= 5 \times 33$

$= 165$

Exercise 5.23

(i) Find the sum of first 15 multiple of 8.

(ii) Find the sum of first 20 multiple of 3.

(iii) Find the sum of first 100 multiple of 7.

85. The first term of an A.P. is 5 and the last term is 45. If the sum of all the terms is 400, the number of terms is. **(2020)**

(a) 20 (b) 8

(c) 10 (d) 16

Solution:

Here $a = 5$

$a_n = 45$

$S_n = 400$

$\therefore \frac{n}{2}(a + a_n) = 400$

$\Rightarrow = \frac{n}{2}(5 + 45) = 400$

$\Rightarrow = \frac{n}{2} \times 50 = 400$

$\therefore n = 16$

86. In an A.P. if the first term $a = 7$, n^{th} term $a_n = 84$ and the sum of first n terms $S_n = \frac{2093}{2}$, then n is equal to : **(2024)**

Solution: (c)

$S_n = \frac{n}{2}(a + a_n)$

$\Rightarrow \frac{2093}{2} = \frac{n}{2}(7 + 84)$

$\Rightarrow 91n = 2093$

$\Rightarrow n = 23$

87. In an A.P., $a = -3$ and $s_{17} = 357$. The value of a_{17} is: **(2026)**

(a) 47 (b) 39

(c) 45 (d) 42

Solution: (c)

$s_{17} = 357$

$\Rightarrow \frac{17}{2}\{2a + (17 - 1)d\} = 357$

$\Rightarrow 2 \times (-3) + 16d = 42$

$\Rightarrow -6 + 16d = 42$

$\Rightarrow 16d = 48$

$\therefore d = 3$

$\therefore a_{17} = a + 16d$

$= (-3) + 16 \times 3$

$= 45$

Exercise 5.24

(i) The first term of an A.P. is 5 and the last term is 74. If the sum of all the terms is 948, the number of terms is

(a) 24 (b) 20

(c) 28 (d) 32

(ii) The first term of an A.P. is 110 and the last term is 990. If the sum of all the terms is 44550, the number of terms is

(a) 62 (b) 66

(c) 81 (d) 102

(iii) The first term of an A.P. is 2 and the last term is 56. If the sum of all the terms is 551, the number of terms is

(a) 19 (b) 10

(c) 12 (d) 22

(iv) In an A.P., if the first term $a = 7$, n^{th} term $a_n = 210$ and sum of first n terms $S_n = 4620$, then n is equal to

(a) 20 (b) 40

(c) 10 (d) 18

(v) In an A.P., if the first term $a = -4$, n^{th} term $a_n = 150$ and sum of first n terms $S_n = 1387$, then n is equal to

(a) 19 (b) 21

(c) 23 (d) 20

88. If the sum of the first n terms of an A.P. be $3n^2 + n$ and its common difference is 6, then its first term is **(2023)**

(a) 2 (b) 3

(c) 21 (d) 4

Solution: (d)

$S_n = 3n^2 + n$

$\Rightarrow \frac{n}{2}(2a + (n-1)d) = 3n^2 + n$

$\Rightarrow \frac{n}{2}(2a + (n-1)6) = 3n^2 + n$

$\Rightarrow 2a + 6(n-1) = 6n + 2 \quad [\because n \neq 0]$

$\Rightarrow 2a + 6n - 6 = 6n + 2$

$\Rightarrow 2a = 8$

$\therefore a = 4$

Exercise 5.25

(i) If the sum of the first n terms of an A.P. be $3n^2 - 4n$ and its common difference is 6, then its first term is

(a) -1 (b) 1

(c) 2 (d) -3

(ii) If the sum of the first n terms of an A.P. be $\frac{3n^2}{2} + \frac{5n}{2}$ and its common difference is 7, then its first term is

(a) 2 (b) 4

(c) -2 (d) 3

(iii) If the sum of the first n terms of an A.P. be $2n^2 + 5n$ and its common difference is 4, then its first term is

(a) 2 (b) -2

(c) 3 (d) 4

89. If the sum of frist m terms of an AP is $2m^2 + 3m$, then its second term is: **(2025)**

(a) 10 (b) 9

(c) 12 (d) 4

Solution: (b)

$S_m = 2m^2 + 3m$

$\therefore S_1 = a_1 = 2 \times 1^2 + 3 \times 1 = 5$

$S_2 = a_1 + a_2 = 2 \times 2^2 + 3 \times 2 = 14$

$\therefore 5 + a_2 = 14$

$\Rightarrow a_2 = 9$

90. In an A.P., if the first term $(a) = -16$ and the common difference $(d) = -2$, then the sum of first 10 terms is. **(2024)**

(a) -200 (b) -70

(c) -250 (d) 250

Solution: (c)

$S_n = \frac{n}{2}\{2a + (n-1)d\}$

$\therefore S_{10} = \frac{10}{2}\{2 \times (-16) + (10-1)(-2)\}$

$= 5(-32 - 18)$

$= 5 \times (-50) = -250$

91. If the first term of an AP, is -12 and the common difference is 4, then the sum of its first 7 term is: **(2025)**

(a) -24 (b) -48

(c) 0 (d) 48

Solution: (c)

$a = -12$

$d = 4$

$S_n = \frac{n}{2}\{2a + (n-1)d\}$

$S_7 = \frac{7}{2}\{2 \times (-12) + (7-1) \times 4\}$

$= \frac{7}{2}(-24 + 24)$

$= 0$

92. The sum of three terms in A.P. is 30. If the greatert term is 13, then the common difference is **(2025)**

(a) 2 (b) 3

(c) –2 (d) –3

Solution: (b)

Let, the terms of the A.P are

$13 - 2d, 13 - d, 13$

$\therefore (13 - 2d) + (13 - d) + 13 = 30$

$\Rightarrow 39 - 3d = 30$

$\Rightarrow -3d = -9$

$\therefore d = 3$

Direction: Question number **93 to 96** are Assertion and Reason based question. Two statements are given, one labelled on Assertion (A) and the other labelled an Reason (R). Select the correct answer to these questions from the codes (A), (B), (C) and (D) as given below.

(a) Both Assertion (A) and Reason (R) are true and Reason (R) is the correct explanation of Assertion (A).

(b) Both Assertion (A) and Reason (R) are true, but Reason (R) is not the correct explanation of Assertion (A).

(c) Assertion (A) is true, but reason (R) is false.

(d) Assertion (A) is false, but reason (R) is true.

93. Assertion (A) : a, b, c are in A.P. if and only if $2b = a + c$

Reason (R) : The sum of first n odd natural numbers is n^2 **(2023)**

Solution: (b)

Both assertion (A) and reason (R) are true and reason (R) is the correct explanation of assertion (A).

94. Assertion (A) : The sum of the first fifteen terms of the AP 21, 18, 15, 12,..........is zero

Reason (R): The sum of the first n terms of an AP with first term 'a' and common difference 'd' is given by $S_n = \frac{n}{2}[2a + (n-1)d]$. **(2024)**

Solution: (a)

$S_{15} = \frac{15}{2}[2 \times 21 + (15-1)(-3)]$

$= 0$

95. Assertion (A): Common difference of the AP: 5, 1, –3, –7,...... is 4.

Reason (R): Common difference of the AP: a_1, a_2, a_3, ..., a_n is obtained by $d = a_n - a_{n-1}$ **(2025)**

Solution: (d)

Common difference $= 1 - 5 = -4$

96. Assertion (A): For an A.P 3, 6, 9, ..., 198, 10th term from the end is 168.

Reason (R): If 'a' and 'ℓ' are the first term and last term of an AP with common difference 'd' then n^{th} term from the end of the given A.P is $\ell - (n-1)d$. **(2025)**

Solution: (d)

10th term from end of the AP is

$a_n = \ell - (n-1)d$

$\therefore a_{10} = 198 - (10-1) \times 3$

$= 198 - 27$

$= 171$

Exercise 5.26

Direction: Question number **i to viii** are Assertion and Reason based question. Two statements are given, one labelled on Assertion

(A) and the other labelled an Reason (R). Select the correct answer to these questions from the codes (A), (B), (C) and (D) as given below.

(a) Both Assertion (A) and Reason (R) are true and Reason (R) is the correct explanation of Assertion (A).

(b) Both Assertion (A) and Reason (R) are true, but Reason (R) is not the correct explanation of Assertion (A).

(c) Assertion (A) is true, but reason (R) is false.

(d) Assertion (A) is false, but reason (R) is true.

(i) Assertion (A) : If $(2p + 1)$, 13, $(5p - 3)$ are three consecutive terms of an A.P. then $p = 4$.

Reason (R) : if a, b, c are in A.P. then $2b = a + c$

(ii) Assertion (A) : Sum of first 10 natural numbers is 55.

Reason (R) : Sum of first n natural numbers is $\frac{n(n+1)}{2}$

(iii) Assertion (A) : if the nth term of an A.P. is $(3n + 2)$, then sum of first n terms of the A. P. is $\frac{n(3n+7)}{2}$

Reason (R) : if a, b, c are in A.P. then $2b = a + c$

(iv) Assertion (A): The 10th term of an AP is equal to the 5th term.

Reason (R): The common difference of the AP is zero.

(v) Assertion: The 11th term of an AP 7, 9, 11, 13, ... is 67

Reason: if S_n is the sum of first n terms of an AP then its nth term a_n is given by an $= S_n - S_{n-1}$

(vi) Assertion (A): The sequence 3, 7, 11, 15, ... is an Arithmetic Progression.

Reason (R): The difference between consecutive terms is constant.

(vii) Assertion (A): The nth term of the AP 3, 7, 11, __ is given by $a_n = 4n - 1$.

Reason (R): The general term of an AP is given by $a_n = a + (n - 1)d$

(viii) Assertion (A): If the common difference of an AP is negative, the terms will decrease.

Reason (R): In an AP, if $d < 0$, then each term is smaller than the previous one.

ANSWER KEYS

EXERCISE – 5.1

(i) (c) $\frac{1}{2}$ **(ii)** (a) $\sqrt{2}$ **(iii)** $\frac{1}{a}$

EXERCISE – 5.2

(i) $2\sqrt{2}$ **(ii)** $2\sqrt{3}$ **(iii)** $-5\sqrt{5}$

EXERCISE – 5.3

(i) –1 **(ii)** –6 **(iii)** 3

EXERCISE – 5.4

(i) 5 **(ii)** 5 **(iii)** 4

EXERCISE – 5.5

(i) (b) 4 **(ii)** (c) –3 **(iii)** (a) 4

EXERCISE – 5.6

(i) (b) 2 **(ii)** (a) 7 **(iii)** (c) –2

EXERCISE – 5.7

(i) (b) 2 **(ii)** (a) 5 **(iii)** (d) 9

EXERCISE – 5.8

(i) 35 **(ii)** $26a - 23$ **(iii)** –77

(iv) 57 **(v)** $13\frac{1}{2}$ **(vi)** –15

EXERCISE – 5.9

(i) (b) 6 **(ii)** (a) 0 **(iii)** (d) 100

(iv) 3 **(v)** 40 **(vi)** –12

EXERCISE – 5.10

(i) $4\sqrt{3}$ **(ii)** $7\sqrt{2}$ **(iii)** $7\sqrt{3}$

EXERCISE – 5.11

(i) 105 **(ii)** 14 **(iii)** 11

EXERCISE – 5.12

(i) 34 **(ii)** 20 **(iii)** 18

EXERCISE – 5.13

(i) 10 **(ii)** 30 **(iii)** 20

EXERCISE – 5.14

(i) 8 **(ii)** 101 **(iii)** 183

EXERCISE – 5.15

(i) 10 **(ii)** $\frac{1}{3}$ **(iii)** $\frac{1}{15}$

(iv) (b) 5 **(v)** (c) 1 **(vi)** (d) 8

EXERCISE – 5.16

(i) (a) 0 **(ii)** (d) –4a –4d **(iii)** (c) –2a –2d

EXERCISE – 5.17

(i) 49 **(ii)** 12 **(iii)** 19

EXERCISE – 5.18

(i) (d) **(ii)** (d) **(iii)** (b)

EXERCISE – 5.19

(i) 1275 **(ii)** 210 **(iii)** 465

EXERCISE – 5.20

(i) (b) 900 **(ii)** (b) 676 **(iii)** (c) 10000

EXERCISE – 5.21

(i) (a) 2550 **(ii)** (c) 930 **(iii)** (c) 40200

EXERCISE – 5.22

(i) (b) 860 **(ii)** (a) 8 **(iii)** 406

EXERCISE – 5.23

(i) 960 **(ii)** 630 **(iii)** 35250

EXERCISE – 5.24

(i) (a) 24 **(ii)** (c) 81 **(iii)** (a) 19
(iv) (b) 40 **(v)** (a) 19 **(vi)** (b) 24

EXERCISE – 5.25

(i) (a) –1 **(ii)** (b) 4 **(iii)** (b) 4

EXERCISE – 5.26

(i) (a) **(ii)** (a) **(iii)** (b)
(iv) (a) **(v)** (d) **(vi)** (a)
(vii) (a) **(viii)** (a)

6 Coordinate Geometry

Key Point Distance of point $A\ (a, b)$ from x-axis is b unit

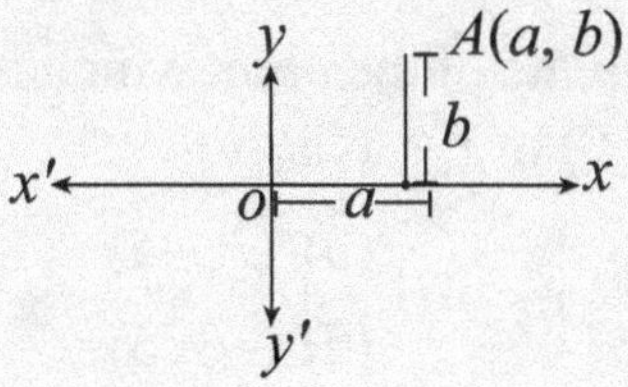

Distance of point $B\ (p, q)$ from y-axis is p unit

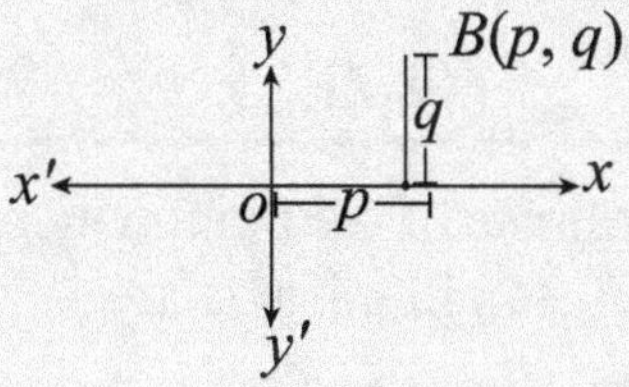

1. Distance of point $P\ (3, 4)$ from x-axis is: **(2020)**

(a) 3 unit (b) 4 unit
(c) 5 unit (d) 1 unit

Solution: (b)

Required distance = 4 unit

2. The distance of the point $(-1, 7)$ from x-axis is: **(2023)**

(a) -1 (b) 7
(c) 6 (d) $\sqrt{50}$

Solution: (b)

Required distance = 7

3. The distance of the point $(-6, 8)$ from x-axis is **(2023)**

(a) 6 units (b) -6 units
(c) 8 units (d) 10 units

Solution: (c)

Required distance = 8 units

4. The distance of the point $(-4, 3)$, from y axis is **(2023)**

(a) -4 (b) 4
(c) 3 (d) 5

Solution: (b)

Required distance = 4 unit

5. Distance of the point $(6, 5)$ from the y-axis is **(2023)**

(a) 6 units (b) 5 units
(c) $\sqrt{61}$ units (d) 0 unit

Solution: (a)

Required distance = 6 units

6. The distance of the point $(4, 5)$ from x-axis is: **(2024)**

(a) 5 (b) 4
(c) 9 (d) 1

7. The distance of the point $P\ (3, -5)$ from x-axis is: **(2024)**

(a) 3 units (b) 5 units
(c) -5 units (d) $\sqrt{34}$ units

Solution: (b)

Required distance = 5 units

8. The distance of the point $A(-3, -4)$ from x-axis is **(2025)**

(a) 3 (b) 4
(c) 5 (d) 7

Solution: (b)

Required distance = 4 unit

9. The distance of point $(a, -b)$ from x-axis is **(2025)**

(a) a (b) $-a$
(c) b (d) $-b$

Solution: (c)

b

10. The distance of point $P(3, 4)$ from the x-aixs is **(2025)**

(a) 3 units (b) 4 units
(c) 5 units (d) 7 units

Solution: (b)

4 units

11. The distance of point $P(1, -1)$ from x-axis is: **(2025)**

(a) 1 (b) –1
(c) 0 (d) $\sqrt{2}$

Solution: (a)

1

12. The distance of the point A $(4a, 3a)$ from $x-axis$ is **(2026)**

(a) $3a$ (b) $-3a$
(c) $4a$ (d) $-4a$

Solution: (a)

The required distance is $3a$.

Exercise 6.1

(i) The distance of the point (8, 6) from x axis is

(a) 8 units (b) 6 units
(c) 10 units (d) 2 units

(ii) The distance of the point (5, 9) from y axis is

(a) 5 units (b) 9 units
(c) 4 units (d) 14 units

(iii) The distance of the point (–3, –7) from x-axis is

(a) –3 units (b) –7 units
(c) 3 units (d) 7 units

(iv) The distance of the point (–6, 0) from y-axis is

(a) 0 units (b) –6 units
(c) 6 units (d) $\sqrt{5}$ unit

(v) The distance of the point (–2, 3) from y axis is

(a) –2 units (b) 3 units
(c) 2 units (d) –3 units

(vi) The distance of the point (8, –10) from x-axis is

(a) 8 units (b) –10 units
(c) –8 units (d) 10 units

Key Point **(i)** Reflection of a point (x, y) over x-axis is $(x, -y)$

(ii) Reflection of a point (x, y) over y axis is $(-x, y)$

13. The coordinate of the point which is reflection of point (–3, 5) in x – axis are **(2020)**

(a) (3, 5) (b) (3, –5)
(c) (–3, –5) (d) (–3, 5)

Solution: (c)

The reflection of (–3, 5) along x – axis is (–3, –5)

Exercise 6.2

(i) The Coordinate of the point which is reflection of point (3, –2) in x – axis are

(a) (3, –2) (b) (–3, –2)
(c) (–3, 2) (d) (3, 2)

(ii) The Coordinate of the point which is reflection of point (–3, 5) in x – axis are

(a) (–3, 5) (b) (–3, –5)
(c) (3, 5) (d) (3,–5)

(iii) The Coordinate of the point which is reflection of point (3, –2) in y axis are

(a) (3, –2) (b) (–3, –2)
(c) (3, 2) (d) (–3, 2)

(iv) The Coordinate of the point which is reflection of point (–4, –7) in y axis are

(a) (–4, 7) (b) (–4, –7)
(c) (4, 7) (d) (4, –7)

Key Point **(i)** The Coordinates of a point A on y-axis at a distance of 'a' units from x axis and above it is $(0, a)$

(ii) The Coordinates of a point A on y axis at a distance of 'a' unit from x – axis and below it is $(0, -a)$

14. The coordinates of a point A on y – axis at a distance of 4 units from x – axis and below it, are **(2020)**

(a) (4, 0) (b) (0, 4)

(c) (–4, 0) (d) (0, –4)

Solution: (d)

The required Coordinates of A are (0, –4)

15. The distance of point $P(3a, 4a)$ from y-axis is **(2025)**

(a) $3a$ (b) $-3a$

(c) $4a$ (d) $-4a$

Solution: (a)

Required distance = $3a$

16. The distance of a point $P(3, -7)$ from y-aixs is **(2025)**

(a) 3 (b) 7

(c) –7 (d) $\sqrt{58}$

Solution: (a)

3

Exercise 6.3

(i) The coordinates of a point A on y – axis at a distance of 6 unit from x – axis and above it, are

(a) (0, 6) (b) (6, 0)

(c) (0, –6) (d) (–6, 0)

(ii) The coordinates of a point A on y – axis at a distance of 3 units from x – axis and above it, are

(a) (–3, 0) (b) (0, –3)

(c) (3, 0) (d) (0, 0)

17. The equation of a line parallel to the x-axis and at a distance of 3 units below x-axis is **(2025)**

(a) $x = 3$ (b) $x = -3$

(c) $y = -3$ (d) $y = 3$

Solution: (c)

$y = -3$

18. The eqution of a line parallel to y-axis and at a distance of 5 unit to the right of y axis is: **(2025)**

(a) $x = 5$ (b) $x = -5$

(c) $y = 5$ (d) $y = -5$

Solution: (a)

$x = 5$

19. The line represented by the eqution $x - y = 0$ is **(2025)**

(a) Parallel to x-axis

(b) Parallel to y-axis

(c) Passing through the origin

(d) Passing through the point (3, 2)

Solution: (c)

Passing through the origin

20. The distance of a point A from x-axis is 3 units. Which of the following cannot be coordinate of the point A ? **(2025)**

(a) (1, 3) (b) (–3, –3)

(c) (–3, 3) (d) (3, 1)

Solution: (d)

(3, 1)

21. For a point (3, –5), the value of (abscissa – ordinate) is: **(2025)**

(a) –8 (b) –2

(c) 2 (d) 8

Solution: (d)

$3 - (-5) = 3 + 5 = 8$

22. For a point $X(a, b)$ where $(b > a > 0)$, the value of its (distance from x-axis – distance from y-axis) is: **(2025)**

(a) $a - b$ (b) $b - a$

(c) $a^2 - b^2$ (d) $b^2 - a^2$

Solution: (b)

Required distance = $b - a$

Key Point If a point (x_1, y_1) lies on a line represented by $ax + by + c = 0$, then $ax_1 + by_1 + c = 0$

23. Find the value of 'a' so that the point (3, a) lies on the line represented by $2x - 3y = 5$. **(2019)**

Solution:

$2 \times 3 - 3a = 5$

$\Rightarrow 6 - 3a = 5$

$\Rightarrow -3a = -1$

$\therefore a = \frac{1}{3}$

24. The value of k, if (6, k) lies in the line represented by $x - 3y + 6 = 0$ is **(2023)**

(a) –4 (b) 12
(c) –12 (d) 4

Solution: (d)

$6 - 3k + 6 = 0$

$\Rightarrow -3k = -12$

$\therefore k = 4$

25. The value of 'p' if (–2, p) lies on the line represented by the equation $2x - 3y + 7 = 0$ is **(2023)**

(a) $\frac{-13}{2}$ (b) $\frac{13}{2}$
(c) –1 (d) 1

Solution: (d)

$2(-2) - 3p + 7 = 0$

$\Rightarrow 3 - 3p = 0$

$\therefore 3p = 3$

$\therefore p = 1$

26. The coordinates of the point where the line $2y = 4x + 5$ crosses x – axis is **(2023)**

(a) $\left(0, \frac{-5}{4}\right)$ (b) $\left(0, \frac{5}{2}\right)$
(c) $\left(\frac{-5}{4}, 0\right)$ (d) $\left(\frac{-5}{2}, 0\right)$

Solution: (a)

Let the line intersects the x – axis at (a, 0)

$\therefore 2 \times 0 = 4a + 5$

$\Rightarrow a = -\frac{5}{4}$

$\therefore$ The required point is $\left(\frac{-5}{4}, 0\right)$

27. The point of intersection of the line represented by $3x - y = 3$ and the y-axis is given by **(2023)**

(a) (0, –3) (b) (0, 3)
(c) (2, 0) (d) (–2, 0)

Solution: (a)

Let the line intersects the y axis at (0, a)

$\therefore 3.0 - a = 3$

$\Rightarrow a = -3$

$\therefore$ The required point is (0, –3)

Exercise 6.4

(i) Find the value of 'a' so that the point (–2, a) lies on the line represented by $3x - 5y = 14$

(a) 4 (b) –4
(c) 0 (d) 8

(ii) The value of k, if (a, –2) lies on the line represented by $3x + 2y + 7 = 0$, is

(a) 1 (b) 0
(c) –3 (d) –1

(iii) The value of 'p', if (1, $-p$) lies on the line represented by the equation $4x - 3y = 10$, is

(a) 2 (b) –2
(c) 1 (d) 3

(iv) The Coordinates of the point where the line $3y = 2x + 6$ crosses x – axis is

(a) (2, 0) (b) (3, 0)
(c) (–1, 0) (d) (–3, 0)

(v) The point of intersection of the line represented by $2x - y = 6$ and y – axis is given by

(a) (–1, 0) (b) (0, –1)
(c) (0, –6) (d) (0, 6)

Key Point **(i)** The distance between two points $P(x_1, y_1)$ and $Q(x_2, y_2)$ is $\sqrt{(x_2 - x_1)^2 + (y_2 - y_1)^2}$

(ii) The distance between two points $P(-x_1, -y_1)$ and $Q(x_2, y_2)$ is

$$\sqrt{\{x_2-(-x_1)\}^2+\{y_2-(-y_1)\}^2}$$
$$=\sqrt{(x_2+x_1)^2+(y_2+y_1)^2}$$

(iii) The distance between $P(x, y)$ and origin is

$$\sqrt{(x-0)^2+(y-0)^2}=\sqrt{x^2+y^2}$$

28. Find the distance of a point $P(x, y)$ from the origin. **(2018)**

Solution:

$$OP=\sqrt{(x-0)^2+(y-0)^2}=\sqrt{x^2+y^2}$$

29. The distance between the points (a, b) and $(-a, -b)$ is ________ **(2020)**

Solution:

$$\text{Required distance}=\sqrt{\{(a-(-a)\}^2+\{b-(-b)\}^2}$$
$$=\sqrt{(a+a)^2+(b+b)^2}$$
$$=\sqrt{(2a)^2+(2b)^2}$$
$$=\sqrt{4(a^2+b^2)}$$
$$=2\sqrt{a^2+b^2}\text{ unit}$$

30. The distance between the points $(m, -n)$ and $(-m, n)$ is **(2020)**

(a) $\sqrt{m^2+n^2}$ (b) $m+n$

(c) $2\sqrt{m^2+n^2}$ (d) $\sqrt{2m^2+2n^2}$

Solution: (c)

The required distance

$$=\sqrt{\{-m-m\}^2+\{n-(-n)\}^2}$$
$$=\sqrt{(-2m)^2+(n+n)^2}$$
$$=\sqrt{4m^2+4n^2}$$
$$=2\sqrt{m^2+n^2}$$

31. The distance of the point $(-12, 5)$ from the origin is **(2020)**

(a) 12 (b) 5

(c) 13 (d) 169

Solution: (c)

$$\text{Required distance}=\sqrt{(-12-0)^2+(5-0)^2}$$
$$=\sqrt{144+25}$$
$$=\sqrt{169}$$
$$=13$$

32. The distance between the points A (2, –3) and B (2, 2) is **(2020)**

(a) 2 units (b) 3 units

(c) 4 units (d) 5 units

Solution: (d)

$$AB=\sqrt{\{2-2\}^2+\{2-(-3)\}^2}$$
$$=\sqrt{0+(2+3)^2}$$
$$=\sqrt{5^2}$$
$$=5\text{ units.}$$

33. Find the distance between the points (a, b) and $(-a, -b)$. **(2019)**

Solution:

$$\text{Required distance}=\sqrt{(-a-a)^2+(-b-b)^2}$$
$$=\sqrt{(-2a)^2+(-2b)^2}$$
$$=\sqrt{4a^2+4b^2}$$
$$=2\sqrt{a^2+b^2}$$

34. Distance between $(a, -b)$ and (a, b) is ________ **(2020)**

Solution:

$$\text{Required distance}=\sqrt{(a-a)^2+\{b-(-b)\}^2}$$
$$=\sqrt{o^2+(2b)^2}$$
$$=\sqrt{4b^2}$$
$$=2b$$

35. The distance between the points (–1, –3) and (5, –2) is **(2020)**

(a) $\sqrt{61}$ unit (b) $\sqrt{37}$ unit

(c) 5 unit (d) $\sqrt{17}$ unit

Solution:

$$\text{Required distance}=\sqrt{\{5-(-1)\}^2+\{(-2)-(-3)\}^2}$$

$=\sqrt{(5+1)^2+(-2+3)^2}$

$=\sqrt{36+1}$

$=\sqrt{37}$ unit

36. The distance between the points $(-a, a)$ and $(-a, -a)$ is________. **(2020)**

Solution:

Required distance $=\sqrt{\{(-a)-(-a)^2\}^2+\{(-a)-a\}^2}$

$=\sqrt{(-a+a)^2+(-a-a)^2}$

$=\sqrt{0+(-2a)^2}$

$=\sqrt{4a^2}$

$=2a$

37. The distance between the points (3, –2) and (–3, 2) is **(2020)**

(a) $\sqrt{52}$ units (b) $4\sqrt{10}$ units

(c) $2\sqrt{10}$ units (d) 40 units

Solution: (a)

Required distance $=\sqrt{\{(-3)-3\}^2+\{2-(-2)\}^2}$

$=\sqrt{(-3-3)^2+(2+2)^2}$

$=\sqrt{(-6)^2+(4)^2}$

$=\sqrt{36+16}$

$=\sqrt{52}$ units.

38. The distance between the points $(0, 2\sqrt{5})$ and $(-2\sqrt{5}, 0)$ is **(2023)**

(a) $2\sqrt{10}$ unit (b) $4\sqrt{10}$ unit

(c) $2\sqrt{20}$ unit (d) 0

Solution: (a)

Requireddistance$=\sqrt{\{(-2\sqrt{5})-0\}^2+\{0-2\sqrt{5}\}^2}$

$=\sqrt{20+20}$

$=\sqrt{40}\;=2\sqrt{10}$ unit

39. The distance between the points $P\left(\frac{-11}{3},5\right)$ and $Q\left(\frac{-2}{3},5\right)$ is: **(2023)**

(a) 6 units (b) 4 units

(c) 2 units (d) 3 units

Solution: (d)

$PQ=\sqrt{\left\{\left(\frac{-2}{3}\right)-\left(\frac{-11}{3}\right)\right\}^2+(5-5)^2}$

$=\sqrt{\left(\frac{-2}{3}+\frac{11}{3}\right)^2+0^2}$

$=\sqrt{\left(\frac{9}{3}\right)^2}$

$=\sqrt{9}$

$=3$ unit

40. The distance between the points (0, 5) and (–3, 1) is **(2023)**

(a) 8 units (b) 5 units

(c) 3 units (d) 25 units

Solution: (b)

Required distance $=\sqrt{(-3-0)^2+(1-5)^2}$

$=\sqrt{9+16}$

$=\sqrt{25}$

$=5$ units

41. The distance between the points (3, 0) and (0, –3) is **(2023)**

(a) $2\sqrt{3}$ units (b) 6 units

(c) 3 units (d) $3\sqrt{2}$ units

Solution: (d)

Required distance $=\sqrt{(0-3)^2+(-3-0)^2}$

$=\sqrt{9+9}$

$=\sqrt{18}$

$=3\sqrt{2}$ units

42. The distance between the points $\left(\frac{-5}{2},7\right)$ and $\left(-\frac{1}{2},7\right)$ is **(2023)**

(a) 3 (b) 2

(c) 4 (d) 9

Solution: (b)

Required distance $= \sqrt{\left(-\frac{1}{2}+\frac{5}{2}\right)^2+(7-7)^2}$

$= \sqrt{2^2+0^2}$

$= 2$

43. The distance between the points A (0, 6) and B(–6, 2) is **(2023)**

(a) 6 units (b) $2\sqrt{6}$ units

(c) $2\sqrt{13}$ units (d) $13\sqrt{2}$ units

Solution: (c)

$AB = \sqrt{(-6-0)^2+(2-6)^2}$

$= \sqrt{36+16}$

$= \sqrt{52}$

$= 2\sqrt{13}$ units.

44. The distance between the points (6, 2) and (–6, 2) is: **(2023)**

(a) $6\sqrt{2}$ units (b) 12 units

(c) $2\sqrt{6}$ units (d) 6 units

Solution: (b)

Required distance $= \sqrt{(-6-6)^2+(2-2)^2}$

$= \sqrt{144}$

$= 12$ units.

45. The distance between the points (5, 0) from the origin is **(2023)**

(a) 0 (b) 5

(c) $\sqrt{5}$ (d) 5^2

Solution: (b)

Required distance $= \sqrt{(5-0)^2+(0-0)^2}$

$= \sqrt{25}$

$= 5$

46. The distance between the points (a cosθ, – a sinθ) and (a sinθ, a cosθ) is **(2024)**

(a) a (b) $a\sqrt{2}$

(c) 0 (d) 2a

Solution:

Required distance =

$\sqrt{(a\sin\theta-a\cos\theta)^2+(a\cos\theta+a\sin\theta)^2}$

$= \sqrt{a^2(\sin\theta-\cos\theta)^2+a^2(\sin\theta+\cos\theta)^2}$

$= a\sqrt{(\sin\theta+\cos\theta)^2+(\sin\theta-\cos\theta)^2}$

$= a\sqrt{2(\sin^2\theta+\cos^2\theta)}$

$= a\cdot\sqrt{2\cdot 1}$

$= \sqrt{2}\,a$

47. The distance between the points (2, –3) and (–2, 3) is **(2024)**

(a) $2\sqrt{13}$ units (b) 5 units

(c) $13\sqrt{2}$ units (d) 10 units

Solution: (a)

Required distance $= \sqrt{(-2-2)^2+(3+3)^2}$

$= \sqrt{(-4)^2+(6)^2}$

$= \sqrt{16+36}$

$= \sqrt{52}$

$= 2\sqrt{13}$ units

48. The distance of the point (5, 4) from the origin is **(2024)**

(a) 41 (b) $\sqrt{41}$

(c) 3 (d) 9

Solution: (b)

Required distance $= \sqrt{(5-0)^2+(4-0)^2}$

$= \sqrt{25+16}$

$= \sqrt{41}$

49. The distance of the point (3, 4) from the origin is **(2024)**

(a) 25 (b) 5

(c) 7 (d) 1

Solution:

Required distance $= \sqrt{(3-0)^2+(4-0)^2}$

$= \sqrt{9+16}$

$= \sqrt{25}$

$= 5$

50. The distance between the points (2, –1) and (–1, –5) is **(2024)**

(a) 15 units (b) 5 units

(c) 25 units (d) 41 units

Solution:

Required distance $= \sqrt{(-1-2)^2 + (-5+1)^2}$

$= \sqrt{(-3)^2 + (-4)^2}$

$= \sqrt{9+16}$

$= \sqrt{25}$

$= 5$ units

51. The distance between the points *A* (0, –1) and *B* (0, –9) is: **(2024)**

(a) 6 (b) 8

(c) 4 (d) 2

Solution:

Required distance $= \sqrt{(0-0)^2 + (-9+1)^2}$

$= \sqrt{0+64}$

$= 8$

52. The distance between the points (a cosθ, + b sinθ, 0) and (0, a sinθ, – b cosθ) is **(2020)**

(a) $a^2 + b^2$ (b) $a^2 - b^2$

(c) $\sqrt{a^2 + b^2}$ (d) $\sqrt{a^2 - b^2}$

Solution: (c)

Required distance =

$$\sqrt{\{0-(a\cos\theta + b\sin\theta)\}^2 + \{(a\sin\theta - b\cos\theta) - 0\}^2}$$

$$= \sqrt{a^2\cos^2\theta + 2ab\cos\theta\sin\theta + b^2\sin^2\theta + a^2\sin^2\theta - 2ab\sin\theta\cos\theta + b^2\cos^2\theta}$$

$$= \sqrt{a^2\left(\sin^2\theta + \cos^2\theta\right) + b^2\left(\sin^2\theta + \cos^2\theta\right)}$$

$$= \sqrt{a^2.1 + b^2.1}$$

$$= \sqrt{a^2 + b^2}$$

53. $\sqrt{2}$ times the distance between (0, 5) and (–5, 0) is________. **(2020)**

Solution:

Distance between (0, 5) and (–5, 0) is

$\sqrt{(-5-0)^2 + (0-5)^2}$

$= \sqrt{25+25}$

$= \sqrt{50}$

$= 5\sqrt{2}$

$\therefore \sqrt{2} \times 5\sqrt{2} = 10$

54. Distance of the point (a cosθ, a sinθ) from origin is: **(2020)**

Solution:

Required distance

$$= \sqrt{(a\cos\theta - 0)^2 + (a\sin\theta - 0)^2}$$

$$= \sqrt{a^2\cos^2\theta + a^2\sin^2\theta}$$

$$= \sqrt{a^2\left(\cos^2\theta + \sin^2\theta\right)}$$

$$= \sqrt{a^2 \cdot 1}$$

$$= \sqrt{a^2} = a$$

55. The distance of the point (–6, 8) from origin is **(2023)**

(a) 6 (b) –6

(c) 8 (d) 10

Solution: (d)

Required distance $= \sqrt{(-6-0)^2 + (8-0)^2}$

$= \sqrt{36+64}$

$= \sqrt{100} = 10$

56. *ABCD* is a rectangle whose three vertices are *B*(4, 0), *C*(4, 3) and *D*(0, 3). The length of one of its diagonals is **(2014)**

(a) 5 (b) 4

(c) 3 (d) 25

Solution: (d)

$BD = \sqrt{(0-4)^2 + (3-0)^2}$

$= \sqrt{16+9}$

$= \sqrt{25} = 5$

57. *AOBC* is a rectangle whose three vertices are *A* (0, –3), *O* (0, 0) and *B* (4, 0). The length of its diagonals is_______ **(2020)**

Solution:

$AB = \sqrt{(4-0)^2 + (0+3)^2}$

$= \sqrt{16+9}$

$= \sqrt{25} = 5\,\text{units.}$

58. *XOYZ* is a rectangle with vertices *X* (–3, 0), O (0,0), *Y* (0, 4) and Z (*x*, *y*). The length of its each diagonal is **(2024)**

(a) 5 units (b) $\sqrt{5}$ units

(c) $x^2 + y^2$ units (d) 4 units

Solution: (a)

$XY = \sqrt{(0+3)^2 + (4-0)^2}$

$= \sqrt{9+16}$

$= \sqrt{25} = 5\,\text{units}$

59. If the distance between the points (4, *k*) and (1, 0) is 5, then what can be the possible values of *k*? **(2017)**

Solution:

$\sqrt{(1-4)^2 + (0-k)^2} = 5$

$\Rightarrow \sqrt{(-3)^2 + (-k)^2} = 5$

$\Rightarrow 9 + k^2 = 25$

$\Rightarrow k^2 = 16$

$\therefore k = \pm 4$

60. Find the value (*s*) of *x*, if the distance between the points *A* (0, 0) and *B* (*x*, –4) is 5 units **(2019)**

Solution:

$AB = 5$

$\Rightarrow \sqrt{(x-0)^2 + (-4-0)^2} = 5$

$\Rightarrow x^2 + 16 = 25$

$\Rightarrow x^2 = 9$

$\Rightarrow \therefore x = \pm 3$

61. If the distance between the points *A* (4, *p*) and *B* (1, 0) is 5 units, then the value (*s*) of *p* is (are) **(2020)**

(a) 4 only (b) –4 only

(c) ±4 (d) 0

Solution: (c)

$AB = 5$

$\Rightarrow \sqrt{(1-4)^2 + (0-p)^2} = 5$

$\Rightarrow 9 + p^2 = 25$

$\Rightarrow p^2 = 16$

$\therefore p = \pm 4$

62. If the distance between the points (3, –5) and (*x*, –5) is 15 units, the value of *x* are **(2024)**

(a) 12, –18 (b) –12, 18

(c) 18, 5 (d) –9, –12

Solution: (b)

$\sqrt{(x-3)^2 + (-5+5)^2} = 15$

$\Rightarrow (x-3)^2 = (15)^2$

$\Rightarrow x - 3 = \pm 15$

$\Rightarrow x = 3 \pm 15$

$\therefore x = 3 + 15, 3 - 15$

$= 18, -12$

63. Find the value of *x* for which the distance between the points *A* (*x*, 2) and *B* (9, 8) is 10 units. **(2019)**

Solution:

$AB = 10$

$\Rightarrow \sqrt{(x-9)^2 + (2-8)^2} = 10$

$\Rightarrow (x-9)^2 + 36 = 100$

$\Rightarrow (x-9)^2 = 64$

$\Rightarrow x - 9 = \pm 8$

$\Rightarrow x = 9 \pm 8$

$\Rightarrow x = 9 + 8, \text{ or } 9 - 8$

$\therefore x = 17, \text{ or } 1$

64. Write the coordinate of a point *P* on *x* – axis which is equidistance from the points *A* (–2, 0) and *B* (6, 0). **(2019)**

Solution:

Let the coordinates of the point *P* be (*a*, 0)

$\therefore \sqrt{(a+2)^2 + (0-0)^2} = \sqrt{(a-6)^2 + (0-0)^2}$

$\Rightarrow (a+2)^2 = (a-6)^2$

$\Rightarrow a^2 + 4a + 4 = a^2 - 12a + 36$

$\Rightarrow 16a = 32$

$\therefore a = 2$

$\therefore$ The coordinates of P are (2, 0)

65. The point P on x – axis equidistance from the points A (–1, 0) and B (5, 0) is **(2020)**

(a) (2, 0) (b) (0, 2)

(c) (3, 0) (d) (2, 2)

Solution: (a)

Let the coordinates of the point P are $(a, 0)$

$\therefore \sqrt{(a+1)^2+(0-0)^2} = \sqrt{(a-5)^2+(0-0)^2}$

$\Rightarrow (a+1)^2 = (a-5)^2$

$\Rightarrow a^2+2a+1 = a^2-10a+25$

$\Rightarrow 12a = 24$

$\therefore a = 2$

$\therefore$ coordinates of P are (2, 0)

66. The point on the x-axis which is equidistance from (–4, 0) and (10, 0) is **(2020)**

(a) (7, 0) (b) (5, 0)

(c) (0, 0) (d) (3, 0)

Solution:

$\sqrt{(a+4)^2+(0-0)^2} = \sqrt{(a-10)^2+(0-0)^2}$

$\Rightarrow (a+4)^2 = (a-10)^2$

$\Rightarrow a^2 + 8a + 16 = a^2 - 20a + 100$

$\Rightarrow 28a = 84$

$\Rightarrow a = 3$

$\therefore$ The coordinates of the point are (3, 0)

67. The end points of a diameter of a circle are (2, 4) and (–3, –1). The radius of the circle is **(2023)**

(a) $2\sqrt{5}$ (b) $\frac{5}{2}\sqrt{5}$

(c) $\frac{5}{2}\sqrt{2}$ (d) $5\sqrt{2}$

Solution: (c)

Length of the diameter $= \sqrt{(-3-2)^2+(-1-4)^2}$

$= \sqrt{(-5)^2+(-5)^2}$

$= \sqrt{25+25}$

$= \sqrt{50} = 5\sqrt{2}$

$\therefore$ Radius of the circle $= \frac{5\sqrt{2}}{2}$

68. If end points of a diameter of a circle are (–5, 4) and (1, 0), then the radius of the circle is: **(2023)**

(a) $2\sqrt{13}$ units (b) $\sqrt{13}$ units

(c) $4\sqrt{2}$ units (d) $2\sqrt{2}$ units

Solution:

Length of the diameter $= \sqrt{(1+5)^2+(0-4)^2}$

$= \sqrt{36+16}$

$= \sqrt{52}$

$= 2\sqrt{13}$ units

$\therefore$ Radius of the circle $= \frac{2\sqrt{13}}{2} = \sqrt{13}$ units

69. The points (–4, 0), (4, 0) and (0, 3) are the vertices of a : **(2023)**

(a) right triangle

(b) isoscles triangle

(c) equilateral triangle

(d) scalene triangle

Solution:

Let the coordinates of the points A, B and C are (–4, 0), (4, 0) and (0, 3)

$AB = \sqrt{(4+4)^2+(0-0)^2}$

$= \sqrt{64}$

$= 8$ units

$BC = \sqrt{(0-4)^2+(3-0)^2}$

$= \sqrt{25}$

$= 5$ units

$CA = \sqrt{(0+4)^2+(3-0)^2}$

$= \sqrt{25}$

$= 5$ units

$\therefore$ In ΔABC

$BC = CA$

$\therefore ABC$ is a isoscles triangle.

70. The diameter of a circle is of length 6 cm. If one end of the diameter is (–4, 0), the other end on x-axis is at **(2024)**

(a) (0, 2) (b) (6,0)

(c) (2, 0) (d) (4, 0)

Solution:

Let the coordinates of other end be (a, 0)

$\therefore \sqrt{(a+4)^2+(0-0)^2}=6$

$\Rightarrow (a+4)^2=36$

$\Rightarrow a+4=6$

$\Rightarrow a=2$

∴ The other end of the diameter is at (2, 0)

71. The end points of a diameter of circle are (2, 4) and (–3, –1). The length is **(2025)**

(a) $\frac{5\sqrt{2}}{2}$ units (b) $5\sqrt{2}$ units

(c) $3\sqrt{2}$ units (d) $\pm\frac{5\sqrt{2}}{2}$ units

Solution: (a)

Length of diameter $=\sqrt{(-3-2)^2+(-1-4)^2}$

$=\sqrt{25+25}$

$=\sqrt{50}$

$=5\sqrt{2}$ units

∴ Length of radian $=\frac{5\sqrt{2}}{2}$ units

72. The coordinates of the end points of a diameter of a circle are (5, –2) and (5, 2). The length of the radius of the circle is **(2025)**

(a) ±2 (b) ±4

(c) 4 (d) 2

Solution: (d)

Length of the diameter $=\sqrt{(5-5)^2+(2+2)^2}$

= 4 units

∴ Length of the radius = 28 units

73. The points (–5, 0), (5, 0) and (0, 4) are the vertion of a triangle which is a/are: **(2025)**

(a) Right–angled triangle

(b) Isoscolen triangle

(c) Equilateral

(d) Scalene triangle

Solution: (b)

Let the vertion of the triangle or $A(-5, 0)$, $B(5, 0)$ and $C(0, 4)$

$\therefore AB=\sqrt{(5+5)^2+(0-0)^2}$

10 units

$BC=\sqrt{(0-5)^2+(4-0)^2}$

$=\sqrt{25+16}$

$=\sqrt{41}$ units

$AC=\sqrt{(0+5)^2+(4-0)^2}$

$=\sqrt{25+16}$

$=\sqrt{41}$

∴ In ΔABC

$BC=AC$

∴ ABC, is a isoscelen triangle.

74. The perimeter of the triangle formed by the vertices (0, 0), (2, 0) and (0, 2) is **(2025)**

(a) 4 units (b) 6 units

(c) $6\sqrt{2}$ units (d) $(4+2\sqrt{2})$ units

Solution: (d)

Let the coordinates of the vertices of the triangle are $A(0, 0)$, $B(2, 0)$ and $C(0, 2)$

$\therefore AB=\sqrt{(2-0)^2+(0-0)^2}$

= 2 units

$BC=\sqrt{(0-2)^2+(2-0)^2}$

$=\sqrt{4+4}$

$=\sqrt{8}$

$2\sqrt{2}$ units

$AC=\sqrt{(0-0)^2+(2-0)^2}$

= 2 units

∴ Required Perimeter $=2+2\sqrt{2}+2$

$=(4+2\sqrt{2})$ units

75. The distance of which of the following points from origin is less than 5 units ? **(2025)**

(a) (3, 4) (b) (2, 6)
(c) (–3, –4) (d) (1, 4)

Solution: (d)

The distance between origin and (1, 4) is $\sqrt{(1-0)^2+(4-1)^2}$

$=\sqrt{1+16}$

$=\sqrt{17}<5$

76. The distance between the points (4cos θ + 3sin θ, 0) and (0, 4sin θ – 3cos θ) is **(2025)**

(a) 25 (b) 7
(c) 5 (d) $\sqrt{7}$

Solution: (c)

Required distance =

$\sqrt{\{0-(4\cos\theta+3\sin\theta)\}^2+\{(4\sin\theta-3\cos\theta)-0\}^2}$

$=\sqrt{(4\cos\theta+3\sin\theta)^2+(4\sin\theta-3\cos)^2}$

$=\sqrt{(16\cos^2\theta+24\sin\theta\cos\theta+9\sin^2\theta}$

$+16\sin^2\theta-24\sin\theta\cos\theta+9\cos^2\theta)$

$=\sqrt{16(\sin^2\theta+\cos^2\theta)+9(\sin^2\theta+\cos^2\theta)}$

$=\sqrt{16.1+9.1}$

$=\sqrt{25}$

$=5$

77. The distance of the point (2, 3) from the origin is: **(2025)**

(a) 2 (b) 3
(c) 5 (d) $\sqrt{13}$

Solution: (d)

Required distance $=\sqrt{(2-0)^2+(3-0)^2}$

$=\sqrt{4+9}$

$=\sqrt{13}$

78. The distance between the points (2, 3) and (–2, –3) is **(2025)**

(a) $4\sqrt{13}$ (b) $\sqrt{40}$
(c) $2\sqrt{13}$ (d) 5

Solution: (c)

Required distance $=\sqrt{(-2-2)^2+(-3-3)^2}$

$=\sqrt{16+36}$

$=\sqrt{52}$

$=2\sqrt{13}$

79. The distance between the points (–6, 9) and (2, 7) is: **(2025)**

(a) $2\sqrt{17}$ (b) $4\sqrt{17}$
(c) $2\sqrt{5}$ (d) $2\sqrt{15}$

Solution: (a)

Required distance $=\sqrt{(2+6)^2+(7-9)^2}$

$=\sqrt{64+4}$

$=\sqrt{68}$

$=2\sqrt{17}$

80. If the distance between the points (3, 0) and (2, *y*) is $\sqrt{5}$, then the value (s) of *y* is **(2025)**

(a) 2, –2 (b) 2, 0
(c) 2, 1 (d) –2, 0

Solution: (a)

$\sqrt{(2-3)^2+(y-0)^2}=\sqrt{5}$

$\Rightarrow 1+y^2=5$

$\Rightarrow y^2=4$

$\therefore y=\pm 2$

81. The distance between the points $(-4,5)$ and $(-1,2)$ is **(2026)**

(a) 5 (b) $3\sqrt{2}$
(c) 6 (d) $2\sqrt{3}$

Solution: (b)

The required distance

$=\sqrt{(-1+4)^2+(2-5)^2}$

$=\sqrt{9+9}$

$=\sqrt{18}$

$=3\sqrt{2}$

82. The distance between the points $(-2,5)$ and $(5,-2)$ is **(2026)**

(a) $7\sqrt{2}$ (b) 14
(c) $2\sqrt{7}$ (d) 7

Solution:

The required distance

$= \sqrt{(5+2)^2 + (-2-5)^2}$

$= \sqrt{49+49}$

$= \sqrt{98}$

$= 7\sqrt{2}$

83. The coordinates of opposite vertices of the square *ABCD* are A $(-5,0)$ and C $(0,5)$. The length of diagonal of the square *ABCD* is **(2026)**

(a) 5 units (b) 10 units
(c) $\sqrt{10}$ units (d) $\sqrt{50}$ units

Solution: (d)

$AC = \sqrt{(0+5)^2 + (5-0)^2}$

$= \sqrt{25+25}$

$= \sqrt{50}$

84. A circle centred at $(-1,2)$ passes through the point $(0,3)$. Radius of the circle is **(2026)**

(a) $2\sqrt{2}$ (b) $\sqrt{2}$
(c) $\sqrt{26}$ (d) 1

Solution: (b)

The required radius is

$\sqrt{(0+1)^2 + (3-2)^2}$

$= \sqrt{1+1}$

$= \sqrt{2}$

85. In the given figure, a circle is centred at $(1,2)$ the diameter of the circle is **(2026)**

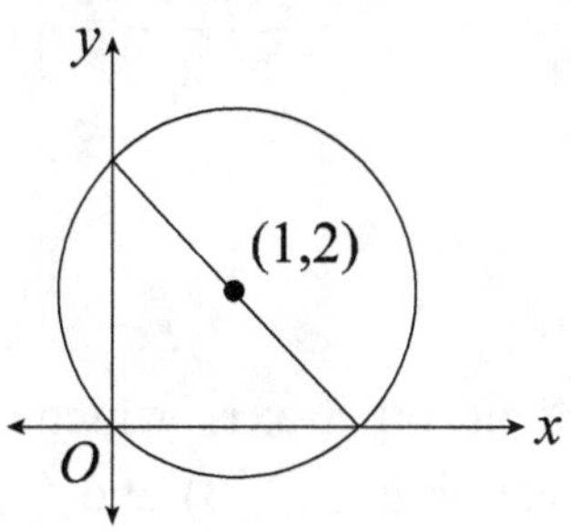

(a) 4 (b) $2\sqrt{2}$
(c) $\sqrt{5}$ (d) $2\sqrt{5}$

Solution: (d)

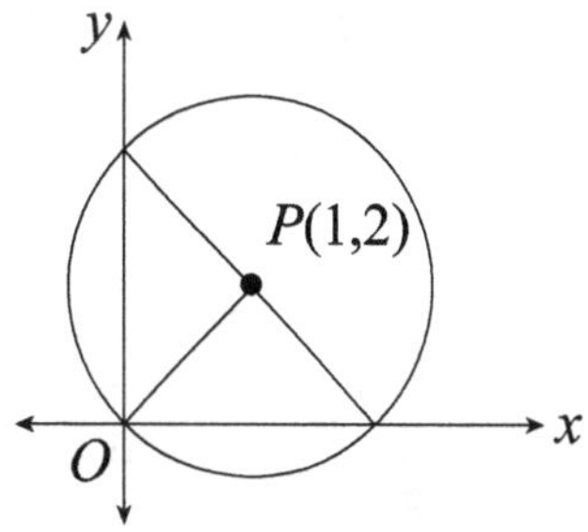

$OP = \sqrt{(1-0)^2 + (2-0)^2}$

$= \sqrt{1+4}$

$= \sqrt{5}$

$\therefore$ diameter of the circle $= 2 \times OP = 2\sqrt{5}$

Exercise 6.5

(i) The distance between (0, 0) and (2, 11) is:

(a) $5\sqrt{2}$ (b) $5\sqrt{3}$
(c) $3\sqrt{5}$ (d) $5\sqrt{5}$

(ii) The distance between (5, –2) and (–1, 3) is:

(a) $\sqrt{61}$ (b) $\sqrt{67}$
(c) $\sqrt{41}$ (d) $\sqrt{37}$

(iii) The point on the *x*-axis that is equidistant from (2, 3) and (4, 5) is:

(a) (3, 0) (b) (–4, 0)
(c) (5, 0) (d) (0, –1)

(iv) If the distance between (4, *k*) and (1, 0) is 5, then the possible value (s) of *k* is/are:

(a) 4
(b) –4
(c) Both (a) and (b)
(d) None of these

(v) If the distance between (5, 11) and (2, *x*) is 10, the possible value (s) of *x* is/are.

(a) $-11+\sqrt{91}, -11-\sqrt{91}$
(b) $11+\sqrt{91}, 11-\sqrt{91}$
(c) $11+\sqrt{91}, 11+\sqrt{91}$
(d) $-11+\sqrt{91}, 11-\sqrt{91}$

(vi) What will be the point of y-axis which will be equidistant from the points (–1, 0) and (3, 9)

(a) $\left(5, \frac{80}{18}\right)$ (b) $\left(1, \frac{89}{18}\right)$

(c) $\left(16, \frac{89}{18}\right)$ (d) $\left(0, \frac{89}{18}\right)$

(vii) If the point P (a, b) is equidistant from the points (3, 1) and (2, 0), then____

(a) $a+b+3=0$

(b) $a-b+3=0$

(c) $a+b=3$

(d) $a-b=3$

(viii) The point on y-axis which is at a distance 5 unit from the point (–5, 7) is

(a) (7, 0) (b) (0, 7)

(c) (1, 7) (d) (0, 8)

(ix) The points (1, 1), (–2, 7) and (3, –3) are

(a) Vertices of an equilateral triangle

(b) Collinear

(c) Vertices of an isosceles triangle

(d) none of these.

(x) $(2a, 4a)$, $(2a, 6a)$ and $(2a + 3a, 5a)$ are the vertices of

(a) Scalene triangle

(b) Isosceles triangle

(c) Equilateral triangle

(d) None of these

(xi) The perimeter of the triangle with vertices (0, 4), (0, 0) and (3, 0) is

(a) $2+\sqrt{5}$ (b) 10

(c) 12 (d) 24

(xii) The distance between the points (a cos20°,0) and (0, a cos 70°) is

(a) a (b) $2a$

(c) $3a$ (d) $4a$

Key Point The coordinates of the point P (x, y) which divides the line segment joining A (x_1, y_1) and B (x_2, y_2) internally in the ratio $m : n$ are given by $x = \frac{mx_2 + nx_1}{m+n}$, $y = \frac{my_2 + ny_1}{m+n}$

86. The coordinates of the point P dividing the line segment joining the points A (1, 3) and B (4, 6) in the ratio 2 : 1 are: **(2012)**

(a) (2, 4) (b) (3, 5)

(c) (4, 2) (d) (5, 3)

Solution: (b)

The coordinates of point P are

$$\left(\frac{2\times4+1\times1}{2+1}, \frac{2\times6+1\times3}{2+1}\right)$$

$$=\left(\frac{9}{3}, \frac{15}{3}\right)$$

$$= (3, 5)$$

87. The coordinate of the point dividing the line segment joining the points A (1, 3) and B (4, 6) in the ratio 2 : 1 is: _________ **(2020)**

Solution:

The required coordinates of the point are

$$\left(\frac{2\times4+1\times1}{2+1}, \frac{2\times6+1\times3}{2+1}\right)$$

$$= (3, 5)$$

88. Find the coordinates of the point on x-axis which divides the line segment joining the points (2, 3) and (5, –6) in the ratio 1 : 2. **(2020)**

Solution:

Required coordinates of the point on x–axis are

$$\left(\frac{1\times5+2\times2}{1+2}, \frac{1\times(-6)+2\times3}{1+2}\right)$$

$$\left(\frac{9}{3}, \frac{0}{3}\right)$$

$$= (3, 0)$$

89. The point on x-axis which divides the line segment joining (2, 3) and (6, –9) in the ratio 1 : 3 is **(2020)**

(a) (4, –3) (b) (6, 0)

(c) (3, 0) (d) (0, 3)

Solution: (c)

Required coordinates of the point on x-axis are

$$\left(\frac{1\times6+3\times2}{1+3},\frac{1\times(-9)+3\times3}{1+3}\right)$$

$$=\left(\frac{12}{4},\frac{0}{4}\right)$$

$= (3, 0)$

90. Point P divides the line segment joining the points A (4, –5) and B (1, 2) in the ratio 5 : 2. Coordinates of point P are **(2024)**

(a) $\left(\frac{5}{2},\frac{-3}{2}\right)$ (b) $\left(\frac{11}{7},0\right)$

(c) $\left(\frac{13}{7},0\right)$ (d) $\left(0,\frac{13}{7}\right)$

Solution: (c)

Coordinates of P are

$$\left(\frac{5\times1+2\times4}{5+2},\frac{5\times2+2\times(-5)}{5+2}\right)$$

$$=\left(\frac{13}{7},0\right)$$

91. The point P which divides the line segment joining the points A (2, –5) and B (5, 2) in the ratio 2 : 3 lie in the quadrant **(2011)**

(a) I (b) II

(c) III (d) IV

Solution: (d)

Coordinates of P are

$$\left(\frac{2\times5+3\times2}{2+3},\frac{2\times2+3\times(-5)}{2+3}\right)$$

$$=\left(\frac{16}{5},\frac{-11}{5}\right)$$

$\therefore$ P lies on IV th quadrant.

92. If the point P (k, 0) divides the line segment joining the points A (2, –2) and B (–7, 4) in the ratio 1 : 2, then the value of k is **(2020)**

(a) 1 (b) 2

(c) –2 (d) –1

Solution: (d)

$$k=\frac{1\times(-7)+2\times2}{1+2} \qquad \left[\because x=\frac{mx_2+nx_1}{m+n}\right]$$

$$\Rightarrow k=\frac{-3}{3}$$

$$\therefore k=-1$$

93. If the point P (6, 2) divides the line segment joining the points A (6, 5) and B (4, y) in the ratio 3 : 1, then the value of y is **(2020)**

(a) 4 (b) 3

(c) 2 (d) 1

Solution: (d)

$$2=\frac{3y+1\times5}{3+1} \qquad \left[\because y=\frac{my_2+ny_1}{m+n}\right]$$

$$\Rightarrow 2=\frac{3y+5}{4}$$

$$\Rightarrow 3y+5=8$$

$$\Rightarrow 3y=3$$

$$\therefore y=1$$

94. If the point C (k, 4) divides the line segment joining the points A (2, 6) and B (5, 1) in the ratio 2 : 3, the value of k is **(2020)**

Solution:

$$k=\frac{2\times5+3\times2}{2+3} \qquad \left[\because x=\frac{mx_2+nx_1}{m+n}\right]$$

$$\Rightarrow k=\frac{16}{5}$$

95. In what ratio, does x-axis divide the line segment joining the points A (3, 6) and B (–12, –3) ? **(2023)**

(a) 1 : 2 (b) 1 : 4

(c) 4 : 1 (d) 2 : 1

Solution: (d)

Let the ratio be $m : n$

$$\because 0=\frac{m(-3)+n\times6}{m+n} \qquad \left[\because y=\frac{mx_2+nx_1}{m+n}\right]$$

$$\Rightarrow -3m+6n=0$$

$$\Rightarrow -3m=-6n$$

$$\Rightarrow \frac{m}{n}=2$$

$$\therefore m:n=2:1$$

96. The ratio in which the x-axis divides the line segment joining the points (–2, 3) and (6, –7) is **(2023)**

(a) 1 : 3 (b) 3 : 7
(c) 7 : 3 (d) 1 : 2

Solution: (b)

$$\frac{m\times(-7)+n\times 3}{m+n}=0 \qquad \left[\because y=\frac{my_2+ny_1}{m+n}\right]$$

$\Rightarrow -7m + 3n = 0$

$\Rightarrow -7m = -3n$

$\Rightarrow \frac{m}{n}=\frac{3}{7}$

$\therefore m : n = 3 : 7$

97. In what ratio, does x-axis divide the line segment joining the points A (2, –3) and B (5, 6) ? **(2023)**

(a) 2 : 3 (b) 2 : 1
(c) 3 : 4 (d) 1 : 2

Solution: (d)

$$\frac{m\times 6+n\times(-3)}{m+n}=0 \qquad \left[\because y=\frac{my_2+ny_1}{m+n}\right]$$

$\Rightarrow 6m - 3n = 0$

$\Rightarrow 6m = 3n$

$\Rightarrow \frac{m}{n}=\frac{1}{2}$

$\therefore m : n = 1 : 2$

98. y-axis divides the line segment joining the points (–6, 2) and (2, –6) in the ratio: **(2023)**

(a) 1 : 3 (b) 3 : 2
(c) 3 : 1 (d) 2 : 3

Solution:

$$\frac{2m-6n}{m+n}=0 \qquad \left[\because x=\frac{mx_2+nx_1}{m+n}\right]$$

$\Rightarrow 2m - 6n = 0$

$\Rightarrow 2m = 6n$

$\Rightarrow \frac{m}{n}=\frac{6}{2}$

$\therefore m : n = 3 : 1$

99. The ratio in which the line segment joining the points A(–4, 8) and B(5, –10) divided by the point p(–2, 4) is **(2025)**

(a) 1 : 3 (b) 3 : 4
(c) 2 : 7 (d) 2 : 5

Solution: (c)

Let the ratio be $m : n$

$\therefore -2=\frac{5m-4n}{m+n}$

$\Rightarrow -2m - 2n = 5m - 4n$

$\Rightarrow -7m = -2n$

$\Rightarrow \frac{m}{n}=\frac{2}{7}$

$\therefore m : n = 2 : 7$

100. The point (x, 0) divides the line segment joining the points (–4, 5) and (0, –10) is the ratio **(2025)**

(a) 1 : 3 (b) 2 : 1
(c) 1 : 1 (d) 1 : 2

Solution: (d)

Let the ratio be $m : n$

$\therefore 0=\frac{m(-10)+n5}{m+n}$

$\Rightarrow -10m + 5n = 0$

$\Rightarrow -10m = -5n$

$\Rightarrow \frac{m}{n}=\frac{5}{10}$

$\therefore m : n = 1 : 2$

101. The point P divides the line segment AB in the ratio 3 : 1 as shown below: **(2025)**

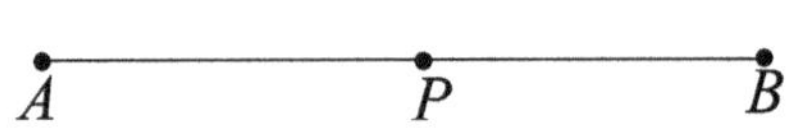

The value of $\frac{AB}{PB}$ is

(a) 3 (b) $\frac{1}{4}$
(c) 4 (d) $\frac{1}{3}$

Solution: (c)

$\frac{AP}{PB}=\frac{3}{1}$

$\Rightarrow AP = 3\ PB$

$AB = AP + PB$

$\Rightarrow AB = 3\ PB + PB$

$\Rightarrow AB = 4\ PB$

$\frac{AB}{PB} = 4$

102. In the following figure P and Q are points of trisection of line segment AB; the value of $\frac{AB}{PB} =$ **(2025)**

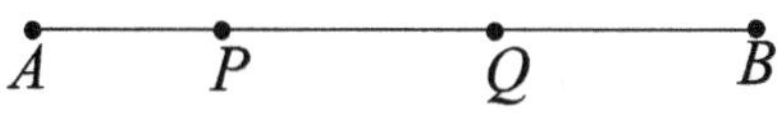

(a) 1 (b) 1.5

(c) $\frac{2}{3}$ (d) 2

Solution: (b)

$AP = PQ = BQ$

$\therefore \frac{AB}{PB} = \frac{3AP}{2AP} = \frac{3}{2} = 1.5$

103. In the figure given below, points P, Q, R divides the line segment AB in four equal parts

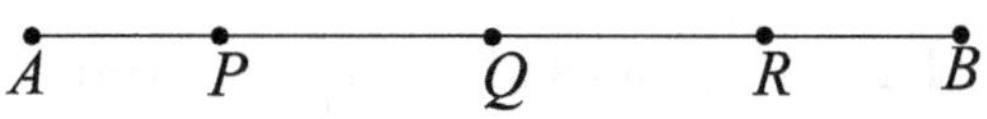

The point Q divides PB in the ratio: **(2025)**

(a) 1 : 3 (b) 2 : 3

(c) 1 : 2 (d) 1 : 1

Solution: (c)

$PQ : BQ$

$= 1 : 2$

104. The line segment joining the points $P(-4,-2)$ and $Q(10,4)$ is divided by $y-axis$ in the ratio **(2026)**

(a) 2:5 (b) 1:2

(c) 2:1 (d) 5:2

Solution:

Let the required ratio be $m : n$

$\therefore \frac{4m-2n}{m+n} = 0$

$\Rightarrow 4m - 2n = 0$

$\Rightarrow 4m = 2n$

$\Rightarrow \frac{m}{n} = \frac{1}{2}$

$\therefore m : n = 1 : 2$

Exercise 6.6

(i) Find the ratio in which the line joining the points (6, 4) and (1, –7) is divided by x-axis

(a) 3 : 1 (b) 2 : 3

(c) 4 : 7 (d) 9 : 4

(ii) The point which divides the line segment joining the points (7, –6) and (3, 4) in ratio 1 : 2 internally lies in the

(a) I quadrant (b) II quadrant

(c) III quadrant (v) IV quadrant

(iii) The ratio in which the point $P\left(\frac{3}{4}, \frac{5}{12}\right)$ divides the line segment joining the points $A\left(\frac{1}{2}, \frac{3}{2}\right)$ and B (2, –5) is:

(a) 1 : 5 (b) 5 : 1

(c) 2 : 3 (d) 3 : 2

(iv) The points which divides the line segment of points P (–1, 7) and Q (4, –3) in the ratio of 2 : 3 is:

(a) (–1, 3) (b) (–1, –3)

(c) (1, –3) (d) (1, 3)

(v) The ratio in which the line segment joining the points P (–3, 10) and Q (6, –8) is divided by R (–1, 6) is

(a) 1 : 3 (b) 3 : 4

(c) 2 : 7 (d) 2 : 5

(vi) The ratio in which the point P (–3, a) divides the line segment joining the points A (–5, 4) and B (–2, 3) is

(a) 2 : 1 (b) 1 : 2

(c) 3 : 2 (d) 4 : 5

(vii) The ratio in which the line segment joining A (3, 4) and B (–2, 1) is divided by the y–axis is

(a) 2 : 3 (b) 3 : 2

(c) 2 : 5 (d) 4 : 3

(viii) The line segment joining the points (3, –1) and (–6, 5) is trisected. The coordinates of point of trisection are

(a) (1, 1) (b) (–3, 3)
(c) (0, 2) (d) (2, –2)

(ix) The ratio in which the point (1, 2) divides the line segment joining (0, 0) and (3, 6) is:

(a) 1 : 2 (b) 3 : 1
(c) 3 : 2 (d) 2 : 3

(x) The ratio in which the *y*-axis divides the line segment joining (2, –3) and (–5, 6) is:

(a) 2 : 3 (b) 3 : 4
(c) 2 : 5 (d) 5 : 7

Key Point The mid point of the join of $A\ (x_1, y_1)$ and $B\ (x_2, y_2)$ is given by $C\left(\frac{x_1+x_2}{2}, \frac{y_1+y_2}{2}\right)$.

105. If *P* (2, p) is the mid-point of the line segment joining the points *A* (6, –5) and *B* (–2, 11), find the value of *p*. **(2010)**

Solution:

$$p=\frac{-5+11}{2} \qquad \left[\because y=\frac{y_1+y_2}{2}\right]$$

$\Rightarrow p=3$

106 The mid-point of segment *AB* is the point *P* (0, 4). If the coordinates of *B* are (–2, 3), then the coordinates of A are **(2011)**

(a) (2, 5) (b) (–2, –5)
(c) (2, 9) (d) (–2, 11)

Solution: (a)

Let the coordinates of *A* are (*a*, *b*)

$\therefore \left(\frac{a-2}{2}, \frac{b+3}{2}\right)=(0,4)$

$\therefore \frac{a-2}{2}=0$

$\Rightarrow a-2=0$

$\Rightarrow a=2$

and $\frac{b+3}{2}=4$

$\Rightarrow b+3=8$

$\therefore b=5$

∴ Coordinates of *A* are (2, 5)

107. If the coordinates of one end of a diameter of a circle are (2, 3) and the coordinates of its centre are (–2, 5), then the coordinates of the other end of the diameter are **(2012)**

(a) (–6, 7) (b) (6, –7)
(c) (6, 7) (d) (–6, –7)

Solution: (a)

Let the coordinates of other end be (*a*, *b*)

$\therefore \left(\frac{a+2}{2}, \frac{b+3}{2}\right)=(-2,5)$

$\therefore \frac{a+2}{2}=-2$

$\Rightarrow a+2=-4$

$\Rightarrow a=-6$

and $\frac{b+3}{2}=5$

$\Rightarrow b+3=10$

$\Rightarrow b=7$

∴ The coordinates of other end is (–6, 7)

108. Find the coordinates of a point *A*, where *AB* is diameter of a circle whose centre is (2, –3) and *B* is the point (1, 4). **(2019)**

Solution:

Let the coordinates of *A* be (*a*, *b*)

$\therefore \left(\frac{a+1}{2}, \frac{b+4}{2}\right)=(2,-3)$

$\therefore \frac{a+1}{2}=2$

$\Rightarrow a+1=4$

$\Rightarrow a=3$

and $\frac{b+4}{2}=-3$

$\Rightarrow b+4=-6$

$\Rightarrow b=-10$

∴ The coordinate of *B* are (3, –10)

109. Find the coordinates of a point *A*, where *AB* is diameter of the circle with centre (–2, 2) and *B* is the point with coordinate (3, 4) **(2019)**

Solution:

Let the coordinates of A be (a, b)

$\therefore \frac{a+3}{2} = -2$

$\Rightarrow a + 3 = -4$

$\Rightarrow a = -7$

and $\frac{b+4}{2} = 2$

$\Rightarrow b + 4 = 4$

$\Rightarrow b = 0$

$\therefore$ The coordinates of A is $(-7, 0)$

110. If (a, b) is the mid-point of the line segment joining the points A $(10, -6)$ and B $(k, 4)$ and $a - 2b = 18$, the value of k is **(2020)**

(a) 30 (b) 22

(c) 4 (d) 40

Solution: (b)

$a = \frac{10+k}{2}$ and $b = \frac{-6+4}{2} = -1$

Now, $a - 2b = 18$

$\Rightarrow \frac{10+k}{2} - 2(-1) = 18$

$\Rightarrow \frac{10+k}{2} + 2 = 18$

$\Rightarrow \frac{10+k}{2} = 16$

$\Rightarrow 10 + k = 32$

$\therefore k = 22$

111. Point $P\left(\frac{a}{8}, 4\right)$ is the mid-point of the line segment joining the points A $(-5, 2)$ and B $(4, 6)$, the value of 'a' is **(2020)**

(a) –4 (b) 4

(c) –8 (d) –2

Solution: (a)

$\frac{a}{8} = \frac{-5+4}{2}$, $\left[\because x = \frac{x_1 + x_2}{2}\right]$

$\Rightarrow \frac{a}{8} = \frac{-1}{2}$

$\Rightarrow a = -4$

112. The centre of a circle whose end points of a diameter are $(-6, 3)$ and $(6, 4)$ is **(2020)**

(a) $(8, -1)$ (b) $(4, 7)$

(c) $\left(0, \frac{7}{2}\right)$ (d) $\left(4, \frac{7}{2}\right)$

Solution: (c)

The centre of the circle $= \left(\frac{-6+6}{2}, \frac{3+4}{2}\right)$

$= \left(0, \frac{7}{2}\right)$

113. If the centre of a circle is $(3, 5)$ and end points of a diameter are $(4, 7)$ and $(2, y)$, then the value of y is **(2020)**

(a) 3 (b) –3

(c) 7 (d) 4

Solution: (a)

$\frac{7+y}{2} = 5$ $\left[\because y = \frac{y_1 + y_2}{2}\right]$

$\Rightarrow 7 + y = 10$

$\Rightarrow y = 3$

114. The mid-point of the line segment AB is P $(0, 4)$. If the coordinates of B are $(-2, 3)$ then the coordinates of A are **(2020)**

(a) $(2, 5)$ (b) $(-2, -5)$

(c) $(2, 9)$ (d) $(-2, 11)$

Solution: (a)

Let the coordinate of A are (a, b)

$\therefore \left(\frac{a-2}{2}, \frac{b+3}{2}\right) = (0, 4)$

$\therefore \frac{a-2}{2} = 0$

$\Rightarrow a - 2 = 0$

$\Rightarrow a = 2$

$\frac{b+3}{2} = 4$

$\Rightarrow b + 3 = 8$

$\Rightarrow b = 5$

$\therefore$ The coordinate of A are $(2, 5)$

115. If $P(-1, 1)$ is the mid point of the line segment joining $A(-3, b)$ and $B(1, b+4)$, then b is equal to **(2020)**

(a) 1 (b) –1

(c) 2 (d) 0

Solution: (b)

$$\frac{b+(b+4)}{2}=1 \qquad \left[\because y=\frac{y_1+y_2}{2}\right]$$

$$\Rightarrow 2b+4=2$$

$$\Rightarrow 2b=-2$$

$$\therefore b=-1$$

116. The mid – point of the line segment joining the points (–3, –3) and (–3, 3) is_______ **(2020)**

Solution:

$$\text{Mid-point}=\left(\frac{-3-3}{2},\frac{-3+3}{2}\right)$$

$$=(-3, 0)$$

117. If (3, –6) is the mid point of the line segment joining (0, 0) and (x, y) then the point (x, y) is **(2020)**

(a) (–3, 6) (b) (6, –6)

(c) (6, –12) (d) $\left(\frac{3}{2},-3\right)$

Solution: (c)

$$(3,-6)=\left(\frac{0+x}{2},\frac{0+y}{2}\right)$$

$$\Rightarrow (3,-6)=\left(\frac{x}{2},\frac{y}{2}\right)$$

$$\therefore \frac{x}{2}=3$$

$$\Rightarrow x=6$$

$$\frac{y}{2}=-6$$

$$\Rightarrow y=-12$$

118. If (2, 4) is the mid-point of the line segment joining (6, 3) and $(a, 5)$, then the value of a is **(2023)**

(a) 2 (b) 4

(c) –4 (d) –2

Solution: (d)

$$2=\frac{6+a}{2} \qquad \left[x=\frac{x_1+x_2}{2}\right]$$

$$\Rightarrow 6+a=4$$

$$\Rightarrow a=-2$$

119. The mid-point of the line segment joining the points (–1, 3) and $\left(8,\frac{3}{2}\right)$ is: **(2024)**

(a) $\left(\frac{7}{2},\frac{-3}{4}\right)$ (b) $\left(\frac{7}{2},\frac{9}{2}\right)$

(c) $\left(\frac{9}{2},\frac{-3}{4}\right)$ (d) $\left(\frac{7}{2},\frac{9}{4}\right)$

Solution: (d)

$$\text{Mid-point}=\left(\frac{-1+8}{2},\frac{3+\frac{3}{2}}{2}\right)$$

$$=\left(\frac{7}{2},\frac{9}{4}\right)$$

120. If $C(1,-1)$ is the mid-point of the line segment AB joining points $A(4, x)$ and $B(-2, 4)$, then value of x is **(2024)**

(a) 5 (b) –5

(c) 6 (d) –6

Solution: (d)

$$\frac{x+4}{2}=-1 \qquad \left[\because x=\frac{x_1+x_2}{2}\right]$$

$$\Rightarrow x+4=-2$$

$$\Rightarrow x=-6$$

121. The mid-point of the line segment AB joining $A(-2, 8)$ and $B(-6, 4)$ is: **(2024)**

(a) (2, 6) (b) (–4, 12)

(c) (–4, 6) (d) (4, –6)

Solution: (c)

$$\text{Mid-point}=\left(\frac{-2-6}{2},\frac{8+4}{2}\right)$$

$$=(-4, 6)$$

122. If (a, 6) is the mid-point of the line segment joining the points A (–7, 7) and B (–3, 5), then the value of 'a' is: **(2024)**

(a) 5 (b) –4

(c) 2 (d) –5

Solution: (d)

$a = \frac{-7-3}{2}$ $\left[\because x = \frac{x_1 + x_2}{2}\right]$

$\Rightarrow a = -5$

123. If A (1, 2), B (4, 3) and C (6, 6) are the three vertices of a parallelogram $ABCD$, find the coordinate of the fourth vertex D. **(2010)**

Solution:

Let, the coordinate of D are (a, b)

$\because$ Mid point of AC and mid point of BD is same.

$\therefore \left(\frac{1+6}{2}, \frac{2+6}{2}\right) = \left(\frac{4+a}{2}, \frac{3+b}{2}\right)$

$\therefore \frac{4+a}{2} = \frac{7}{2}$

$\Rightarrow a + 4 = 7$

$\therefore a = 3$

and $\frac{3+b}{2} = \frac{2+6}{2}$

$\Rightarrow \frac{3+b}{2} = 4$

$\Rightarrow 3 + b = 8$

$\Rightarrow b = 5$

124. The coordinates of the vertex A of a rectangle $ABCD$ whose three vertices are given is B (0, 0), C (3, 0) and D (0, 4) are: **(2023)**

(a) (4, 0) (b) (0, 3)

(c) (3, 4) (d) (4, 3)

Solution: (c)

Mid point of AB = mid point DC

$\Rightarrow \left(\frac{a+0}{2}, \frac{b+0}{2}\right) = \left(\frac{3+0}{2}, \frac{0+4}{2}\right)$

$\Rightarrow \left(\frac{a}{2}, \frac{b}{2}\right) = \left(\frac{3}{2}, 2\right)$

$\therefore \frac{a}{2} = \frac{3}{2}$

$\Rightarrow a = 3$

and $\frac{b}{2} = 2$

$\Rightarrow b = 4$

$\therefore$ The coordinates of A are (3, 4)

125. The fourth vertex D of a parallelogram $ABCD$ whose three vertices are A (–2, 3), B (6, 7) and C (8, 3) is: **(2024)**

(a) (0, 1) (b) (0, –1)

(c) (–1, 0) (d) (1, 0)

Solution: (b)

Mid point of AC = mid point of BD

$\Rightarrow \left(\frac{-2+8}{2}, \frac{3+3}{2}\right) = \left(\frac{6+a}{2}, \frac{7+b}{2}\right)$

$\Rightarrow (3,3) = \left(\frac{6+a}{2}, \frac{7+b}{2}\right)$

$\therefore \frac{6+a}{2} - 3$

$\Rightarrow 6 + a = 6$

$\Rightarrow a = 0$

and $\frac{7+b}{2} = 3$

$\Rightarrow 7 + b = 6$

$\Rightarrow b = -1$

$\therefore$ The coordinates of D are (0, –1)

126. The centre of a circle is at (2, 0). If one end of a diameter is at (6, 0), then the other end is at: **(2024)**

(a) (0, 0) (b) (4, 0)

(c) (–2, 0) (d) (–6, 0)

Solution:

Let the coordinate of other end be (a, b)

$\therefore \left(\frac{a+6}{2}, \frac{b+0}{2}\right) = (2,0)$

$\therefore \frac{a+6}{2} = 2$

$\Rightarrow a + 6 = 4$

$\Rightarrow a = -2$

and $\frac{b}{2} = 0$

$\Rightarrow b = 0$

$\therefore$ The coordinate of the other end is (–2, 0)

127. *AD* is a median of ΔABC with vertices *A* (5, –6), *B* (6, 4) and *C* (0, 0). Length *AD* is equal to: **(2024)**

(a) $\sqrt{68}$ units (b) $2\sqrt{15}$ units
(c) $\sqrt{101}$ units (d) 10 units

Solution: (a)

Coordinates of $D = \left(\frac{6+0}{2}, \frac{4+0}{2}\right)$
$= (3, 2)$

$\therefore AD = \sqrt{(3-5)^2 + (2+6)^2}$
$= \sqrt{4+64}$
$\sqrt{68}$ units

128. The mid-point of the line segment joining the points *P*(–4, 5) and *Q*(4, 6) lies on: **(2025)**

(a) *x*-axis
(b) *y*-axis
(c) Origin
(d) Neither *x*-axis nor *y*-axis

Solution: (b)

Mid point of *PQ* is $\left(\frac{-4+4}{2}, \frac{5+6}{2}\right)$
$= \left(0, \frac{11}{2}\right)$

Which is lie on *y*-axis

129. The mid-point of the line segment joining points (1, 3) and (1, –3) lies: **(2025)**

(a) at the origin
(b) in the second quadrant
(c) on *x*-axis
(d) on *y*-axis

Solution: (c)

Mid-point of (1, 3) and (1, –3) is $\left(\frac{1+1}{2}, \frac{3-3}{2}\right)$
$= (1, 0)$

130. The coordinates of the mid-points of The line segment joining points *A* $(p-1, q+1)$ and *B* $(p+1, q-1)$ are given by : **(2026)**

(a) (p, q) (b) $(2p, 2q)$
(c) $\left(\frac{p+1}{2}, \frac{q+1}{2}\right)$ (d) $\left(\frac{p-1}{2}, \frac{q-1}{2}\right)$

Solution: (a)

Mid point of AB is $\left(\frac{p-1+p+1}{2}, \frac{q+1+q-1}{2}\right)$
$= \left(\frac{2p}{2}, \frac{2q}{2}\right)$
$= (p, q)$

131. The mid-point of the segment joining the points $(5, -4)$ and $(6, 4)$ has on: **(2026)**

(a) *x*-axis
(b) *y*-axis
(c) origin
(d) neither *x*-axis is nor *y*-axis

Solution: (a)

Mid point of the line segment is

$\left(\frac{5+6}{2}, \frac{-4+4}{2}\right) = \left(\frac{11}{2}, 0\right)$ which is lie on *x*-axis.

Exercise 6.7

(i) The mid point of the line segment joining the points *A* (2, 8) and *B* (–6, 4) is

(a) (2, 2) (b) (–2, 6)
(c) (2, –6) (d) (0, 4)

(ii) If *P* (2, –2) is the mid point of the line segment joining *A* (0, *b*) and *B* (4, *b* + 2), then *b* = ?

(a) 3 (b) 0
(c) –2 (d) –3

(iii) If $P\left(\frac{a}{5}, 2\right)$ is the midpoint of the line segment joining *A* (4, 6) and *B* (2, –2), then *a* = ?

(a) 5 (b) 10
(c) 12 (d) 15

(iv) If *A* (–1, 0), *B* (5, –2) and *C* (8, 2) be the vertices of ΔABC and *AD* is a median, then the coordinates of *D* are

(a) $\left(\frac{5}{2}, 0\right)$ (b) $(4, 0)$
(c) $\left(\frac{13}{2}, 0\right)$ (d) $(-4, 0)$

(v) The three vertices of a parallelogram $ABCD$ are A (9, 0), B (9, 6) and C(–9, 6). The fourth vertex D is

(a) (9, 0) (b) (–9, 0)
(c) (0, 9) (d) (0, –9)

(vi) If the points A (2, 3) and B (x, –4) lie on a circle with the centre O (–1, 2), then x =

(a) 4 (b) –4
(c) 2 (d) –3

(vii) Find the coordinates of the points which divide the join of A (–4, 0) and B (0, 6) in equal parts

(a) (–2, 0) (b) (–2, 3)
(c) (–4, 6) (d) (2, 3)

(viii) The mid point of the line segment joining (a, 2) and B (–3, b + 1) is C (4, a – 2). Find the values of a and b

(a) $a = 11, b = 18$
(b) $a = 11, b = 15$
(c) $a = 9, b = 11$
(d) $a = 10, b = 12$

(ix) BE is a median of ΔABC with vertices A (0, –1), B (2, 1) and C(0, 0). Length BE is equal to

(a) 2 units (b) $2\sqrt{11}$ units
(c) $\sqrt{10}$ units (d) 3 units

(x) If (1, 2), (4, y), (x, 6) and (3, 5) are the vertices of a parallelogram taken in order, find x and y

(a) $x = 6, y = 3$ (b) $x = 3, y = 6$
(c) $x = 4, y = 4$ (d) $x = 0, y = -1$

Key Point (i) The area of a ΔABC with vertices A (x_1, y_1), B (x_2, y_2) and C (x_2, y_2) is given by ar (ΔABC)

$$= \left|\frac{1}{2}\{x_1(y_2 - y_3) + x_2(y_3 - y_1) + x_3(y_1 - y_2)\}\right|$$

(ii) Three points A (x_1, y_1), B (x_2, y_2) and C $(x_3, y_3$ are collinear, if ar (ΔABC) = 0

i. e $x_1(y_2 - y_3) + x_2(y_3 - y_1) + x_3(y_1 - y_2) = 0$

132. The value of p, for which the points A (3, 1), B (5, p) and C (7, –5) are collinear is **(2020)**

(a) –2 (b) 2
(c) –1 (d) 1

Solution: (a)

The given points A, B, C are collinear

$$\Rightarrow x_1(y_2 - y_3) + x_2(y_3 - y_1) + x_3(y_1 - y_2) = 0$$
$$\Rightarrow 3(p+5) + 5(-5-1) + 7(1-p) = 0$$
$$\Rightarrow 3p + 15 - 30 + 7 - 7p = 0$$
$$\Rightarrow -4p - 8 = 0$$
$$\Rightarrow -4p = 8$$
$$\therefore p = -2$$

133. The value of k for which the points A (0, 1), B (2, k) and C (4, –5) are collinear is **(2020)**

(a) 2 (b) –2
(c) 0 (d) 4

Solution: (b)

The given points A, B, C are collinear

$$\Rightarrow x_1(y_2 - y_3) + x_2(y_3 - y_1) + x_3(y_1 - y_2) = 0$$
$$\Rightarrow 0(k+5) + 2(-5-1) + 4(1-k) = 0$$
$$\Rightarrow -12 + 4 - 4k = 0$$
$$\Rightarrow -4k = 8$$
$$\therefore k = -2$$

134. It is being given that the points A (1, 2), B (0, 0) and C (a, b) are collinear, which of the following relations between a and b is true? **(2020)**

(a) $a = 2b$ (b) $2a = b$
(c) $a + b = 0$ (d) $a - b = 0$

Solution: (b)

The given points A, B, C are collinear

$$\Rightarrow x_1(y_2 - y_3) + x_2(y_3 - y_1) + x_3(y_1 - y_2) = 0$$
$$\Rightarrow 1(0-b) + 0(b-2) + a(2-0) = 0$$
$$\Rightarrow -b + 2a = 0$$
$$\Rightarrow 2a = b$$

135. If points A (–3, 12), B (7, 6) and C (x, 9) are collinear, then the value of x is ______ **(2020)**

Solution:

The given points A, B, C are collinear

$$\Rightarrow x_1(y_2 - y_3) + x_2(y_3 - y_1) + x_3(y_1 - y_2) = 0$$

$\Rightarrow -3(6-9)+7(9-12)+x(12-6)=0$

$\Rightarrow 9-21+6x=0$

$\Rightarrow 6x-12=0$

$\Rightarrow 6x=12$

$\therefore x=2$

136. The area of a triangle with vertices A (5, 0), B (8, 0) and C (8, 4) in square unit is **(2020)**

(a) 20 (b) 12

(c) 6 (d) 16

Solution: (c)

$ar(\Delta ABC)=\frac{1}{2}[x_1(y_2-y_3)+x_2(y_3-y_1)+x_3(y_1-y_2)]$

$=\frac{1}{2}[5(0-4)+8(4-0)+8(0-0)]$

$=\frac{1}{2}(-20+32)$

$=\frac{1}{2}\times 12=6$

137. The area of a triangle formed with the origin and the points (4, 0) and (0, 6) is______ **(2020)**

Solution:

Required area

$=\frac{1}{2}[x_1(y_2-y_3)+x_2(y_3-y_1)+x_3(y_1-y_2)]$

$=\frac{1}{2}[0(0-6)+4(6-0)+0(0-0)]$

$=\frac{1}{2}\times 24$

= 12 sq. unit.

Exercise 6.8

(i) The value of p for which the points (–5, 1), (1, p) and (4, –2) are collinear.

(a) –2 (b) –1

(c) 0 (d) 1

(ii) The value of p for which the points (–1, 3), (2, p) and (5, –1) are collinear is

(a) 1 (b) –1

(c) 2 (d) 3

(iii) If the points (a, b), (–5, 7) and (–4, 5) are collinear, then

(a) $2a+b=0$

(b) $b=2a$

(c) $2a+b+1=0$

(d) $2a+b+3=0$

(iv) If the points A (a, 0), B (0, b) and C (1, 1) are collinear, then

(a) $a+b=1$ (b) $a^2+b^2=2$

(c) $ab+b=a$ (d) $a+b=ab$

(v) The area of ΔABC whose vertices are A (–5, 7), B (–4, –5) and C (4, 5) is (in sq. unit)

(a) 64 (b) 58

(c) 53 (d) 48

138. The area of the triangle from by the line $\frac{x}{a}+\frac{y}{b}=1$ with the coordinate axes is: **(2020)**

(a) ab (b) $\frac{1}{2}ab$

(c) $\frac{1}{4}ab$ (d) $2ab$

Solution: (b)

The given line intersects the coordinate axises at (a, 0) and (0, b)

$\therefore$ The area of the triangle $=\frac{1}{2}ab$ sq. unit.

139. The lines represented by the linear equalties $y = x$ and $x = 4$ intersect at P. The coordinates of the point P are: **(2023)**

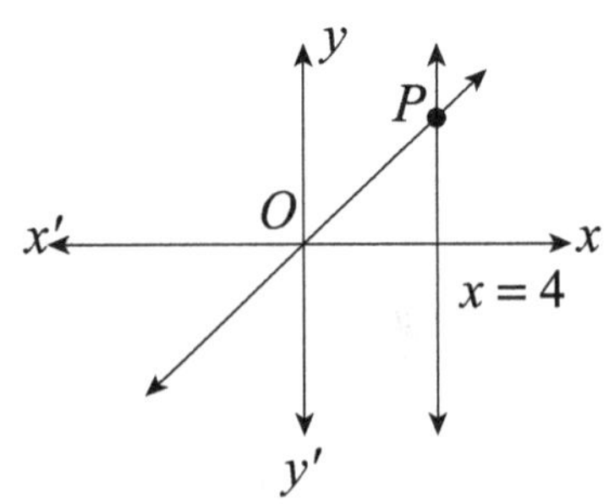

(a) (4, 0) (b) (4, 4)

(c) (0, 4) (d) (–4, 4)

Solution: (b)

The coordinate of P are (4, 4)

140. The perpendicular bisector of the line segment joining the points $A(-1, 3)$ and $B(2, 4)$ cuts the y-axis at: **(2024)**

(a) (0, 5) (b) (0, –5)
(c) (0, 4) (d) (0, –4)

Solution: (a)

Let CP be the perpendicular bisector of AB

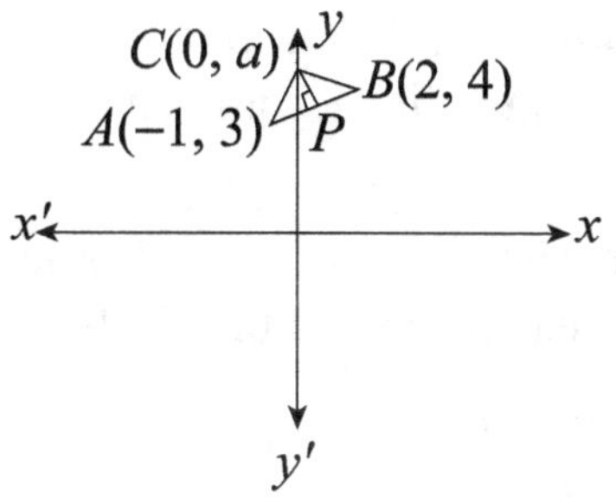

Now, $CA = CB$

$\Rightarrow \sqrt{(0+1)^2+(a-3)^2} = \sqrt{(0-2)^2+(a-4)^2}$

$\Rightarrow 1+(a-3)^2 = 4+(a-4)^2$

$\Rightarrow 1+a^2-6a+9 = 4+a^2-8a+16$

$\Rightarrow 2a = 10$

$\Rightarrow a = 5$

$\therefore$ Coordinate of C are (0, 5)

141. In Fig.3, the area of triangle ABC (in sq. units) is: **(2013)**

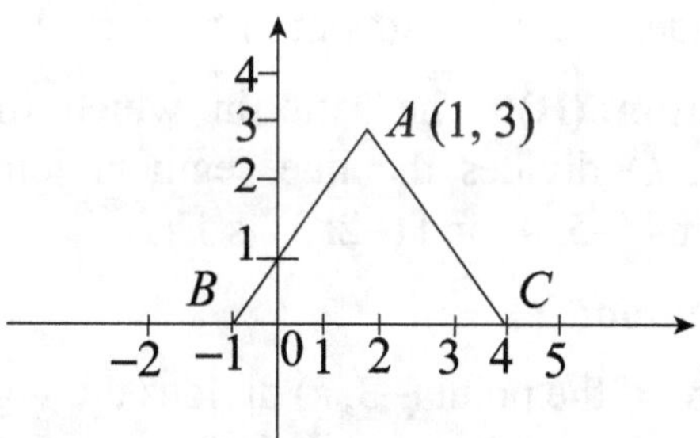

Fig. 3

(a) 15 (b) 10
(c) 7.5 (d) 2.5

Solution: (c)

Height of the triangle ABC = 3 units. Base of the triangle ABC = 5 units

$\therefore$ ar $(\Delta ABC) = \frac{1}{2}\times 3\times 5$

$= 7.5$ sq. units

142. In the given figure, ΔABC in an equilateral triangle AD is a median of the triangle joining the points A $\left(0, \frac{5\sqrt{3}}{2}\right)$, D $(0,0)$. Points B and C are (in same order): **(2026)**

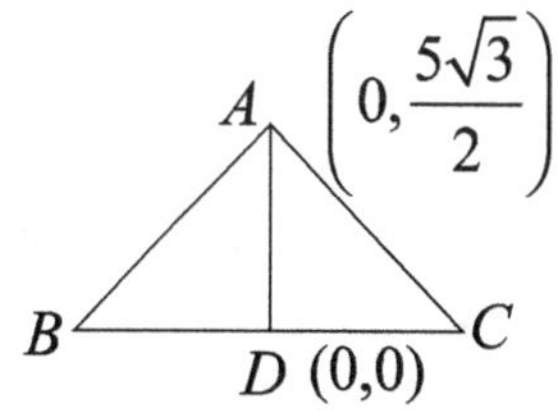

(a) $(-5,0),(5,0)$ (b) $\left(\frac{-5}{2},0\right),\left(\frac{5}{2},0\right)$

(c) $(-10,0),(10,0)$ (d) $(-5\sqrt{3},0),(5\sqrt{3},0)$

Solution:

$AD = \frac{\sqrt{3}}{2}BC$

$\Rightarrow \frac{5\sqrt{3}}{2} = \frac{\sqrt{3}}{2}BC$

$\therefore BC = 5$ unit

BC lies on x-axis

$\therefore$ coordinate of B is $\left(-\frac{5}{2},0\right)$ and coordinate of C is $\left(\frac{5}{2},0\right)$

Exercise 6.9

(i) The perpendicular bisector of the line segment joining the points $A(1, 5)$ and $B(4, 6)$ cuts the y axis at:

(a) (0, 8) (b) (0, 11)
(c) (0, –3) (d) (0, 13)

(ii) The lines represented by the linear equations $y = x$ and $x = 6$ intersect at P. The coordinate of the point P are:

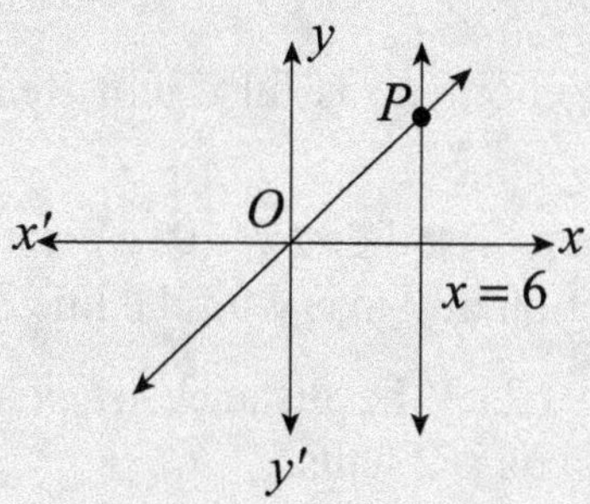

(a) (0, 0) (b) (0, 6)
(c) (6, 0) (d) (6, 6)

(iii) The area of the triangle from by the line $\frac{x}{6}+\frac{y}{8}=1$ with the coordinate axises is

(a) 48 (b) 64
(c) 24 (d) 12

(iv) In the given figure, the area of the triangle ABC (is sq. units) is

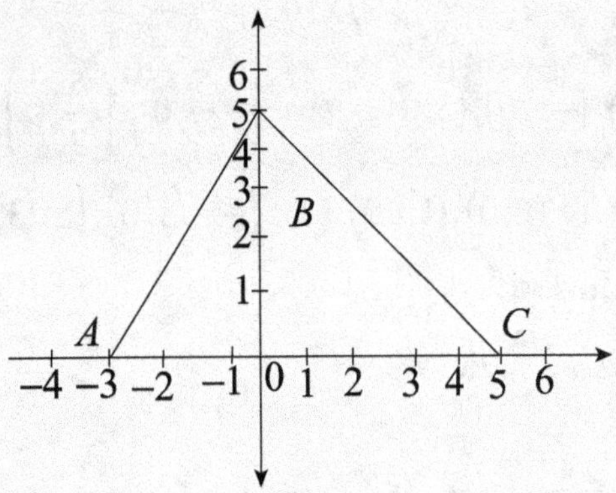

(a) 22.5 (b) 24
(c) 45 (d) 64

Question number **143 to 146** are Assertion and Reason based questions. Two statements are given, one labelled as Assertion (A) and the other labelled as Reason (R). Select the correct answer to these questions from the codes (A), (B), (C) and (D) as given below.

(a) Both Assertion (A) and Reason (R) are true and Reason (R) is the correct explanation of Assertion (A).

(b) Both Assertion (A) and Reason (R) are true and Reason (R) is not the correct explanation of Assertion (A).

(c) Assertion (A) is true, but Reason (R) is false.

(d) Assertion (A) is false, but Reason (R) is true.

143. Assertion (A): Point P (0, 2) is the point of intersection of y-axis with the line $3x + 2y = 4$

Reason (R): The distance of point P (0, 2) from x-axis is 2 units. **(2023)**

Solution: (b)

$3.0 + 2.y = 4$

$\Rightarrow 2y = 4$

$\Rightarrow y = 2$

∴ the line intersects the y axis at (2, 0) point

The distance of the point P (0, 2) from the x-axis is 2 unit.

144. Assertion (A): If the points A (4, 3) and B (x, 5) lie on a circle with centre O (2, 3), then the value of x is 2.

Reason (R): Centre of the circle is the mid-point of each chord of the circle. **(2023)**

Solution: (c)

$OA = OB$

$\Rightarrow \sqrt{(4-2)^2+(3-3)^2} = \sqrt{(x-2)^2+(5-3)^2}$

$\Rightarrow 4+0=(x-2)^2+4$

$\Rightarrow (x-2)^2=0$

$\Rightarrow x-2=0$

$\Rightarrow x=2$

Centre of the circle is only the mid point of tangent chord (diameter) of the circle.

145. Assertion (A): Mid-point of a line segment divides the line segment in the ratio 1 : 1.

Reason (R): The ratio in which the point (–3, k) divides the line segment joining the points (–5, 4) and (–2, 3) is 1 : 2. **(2024)**

Solution: (c)

Here, if the point (–3, k) divides the segment in the ratio 1 : 2 between (–5, 4) and (–2, 3), then.

$$-3=\frac{1\times(-2)+2\times(-5)}{1+2}$$

$$\Rightarrow -3=\frac{-2-10}{3}=-4$$

This is not true. Therefore, the x-coordinate does not satisty the 1 : 2 division ratio.

146. Assertion (A): The point (0,4) lies on y-axis

Reason (R): The x-coordinate of a point, lying on y-axis is zero. **(2024)**

Solution: (a)

The x-coordinate of a point lying on y-axis is always zero.

Direction: Question number **i to x** are Assertion and Reason based question. Two statements are given, one labelled on Assertion (A) and the other labelled an Reason (R). Select the correct answer to these questions from the codes (A), (B), (C) and (D) as given below.

(a) Both Assertion (A) and Reason (R) are true and Reason (R) is the correct explanation of Assertion (A).

(b) Both Assertion (A) and Reason (R) are true, but Reason (R) is not the correct explanation of Assertion (A).

(c) Assertion (A) is true, but reason (R) is false.

(d) Assertion (A) is false, but reason (R) is true.

Exercise 6.10

(i) Assertion (A): The distance between the points (1, 2) and (5, 4) is $2\sqrt{5}$ units.

Reason (R): The distance between two points (x_1, y_1) and (x_2, y_2) is given by $\sqrt{(x_2-x_1)^2+(y_2-y_1)^2}$

(ii) Assertion (A): The point (3, 0) lies on the x-axis.

Reason (R): Any point with its y-coordinate zero lies on the x axis.

(iii) Assertion (A): The midpoint of the line segment joining (4, 6) and (–8, –4) is (–2, 1).

Reason (R): The midpoint of the line segment joining (x_1, y_1) and (x_2, y_2) is $\left(\frac{x_1+x_2}{2}, \frac{y_1+y_2}{2}\right)$

(iv) Assertion (A): The points (1, 1), (2, 2) and (3, 3) are collinear.

Reason (R): Three points are collinear if the area of the triangle formed by them is zero.

(v) Assertion (A): The value of a is 6, for which the distance between the points $A(2, -3)$ and $B(10, a)$ is 10

Reason (R): Distance between two given points $A(x_1, y_1)$ and $B(x_2, y_2)$ is given by $AB=\sqrt{(x_2-x_1)^2+(y_2-y_1)^2}$

(vi) Assertion (A): The coordinate of the point which divides the join of $A(4, -3)$ and $B(9, 7)$ in the ratio 3 : 2.

Reason (R): If $P(x, y)$ divides the join of $A(x_1, y_1)$ and $B(x_2, y_2)$ in the ratio $m : n$, then $x = \frac{mx_2+nx_1}{m+n}, y=\frac{my_2+ny_1}{m+n}$

(vii) Assertion (A): The points $(a, 2-2a)$, $(-a+1, 2a)$ and $(-4-a, 6-2a)$ are collinear if $a=\frac{1}{2}$.

Reason (R): Three points A, B and C are collinear in same line, if $AB + BC = AC$.

(viii) Assertion (A): If the points $P(4, 3)$ and $Q(a, 5)$ lies on a circle with the centre $O(2, 3)$, then the value of a is 2.

Reason (R): The mid-point of the line segment joining the points $A(x_1, y_1)$ and $B(x_2, y_2)$ is $\left(\frac{x_1+x_2}{2}, \frac{y_1+y_2}{2}\right)$

(ix) Assertion (A): If the points A (–3, 9), B (2, p) and C (4, –5) are collinear, then $p = -1$.

Reason (R): Three points (x_1, y_1), (x_2, y_2) and (x_3, y_3) are collinear only when $x_1(y_2-y_3)+x_2(y_3-y_1)+x_3(y_1-y_2)=0$

(x) Assertion (A): The points A (–1, 0), B (3, 1), C (2, 2) and D (–2, 1) are the vertices of a parallelogram.

Reason (R): The coordinates of the mid points of both the diagonals AC and BD are $\left(\frac{1}{2}, 1\right)$

ANSWER KEYS

Exercise – 6.1

(i) (b) 6 units **(ii)** (a) 5 units **(iii)** (d) 7units
(iv) (c) 6 units **(v)** (c) 2 units **(vi)** (d) 10 units

Exercise – 6.2

(i) (d) (3, 2) **(ii)** (b) (–3, –5)
(iii) (b) (–3, –2) **(iv)** (a) (–4, 7)

Exercise – 6.3

(i) (a) (0, 6) **(ii)** (0, –3)

Exercise – 6.4

(i) (b) –4 **(ii)** (d) –1 **(iii)** (a) 2
(iv) (d) (–3, 0) **(v)** (c) (0, –6)

Exercise – 6.5

(i) (d) $5\sqrt{5}$ **(ii)**(a) $\sqrt{61}$ **(iii)** (c) (5, 0)
(iv) (c) Both (a) and (b) **(v)** (b)$11+\sqrt{91}, 11-\sqrt{91}$
(vi) (d) $\left(0, \frac{89}{18}\right)$ **(vii)** (a) a + b +3 = 0
(viii) (b) (0, 7) **(ix)** (b) collinear
(x) (b) Isosceles triangle **(xi)**(c) 12
(xii) (a) a

Exercise – 6.6

(i) (c) 4 : 7 **(ii)** (d) IV quadrant **(iii)** (a) 1 : 5
(iv) (d) (1, 3) **(v)** (c) 2 : 7 **(vi)** (a) 2 : 1
(vii) (b) 3 : 2 **(viii)** (b) (–3, 3) **(ix)** (a) 1 : 2
(x) (c) 2 : 5

Exercise – 6.7

(i) (b) (–2, 6) **(ii)** (d) –3 **(iii)** (d) 15
(iv) $\left(\frac{13}{2}, 0\right)$ **(v)** (b) (–9, 0) **(vi)** (b) –4
(vii) (b) (–2, 3) **(viii)** (b) $a = 11, b = 15$
(ix) (a) 2 units **(x)** (a) $x = 6, y = 3$

Exercise – 6.8

(i) (b) –1 **(ii)** (a) 1 **(iii)** (d) 2a + b + 3 = 0
(iv) (d) $a + b = ab$ **(v)** (c) 53

Exercise – 6.9

(i) (d) (0, 13) **(ii)** (d) (6, 6)
(iii) (c) 24 **(iv)** (a) 22.5

Exercise – 6.10

(i) a **(ii)** a **(iii)** a **(iv)** a **(v)** d
(vi) a **(vii)** a **(viii)** b **(ix)** a **(x)** a

7 Geometry

Key Point If ABC and DEF are two similar triangles, then $\angle A = \angle D$, $\angle B = \angle E$ and $\angle C = \angle F$

1 in ΔLMN, $\angle L = 50°$and $\angle N = 60$. If $\Delta LMN \sim \Delta PQR$, then find $\angle Q$. **(2009)**

Solution:

$\Delta LMN \sim \Delta PQR$

$\therefore \angle L = \angle P = 50°$

$\angle N = \angle R = 60°$

$\therefore \angle M = \angle Q = 180° – (50° + 60°) = 70°$

2. In $\Delta ABC \sim \Delta PQR$ with $\angle A = 32°$ and $\angle R = 65°$, then the measure of $\angle B$ is **(2023)**

(a) 32° (b) 65°
(c) 83° (d) 97°

Solution: (c)

$\Delta ABC \sim \Delta PQR$

$\therefore \angle A = \angle P = 32°$

$\angle C = \angle R = 65°$

$\therefore \angle B = \angle Q = 180° - (32° + 65°) = 180° - 97° = 83°$

3. If $\Delta ABC \sim \Delta DEF$ and $\angle A = 47°$, $\angle E = 83°$, then $\angle C$ is equal **(2023)**

(a) 47° (b) 50°
(c) 83° (d) 130°

Solution: (b)

$\Delta ABC \sim \Delta DEF$

$\therefore \angle A = \angle D = 47°$

$\angle B = \angle E = 83°$

$\therefore \angle C = \angle F = 180° – (47° + 83°) = 180° – 130° = 50°$

4. It is given that $\Delta ABC \sim \Delta DEF$. if $\angle A = 55°$, $\angle E = 45°$, then $\angle C$ is **(2024)**

(a) 80° (b) 90°
(c) 55° (d) 45°

Solution: (a)

$\Delta ABC \sim \Delta DEF$

$\therefore \angle A = \angle D = 55°$

$\angle B = \angle E = 45°$

$\therefore \angle C = 180° - (55° + 45°) = 180° - 100° = 80°$

5. Given $\Delta ABC \sim \Delta PQR$, $\angle A = 30°$ and $\angle Q = 90°$. The value of $(\angle R + \angle B)$ is **(2025)**

(a) 90° (b) 120°
(c) 150° (d) 180°

Solution:

$\Delta ABC \sim \Delta PQR$

$\therefore \angle A = \angle P = 30°$

$\angle B = \angle Q = 90°$

$\therefore \angle C = \angle R = 180° - (90° + 30°) = 60°$

$\therefore \angle R + \angle B = 60° + 90° = 150°$

Exercise 7.1

(i) In ΔPDC, $\angle P = 75°$, $\angle D = 70°$. If $\Delta PDC \sim \Delta OBA$, then find $\angle A$.

(ii) In ΔABC, $\angle A = 60°$, $\angle C = 62°$.If $\Delta ABC \sim \Delta DEF$, then find $\angle E$.

(iii) In ΔABC, $\angle A = 80°$, $\angle C = 70°$. If $\Delta ABC \sim \Delta DEF$ then find $\angle E$.

(iv) If $\Delta ABC \sim \Delta DEF$, $\angle A = 80°$, $\angle E = 70°$, then $\angle C$ is equal

(a) 50° (b) 30°
(c) 80° (d) 100°

(v) If $\Delta ABC \sim \Delta DEF$, $\angle A = 85°$, $\angle E = 45°$, then $\angle C$ is equal

(a) 45° (b) 85°
(c) 58° (d) 25°

(vi) If $\Delta ABC \sim \Delta DEF$ and $\angle A = 53°$, $\angle E = 48°$, then $\angle C$ is equal

(a) 71° (b) 83°
(c) 99° (d) 79°

Key Point If $\Delta ABC \sim \Delta DEF$ then

$$\frac{AB}{DE} = \frac{BC}{EF} = \frac{AC}{DF}$$

6.

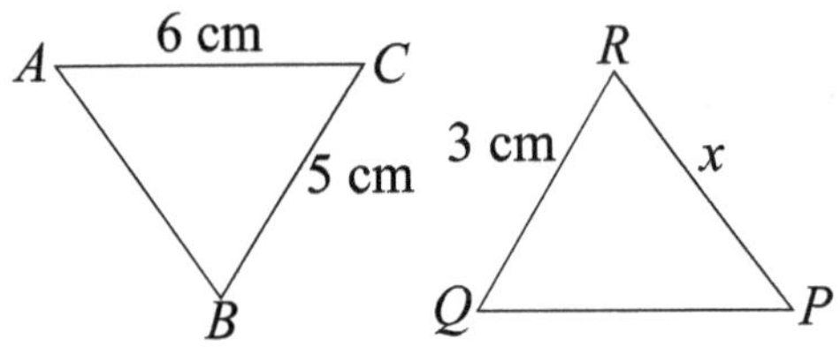

In the given figure, $\Delta ABC \sim \Delta QPR$. If $AC = 6$ cm, $BC = 5$ cm, QR = 3 cm and $PR = x$, then the value of x is **(2023)**

(a) 3.6 cm (b) 2.5 cm
(c) 10 cm (d) 3.2 cm

Solution: (b)

$\Delta ABC \sim \Delta QPR$

$$\therefore \frac{BC}{PR} = \frac{AC}{QR} = \frac{AB}{QP}$$

$$\Rightarrow \frac{BC}{PR} = \frac{AC}{QR}$$

$$\Rightarrow \frac{5}{x} = \frac{6}{3}$$

$x = 2.5$ cm

7. In the given figure, if $\Delta ABC \sim$ ΔQPR, then the Value of x is **(2024)**

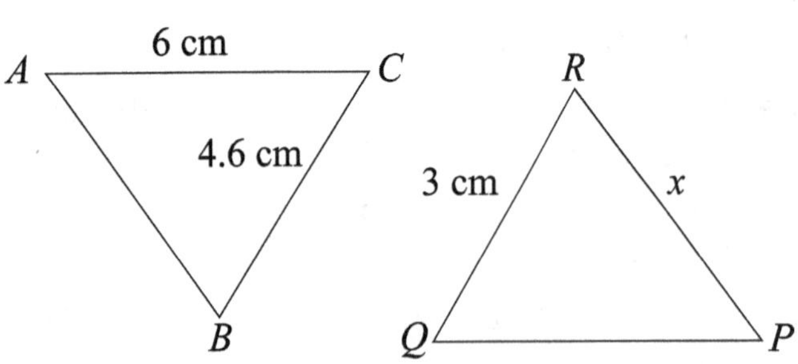

(a) 5.3 cm (b) 4.6 cm
(c) 2.3 cm (d) 4 cm

Solution: (c)

$\Delta ABC \sim \Delta QPR$

$$\therefore \frac{AC}{QR} = \frac{BC}{PR}$$

$$\Rightarrow \frac{6}{3} = \frac{4.6}{x}$$

$x = 2.3$ cm

8. ΔABC and ΔPQR are shown in the adjoining figures. The measure of $\angle C$ is: **(2025)**

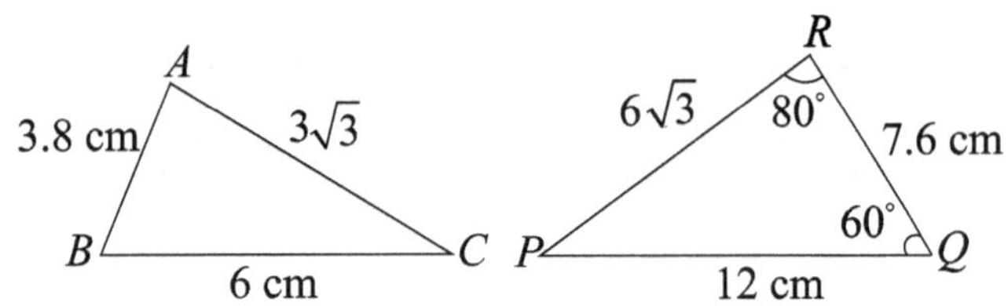

(a) 140° (b) 80°
(c) 60° (d) 40°

Solution: (d)

$$\frac{AB}{RQ} = \frac{1}{2}, \frac{BC}{PQ} = \frac{1}{2}, \frac{AC}{RP} = \frac{1}{2}$$

$\therefore \Delta ABC \sim \Delta RQP$

$\angle C = \angle P = 180° - (80° + 60°) = 40°$

9. If $\Delta ABC \sim \Delta PQR$ such that $AB = 6$ cm, $AC = 7$ cm, $QR = 15$ cm and $PQ = 12$ cm, then the sum of lengths of BC and PR is **(2025)**

Solution: (b)

$\Delta ABC \sim \Delta PQR$

$$\therefore \frac{AB}{PQ} = \frac{BC}{QR} = \frac{AC}{PR}$$

$$\Rightarrow \frac{6}{12} = \frac{BC}{15} = \frac{7}{PR}$$

$$\therefore \frac{BC}{15} = \frac{6}{12}$$

$\Rightarrow BC = 7.5$ cm

and $\dfrac{7}{PR} = \dfrac{6}{12}$

$\Rightarrow PR = 14$ cm

$\therefore BC + PR = 7.5 + 14$

$= 21.5$ cm

10. From the figures given below, which of the following is the measure of $\angle P$? **(2025)**

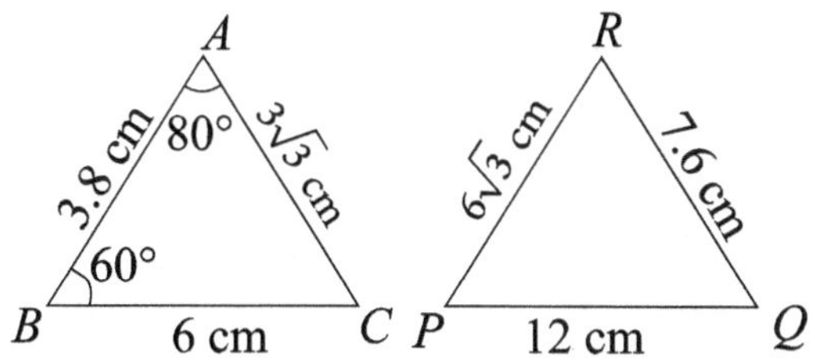

(a) $\angle P = 60°$

(b) $\angle P = 80°$

(c) $\angle P = 40°$

(d) The measure of $\angle P$ can not be determind

Solution: (c)

$$\frac{AB}{RQ} = \frac{AC}{PR} = \frac{BC}{PQ}$$

$\therefore \Delta ABC \sim \Delta RQP$

$\therefore \angle A = \angle R = 80°$

$\angle B = \angle Q = 60°$

$\angle C = \angle P = 180° - (80° + 60°) = 40°$

11. In $\Delta ABC \sim \Delta PQR$, $PQ = 3AB$ and $BC = 2$ cm, then QR is equal to **(2025)**

(a) 2 cm (b) 6 cm

(c) $\frac{2}{3}$ cm (d) 12 cm

Solution: (b)

$\Delta ABC \sim \Delta PQR$

$$\therefore \frac{AB}{PQ} = \frac{BC}{QR} = \frac{AC}{PR}$$

$$\Rightarrow \frac{PQ}{AB} = \frac{QR}{BC}$$

$$\Rightarrow 3 = \frac{QR}{2} \; [\because PQ = 3AB]$$

$\Rightarrow QR = 6$ cm

12. It is given that $\Delta ABC \sim \Delta QRP$ such that $AB = 9$ cm, $BC = 5$ cm and $PR = 2$ cm, length of side QR is **(2026)**

(a) 0.9 cm (b) $\frac{5}{18}$ cm

(c) $\frac{10}{9}$ cm (d) 3.6 cm

Solution: $\Delta ABC \sim \Delta QRP$

$$\therefore \frac{AB}{QR} = \frac{BC}{RP} = \frac{AC}{QP}$$

$$\Rightarrow \frac{9}{QR} = \frac{5}{2}$$

$\Rightarrow QR = 3.6$ cm

13. Observe the given triangles:

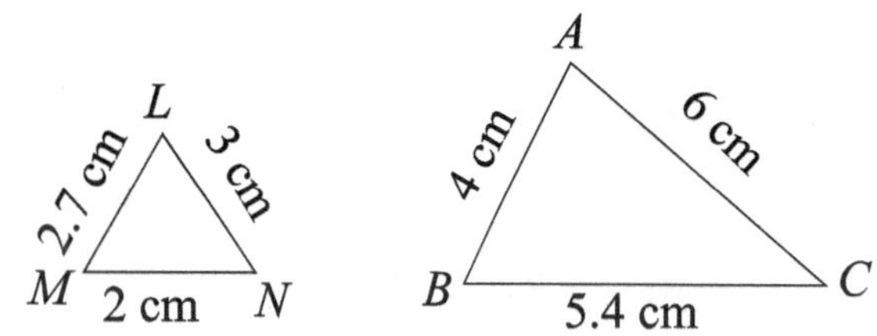

Which of the following statements is true? **(2026)**

(a) $\Delta LMN \sim \Delta ABC$

(b) $\Delta LMN \sim \Delta ACB$

(c) $\Delta NML \sim \Delta ABC$

(d) $\Delta NML \sim \Delta ACD$

Solution: $\frac{MN}{AB} = \frac{ML}{BC} = \frac{LN}{CA}$

$\therefore \Delta NML \sim \Delta ABC$

14. In triangle ABC and PQR, $\angle A = \angle Q$ and $\angle B = \angle R$, then $AB : AC$ is equal to **(2026)**

(a) $PQ : PR$ (b) $PQ : QR$

(c) $QR : QP$ (d) $PR : QR$

Solution:

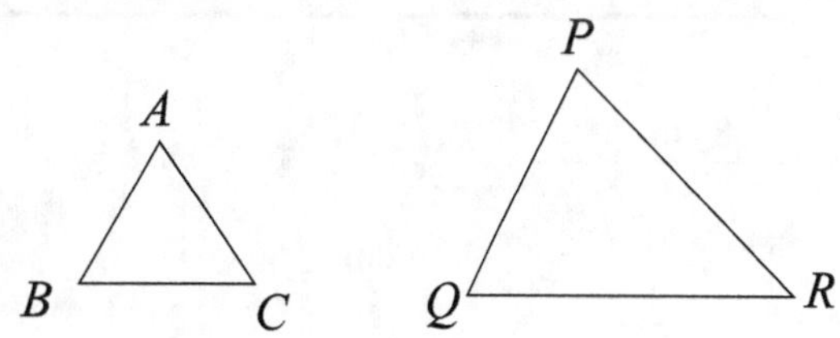

In ΔABC & ΔPQR

$\angle A = \angle Q$ and $\angle B = \angle R$

$\therefore \Delta ABC \sim \Delta QRP$

$$\therefore \frac{AB}{QR} = \frac{BC}{RP} = \frac{AC}{QP}$$

$$\therefore \frac{AB}{QR} = \frac{AC}{QP}$$

$\Rightarrow \frac{AB}{AC} = \frac{QP}{QP}$

15. If ΔABC and ΔDEF are similar such that $2\ AB = DE$ and $BC = 8$ cm, Then EF is equal to: **(2026)**

(a) 4 cm (b) 8 cm
(c) 12 cm (d) 16 cm

Solution: (d)

$\Delta\ ABC \sim \Delta\ DEF$

$\therefore \frac{AB}{DE} = \frac{BC}{EF}$

$\Rightarrow \frac{1}{2} = \frac{8}{EF} [\because 2AB = DE]$

$\Rightarrow EF = 16 \text{cm}$

16. Devansh proved that $\Delta ABC \sim \Delta PQR$ using SAS similarity criteria. If he found $\angle C = \angle R$, which of the following was proved true? **(2026)**

(a) $\frac{AC}{AB} = \frac{PR}{PQ}$ (b) $\frac{BC}{AC} = \frac{PR}{QR}$

(c) $\frac{AC}{BC} = \frac{PR}{PQ}$ (d) $\frac{AC}{BC} = \frac{PR}{QR}$

Solution: (d)

$\frac{AC}{BC} = \frac{PR}{QR}$

Exercise 7.2

(i)

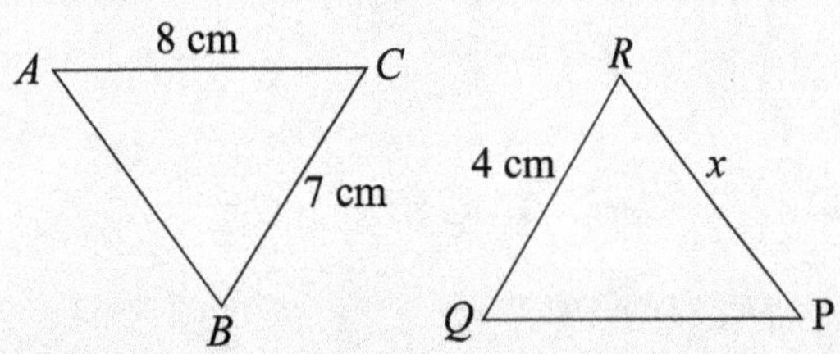

In the given figure, $\Delta ABC \sim \Delta QPR$. If $AC = 8$ cm, $BC = 7$ cm, $QR = 4$ cm and $PR = x$, then the value of x in

(a) 3.5 cm (b) 4 cm
(c) 2 cm (d) 1.5 cm

(ii)

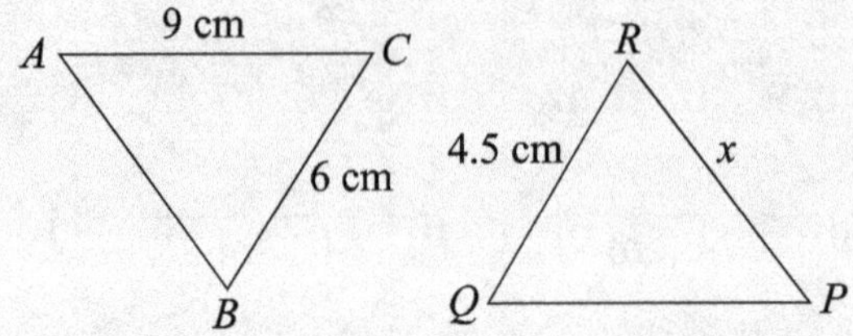

In the given figure, $\Delta ABC \sim \Delta QPR$. If $AC = 9$ cm, $BC = 6$ cm, $QR = 4.5$ cm and $PX = x$, then the value of x is

(a) 8 cm (b) 6 cm
(c) 3 cm (d) 10 cm

(iii)

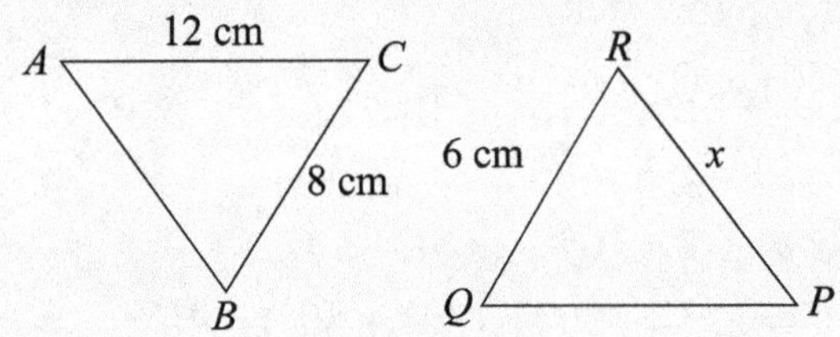

In the given figure, $\Delta ABC \sim \Delta QPR$ .If $AC = 12$ cm, $BC = 8$ cm, $QR = 6$ cm and $PQ = x$, then the value of x is

(a) 4 cm (b) 6 cm
(c) 3 cm (d) 5 cm

Key Point

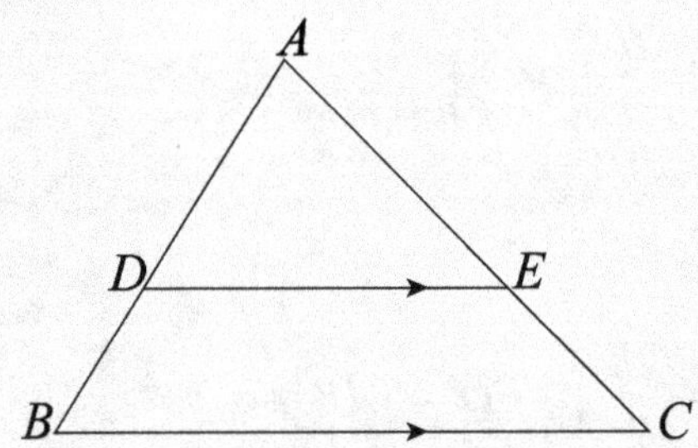

In ΔABC, in which $DE||BC$ and DE intersects AB at D and AC at E, then

(i) $\frac{AD}{DB} = \frac{AE}{EC}$ **(ii)** $\frac{AD}{AB} = \frac{AE}{AC}$

17. In figure 2, $DE||BC$. Find the length of side AD, given that $AE = 1.8$ cm, $BD = 7.2$ cm and $CE = 5.4$ cm. **(2019)**

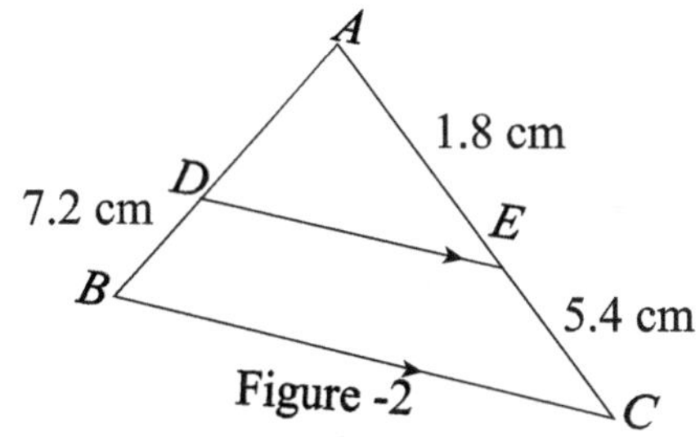

Figure -2

Solution:

in ΔABC $DE \parallel BC$

$\therefore \frac{AD}{DB} = \frac{AE}{EC}$

$\Rightarrow \frac{AD}{7.2} = \frac{1.8}{5.4}$

$\Rightarrow AD = 2.4$ cm

18. In figure –2, $DE \parallel BC$. If $\frac{AD}{DB} = \frac{3}{2}$ and $AE = 2.7$ cm, then EC is equal to **(2020)**

(a) 2.0 cm (b) 1.8 cm

(c) 4.0 cm (d) 2.7 cm

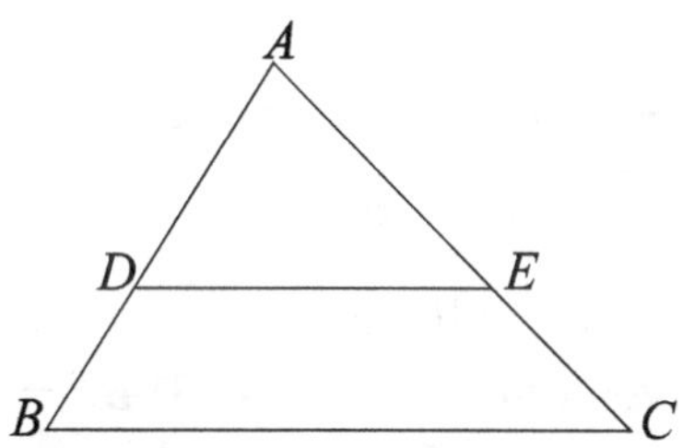

Solution:

(b) In ΔABC, $DE \parallel BC$

$\therefore \frac{AD}{DB} = \frac{AE}{EC}$

$\Rightarrow \frac{3}{2} = \frac{2.7}{EC}$

$\Rightarrow EC = 1.8$ cm

19. In fig.3, in ΔABC, $DE \parallel BC$ such that $AD = 2.4$ cm, $AB = 3.2$ cm and $AC = 8$ cm, Then what is the length of AE ? **(2020)**

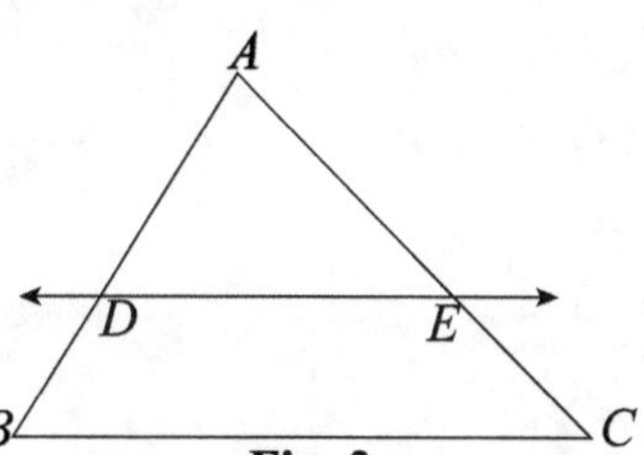

Fig. 3

Solution:

In ΔABC $DE \parallel BC$

$\therefore \frac{AD}{AB} = \frac{AE}{AC}$

$\Rightarrow \frac{2.4}{3.2} = \frac{AE}{8}$

$\Rightarrow AE = 6$ cm

20. In fig. 1, D and E are points on sides AB and AC respectively of a ΔABC such that $DE \parallel BC$. If $AD = 3.6$ cm, $AB = 10$ cm and $AE = 4.5$ cm, find EC and AC **(2020)**

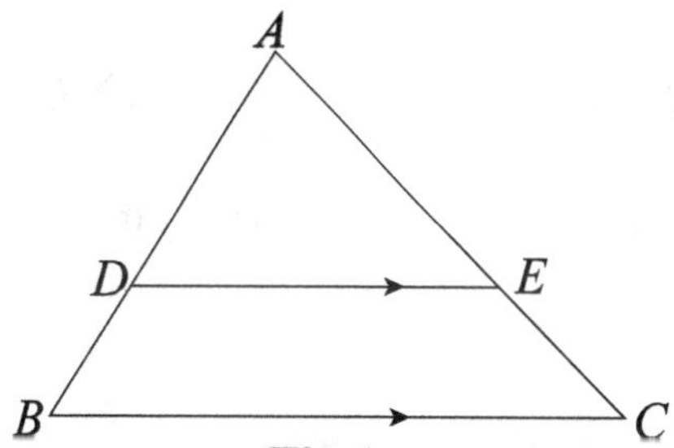

Fig-1

Solution:

$\frac{AD}{AB} = \frac{AE}{AC}$

$\Rightarrow \frac{3.6}{10} = \frac{4.5}{AC}$

$\Rightarrow AC = 12.50$ cm

Now, $EC = AC - AE = 12.50 - 4.50 = 8$ cm

21. In the given figure, $DE \parallel BC$. If $AD = 2$ units, $DB = AE = 3$ Units and $EC = x$ units, then the value of x is: **(2023)**

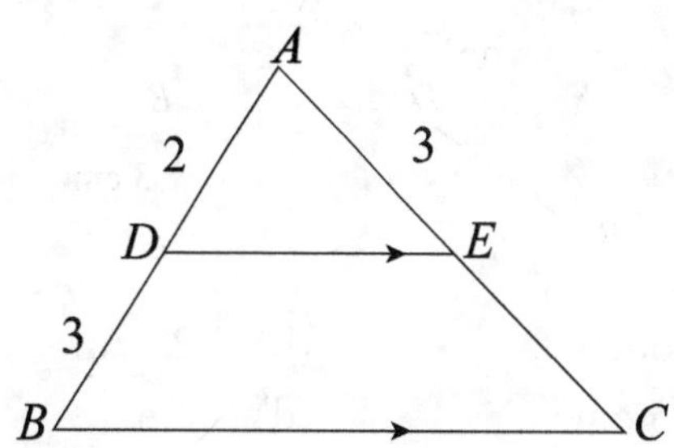

(a) 2 (b) 3

(c) 5 (d) $\frac{9}{2}$

Solution: (d)

In ΔABC , $DE \parallel BC$

$\therefore \frac{AD}{DB} = \frac{AE}{EC}$

$\Rightarrow \frac{2}{3} = \frac{3}{x}$

$\therefore x = \frac{9}{2}$

22. In ΔABC, $PQ \parallel BC$. If $PB = 6$ cm, $AP = 4$ cm, $AQ = 8$ cm, find the length of AC. **(2023)**

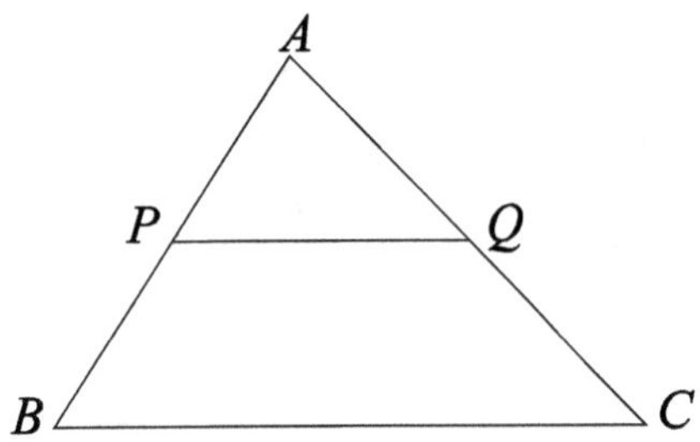

(a) 12 cm (b) 20 cm
(c) 6 cm (d) 14 cm

Solution:

In ΔABC $PQ \parallel BC$

$\therefore \frac{AP}{PB} = \frac{AQ}{QC}$

$\Rightarrow \frac{4}{6} = \frac{8}{QC}$

$\Rightarrow QC = 12$ cm

$\therefore AC = AQ + QC = 8 + 12 = 20$ cm

23. In the given figure, $DE \parallel BC$. If $AD = 3$ cm, $AB = 7$ cm and $BC = 3$ cm, then the length of AE is **(2023)**

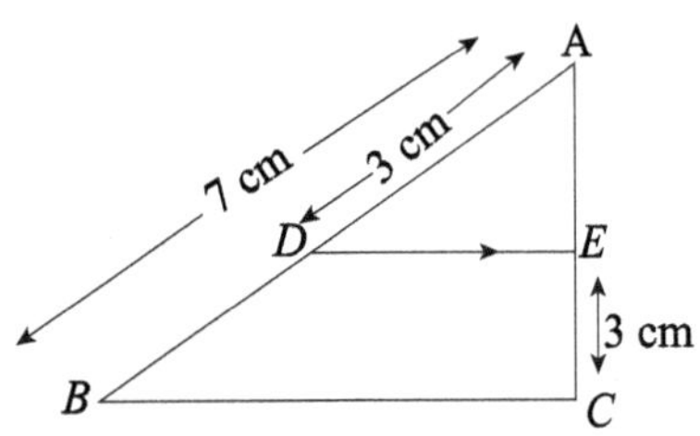

(a) 2 cm (b) 2.25 cm
(c) 3.5 cm (d) 4 cm

Solution: (b)

$BD = AB - AD = 7 - 3 = 4$ cm

In ΔABC $DE \parallel BC$

$\therefore \frac{AD}{DB} = \frac{AE}{EC}$

$\Rightarrow \frac{3}{4} = \frac{AE}{3}$

$\Rightarrow AE = 2.25$ cm

24. In the given figure ABC is a triangle in which $AD = 1.6$ cm, $BD = 4.8$ cm, $AE = 1.1$ cm and $EC = 2.2$ cm, then : **(2023)**

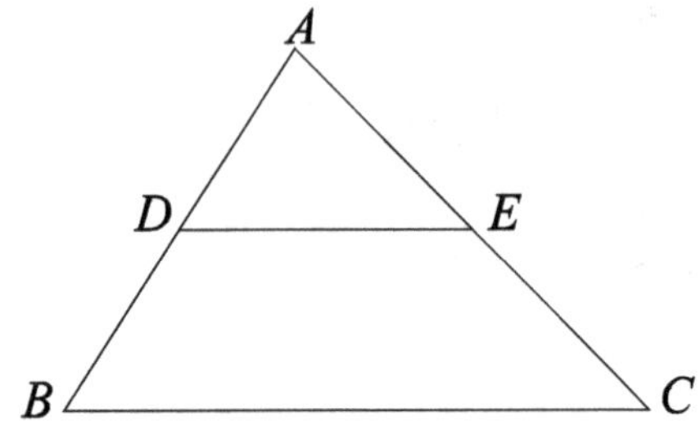

(a) $DE \parallel BC$

(b) $DE = \frac{1}{2} BC$

(c) $DE = BC$

(d) DE is not parallel to BC ($DE \nparallel BC$)

Solution: (d)

$\frac{AD}{DB} = \frac{1.6}{4.8} = \frac{1}{3}$

$\frac{AE}{EC} = \frac{1.1}{2.2} = \frac{1}{2}$

$\therefore \frac{AD}{AB} \neq \frac{AE}{EC}$

$\therefore DE$ is not parallel to BC.

25. In the given figure, $PQ \parallel AC$. If $BP = 4$ cm, $AP = 2.4$ cm and $BQ = 5$ cm, then the length of BC is **(2023)**

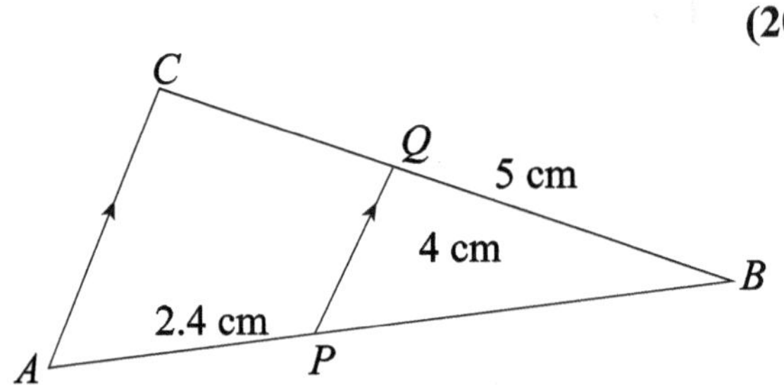

(a) 8 cm (b) 3 cm
(c) 0.3 cm (d) $\frac{25}{3}$ cm

Solution: (a)

In ΔABC, $PQ \parallel AC$

$\therefore \frac{BP}{PA} = \frac{BQ}{QC}$

$\Rightarrow \frac{4}{2.4} = \frac{5}{QC}$

$\Rightarrow QC = 3$ cm

Now, $BC = BQ + QC = 5 + 3 = 8$ cm

26. In ΔABC, $DE \parallel BC$ (as shown in the figure). If $AD = 4$ cm, $AB = 9$ cm and $AC = 13.5$ cm, then the length of EC is **(2024)**

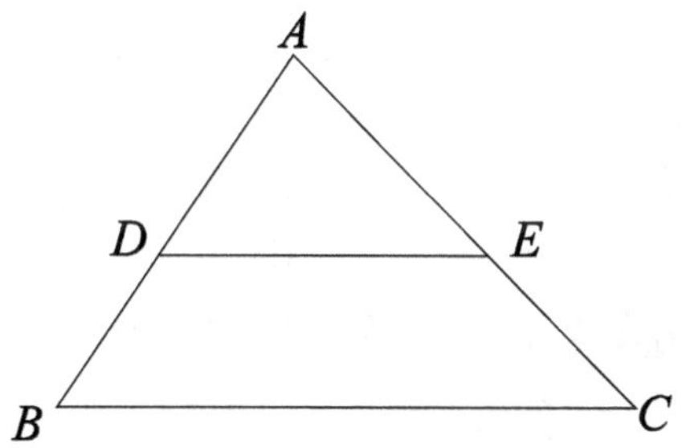

(a) 6 cm (b) 7.5 cm
(c) 9 cm (d) 5.7 cm

Solution:

In ΔABC, $DE \parallel BC$

$\therefore \frac{AD}{AB} = \frac{AE}{AC}$

$\Rightarrow \frac{4}{9} = \frac{AE}{13.5}$

$\Rightarrow AE = 6$ cm

Now, $EC = AC - AE$

$= 13.5 - 6$

$= 7.5$ cm

27. In the given figure, in ΔABC, $DE \parallel BC$. If $AD = 2.4$ cm, $DB = 4$ cm and $AE = 2$ cm then the length of AC is **(2024)**

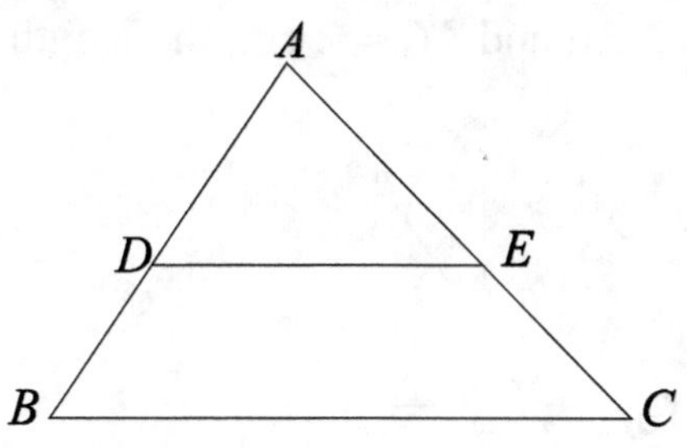

(a) $\frac{10}{3}$ cm (b) $\frac{3}{10}$ cm
(c) $\frac{16}{3}$ cm (d) 1.2 cm

Solution: (c)

In ΔABC, $DE \parallel BC$

$\therefore \frac{AD}{DB} = \frac{AE}{EC}$

$\Rightarrow \frac{2.4}{4} = \frac{2}{EC}$

$\Rightarrow EC = \frac{10}{3}$ cm

$\Rightarrow AC = AE + EC = 2 + \frac{10}{3} = \frac{16}{3}$ cm

28. A line ℓ intersects the sides PQ and PR of a ΔPQR at L and M respectively such that $LM \parallel QR$. If $PL = 5.7$ cm, $PQ = 15.2$ cm and $MR = 5.5$ cm, then the length of PM (in cm) is : **(2024)**

(a) 3 (b) 1.8
(c) 2.5 (d) 3.3

Solution: (d)

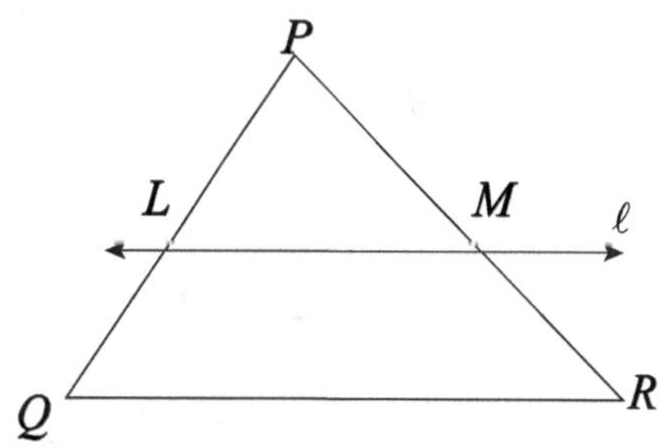

In ΔPQR, $LM \parallel QR$

$\therefore \frac{PL}{PQ} = \frac{PM}{PR}$

$\Rightarrow \frac{PL}{PQ} = \frac{PM}{PM + MR}$

$\Rightarrow \frac{5.7}{15.2} = \frac{PM}{PM + 5.5}$

$\Rightarrow 5.7\,\text{PM} + 5.7 \times 5.5 = 15.2\,\text{PM}$

$\Rightarrow 9.5\ \text{PM} = 5.7 \times 5.5$

$\Rightarrow \text{PM} = 3.3$ cm

29. In the given figure, in ΔABC, $DE \parallel BC$. If $AD = 2x$ cm, $AE = (x+2)$cm, $DB = 4$ cm, $EC = 3$ cm, Then the value of x is **(2023)**

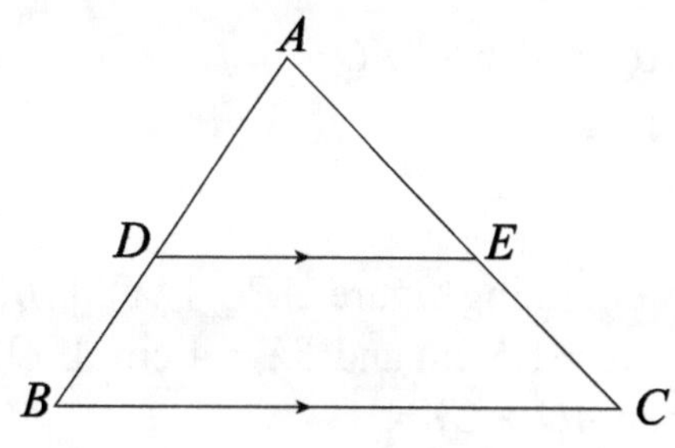

(a) 3 (b) 2
(c) 6 (d) 4

Solution:

In ΔABC, $DE \parallel BC$

$\therefore \frac{AD}{DB} = \frac{AE}{EC}$

$\Rightarrow \frac{2x}{4} = \frac{x+2}{3}$

$\Rightarrow 6x = 4x + 8$

$\Rightarrow 2x = 8$

$\therefore x = 4$

30. In the given figure, $PQ \parallel BC$. If $\frac{AP}{PB} = \frac{4}{13}$ and $AC = 20.4$ cm, then the length of AQ is: **(2025)**

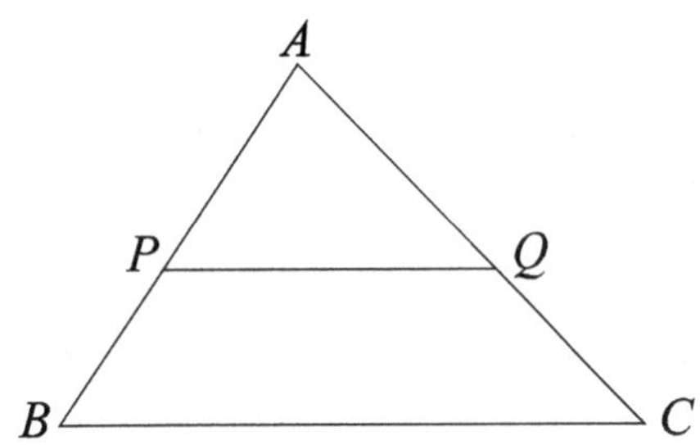

(a) 2.8 cm (b) 5.8 cm
(c) 3.8 cm (d) 4.8 cm

Solution:

$PQ \parallel BC$

$\therefore \frac{AP}{PB} = \frac{AQ}{QC}$

$\Rightarrow \frac{AQ}{QC} = \frac{AP}{PB}$

$\Rightarrow \frac{AQ}{AC - AQ} = \frac{4}{13}$

$\Rightarrow \frac{AQ}{20.4 - AQ} = \frac{4}{13}$

$\Rightarrow 13AQ = 81.6 - 4AQ$

$\Rightarrow 17AQ = 81.6$

$\therefore AQ = 4.8$ cm

31. In the adjoining figure , $PQ \parallel XY \parallel BC$, $AP = 2$ cm, $PX = 1.5$ cm and $BX = 4$ cm. If $QY = 0.75$ cm, then $AQ + CY =$ **(2025)**

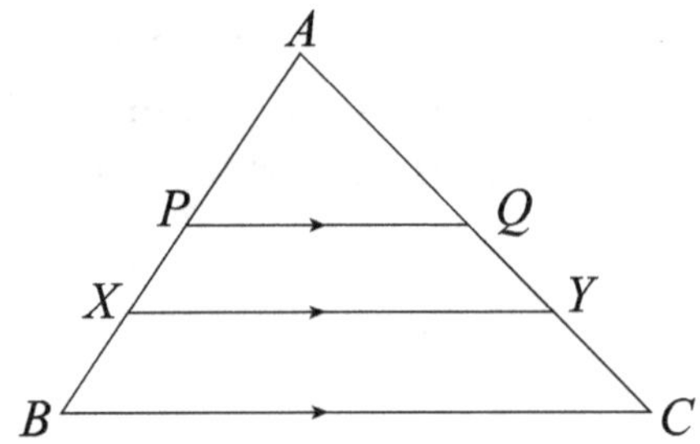

(a) 6 cm (b) 4.5 cm
(c) 3 cm (d) 5.25 cm

Solution: (c)

In ΔAXY, $PQ \parallel XY$

$\therefore \frac{AP}{PX} = \frac{AQ}{QY}$

$\Rightarrow \frac{2}{1.5} = \frac{AQ}{0.75}$

$\Rightarrow AQ = 1$ cm

Now, in ΔABC

$XY \parallel BC$

$\therefore \frac{AX}{BX} = \frac{AY}{CY}$

$\Rightarrow \frac{3.5}{4} = \frac{1.75}{CY}$

$\Rightarrow CY = 2$ cm

$\therefore AQ + CY = 1 + 2 = 3$ cm

32. In the given figure, if $DE \parallel BC$, $AD = 1.5$ cm, $DB = 3$ cm and $EC = 2$ cm, the length of AC is **(2025)**

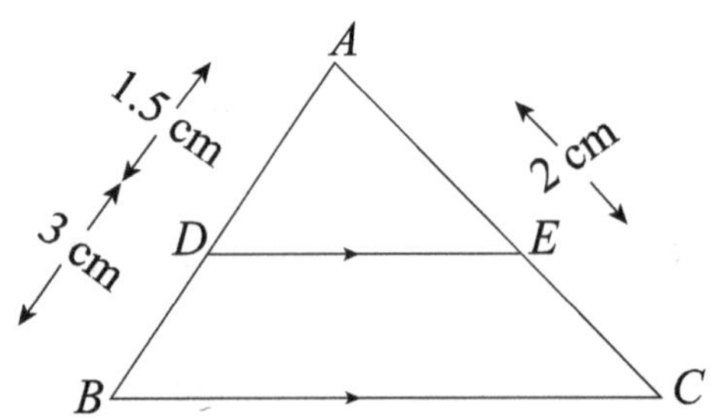

(a) 1.5 cm (b) 3 cm
(c) 3.5 cm (d) 4.5 cm

Solution:

In ΔABC, $DE \parallel BC$

$$\therefore \frac{AD}{DB}=\frac{AE}{EC}$$

$$\Rightarrow \frac{1.5}{3}=\frac{AE}{2}$$

$\Rightarrow AE = 1$ cm

$AC = AE + EC$

$= 1 + 2$

$= 3$ cm

33. Line ST is drawn parallel to the base QR of a ΔPQR, meeting PQ at S and PR at T. If $\frac{PQ}{QS}=3$ and $TR = 3$ cm, then the length of PT is **(2025)**

(a) 9 cm (b) 12 cm

(c) 6 cm (d) 3 cm

Solution:

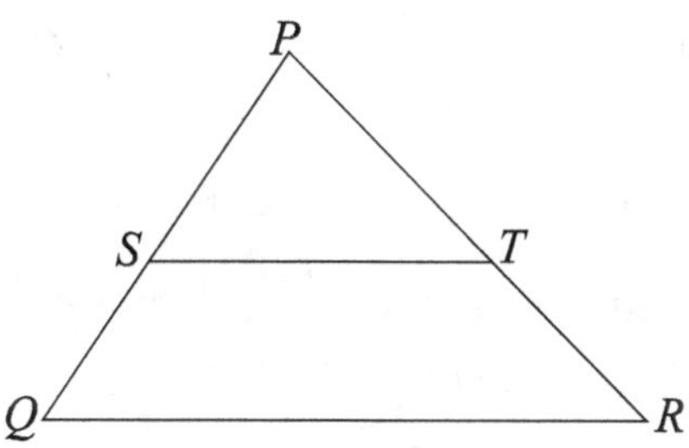

$$\frac{PQ}{QS}=3$$

$\Rightarrow PQ = 3\ QS$

$\Rightarrow PS + QS = 3\ QS$

$\Rightarrow PS = 2\ QS$

$$\Rightarrow \frac{PS}{QS}=2$$

Now, $ST \parallel QR$

$$\therefore \frac{PS}{QS}=\frac{PT}{TR}$$

$$\Rightarrow 2=\frac{PT}{3}$$

$\therefore PT = 6$ cm

34. In $\Delta\ DEF$, $AB \parallel EF$ .The value of x is : **(2026)**

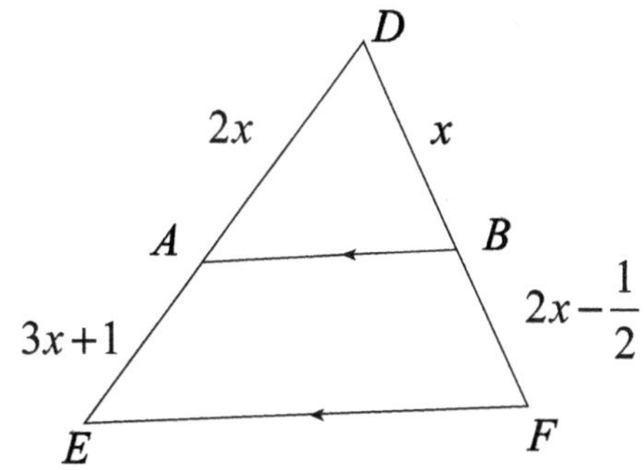

(a) 0, 2 (b) 2 only

(c) −2 (d) 1

Solution: (b)

$$\frac{DA}{AE}=\frac{DB}{DF}$$

$$\Rightarrow \frac{2x}{3x+1}=\frac{x}{2x-\frac{1}{2}}$$

$$\rightarrow 2x\left(2x-\frac{1}{2}\right)-x(3x+1)$$

$\Rightarrow 4x^2 - x = 3x^2 + x$

$\Rightarrow x^2 - 2x = 0$

$\Rightarrow x(x-2) = 0$

$\Rightarrow x - 2 = 0 \quad [\therefore x \neq 0]$

$\Rightarrow x = 2$

35. In the given figure, $DE \parallel BC$, If $AD : AB = 1 : 3$ and $AE = 2.5$ cm, then AC equals. **(2026)**

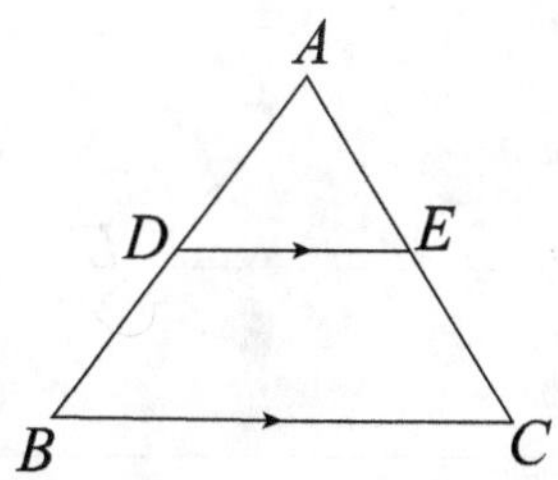

(a) 7.5 cm (b) 5 cm

(c) 10 cm (d) 2.5 cm

Solution: (a)

$$\frac{AD}{AB}=\frac{AE}{AC}$$

$$\Rightarrow \frac{1}{3}=\frac{2.5}{AC}$$

$\Rightarrow AC = 7.5$ cm

36. In the adjoining figure, if $EA \parallel SR$ and $PE = x$ cm, then the value of $5x$ is: **(2026)**

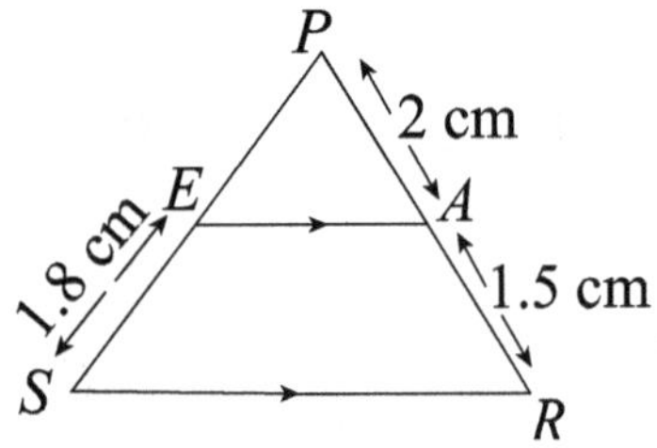

(a) 2.4 cm (b) 12 cm
(c) 1.35 cm (d) 6.75 cm

Solution: (b)

$$\frac{PE}{ES} = \frac{PA}{AR}$$

$$\Rightarrow \frac{x}{1.8} = \frac{2}{1.5}$$

$$\Rightarrow 5x = \frac{3.6}{0.3}$$

$$\Rightarrow 5x = 12$$

Exercise 7.3

(i) In the given figure, $DE \parallel BC$. Find the length of side EC, given that $AD = 6$ cm, $DB = 9$ cm, $AE = 8$ cm.

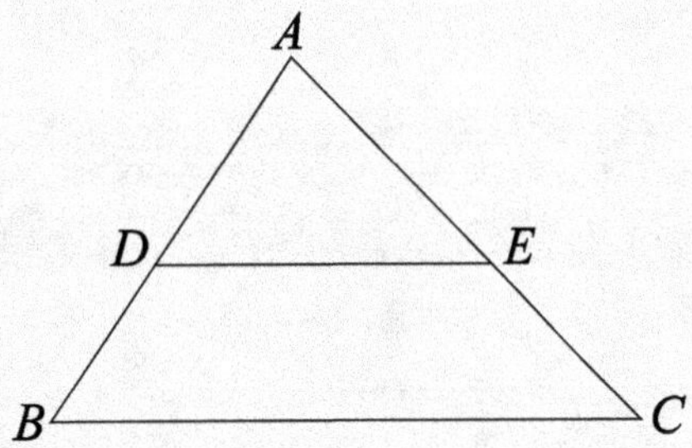

(ii) In the ΔABC, $DE \parallel BC$. Find the length of side AE, given that $AD = 8$ cm, $DB = 6$ cm, $EC = 9$ cm.

(iii) In the ΔABC, $DE \parallel BC$. Find the length of side EC, given that $AD = 16$ cm, $AE = 24$ cm, $DB = 14$ cm.

(iv) In the given figure, $\frac{AD}{DB} = \frac{4}{5}$ and $AE = 1.6$ cm, then EC in equal to

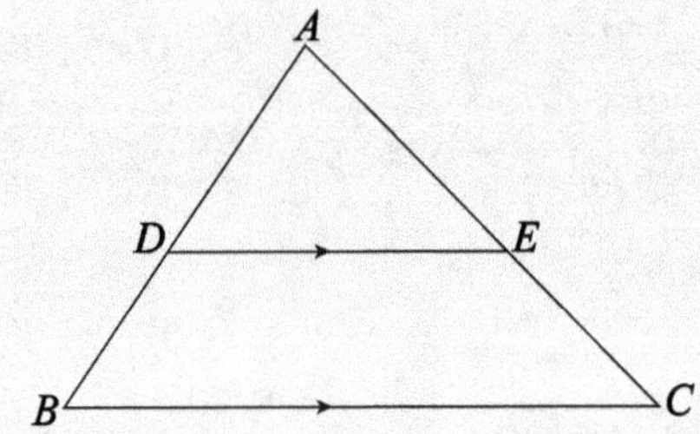

(a) 2 cm (b) 3 cm
(c) 1.5 cm (d) 1.6 cm

(v) In the given figure $\frac{AD}{DB} = \frac{2}{3}$ and $AE = 1.2$ cm, then EC is equal to.

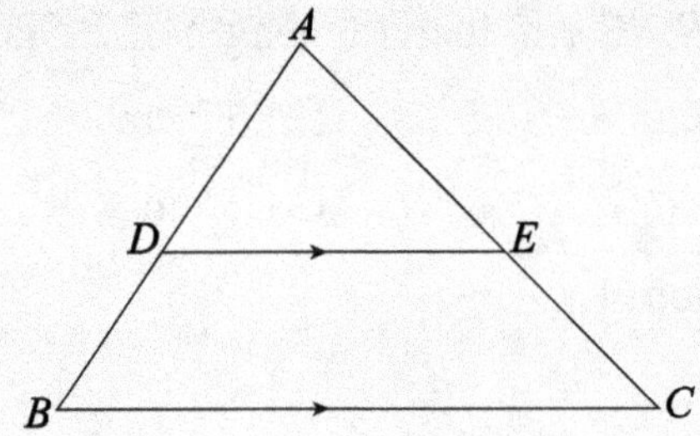

(a) 1.2 cm (b) 1.6 cm
(c) 1.8 cm (d) 2 cm

(vi) In the given figure, $\frac{AD}{DB} = \frac{4}{5}$ and $AE = 4$ cm, then EC is equal to

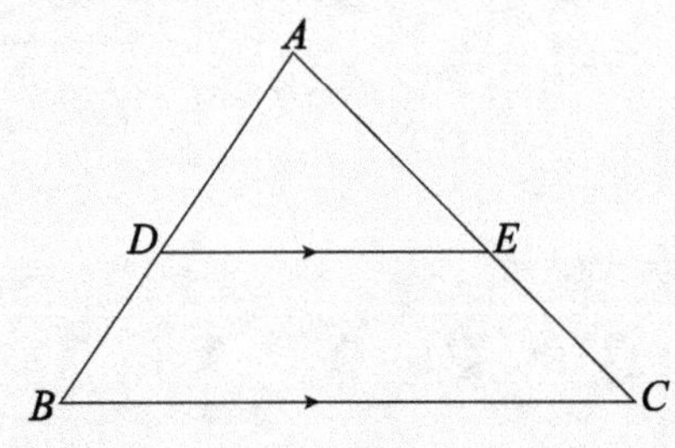

(a) 4 cm (b) 5 cm
(d) 6 cm (d) 10 cm

(vii) In the given figure, in ΔABC, $DE \parallel BC$ such that $AD = 1.5$ cm, $AB = 4.5$ cm, $AC = 6$ cm, then what is the length of AE?

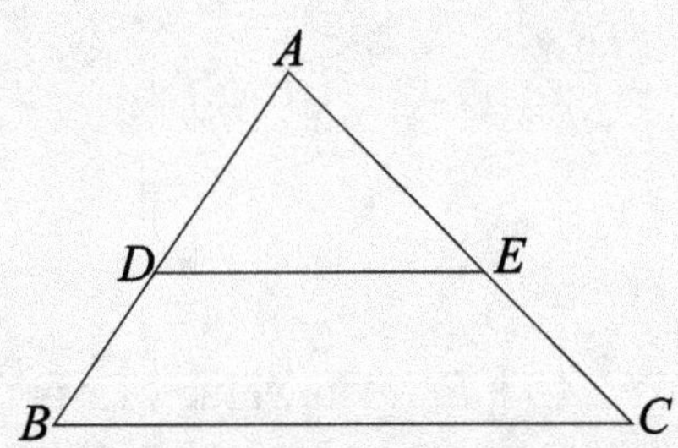

(viii) In the given figure, in ΔABC, $DE \parallel BC$ such that $AD = 6$ cm, $AB = 18$ cm, $AC = 4$ cm, then what is the length of AE ?

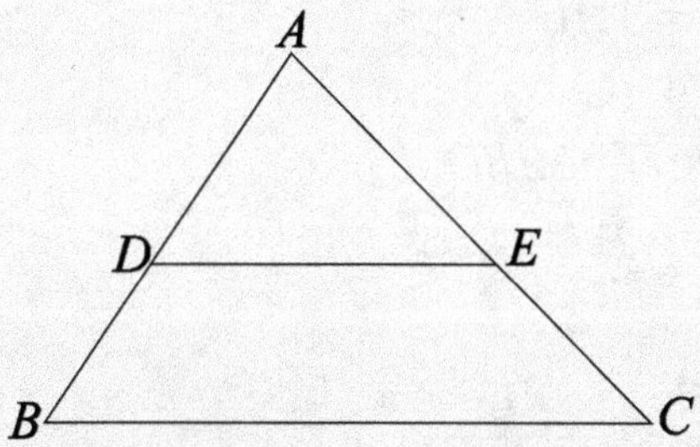

(ix) In the given figure, in ΔABC $DE \parallel BC$. If $AD = (x - 1)$cm, $BD = (x + 1)$cm, $AE = (2x - 5)$cm and $EC = (2x - 3)$cm, then find the value of x.

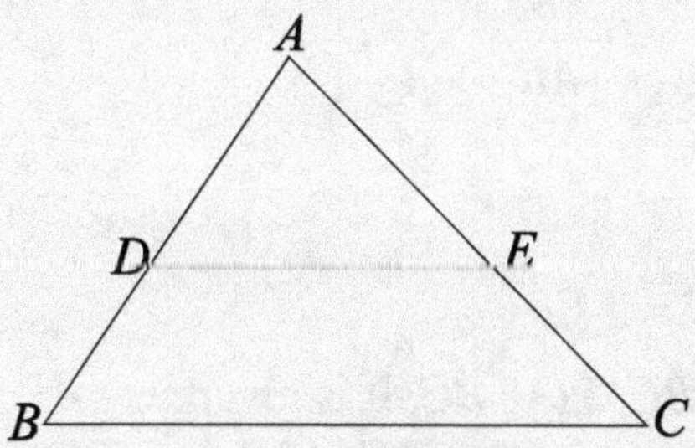

(x) In ΔABC, D and E are two points on sides AB and AC such that $DE \parallel BC$. If $AD = x$ cm, BD = $(x - 2)$cm, $AE = (x + 2)$cm and $EC = (x - 1)$cm, then find the value of x.

(xi) In ΔABC, D and E are two points on sides AB and AC such that $DE \parallel BC$. If $AD = (x–3)$ cm, $DB = (x - 1)$ cm, $AE = (x + 2)$cm and $EC = (x + 5)$ cm, then find the value of x.

Key Point

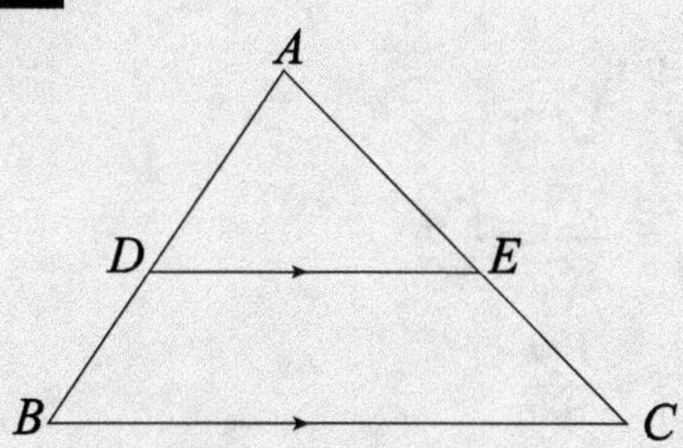

In ΔABC, $DE \parallel BC$, then.

$\Delta ADE \sim \Delta ABC$

$$\therefore \frac{AD}{AB} = \frac{AE}{AC} = \frac{DE}{BC}$$

37. In the given figure, $DE \parallel BC$. Then value of x is **(2023)**

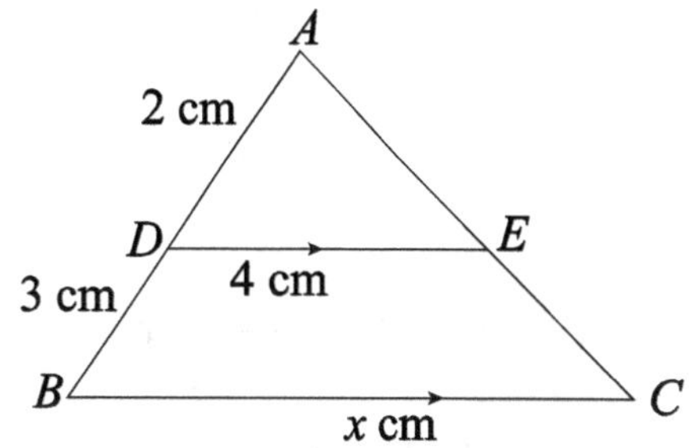

(a) 6 (b) 12.5

(c) 8 (d) 10

Solution: (d)

$DE \parallel BC$

$\therefore \Delta ADE \sim \Delta ABC$

$$\therefore \frac{AD}{AB} = \frac{AE}{AC} = \frac{DE}{BC}$$

$$\therefore \frac{AD}{AB} = \frac{DE}{BC}$$

$$\Rightarrow \frac{AD}{AD + DB} = \frac{DE}{BC}$$

$$\Rightarrow \frac{2}{2+3} = \frac{4}{x}$$

$\Rightarrow 2x = 20$

$\therefore x = 10$

38. In the given figure, $AD = 2$ cm, $DB = 3$ cm, $DE = 2.5$ cm and $DE \parallel BC$. The value of x is **(2023)**

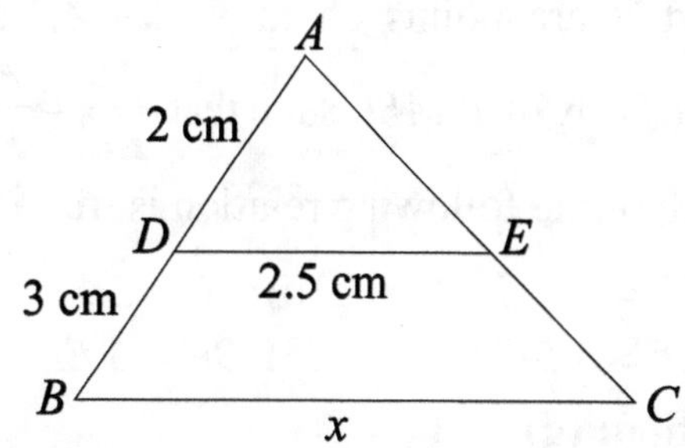

(a) 6 cm (b) 3.75 cm

(d) 6.25 cm (d) 7.5 cm

Solution: (c)

$DE \parallel BC$

$\therefore \Delta ADE \sim \Delta ABC$

$\therefore \frac{AD}{AB} = \frac{DE}{BC}$

$\Rightarrow \frac{2}{2+3} = \frac{2.5}{x}$

$\Rightarrow 2x = 2.5 \times 5$

$\therefore x = 6.25$ cm

39. In the given figure ΔABC is shown, DE is parallel to BC if $AD = 5$ cm, $DB = 2.5$ cm and $BC = 12$ cm then DE is equal to **(2024)**

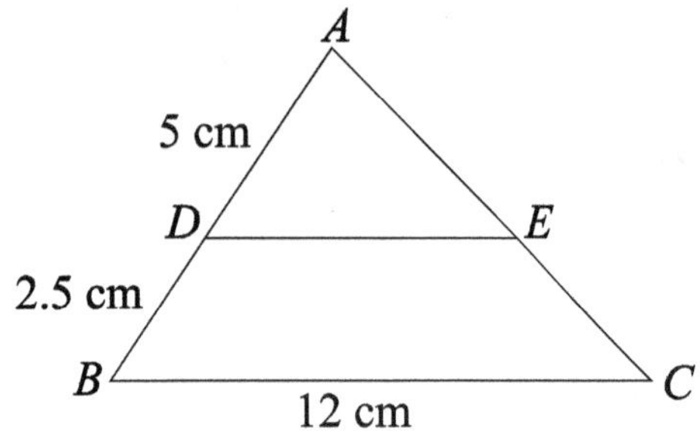

(a) 10 cm (b) 6 cm
(d) 8 cm (d) 7.5 cm

Solution: (c)

$DE \parallel BC$

$\therefore \Delta ADE \sim \Delta ABC$

$\therefore \frac{AD}{AB} = \frac{DE}{BC}$

$\Rightarrow \frac{5}{5+2.5} = \frac{DE}{12}$

$\Rightarrow DE = \frac{60}{7.5} = 8$ cm

40. E and F are points on the sides AB and AC respectively of a ΔABC such that $\frac{AE}{EB} = \frac{AF}{FC} = \frac{1}{2}$. Which of the following relation is true ? **(2025)**

(a) $EF = 2\,BC$ (b) $BC = 2\,EF$
(c) $EF = 3\,BC$ (d) $BC = 3\,EF$

Solution: (d)

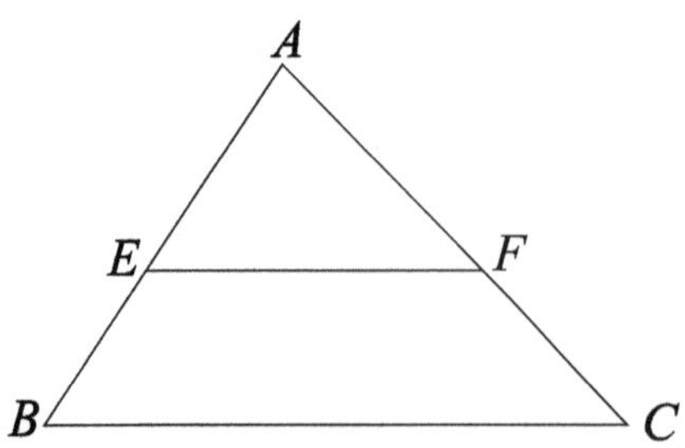

In ΔABC

$\frac{AE}{EB} = \frac{AF}{FC} = \frac{1}{2}$.

$\therefore EB = 2\,AE$

$\therefore EF \parallel BC$

$\therefore \Delta AEF \sim \Delta ABC$

$\therefore \frac{AE}{AB} = \frac{EF}{BC}$

$\Rightarrow \frac{EF}{BC} = \frac{AE}{AE+EB}$

$\Rightarrow \frac{EF}{BC} = \frac{AE}{AE+2AE}$

$\Rightarrow \frac{EF}{BC} = \frac{AE}{3AE}$

$\Rightarrow \frac{EF}{BC} = \frac{1}{3}$

$\Rightarrow BC = 3\,EF$

41. In ΔABC, $PQ \parallel BC$. It is given that $AP = 2.4$ cm, $PB = 3.6$ cm and $BC = 5.4$ cm. PQ is equal to **(2025)**

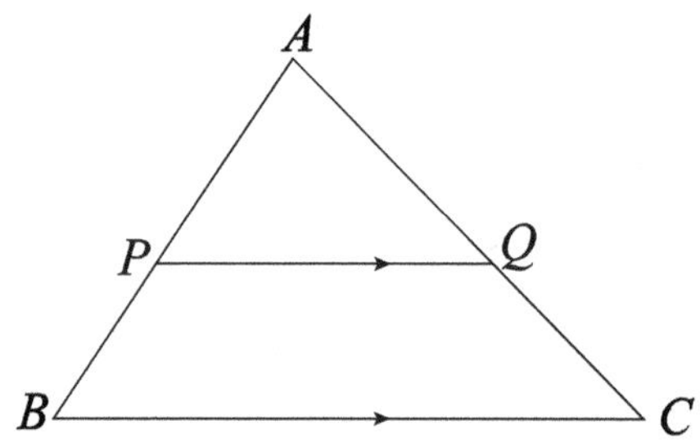

(a) 2.7 cm (b) 1.8 cm
(c) 3.6 cm (d) 2.16 cm

Solution:

$\Delta APQ \sim \Delta ABC$

$\therefore \frac{AP}{AB} = \frac{AQ}{AC} = \frac{PQ}{BC}$

$\Rightarrow \frac{AP}{AB} = \frac{PQ}{BC}$

$\Rightarrow \frac{2.4}{2.4+3.6} = \frac{PQ}{5.4}$

$\Rightarrow \frac{2.4}{6} = \frac{PQ}{5.4}$

$\Rightarrow PQ = 2.16$ cm

42. In The given figure, $AB \parallel EF$. If $AB = 24$ cm, $EF = 36$ cm and $DA = 7$ cm, then AE equals. **(2026)**

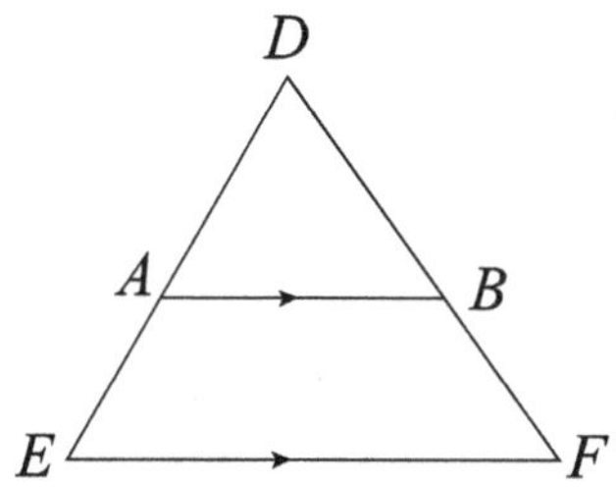

(a) 2.5 cm (b) 10.5 cm

(c) 3.5 cm (d) $\frac{14}{3}$ cm

Solution: (c)

In ΔDEF, $AB \parallel EF$

$\therefore \Delta DAB \sim \Delta DEF$

$\therefore \frac{DA}{DE} = \frac{AB}{EF}$

$\Rightarrow \frac{7}{DA+AE} = \frac{24}{36}$

$\Rightarrow \frac{7}{7+AE} = \frac{2}{3}$

$\Rightarrow 14 + 2AE = 21$

$\Rightarrow 2AE = 7$

$\Rightarrow AE = 3.5$ cm

43. In The given figure, $PQ \parallel YZ$ Such that $XP : PY = 2 : 3$. If $PQ = 5$ cm, Then YZ equals **(2026)**

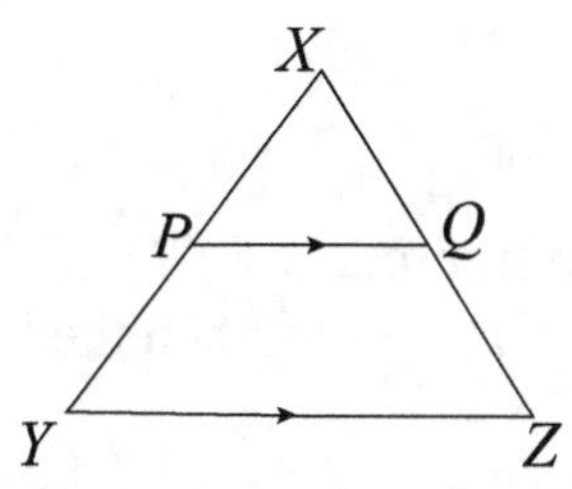

(a) 12.5 cm (b) 10 cm

(c) 15 cm (d) 7.5 cm

Solution: (a)

$PQ \parallel YZ$

$\therefore \Delta XPQ \sim \Delta XYZ$

$\therefore \frac{XP}{XY} = \frac{PQ}{YZ}$

$\Rightarrow \frac{2x}{5x} = \frac{5}{YZ}$

$\Rightarrow YZ = \frac{25}{2} = 12.5\,\text{cm}$

44. In the given figure ΔABC is shown, in which $DE \parallel BC$. If $AD = 5$ cm, $DB = 2.5$ cm and $DE = 8$ cm, then length of BC is : **(2026)**

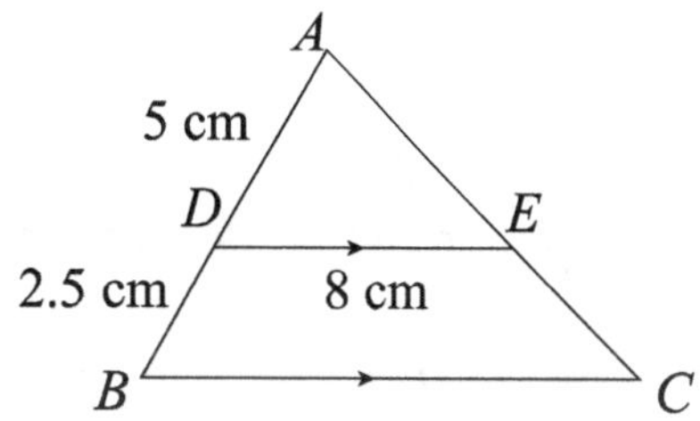

(a) 10 cm (b) 6 cm

(c) 12 cm (d) 7.5 cm

Solution: (c)

$DE \parallel BC$

$\therefore \Delta ADE \sim \Delta ABC$

$\therefore \frac{AD}{AB} = \frac{DE}{BC}$

$\Rightarrow \frac{5}{7.5} = \frac{8}{BC} = [\because AB = AD + DB]$

$\Rightarrow BC = 12$ cm

Exercise 7.4

(i) In the given figure ΔABC is shown, DE is parallel to BC. Then value of x is

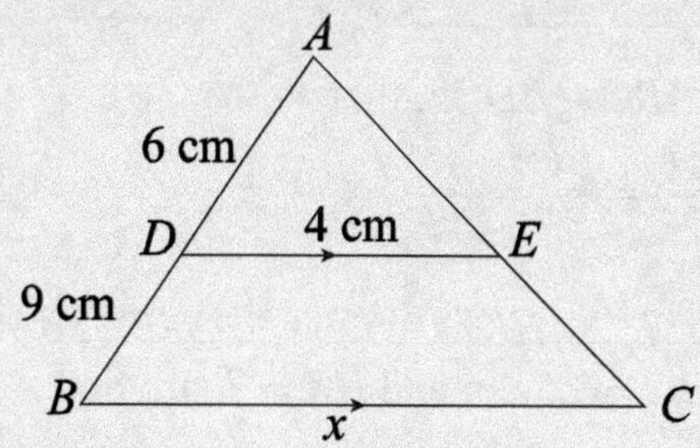

(a) 10 (b) 12

(c) 8 (d) 15

(ii) In the given figure, $DE \parallel BC$. Then value of x is

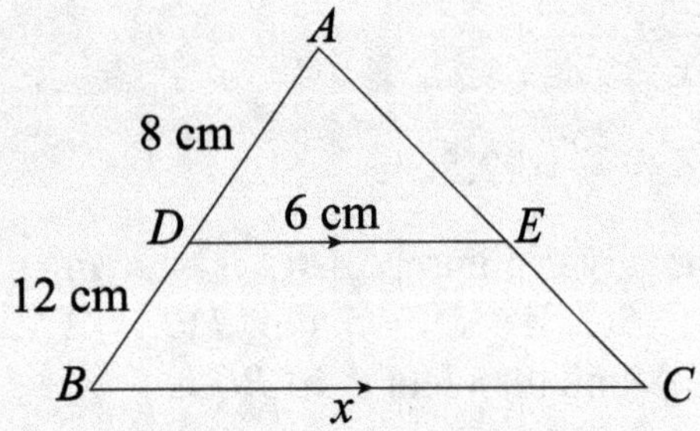

(a) 10 (b) 12
(c) 20 (d) 15

(iii) In the given figure, $DE \parallel BC$. Then value of x is

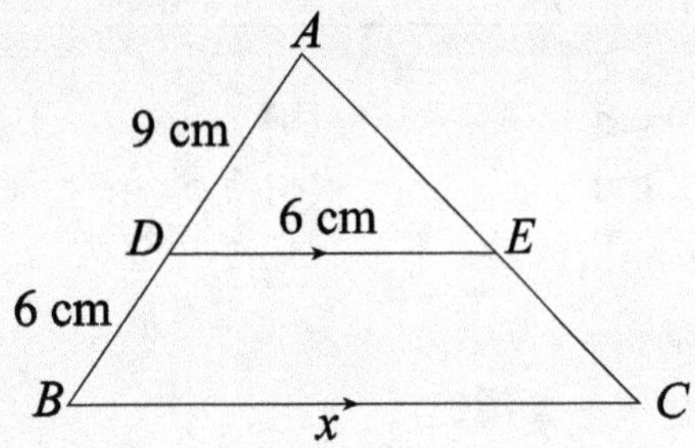

(a) 5 (b) 8
(c) 9 (d) 10

Key Point

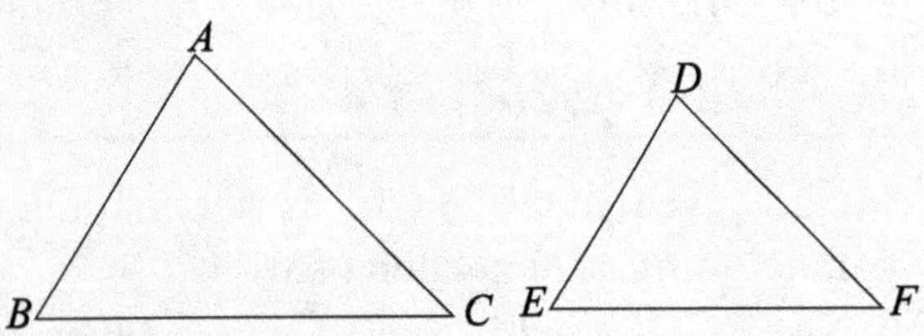

In $\Delta ABC \sim \Delta DEF$, then.

$$\therefore \frac{AB}{DE} = \frac{BC}{EF} = \frac{AC}{DF}$$

45. In fig. 2, ΔAHK is similar to ΔABC. If $AK = 10$ cm, $BC = 3.5$ cm and $HK = 7$ cm, find AC. **(2010)**

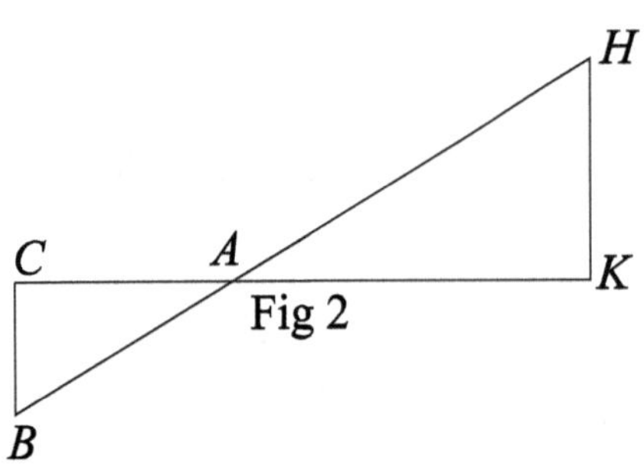

Fig 2

Solution:

$\Delta AHK \sim \Delta ABC$

$$\therefore \frac{HK}{BC} = \frac{AK}{AC}$$

$$\Rightarrow \frac{7}{3.5} = \frac{10}{AC}$$

$$\therefore AC = 5 \text{ cm}$$

46. In the given figure, $AB \parallel PQ$. If $AB = 6$ cm, $PQ = 2$ cm and $OB = 3$ cm, then the length of OP is **(2023)**

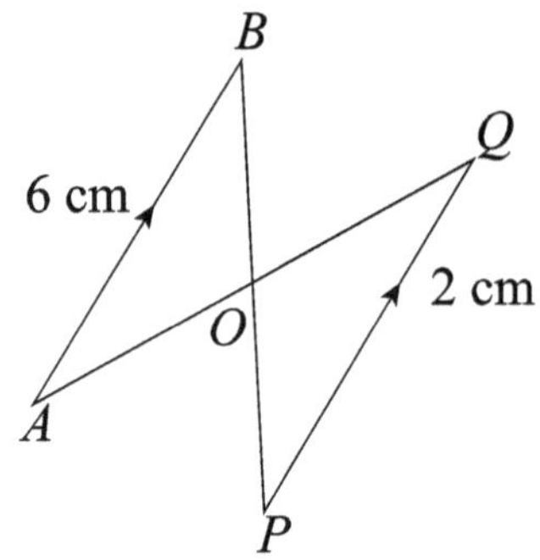

(a) 9 cm (b) 3 cm
(c) 4 cm (d) 1 cm

Solution: (d)

$AB \parallel PQ$

$\therefore \Delta AOB \sim \Delta QOP$

$$\therefore \frac{OB}{OP} = \frac{AB}{PQ} = \frac{OA}{OQ}$$

$$\therefore \frac{OB}{OP} = \frac{AB}{PQ}$$

$$\therefore \frac{3}{OP} = \frac{6}{2}$$

$$\therefore OP = 1 \text{ cm}$$

47. In the given figure, $AB \parallel CD$. If $AB = 5$ cm, $CD = 2$ cm and $OB = 3$ cm, then the length of OC is **(2023)**

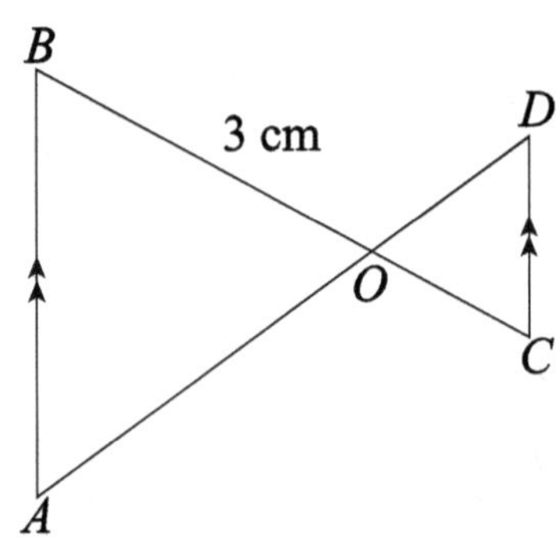

(a) $\frac{15}{2}$ cm (b) $\frac{10}{3}$ cm

(c) $\frac{6}{5}$ cm (d) $\frac{3}{5}$ cm

Solution: (c)

$AB \parallel CD$

$\therefore \Delta AOB \sim \Delta DOC$

$\therefore \frac{OB}{OC} = \frac{AB}{CD} = \frac{OA}{OD}$

$\therefore \frac{OB}{OC} = \frac{AB}{CD}$

$\Rightarrow \frac{3}{OC} = \frac{5}{2}$

$\Rightarrow OC = \frac{6}{5}$ cm

48. In the adjoining figure, $ABCD$ is a trapezium in which $XY \parallel AB \parallel CD$. If $AX = \frac{2}{3}AD$, then CY : YB = **(2025)**

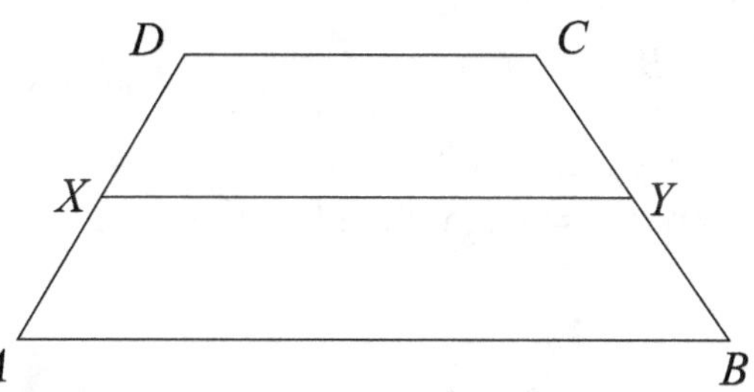

(a) 2 : 3 (b) 3 : 2

(c) 1 : 3 (d) 1 : 2

Solution: (d)

$AX = \frac{2}{3}AD$

$\Rightarrow AD = \frac{3}{2}AX$

Now, $DX = AD - AX$

$= \frac{3}{2}AX - AX$

$= \frac{1}{2}AX$

$\therefore AX : DX = AX : \frac{1}{2}AX$

$= 1 : \frac{1}{2}$

$= 2 : 1$

$\Rightarrow DX : AX = 1 : 2$

Now, In the trapezium $ABCD$, $XY \parallel AB \parallel CD$

$CY : BY = DX : AX$

$\Rightarrow CY : BY = 1 : 2$

49. In the given figure, $OA \times OB = OC \times OD$. Which of the following option is correct? **(2026)**

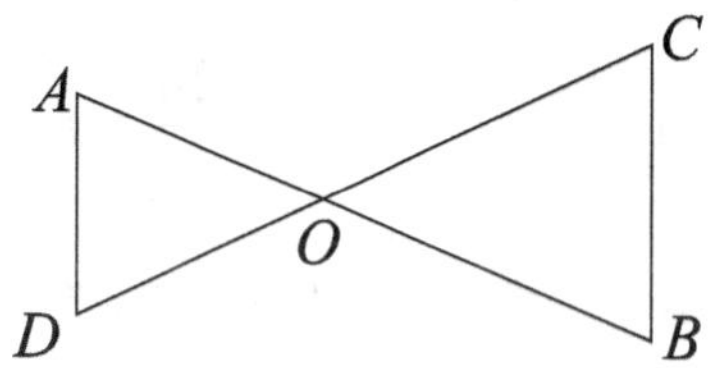

(a) $\angle A = \angle C$

(b) $\angle A = \angle B$

(c) $\angle A = \angle D$

(d) $\Delta OAD \sim \Delta OBC$

Solution:

In ΔAOD and ΔBOC

$\frac{OA}{OC} = \frac{OD}{OB}$ $[\because OA \times OB = OC \times OD]$

and $\angle AOD = \angle BOC$ [vertically opposite angles]

$\therefore \Delta OAD \sim \Delta OCB$

$\therefore \angle A = \angle C$

50. In the adjoining figure, $\Delta PQT \sim \Delta RST$, which of the following is true?

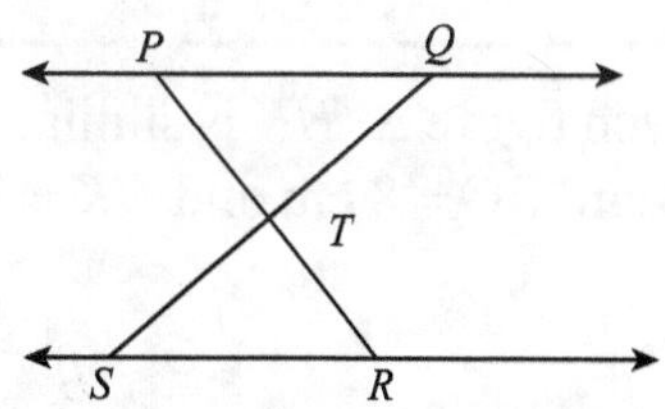

(a) $\frac{PQ}{ST} = \frac{QT}{RS}$ (b) $QT.RT = PT.ST$

(c) $PQ.PT = RS.RT$ (d) $\frac{PQ}{PT} = \frac{RT}{RS}$

Solution: (b)

$\Delta PQT \sim \Delta RST$

$\therefore \frac{PQ}{RS} = \frac{QT}{ST} = \frac{PT}{RT}$

$\therefore \frac{QT}{ST} = \frac{PT}{RT}$

$\Rightarrow QT.RT = PT.ST$

51. $ABCD$ is a parallelogram such that $AF = 7$ cm, $FB = 3$ cm and $EF = 4$ cm length $FD =$ equals **(2026)**

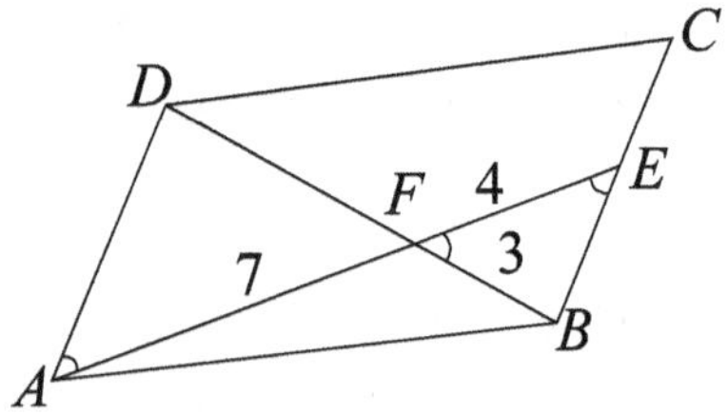

(a) $\dfrac{21}{4}$ cm (b) $\dfrac{28}{3}$ cm

(c) $\dfrac{12}{7}$ cm (d) 5.5 cm

Solution: (b)

In ΔAFD and ΔBFE

$\angle FAD = \angle FEB$

$\angle AFD = \angle BFE$

$\therefore \Delta FDA \sim \Delta FBE$

$\therefore \dfrac{FB}{FA} = \dfrac{FE}{FD}$

$\Rightarrow \dfrac{3}{7} = \dfrac{4}{FD}$

$\therefore FD = \dfrac{28}{3}$ cm

Exercise 7.5

(i) In the given figure ΔAHK is similar to ΔABC.If $AK = 12$ cm, $BC = 8$ cm and $HK = 16$ cm, find AC.

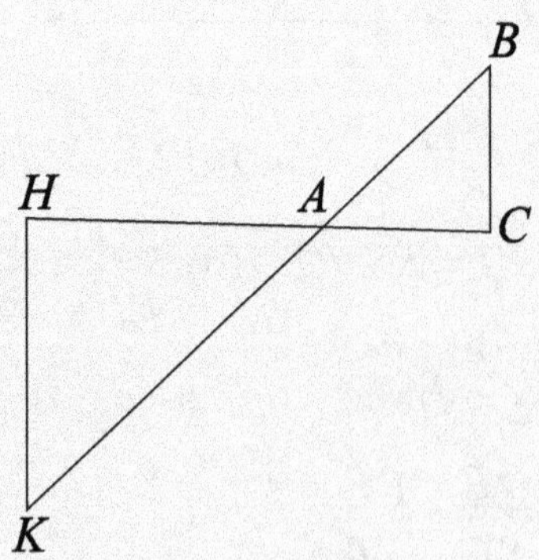

(ii) In the given figure ΔAHK is similar to ΔABC.If $AK = 14$ cm, $BC = 4.5$ cm and $HK = 9$ cm, find AC.

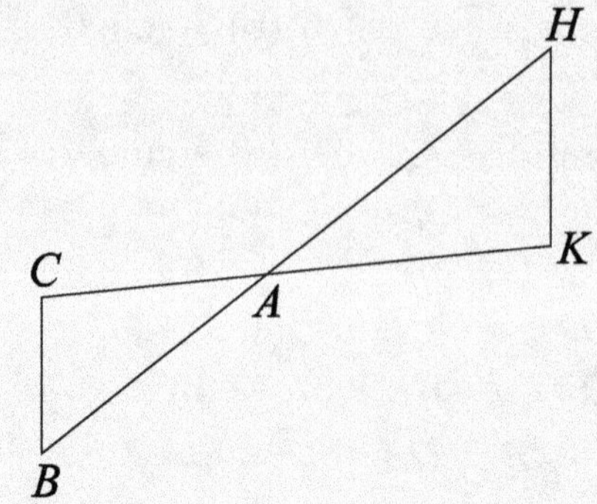

(iii) In the given figure ΔAHK is similar to ΔABC. If $AK = 20$ cm, $BC = 7.5$ cm and $HK = 15$ cm, find AC.

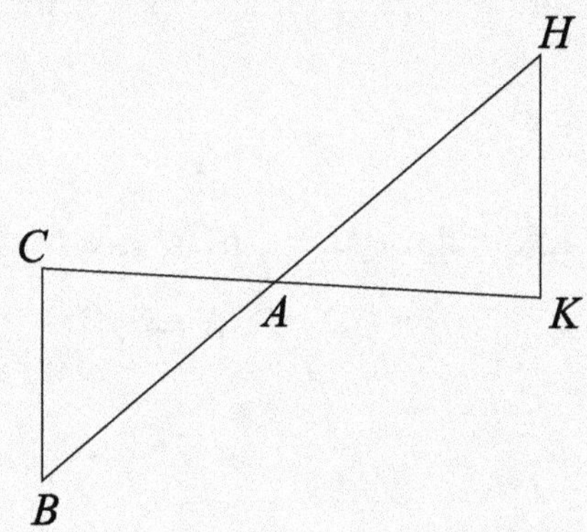

52. In the given figure, $\angle A = \angle C$, $AB = 6$ cm, $AP = 12$ cm, $CP = 4$ cm, then length of CD is **(2023)**

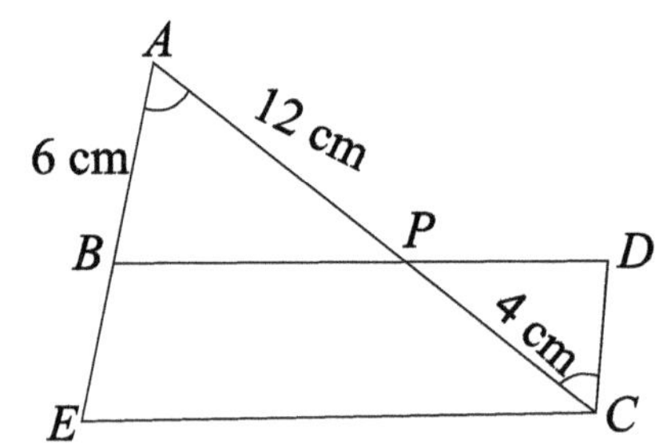

(a) 2 cm (b) 6 cm

(c) 8 cm (d) 18 cm

Solution: (a)

$\Delta APB \sim \Delta CPD$

$\therefore \dfrac{AP}{PC} = \dfrac{AB}{CD}$

$\Rightarrow \dfrac{12}{4} = \dfrac{6}{CD}$

$\Rightarrow CD = 2$ cm.

Exercise 7.6

(i) In the given figure, $\angle A = \angle C$, $AB = 9$ cm. $AP = 6$ cm, $CP = 4$ cm, then the length of CD is

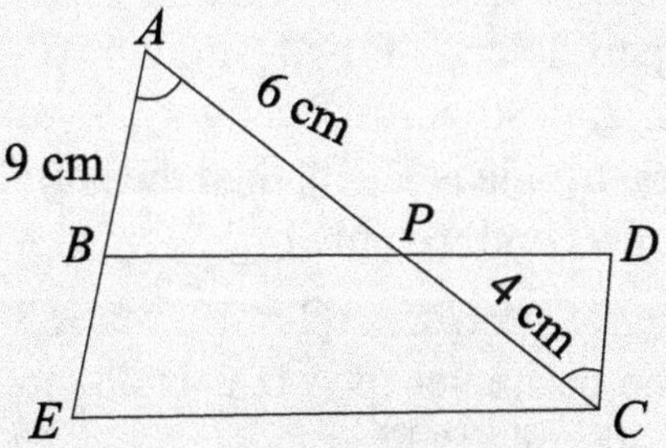

(a) 4 cm (b) 6 cm
(c) 8 cm (d) 2 cm

(ii) In the given figure, $\angle A = \angle C$, $AB = 12$ cm. $AP = 9$ cm, $CP = 6$ cm, then the length of CD is

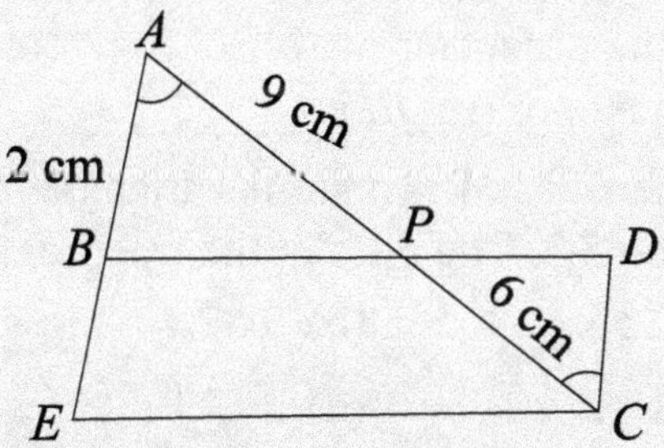

(a) 4 cm (b) 6 cm
(c) 8 cm (d) 12 cm

(iii) In the given figure, $\angle A = \angle C$, $AB = 10$ cm, $AP = 5$ cm, $CP = 7$ cm, then the length of CD is

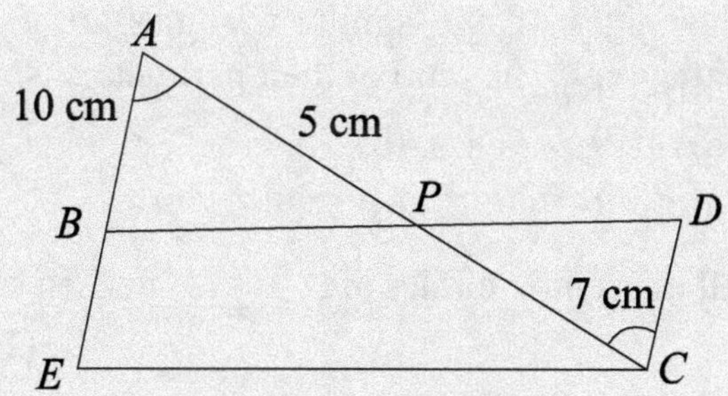

(a) 14 cm (b) 10 cm
(c) 9 cm (d) 8 cm

Key Point

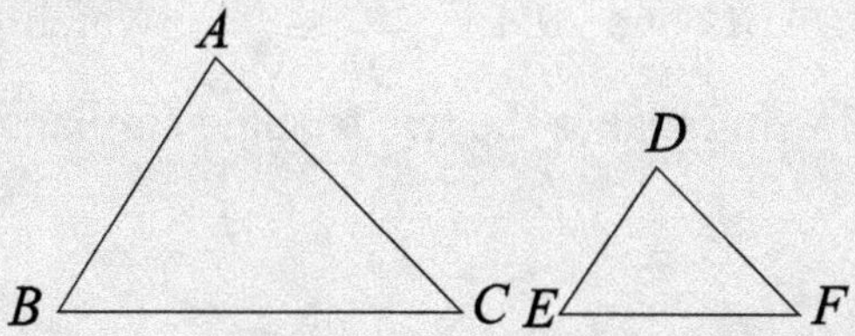

If $\Delta ABC \sim \Delta DEF$ then,

$$\frac{AB}{DE} = \frac{BC}{EF} = \frac{AC}{DF} = \frac{AB+BC+AC}{DE+EF+DF}$$

53. The perimeter of two similar triangles are 25 cm and 15 cm respectively. If one side of the first triangle is 9 cm, then the corresponding side of second triangle is **(2020)**

Solution:

$$\frac{\text{Perimeter of 1st triangle}}{\text{Perimeter of 2nd triangle}} = \frac{\text{Side of first triangle}}{\text{Corresponding side of second triangle}}$$

$$\Rightarrow \frac{25}{15} = \frac{9}{x}$$

$$\Rightarrow x = 5.4 \text{ cm.}$$

54. In ratio of the corresponding side of two similar triangles is 2:3, Then ratio of their perimeter is ________. **(2020)**

Solution:

The ratio of perimeter of two similar triangle is equel to ratio of their corresponding sides = 2:3

55. If $\Delta PQR \sim \Delta ABC$, $PQ = 6$ cm, $AB = 8$ cm and the perimeter of ΔABC is 36 cm, then the perimeter of ΔPQR is **(2023)**

(a) 20.25 cm (b) 27 cm
(c) 48 cm (d) 64 cm

Solution: (b)

$$\Delta PQR \sim \Delta ABC$$

$$\therefore \frac{\text{Perimeter of } \Delta PQR}{\text{Perimeter of } \Delta ABC} = \frac{PQ}{AB}$$

$$\Rightarrow \frac{\text{Perimeter of } \Delta PQR}{36} = \frac{6}{8}$$

$$\Rightarrow \text{Perimeter of } \Delta PQR = 27 \text{ cm.}$$

56. In sides of two similar triangles are in the ratio 4:7. The ratio of their perimeter is **(2023)**

(a) 4:7 (b) 12:21
(c) 16:49 (d) 7:4

Solution: (a)

Ratio of two sides of two triangles =
Ratio of their perimeter = 4 : 7

57. $\Delta ABC \sim \Delta DEF$ and their perimeters are 32 cm and 24 cm respectively. If $AB = 10$ cm, then DE equals **(2023)**

(a) 8 cm (b) 7.5 cm
(c) 15 cm (d) $5\sqrt{3}$cm

Solution: (b)

$\Delta ABC \sim \Delta DEF$

$\therefore \frac{AB}{DE} = \frac{\text{Perimeter of } \Delta ABC}{\text{Perimeter of } \Delta DEF}$

$\Rightarrow \frac{10}{DE} = \frac{32}{24}$

$\therefore DE = 7.5$ cm

58. The Perimeters of two similar triangles ABC and PQR are 56 cm and 48 cm respectively. PQ / AB is equal to **(2024)**

(a) $\frac{7}{8}$ (b) $\frac{6}{7}$
(c) $\frac{7}{6}$ (d) $\frac{8}{7}$

Solution: (c)

$\frac{\text{Perimeter of } \Delta ABC}{\text{Perimeter of } \Delta PQR} = \frac{56}{48} = \frac{7}{6}$

59. It is given that $\Delta ABC \sim \Delta EDF$. Which of the following is not true? **(2026)**

(a) $\frac{\text{Perimeter } \Delta ABC}{\text{Perimeter } \Delta EDF} = \frac{AB}{ED}$

(b) $\frac{AB}{ED} = \frac{AC}{EF}$

(c) $\angle A = \angle D, \angle C = \angle F$

(d) $\frac{AB + BC}{AC} = \frac{DE + DF}{EF}$

Solution: (c)

$\angle A = \angle D, \angle C = \angle F$

Exercise 7.7

(i) The perimeters of two similar triangles are 30 cm and 40 cm respectively. If one side of the first tringle is 12 cm, then the corresponding side of second triangle is__________

(ii) The perimeters of two similar triangles are 36 cm and 42 cm respectively. If one side of the first tringle is 12 cm, then the corresponding side of second triangle is__________

(iii) The perimeters of two similar triangles are 60 cm and 80 cm respectively. If one side of the first tringle is 15cm, then the corresponding side of second triangle is__________

(iv) If $\Delta PQR \sim \Delta ABC$, $PQ = 12$ cm, $AB = 16$ cm and the perimeter of ΔABC is 48 cm, then the perimeter of ΔPQR is __________.

(v) If $\Delta PQR \sim \Delta ABC$; $PQ = 9$ cm $AB = 15$ cm and the perimeter of ΔABC is 45 cm, then the perimeter of ΔPQR is __________.

(vi) If $\Delta PQR \sim \Delta ABC$; $PQ = 11$ cm $AB = 16$ cm and the perimeter of ΔABC is 48 cm, then the perimeter of ΔPQR is __________.

(vii) The sides of two similar triangles are in the ratio 3:2. The ratio of their perimeters is

(a) 2 : 3 (b) 3 : 2
(c) 1 : 2 (d) 2 : 1

(viii) The sides of two similar triangles are in the ratio 8:15, The ratio of their perimeters is

(a) 4 : 3 (b) 3 : 4
(c) 8 : 15 (d) 15 : 8

(ix) The sides of two similar triangles are in the ratio 4 : 9, The ratio of their perimeters is

(a) 4 : 9 (b) 9 : 4
(c) 2 : 3 (d) 3 : 2

60. All concentric circles are _______to each other. **(2020)**

Solution: Similar

61. All square are ________(congruent/similar) **(2020)**

Solution: Similar.

62. In ΔABC and ΔDEF, $\frac{AB}{DE} = \frac{BC}{FD}$, which of the following makes the two triangle similar **(2023)**

(a) $\angle A = \angle D$ (b) $\angle B = \angle D$
(c) $\angle B = \angle E$ (d) $\angle A = \angle F$

Solution: (b)

$\angle B = \angle D$

63.

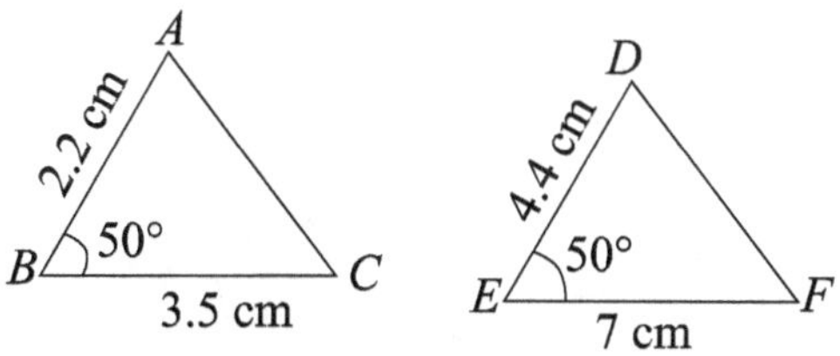

In the above figure, the criteron of similarity by which $\Delta ABC \sim \Delta PQR$ is **(2023)**

(a) SSA (Side - Side - Angle) similarity
(b) ASA (Angle – Side – Angle) similarity
(c) SAS (Side – Angle – Side) similarity
(d) AA (Angle – Angle) similarity

Solution: (c) SAS

64. In two triangles ΔPQR and ΔABC. It is given that $\frac{AB}{BC} = \frac{PQ}{PR}$. For these two triangles to be similar, which of the following should be true ? **(2023)**

(a) $\angle A = \angle P$ (b) $\angle B = \angle Q$
(c) $\angle B = \angle P$ (d) $CA = QR$

Solution: (c)

$\angle B = \angle P$

65. If in triangles ABC and PQR, $\frac{AB}{QR} = \frac{BC}{PR}$; then they will be similar, when **(2024)**

(a) $\angle B = \angle Q$ (b) $\angle A = \angle R$
(c) $\angle B = \angle R$ (d) $\angle C = \angle Q$

Solution: (c)

$\angle B = \angle R$

66. In Figure 1, ABC is an isosceles triangle right, angled at C with $AC = 4$ cm. Find the length of AB. **(2019)**

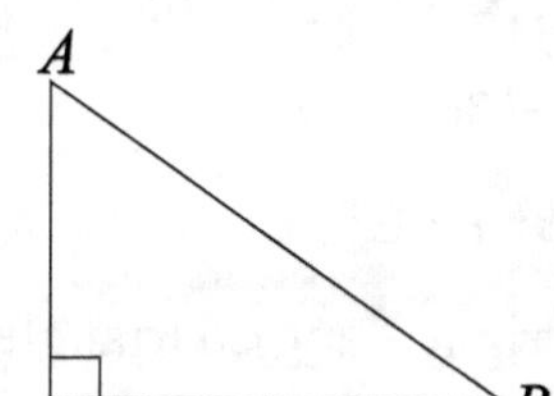

Figure 1

Solution:

In ΔABC $\angle C = 90°$

$\therefore AB^2 = AC^2 + BC^2$

$\Rightarrow AB^2 = 4^2 + 4^2$ [$\because AC = BC = 4$ cm]

$\Rightarrow AB^2 = 32$

$\therefore AB = 4\sqrt{2}$cm

67. In Figure–1, ABC is an isosceles triangle, right - angled at C. Therefore **(2020)**

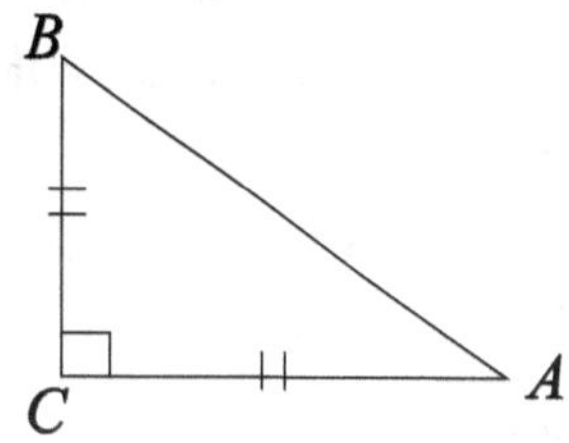

(a) $AB^2 = 2AC^2$ (b) $BC^2 = 2AB^2$
(c) $AC^2 = 2AB^2$ (d) $AB^2 = 4AC^2$

Solution: (a)

In ΔABC $\angle C = 90°$

$\therefore AB^2 = AC^2 + BC^2$

$= AC^2 + AC^2$ [$\because AB = AC$]

$= 2AC^2$

68. ΔABC is an isosceles with $AC = BC$. If $AB^2 = 2AC^2$, then find the measure of $\angle C$. **(2020)**

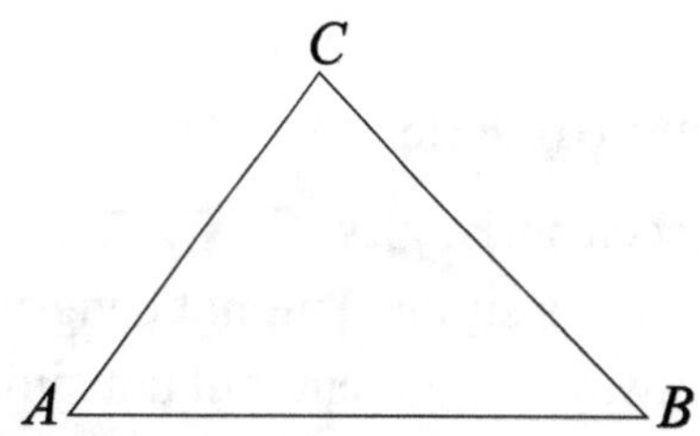

Solution:

$AB^2 = 2AC^2$

$= AC^2 + AC^2$

$= AC^2 + BC^2$ [$\because AC = BC$]

$\therefore ABC$ is a right - angled triangle.

$\therefore \angle C = 90°$

69. In triangle ABC and DEF, $\angle B = \angle E$, $\angle F = \angle C$ and $AB = 3DE$. Then, the two triangles are **(2025)**

(a) Congurent but not similar
(b) Congurent as well as similar

(c) Niether congruent not similar
(d) Smilar but not congurent

Solution: (d)

70. Which of the following statements is incorrect ? **(2025)**

(a) Two congruent figures are always similar.
(b) A square and a rhombus of the same area are always similar.
(c) Two equilateral triangles are always similar.
(d) Two similar triangles need not be congruent.

Solution: (b)

A square and a rhombus of the same area are always similar.

71. Which of the following statements is false ? **(2025)**

(a) Two right triangles are always similar
(b) Two squares are always similar
(c) Two quilateral triangles are always similar
(d) Two circles are always similar

Solution: (a)

Two right triangles are always similar

72. Which of the following is not the oriterion for similarity of triangles ? **(2025)**

(a) AAA (b) SSS
(c) SAS (d) RHS

Solution: (d) RHS

73. Two isosceles triangles: **(2026)**

(a) are always similar, but not congruent
(b) are always congruent, but not similar
(c) may or may not be similar
(c) are always similar and congruent

Solution: (c)

may or may not be similar

74. Which of the following statements is not always true? **(2026)**

(a) Two circles are similar.
(b) Two isosceles right triangles are similar.
`(c) Two rectangles are similar.
(d) Two equilateral triangles are similar.

Solution : (b)

Two isosceles right triangles are similar.

Exercise 7.8

(i) In the given figure, ABC is an isosceles triangle, right – angled at C with $AC = 30$ cm. Find the length of AB.

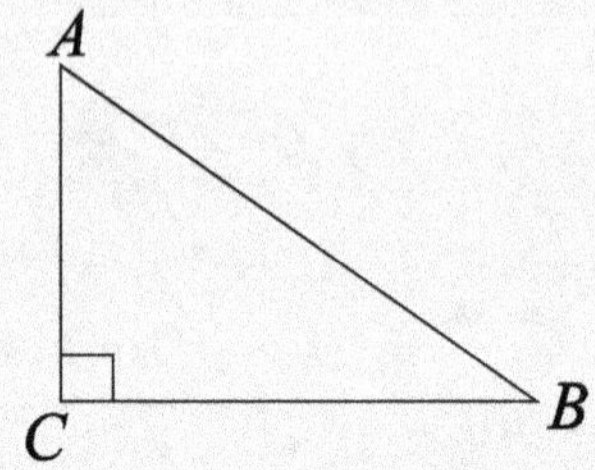

(ii) In the given figure, ABC is an isosceles triangle, right – angled at C, with $AC = 5$ cm, Find the length of AB.

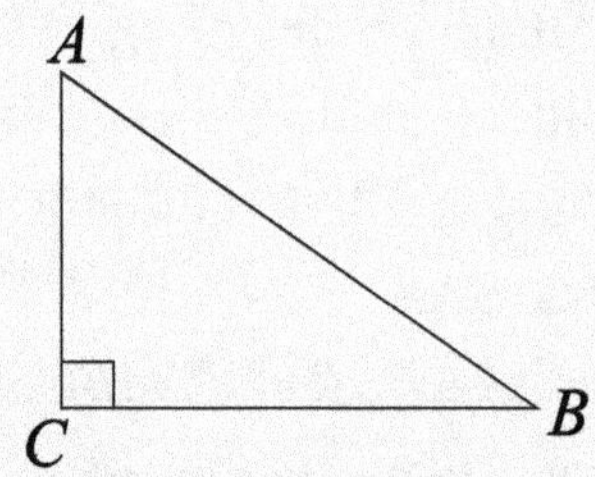

(iii) In the given figure, ABC is an isosceles triangle, right – angled at C, with $AC = a$ cm, Find the length of AB.

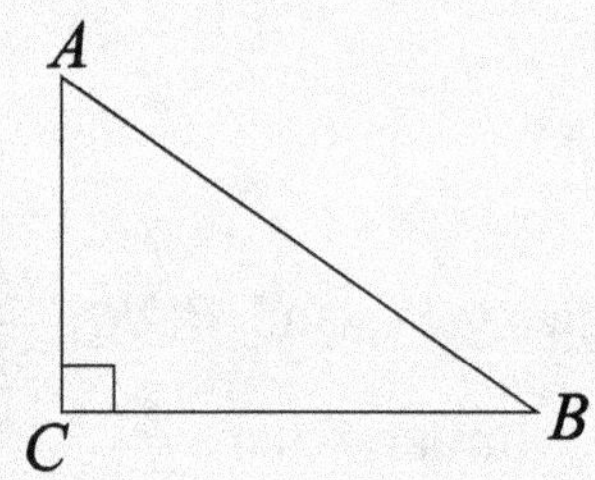

75. In ΔABC, $AB = 6\sqrt{3}$ cm, $AC = 12$ cm and $BC = 6$ cm, Then $\angle B =$ ________ **(2020)**

Solution:

$AC^2 = (12)^2 = 144$

$AB^2 = \left(6\sqrt{3}\right)^2 = 36 \times 3 = 108$

$(BC)^2 = (6)^2 = 36$

$\therefore AC^2 = AB^2 + BC^2$

$\therefore ABC$ is a right – angled triangle and $\angle B = 90°$

Key Point

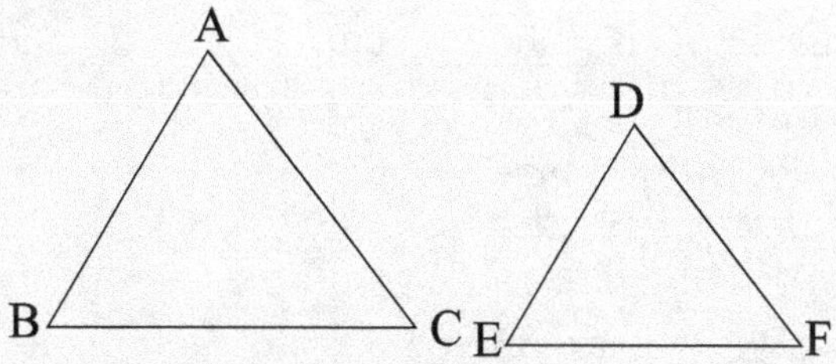

If $\Delta ABC \sim \Delta DEF$, Then

$$\frac{ar(\Delta ABC)}{ar(\Delta DEF)} = \frac{AB^2}{DE^2} = \frac{AC^2}{DF^2} = \frac{BC^2}{EF^2}$$

76. In figure 1, $PQ \parallel BC$ and $AP: PB = 1:2$. Find $\frac{ar(\Delta APQ)}{ar(ABC)}$ **(2008)**

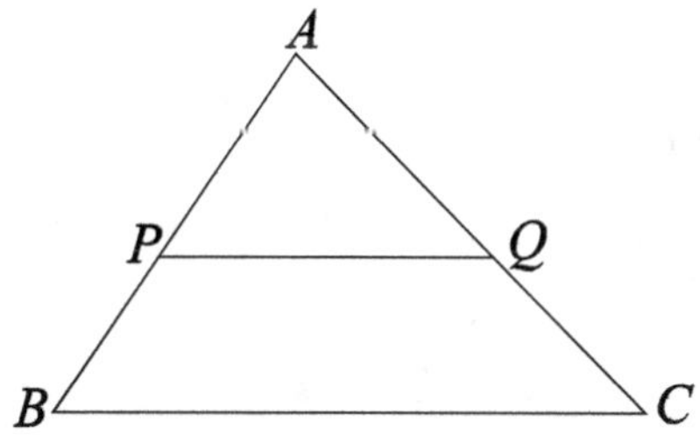

Solution:

$\frac{AP}{PB} = \frac{1}{2}$

$\Rightarrow \frac{PB}{AP} = 2$

$\Rightarrow \frac{PB}{AP} + 1 = 2 + 1$

$\Rightarrow \frac{PB + AP}{AP} = 3$

$\Rightarrow \frac{AB}{AP} = 3$

$\Rightarrow \frac{AP}{AB} = \frac{1}{3}$

Now, $PQ \parallel BC$

$\therefore \Delta APQ \sim \Delta ABC$

$$\therefore \frac{ar(\Delta APQ)}{ar(\Delta ABC)} = \frac{AP^2}{AB^2} = \left(\frac{AP}{AB}\right)^2 = \left(\frac{1}{3}\right)^2 = \frac{1}{9}$$

77. In Fig. 1, S and T are points on the sides PQ and PR respectively of ΔPQR, such that $PT = 2$ cm, $TR = 4$ cm and ST is parallel to QR. Find the ratio of the areas of ΔPST and ΔPQR. **(2010)**

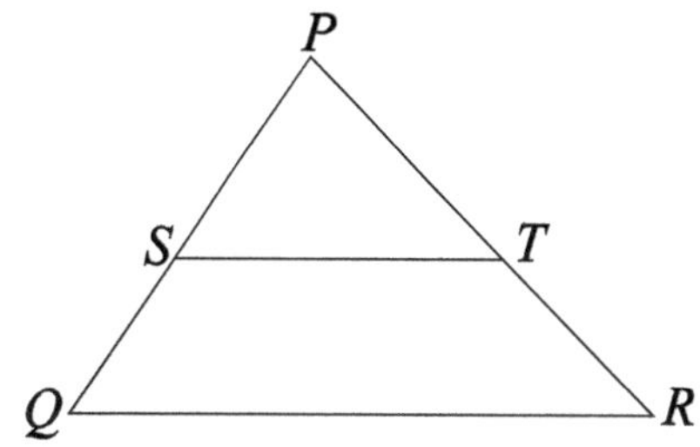

Solution:

$PR = PT + TR = 2 + 4 = 6$ cm

Now, $ST \parallel QR$

$\therefore \Delta PST \sim \Delta PQR$

$$\therefore \frac{ar(\Delta PST)}{ar(\Delta PQR)} = \frac{PT^2}{PR^2} = \left(\frac{PT}{PR}\right)^2 = \left(\frac{2}{6}\right)^2 = \frac{1}{9}$$

78. Given $\Delta ABC \sim \Delta PQR$, If $\frac{AB}{PQ} = \frac{1}{3}$, then find $\frac{ar\Delta ABC}{ar\Delta PQR}$ **(2018, 2019)**

Solution:

$\Delta ABC \sim \Delta PQR$

$$\therefore \frac{ar(\Delta ABC)}{ar(\Delta PQR)} = \frac{AB^2}{PQ^2} = \left(\frac{AB}{PQ}\right)^2 = \left(\frac{1}{3}\right)^2 = \frac{1}{9}$$

79. In Fig. 1, $DE \parallel BC$, $AD = 1$ cm and $BD = 2$ cm. What is the ratio of the ar (ΔABC) to the ar (ΔADE)? **(2019)**

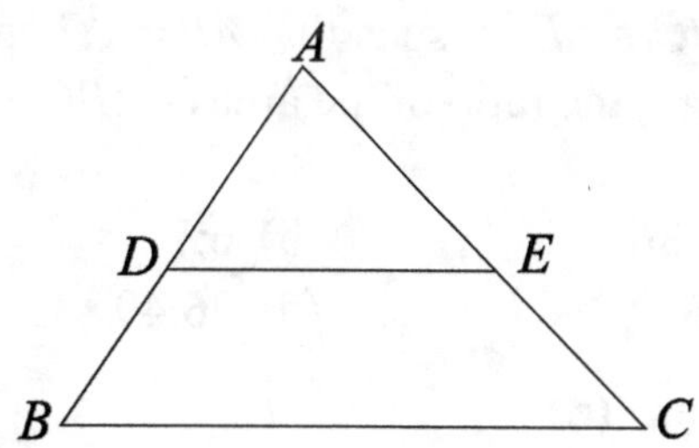

Solution:

$AB = AD + DB = 1 + 2 = 3$ cm

$\Delta ADE \sim \Delta ABC$

$$\therefore \frac{ar(\Delta ADE)}{ar(\Delta ABC)} = \frac{AD^2}{AB^2} = \left(\frac{AD}{AB}\right)^2 = \left(\frac{1}{3}\right)^2 = \frac{1}{9}$$

$\therefore$ ar (ΔADE): ar (ΔABC) = 1:9

80. Let $\Delta ABC \sim \Delta DEF$ and their areas be respectively 64 cm^2 and 121 cm^2. If $EF = 15.4$ cm, find BC. **(2019)**

Solution:

$\Delta ABC \sim \Delta DEF$

$$\therefore \frac{ar(\Delta ABC)}{ar(\Delta DEF)} = \frac{BC^2}{EF^2}$$

$$\Rightarrow \frac{BC^2}{EF^2} = \frac{ar(\Delta ABC)}{ar(\Delta DEF)}$$

$$\Rightarrow \frac{BC^2}{(15.4)^2} = \frac{64}{121}$$

$$\Rightarrow BC^2 = \frac{64}{121} \times (15.4)^2$$

$\therefore BC = 11.2$ cm

81. The area of two similar triangles are 25 sq. cm and 121 sq. cm. Find the ratio of their corresponding sides. **(2019)**

Solution:

$$\frac{\text{area of } \Delta_1}{\text{area of } \Delta_2} = \frac{(\text{side of } \Delta_1)^2}{(\text{side of } \Delta_2)^2}$$

$$\Rightarrow \left(\frac{\text{side of } \Delta_1}{\text{side of } \Delta_2}\right)^2 = \frac{25}{121}$$

$$\Rightarrow \frac{\text{side of } \Delta_1}{\text{side of } \Delta_2} = \frac{5}{11}$$

$\therefore$ side of $\Delta 1$: side of $\Delta 2 = 5:11$

82. If $\Delta ABC \sim \Delta DEF$ such that $AB = 1.2$ cm and $DE = 1.4$ cm, the ratio of the areas ΔABC and ΔDEF is **(2020)**

(a) 49:36 (b) 6:7

(c) 7:6 (d) 36:49

Solution: (d)

$$\frac{ar(\Delta ABC)}{ar(\Delta DEF)} = \frac{(1.2)^2}{(1.4)^2}$$

$$= \frac{1.44}{1.96}$$

$$= \frac{36}{49}$$

$\therefore$ ar (ΔABC): ar (ΔDEF) = 36 : 49

83. Let $\Delta ABC \sim \Delta DEF$ and their areas be respectively 81 cm^2 and 144 cm^2. If $EF = 24$ cm, then length of side BC is ________ cm. **(2020)**

Solution:

$$\frac{ar(\Delta ABC)}{ar(\Delta DEF)} = \frac{BC^2}{EF^2}$$

$$\Rightarrow \left(\frac{BC}{EF}\right)^2 = \frac{\text{ar}(\Delta ABC)}{\text{ar}(\Delta DEF)}$$

$$\Rightarrow \left(\frac{BC}{24}\right)^2 = \frac{81}{144}$$

$$\Rightarrow \frac{BC}{24} = \frac{9}{12}$$

$\therefore BC = 18$ cm.

84. The areas of two similar triangles ABC and PQR are 25 cm^2 and 49 cm^2 respectively. If $QR = 9.8$ cm, find BC. **(2020)**

Solution:

$$\frac{ar(\Delta ABC)}{ar(\Delta PQR)} = \frac{BC^2}{QR^2}$$

$$\Rightarrow \frac{25}{49} = \frac{BC^2}{(9.8)^2}$$

$$\Rightarrow \left(\frac{BC}{9.8}\right)^2 = \left(\frac{5}{7}\right)^2$$

$$\Rightarrow \frac{BC}{9.8} = \frac{5}{7}$$

$\Rightarrow BC = 7$ cm.

85. ABC and BDE are two equilateral triangle such that D is the mid–point of BC. Ratio of the areas of triangles ABC and BDE is ________ **(2020)**

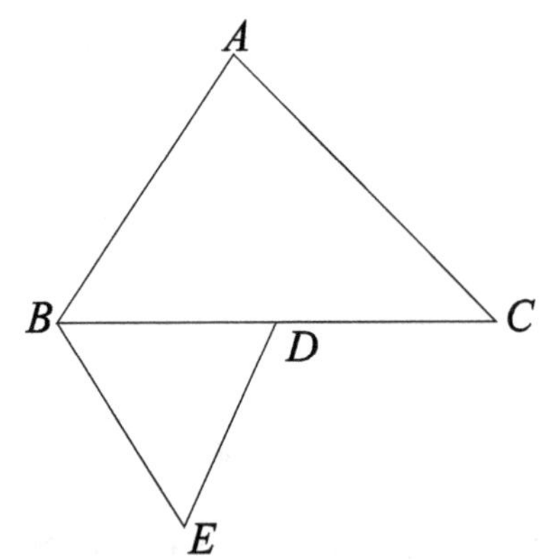

Solution:

$\Delta ABC \sim \Delta BDE$

$\therefore \dfrac{ar(\Delta ABC)}{ar(\Delta BDE)} = \dfrac{BC^2}{BD^2}$

$= \dfrac{(2BD)^2}{BD^2}$

$= 4$

$\therefore$ ar (ΔABC): ar (ΔBDE) = 4:1

86. In fig. 1, $MN \parallel BC$ and $AM : MB = 1:2$, then $\dfrac{ar(\Delta AMN)}{ar(\Delta ABC)} =$ ______ **(2020)**

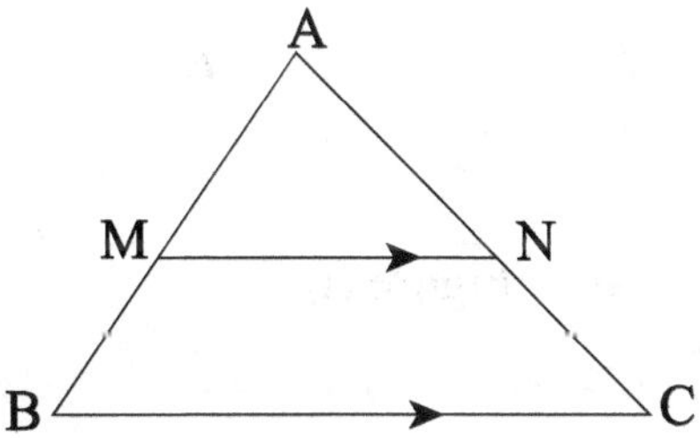

Solution:

$\dfrac{AM}{MB} = \dfrac{1}{2}$

$\Rightarrow \dfrac{MB}{AM} = 2$

$\Rightarrow \dfrac{MB}{AM} + 1 = 2 + 1$

$\Rightarrow \dfrac{MB + AM}{AM} = 3$

$\Rightarrow \dfrac{AB}{AM} = 3$

$\Rightarrow \dfrac{AM}{AB} = \dfrac{1}{3}$

Now, $MN \parallel BC$

$\therefore \Delta AMN \sim \Delta ABC$

$\therefore \dfrac{ar(\Delta AMN)}{ar(\Delta ABC)} = \dfrac{AM^2}{AB^2}$

$= \left(\dfrac{AM}{AB}\right)^2$

$= \left(\dfrac{1}{3}\right)^2$

$= \dfrac{1}{9}$

87. The corresponding sides of two similar triangles are in the ratio 3:4, then the ratio of the area of the triangle is______ **(2020)**

Solution:

The ratio of areas of two similar triangle

= The ratio of square of their corresponding sides

$= (3k)^2 : (4k)^2$

$= 9 : 16$

88. In Fig. 1, $DE \parallel BC$, $AD = 3$ cm and $BD = 2$ cm. Find $\dfrac{ar(\Delta ADE)}{ar(\Delta ABC)}$ **(2020)**

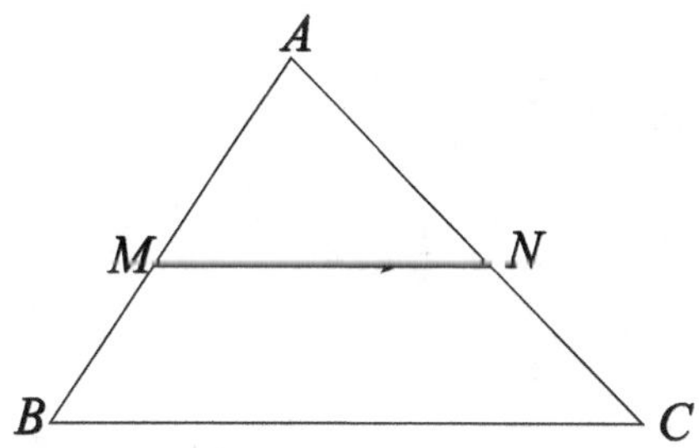

Solution:

$AB = AD + DB = 3 + 2 = 5$ cm

$DE \parallel BC$

$\therefore \Delta ADE \sim \Delta ABC$

$\therefore \dfrac{ar(\Delta ADE)}{ar(\Delta ABC)} = \dfrac{AD^2}{AB^2}$

$= \left(\dfrac{AD}{AB}\right)^2$

$= \left(\dfrac{3}{5}\right)^2 = \dfrac{9}{25}$

89. Sides of two similar triangles are in the ratio 4 : 9. Areas of these triangles are in the ratio : **(2020)**

(a) 4 : 9 (b) 2 : 3

(c) 81 : 16 (d) 16 : 81

Solution:

Ratio of areas of two similar triangles

$= 4^2 : 9^2$

$= 16 : 81$

Exercise 7.9

(i) In the figure 1, $DE \parallel AB$ and $AD{:}DE = 2 : 3$

Find $\dfrac{ar(\Delta ADE)}{ar(\Delta ABC)}$.

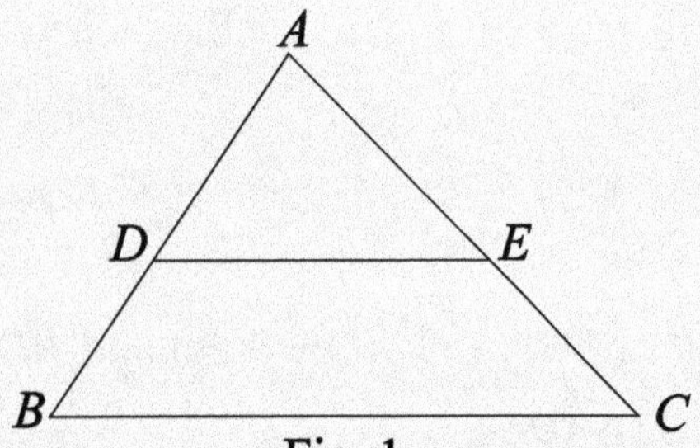

Fig. 1

(ii) In the figure 2, $PQ \parallel PB$ and $AP{:}PB = 1 : 3$,

Find $\dfrac{ar(\Delta APQ)}{ar(\Delta ABC)}$.

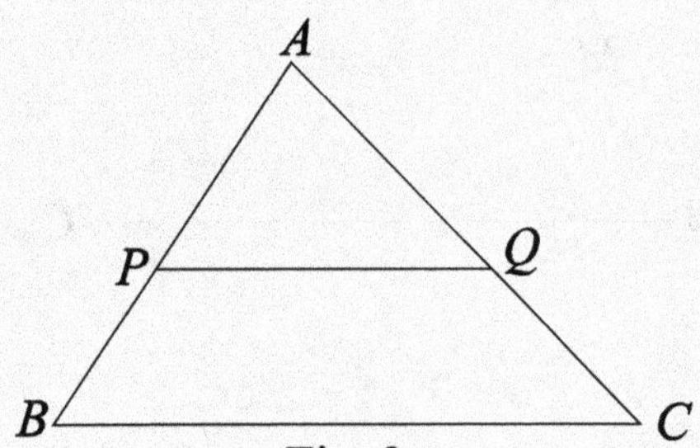

Fig. 2

(iii) In the figure 2, $PQ \parallel BC$ and $AP{:}PB = 4{:}5$, Find $\dfrac{ar(\Delta APQ)}{ar(\Delta ABC)}$.

(iv) Given $\Delta ABC \sim \Delta PQR$, if $\dfrac{AB}{PQ} = \dfrac{1}{4}$, then find $\dfrac{ar(\Delta ABC)}{ar(\Delta PQR)}$.

(v) Given $\Delta ABC \sim \Delta PQR$, If $= \dfrac{AB}{PQ} = \dfrac{2}{3}$, then find $\dfrac{ar(\Delta ABC)}{ar(\Delta PQR)}$.

(vi) Given $\Delta ABC \sim \Delta PQR$, If $\dfrac{AB}{PQ} = \dfrac{3}{5}$, then find $\dfrac{ar(\Delta ABC)}{ar(\Delta PQR)}$.

(vii) In the given figure 1, $DE \parallel BC$, $AD = 3$ cm and $BD = 4$ cm what is the ratio of the ar(ΔABC) to the ar(ΔADE) =?

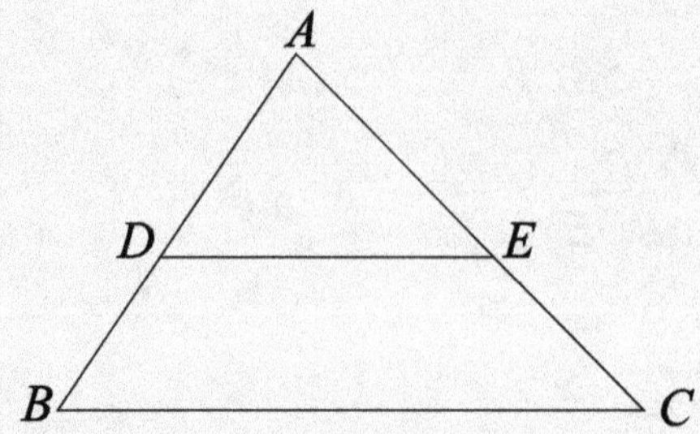

(viii) In the figure (i), $DE \parallel BC$, $AD = 2$ cm and $BD = 3$ cm what is the ratio of the ar(ΔABC) to the ar(ΔADE)?

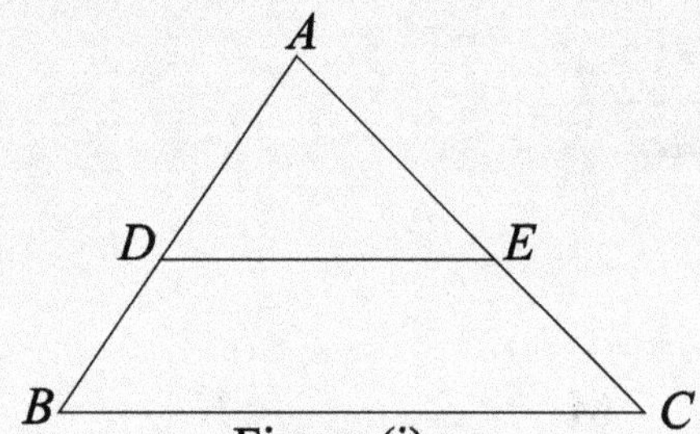

Figure (i)

(ix) In the given figure, $DE \parallel BC$, $AD = 4$ cm and $BD = 5$ cm what is the ratio of the ar(ΔABC) to the ar(ΔADE).

(x) The area of two similar triangles are 64 sq cm and 100 sq cm. Find the ratio of their corresponding sides.

(xi) The area of two similar triangles are 49 sq cm and 100 sq cm. Find the ratio of their corresponding sides.

(xii) The area of two similar triangles are 100 sq cm and 121 sq cm. Find the ratio of their corresponding sides.

(xiii) If $\Delta ABC \sim \Delta DEF$ such that $AB = 1.5$ cm and $DE = 1.8$ cm, the ratio of the areas ΔABC and ΔDEF is ______.

(xiv) If $\Delta ABC \sim \Delta DEF$ such that $AB = 0.6$ cm and $DE = 0.8$ cm, the ratio of the areas ΔABC and ΔDEF is ______.

(xv) If $\Delta ABC \sim \Delta DEF$ such that $AB = 1.1$ cm and $DE = 1.5$ cm, the ratio of the areas ΔABC and ΔDEF is ______.

(xvi) ABC and BDE are two equilateral triangle such that $BD : DC = 2{:}3$. Ratio of the areas of triangles ABC and BDE is ________.

(xvii) ABC and BDE are two equilateral triangle such that $BD : DC = 3 : 2$. Ratio of the areas of triangles ABC and BDE is ________.

(xviii) ABC and BDE are two equilateral triangle such that $BD : DC = 4:5$. Ratio of the areas of triangles ABC and BDE is ________.

Key Point In the given figure, O is the centre of a circle, AB is a chord and AT is tangent at A if $\angle AOB = a°$, then $\angle BAT = \frac{a°}{2}$

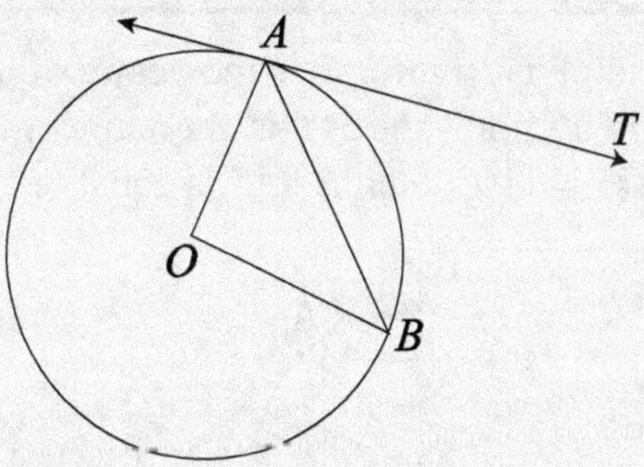

90. In Fig. 1, O is the centre of a circle, AB is a chord and AT is the tangent at A. If $\angle AOB = 100°$, then $\angle BAT$ is equal to **(2011)**

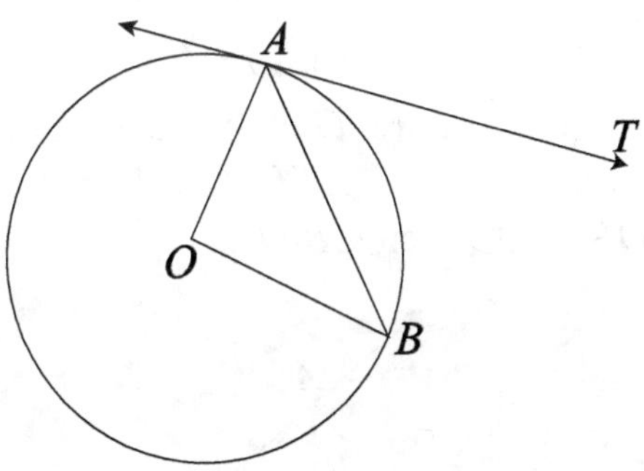

Fig. 1

(a) 100° (b) 40°
(c) 50° (d) 90°

Solution:

In ΔAOB

$OA = OB$ (radius of the circle)

$\therefore \angle OAB = \angle OBA$

and $\angle AOB + \angle OAB + \angle OBA = 180°$

$\Rightarrow 100 + 2\angle OBA = 180°$

$\Rightarrow 2\angle OAB = 80°$

$\therefore \angle OAB = 40°$

Now, $OA \perp AT$

$\therefore \angle OAT = 90°$

$\Rightarrow \angle OAB + \angle BAT = 90°$

$\Rightarrow 40° + \angle BAT = 90°$

$\therefore \angle BAT = 50°$

91. In the given figure, PQ is tangent to the circle centred at O. If $\angle AOB = 95°$, then the measures of $\angle ABQ$ will be. **(2023)**

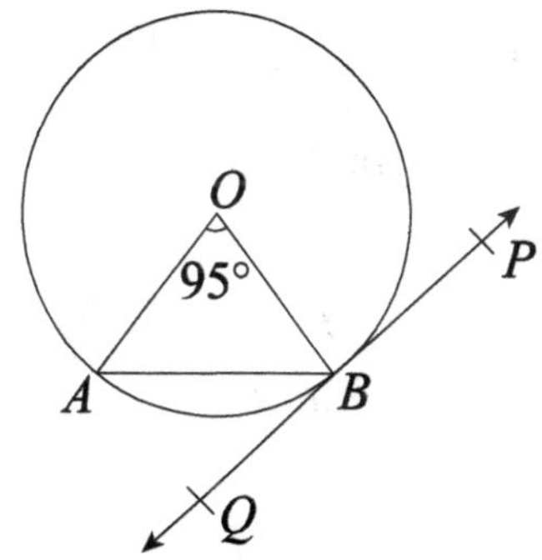

(a) 47.5° (b) 42.5°
(c) 85° (d) 95°

Solution: (a)

$\angle OAB = \angle OBA = \frac{180° - 95°}{2} = 42.5°$

Now, $\angle ABQ = \angle OBQ - \angle OBA$

$= 90° - 42.5°$

$= 47.5°$

92. In the given figure, PT is tangent to a circle with centre O. Chord PQ subtends an angle of 65° at the centre. The measure of $\angle QPT$ is **(2024)**

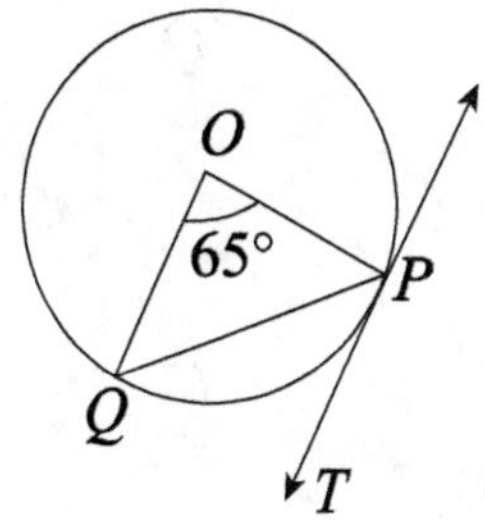

(a) 65° (b) 57.5°
(c) 67.5° (d) 32.5°

Solution: (d)

$\angle OQP = \angle OPQ = \frac{180° - \angle POQ}{2}$

$= \frac{180° - 65°}{2}$

$= 57.5$

$\therefore \angle QPT = \angle OPT - \angle OPQ$

$= 90° - 57.5° = 32.5°$

93. In the figure-2, PT is tangent to a circle with centre at O, at the point B. If $\angle AOB = 100°$, then $\angle ABP$ is equal to **(2020)**

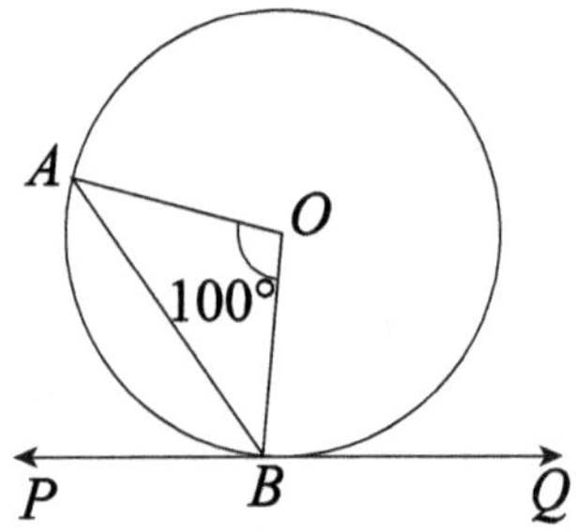

Fig. 2

(a) 50° (b) 40°
(c) 60° (d) 80°

Solution: (a)

$\angle OAB = \angle OBA = \frac{180° - 100°}{2} = 40°$

$\therefore \angle ABP = 90° - \angle OBA = 90° - 40° = 50°$

94. A chord QR subtends an angel of 105° at the center O of the circle. The measure of $\angle RQP$ is **(2026)**

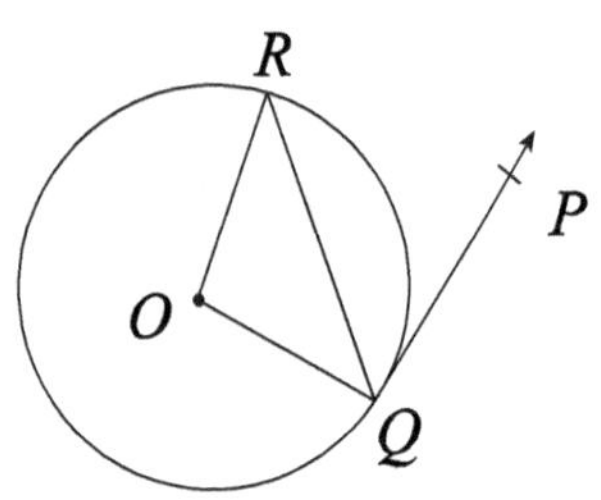

(a) $\frac{75°}{2}$ (b) $\frac{105°}{2}$
(c) 75° (d) 15°

Solution : (b)

In ΔROQ, $OR = OQ$

$\therefore \angle ORQ = \angle OQR = \frac{180° - 105°}{2}$

$= \frac{75°}{2}$

Now, $\angle OQP = 90°$

$\Rightarrow \angle OQR + \angle RQP = 90°$

$\Rightarrow \frac{75°}{2} + \angle RQP = 90°$

$\Rightarrow \angle RQP = 90° - \frac{75°}{2}$

$\Rightarrow \angle RQP = \frac{105°}{2}$

Exercise 7.10

(i) In the given figure, O is the centre of a circle, AB is a chord and AT is the tangent at A. If $\angle AOB = 110°$, then $\angle BAT$ is equal to.

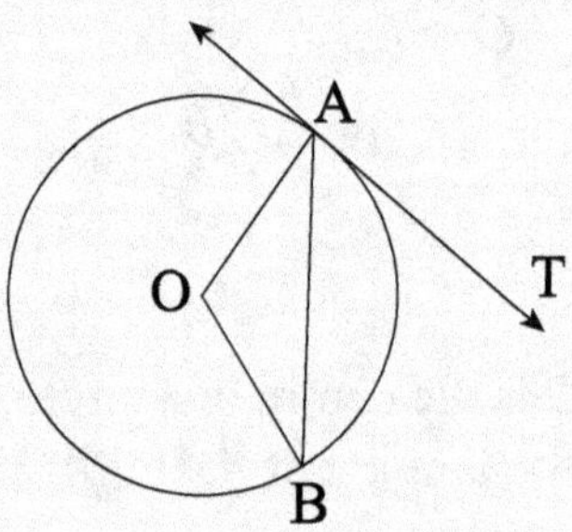

(a) 45° (b) 55°
(c) 65° (d) 30°

(ii) In the given figure, O is the centre of a circle, AB is a chord and AT is the tangent at A. If $\angle AOB = 80°$, then $\angle BAT$ is equal to.

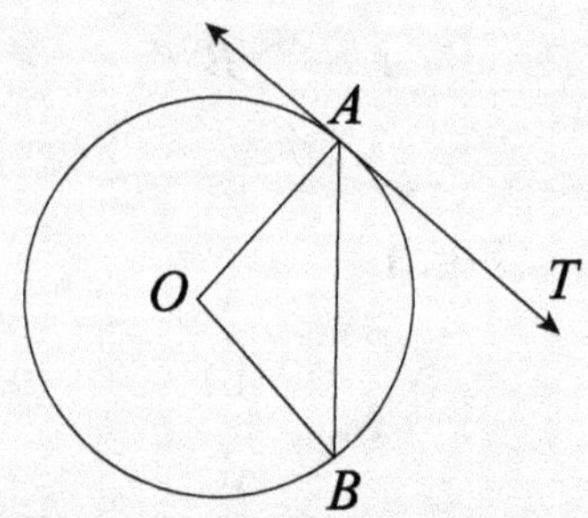

(a) 50° (b) 80°
(c) 40° (d) 70°

(iii) In the given figure, O is the centre of a circle, AB is a chord and AT is the tangent at A. If $\angle AOB = 120°$, then $\angle BAT$ is equal to.

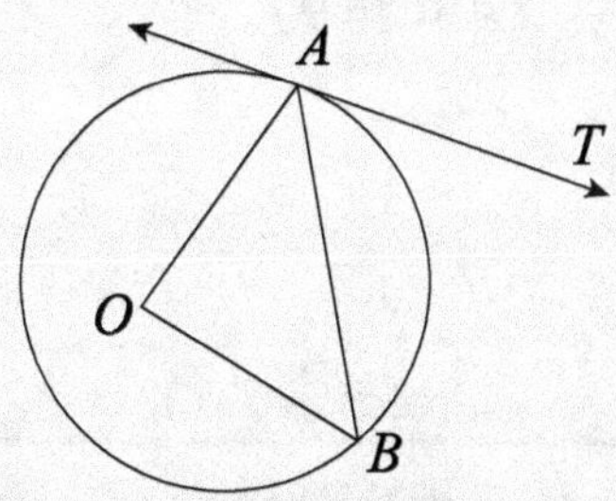

(a) 30° (b) 60°
(c) 50° (d) 40°

(iv) In the given figure, PQ is tangent to the circle centred at O. If $\angle AOB = 100°$ then the measure of $\angle ABQ$ will be

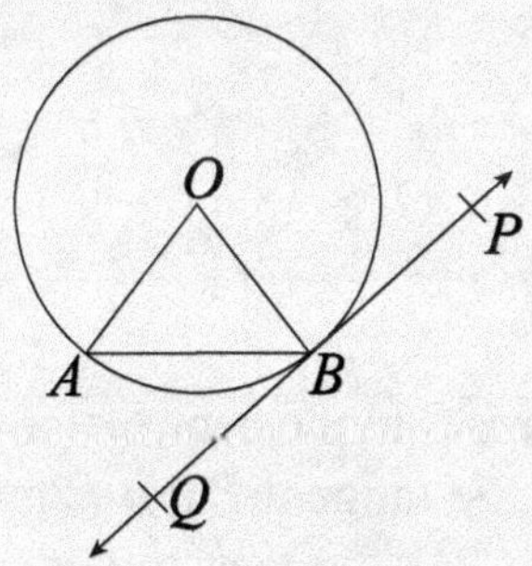

(a) 50° (b) 40°
(c) 60° (d) 90°

(v) In the given figure, PQ is tangent to the circle centred at O. If $\angle AOB = 110°$, then the measure of $\angle BAP$ will be.

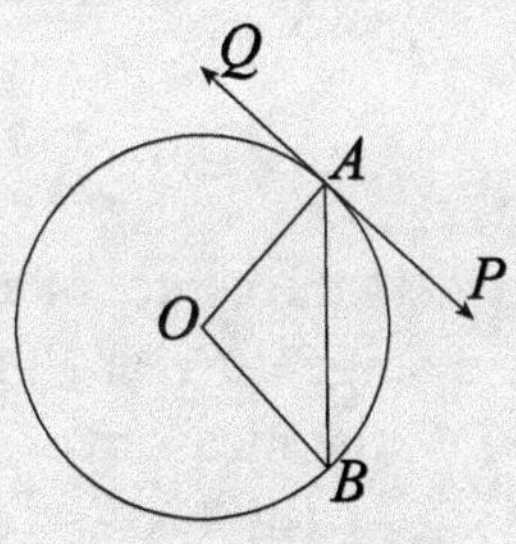

(a) 45° (b) 60°
(c) 55° (d) 40°

(vi) In the given figure, PQ is tangent to the circle with centre at O, at the point B. If $\angle AOB = 100°$, then $\angle ABP$ is equal to ______.

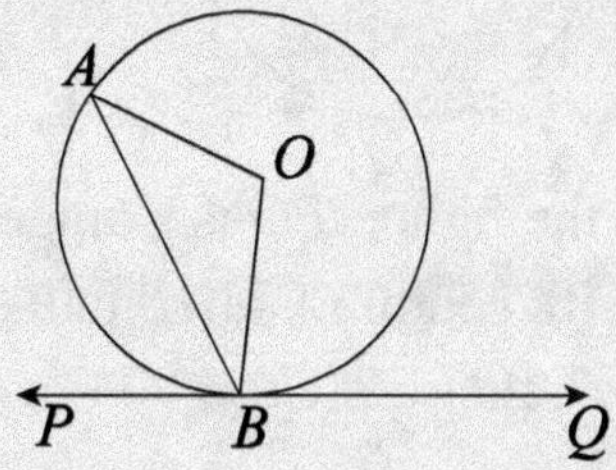

Key Point In the given figure, O is the centre of circle. PQ is a chord and PT is tangent at P which makes on angle of 'a' with PQ, then $\angle POQ = 2a°$

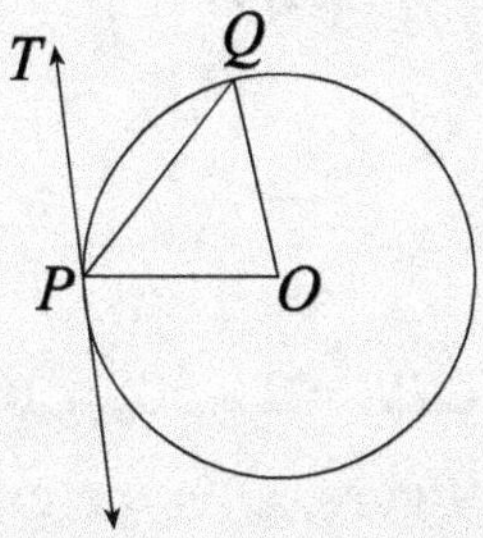

95. In fig 1, O is the centre of circle. PQ is a chord and PT is tangent at P which makes an angel of 50° with PQ, $\angle POQ$ is **(2020)**

(a) 130° (b) 90°
(c) 100° (d) 75°

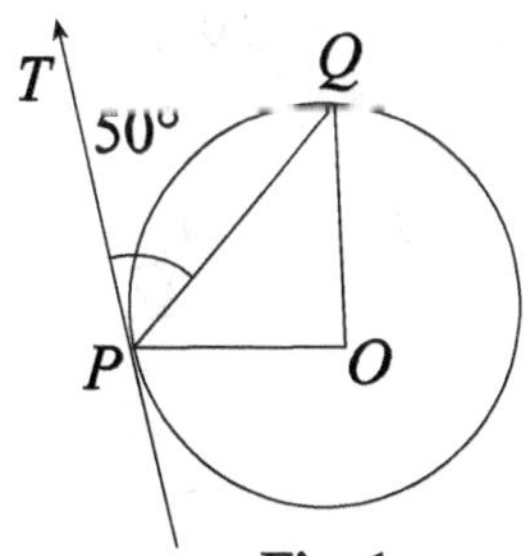

Fig. 1

Solution: (c)

$\angle OPT = 90°$

$\Rightarrow \angle OPQ + \angle QPT = 90°$

$\Rightarrow \angle OPQ + 50° = 90°$

$\Rightarrow \angle OPQ = 40°$ $\therefore \angle OQP = 40°$

Now, in ΔPOQ

$\angle OPQ + \angle OQP + \angle POQ = 180°$

$\Rightarrow 40° + 40° + \angle POQ = 180°$

$\Rightarrow \angle POQ = 100°$

Note: $\angle POQ = 2\angle QPT = 2 \times 50° = 100°$

96. In the given figure, O is the centre of the circle and PQ is the chord. If the tangent PR at P makes an angel of 50° with PQ, then the measure of $\angle POQ$ is **(2023)**

(a) 50° (b) 40°
(c) 100° (d) 130°

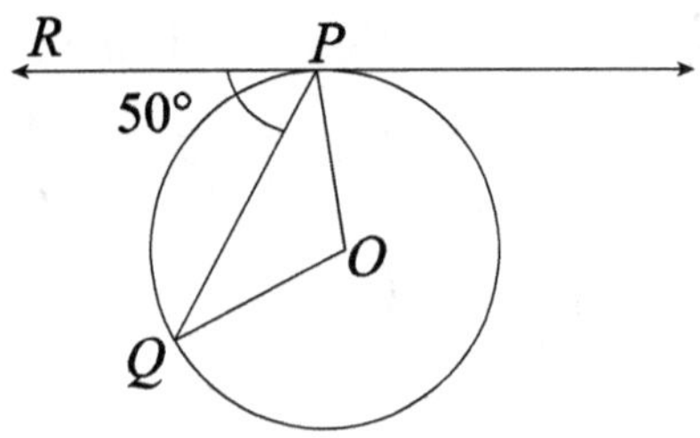

Solution: (c)

$\angle POQ = 2\angle QPR = 2 \times 50° = 100°$

97. In the given figure, O is the centre of a circle, PQ is a chord and the tangent PT at P makes an angel of 50° with PQ, the measure of $\angle POQ$ is **(2023)**

(a) 130° (b) 100°
(c) 90° (d) 75°

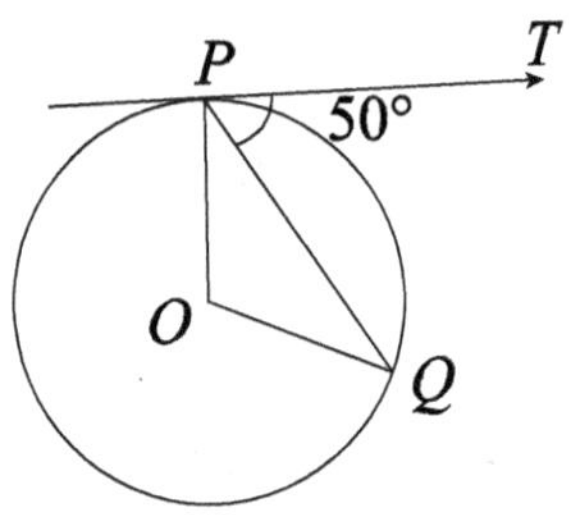

Solution: (b)

$\angle POQ = \angle QPT = 2 \times 50° = 100°$

98. Tangent BC makes an angle of $x°$ with the chord AB of circle centerd at O. If $\angle AOB = 100°$, then value of x is **(2025)**

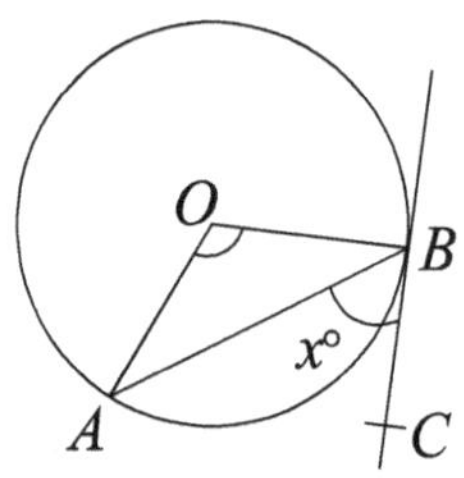

(a) 40 (b) 80
(c) 90 (d) 50

Solution: (d)

In ΔAOB

$\angle AOB = 100°$

$\therefore\ \angle OAB = \angle OBA = \dfrac{180° - 100°}{2} = 40°$

Now $OB \perp BC$

$\therefore\ \angle OBC = 90°$

$\Rightarrow \angle OBA + \angle ABC = 90°$

$\Rightarrow 40° + x° = 90°$

$\therefore\ x° = 50°$

$\therefore\ x = 50$

Exercise 7.11

(i) In the figure, O in the centre of a circle, PQ is a chord and the tangent PT at P makes an angle of 45° with PQ, the measure of $\angle POQ$ is

(a) 45° (b) 60°
(c) 90° (d) 120°

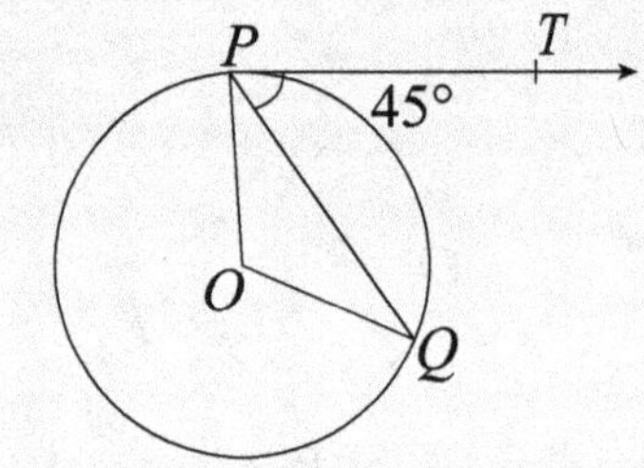

(ii) In the figure, O in the centre of a circle, PQ is a chord and the tangent PT at P makes an angle of 30° with PQ, the measure of $\angle POQ$ is

(a) 30° (b) 60°
(c) 90° (d) 80°

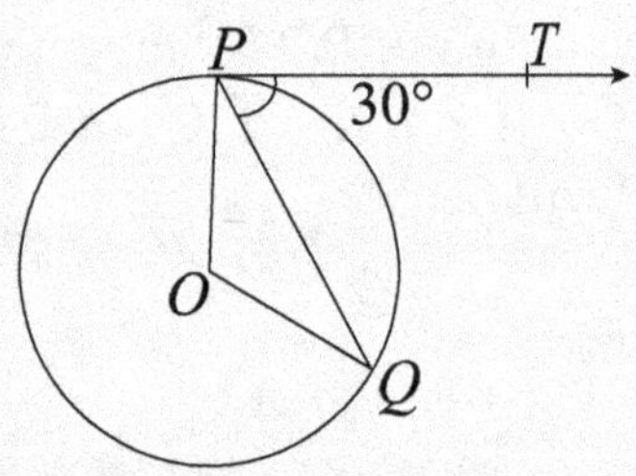

(iii) In the figure, O in the center of a circle, PQ is a chord and the tangent PT at P makes an angle of 62° with PQ, the measure of $\angle POQ$ is

(a) 62° (b) 120°
(c) 124° (d) 126°

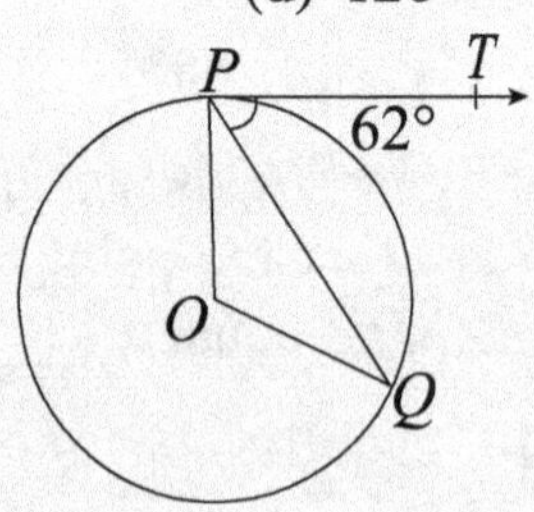

Key Point In the figure, PQ is a tangent at a point C to a circle with centre O. If AB in a diameter and $\angle CAB = a°$, then $\angle ACP = 90 - a°$

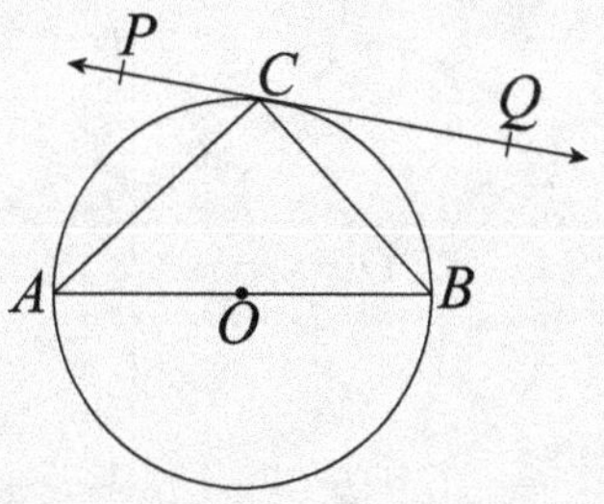

99. In fig.1, PQ is a tangent at a point C to a circle with centre O. If AB is a diameter and $\angle CAB = 30°$, find $\angle PCA$ **(2016)**

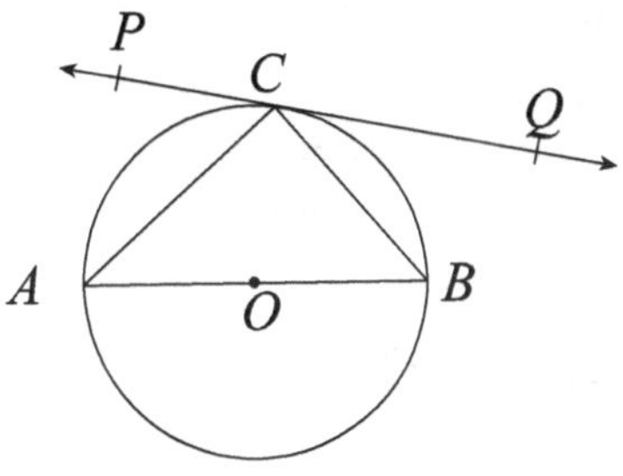

Solution:

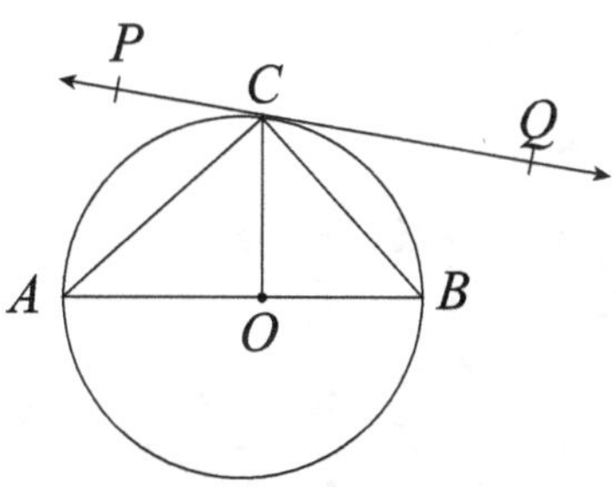

In ΔAOC

$OA = OC$

$\therefore \angle OAC = \angle OCA = 30°$

$OC \perp PQ$

$\therefore \angle OCP = 90°$

$\Rightarrow \angle OCA + \angle PCA = 90°$

$\Rightarrow 30° + \angle PCA = 90°$

$\therefore \angle PCA = 60°$

100. In the given figure, RS is the tangent to the circle at the point L and MN is the diameter. If $\angle NML = 30°$, then $\angle RLM$ is: **(2025)**

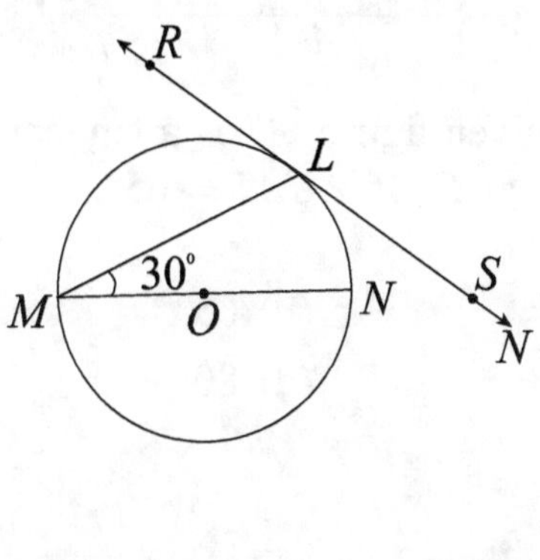

(a) 30° (b) 60°
(c) 90° (d) 120°

Solution: (b)

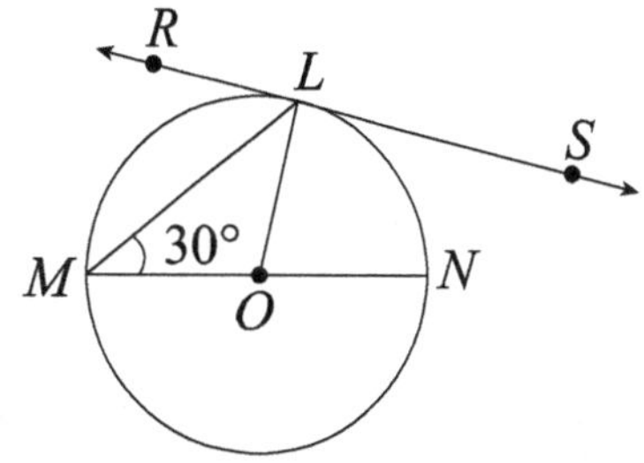

$\therefore$ In ΔMOL

$OM = OL$

$\therefore \angle OML = \angle OLM = 30°$

Now $\angle OLR = 90°$

$\Rightarrow \angle OLM + \angle MLR = 90°$

$\Rightarrow 30° + \angle MLR = 90°$

$\therefore \angle MLR = 60°$

Exercise 7.12

(i) In the given figure, PQ is a tangent at a point C to a circle with centre O. If AB is a diameter and $\angle CAB = 40°$, find $\angle PCA$

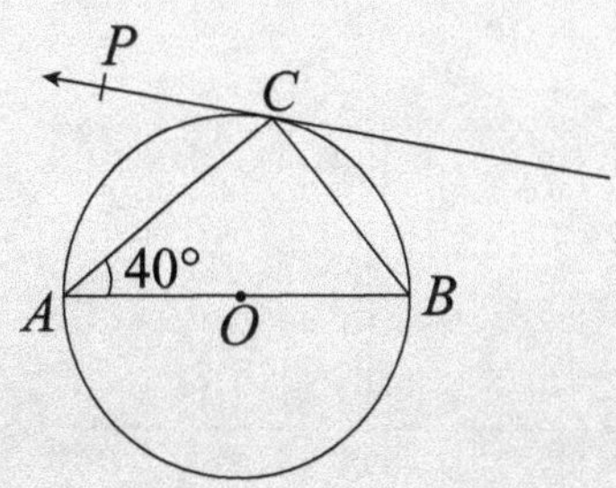

(ii) In the given figure, PQ is a tangent at a point C to a circle with centre O. If AB is a diameter and $\angle CAB = 45°$, find $\angle PCA$

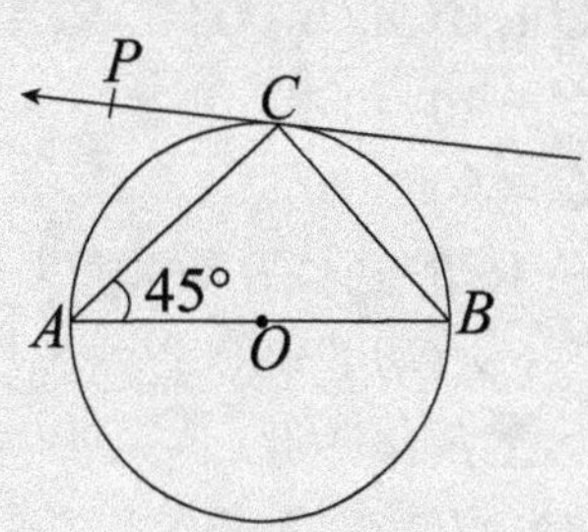

(iii) In the given figure, PQ is a tangent at a point C to a circle with centre O. If AB is a diameter and $\angle CAB = 52°$, find $\angle PCA$

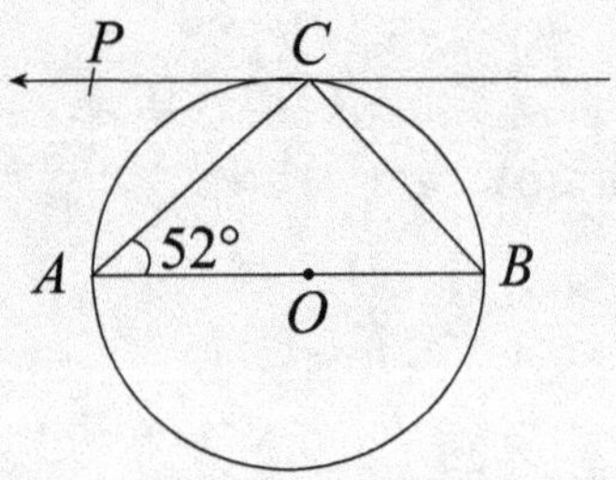

Key Point In a given figure, AT is a tangent to a circle centred at O. If $\angle CAT = a°$ then $\angle CBA = a°$

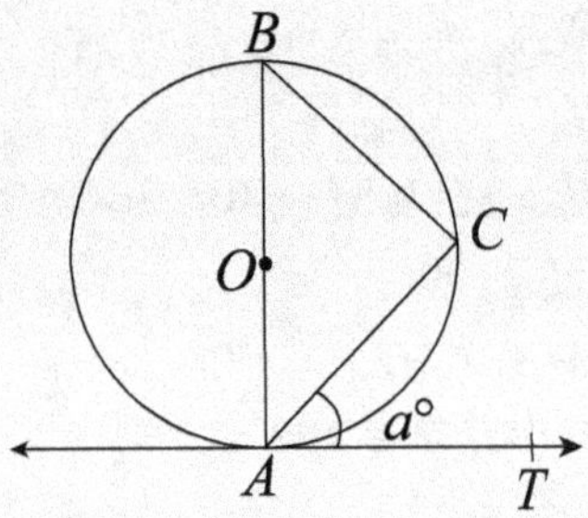

101. In a given figure, AT is tangent to a circle centred at O. If $\angle CAT$ = 40°, then $\angle CBA$ is equal to **(2024)**

(a) 70° (b) 50°
(c) 65° (d) 40°

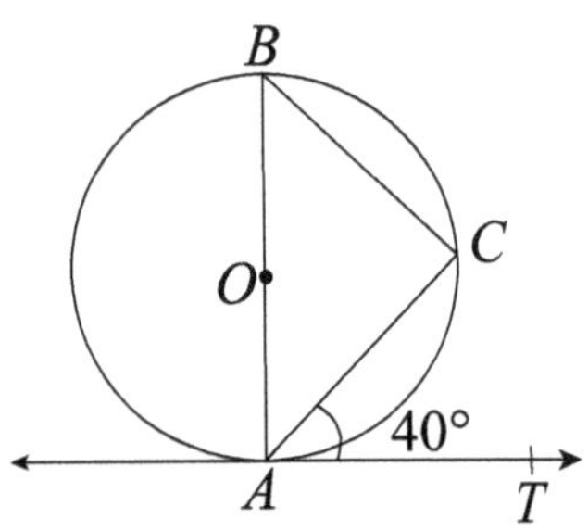

Solution: (d)

$\angle BAT = 90°$

$\Rightarrow \angle BAC + \angle CAT = 90°$

$\Rightarrow \angle BAC + 40° = 90°$

$\Rightarrow \angle BAC = 50°$

Now, in ΔACB, $\angle ACB = 90°$

$\therefore \angle BAC + \angle CBA = 90°$

$\Rightarrow 50° + \angle CBA = 90°$

$\therefore \angle CBA = 40°$

102. In the figure, AB is diameter of the circle with center O. CD is tangent to the circle so that $\angle ACD = 40°$. The value of $\angle CBA$ is **(2025)**

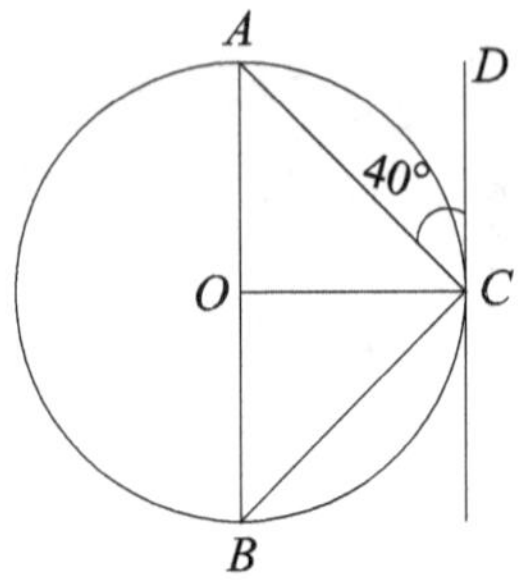

(a) 50° (b) 40°
(c) 80° (d) 45°

Solution:

$\angle OCD = 90°$

$\Rightarrow \angle OCA + \angle ACD = 90°$

$\Rightarrow \angle OCA + 40° = 90°$

$\Rightarrow \angle OCA = 50°$

Now, $\angle ACB = 90°$

$\Rightarrow \angle OCA + \angle OCB = 90°$

$\Rightarrow 50° + \angle OCB = 90°$

$\Rightarrow \angle OCB = 40°$

In ΔBOC

$OB = OC$

$\therefore \angle OCB = \angle OBC = 40°$

$\therefore \angle ABC = 40°$

Exercise 7.13

(i) In the given figure, AT is a tangent to a circle centred at O. If $\angle CAT$ = 50°, then $\angle CBA$ is equal to

(a) 40° (b) 50°
(c) 60° (d) 90°

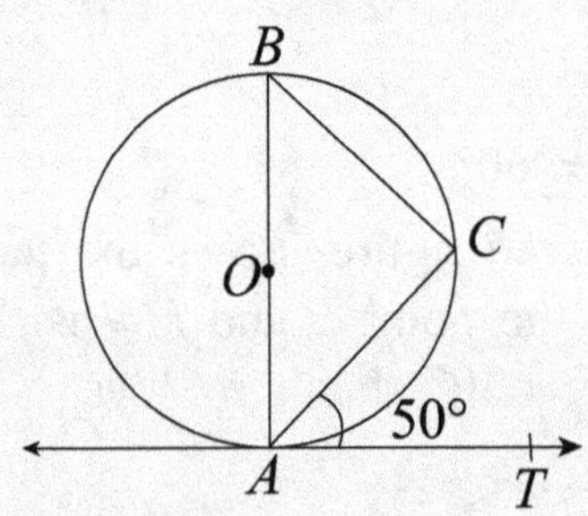

(ii) In the given figure, AT is a tangent to a circle centred at O. If $\angle CAT$ = 65°, then $\angle CBA$ is equal to

(a) 45° (b) 65°
(c) 50° (d) 80°

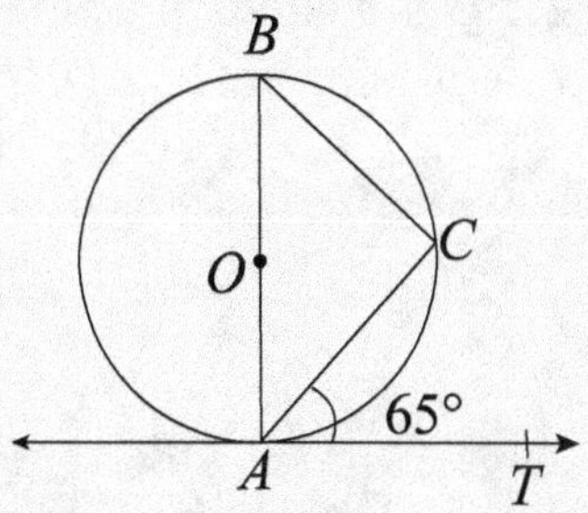

(iii) In the given figure, AT is a tangent to a circle centred at O. If $\angle CAT = 52°$, then $\angle CBA$ is equal to

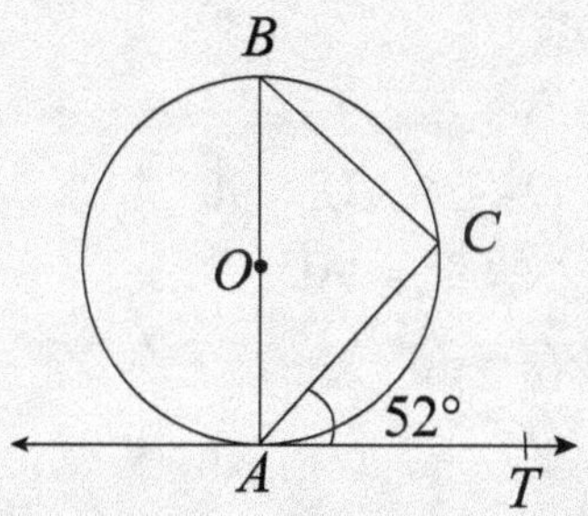

(a) 52° (b) 42°

(c) 48° (d) 60°

Key Point In the given figure, AOB is a diameter of a circle with centre O and AC is a tangent to the circle at A. If $\angle BOC = a°$, then $\angle ACO = (a - 90)°$

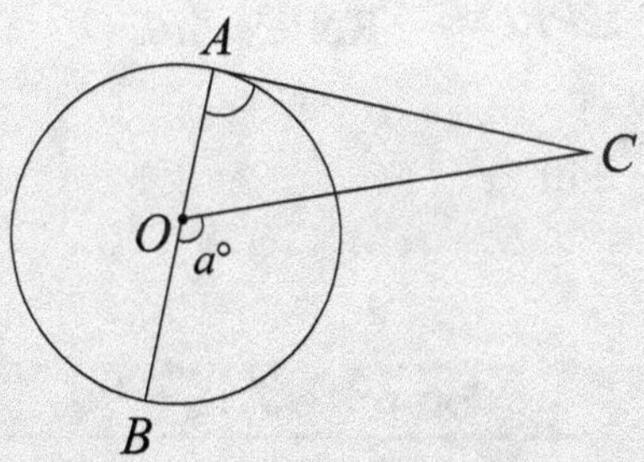

103. In fig-1, AOB is a diameter of a circle with circle O and AC is a tangent to the circle at A. If $\angle BOC = 130°$, then find $\angle ACO$. **(2016)**

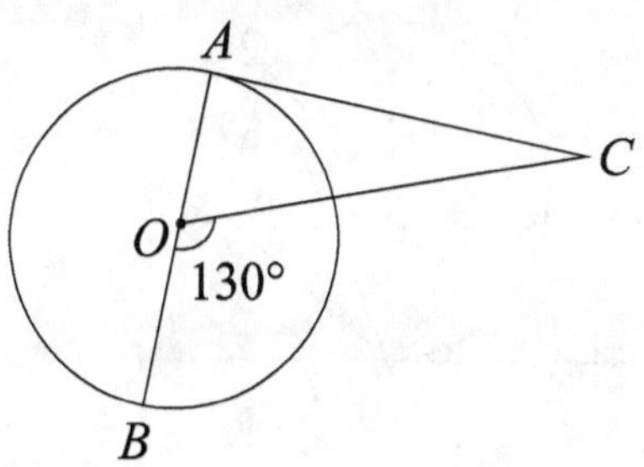

Solution:

$\angle AOC = 180° - \angle BOC = 180° - 130° = 50°$

Now, in ΔAOC, $\angle OAC = 90°$

$\therefore \angle AOC + \angle ACO = 180°$

$\Rightarrow 50° + \angle ACO = 90°$

$\therefore\ \angle ACO = 40°$

104. PQ is a tangent drawn from an external point P to a circle with center O, QOR is the diameter of the circle. If $\angle POR = 120°$, what is the measure of $\angle OPQ$? **(2017)**

Solution:

$\angle OPQ = 120° - 90° = 30°$

105. In the given figure, PA is tangent from an external point P to a circle with centre O. If $\angle POB = 115°$, then $\angle APO$ is equal to **(2025)**

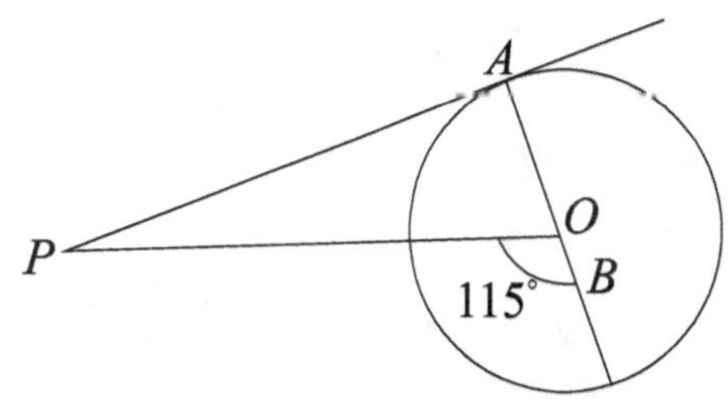

(a) 25° (b) 65°
(c) 90° (d) 35°

Solution: (a)

$\angle APO = 115° - 90° = 25°$

106. In the given figure, PA is a tangent from an external point P to a circle with center O. If $\angle 125°$, then $\angle APO$ is equal to: **(2026)**

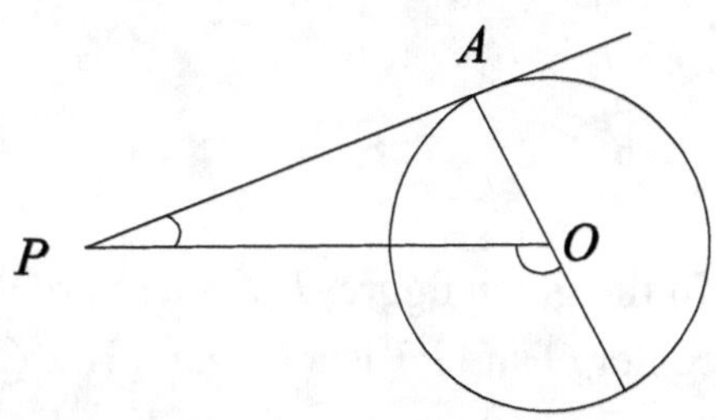

(a) 25° (b) 65°
(c) 90° (c) 35°

Solution : (d)

$\angle POB = \angle PAO + \angle APO$

$\Rightarrow 125° = 90° + \angle APO$

$\therefore\ \angle APO = 35°$

Exercise 7.14

(i) In the given figure, AOB is a diameter of a circle with circle centre O and AC is a tangent to the circle at A if $\angle BOC = 120°$, then find $\angle ACO$

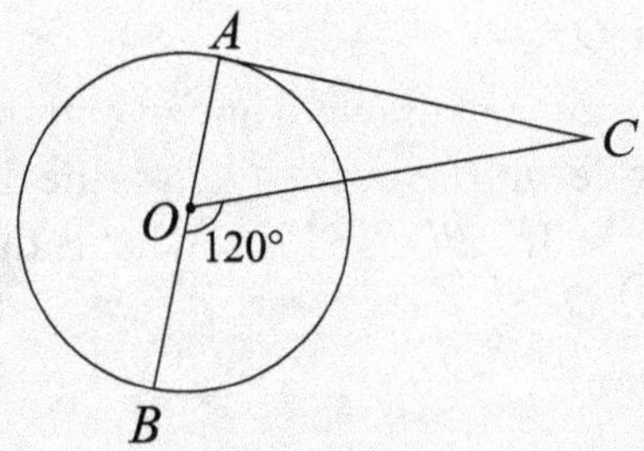

(ii) In the given figure, AOB is a diameter of a circle with center O and AC is a tangent to the circle at A. If $\angle BOC = 110°$, then find $\angle ACO$

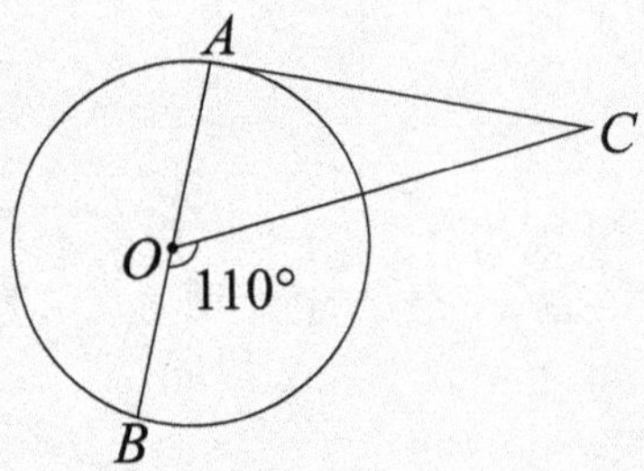

(iii) In the given figure, AOB is a diameter of a circle with center O and AC is a tangent to the circle at A. If $\angle BOC = 140°$, then find $\angle ACO$

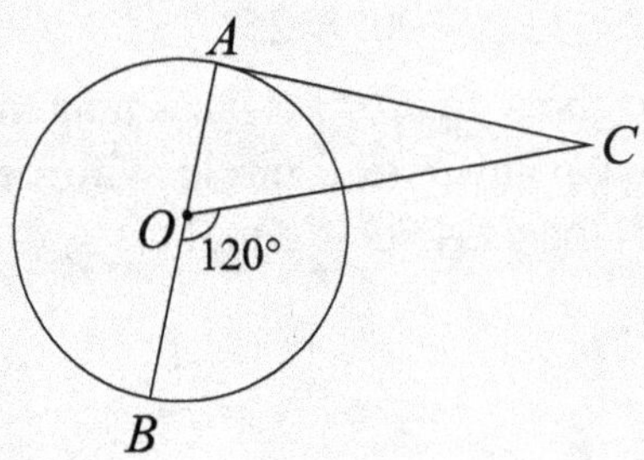

Key Point In the given figure, PQ is a chord of a circle with center O and PT is a tangent. If $\angle QPT = a°$, then $\angle PRQ = (180 - a)°$

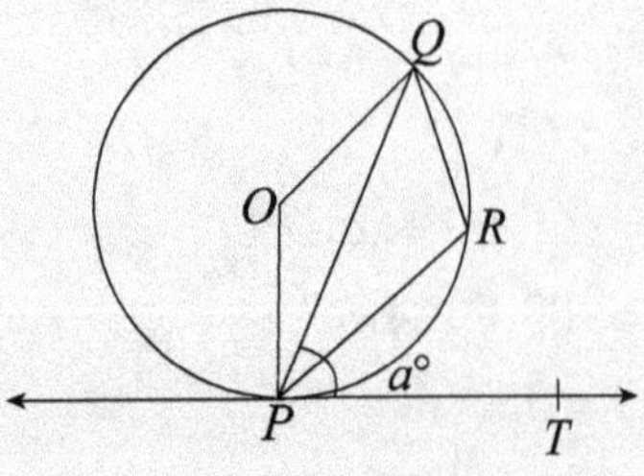

107. In figure 2, PQ is a chord of a circle with center O and PT is a tangent. If $\angle QPT = 60°$, find $\angle PRQ$ **(2015, 2020)**

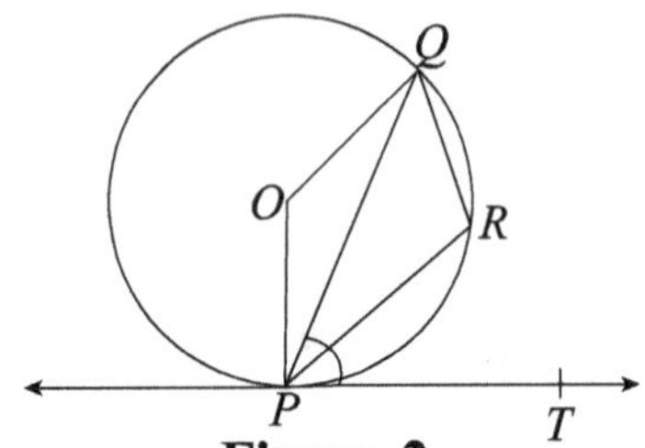

Figure-2

Solution:

$\angle OPT = 90°$

$\Rightarrow \angle OPQ + \angle QPT = 90°$

$\Rightarrow \angle OPQ + 60° = 90°$

$\Rightarrow \angle OPQ = 30°$

In $\Delta\, POQ$

$OP = OQ$

$\therefore\ \angle OPQ = \angle OPQ = 30°$

$\therefore\ \angle POQ = 180° - 2\angle OPQ$

$= 180° - 2 \times 30°$

$= 120°$

$\therefore$ Reflex $\angle POQ = 360° - 120°$

$= 240°$

Now, $\angle PRQ = \frac{1}{2}$ Reflex $\angle POQ$

$= \frac{1}{2} \times 240°$

$= 120°$

Exercise 7.15

(i) In the given figure, PQ is a chord of a circle with center O and PT is a tangent. If $\angle QPT = 50°$, find $\angle PRQ$

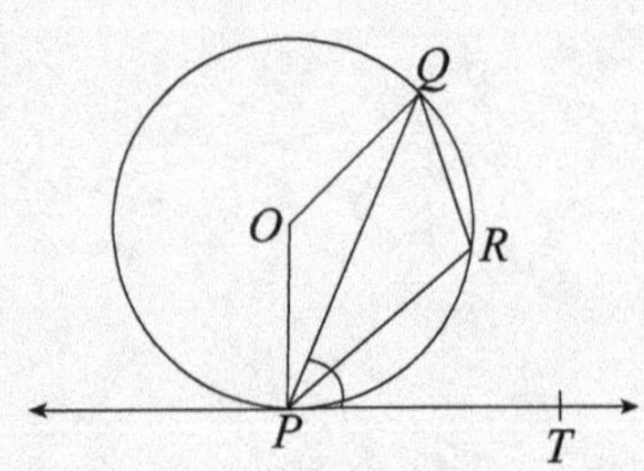

(ii) In the given figure, PQ is a chord of a circle with center O and PT is a tangent. If $\angle QPT = 40°$, find $\angle PRQ$

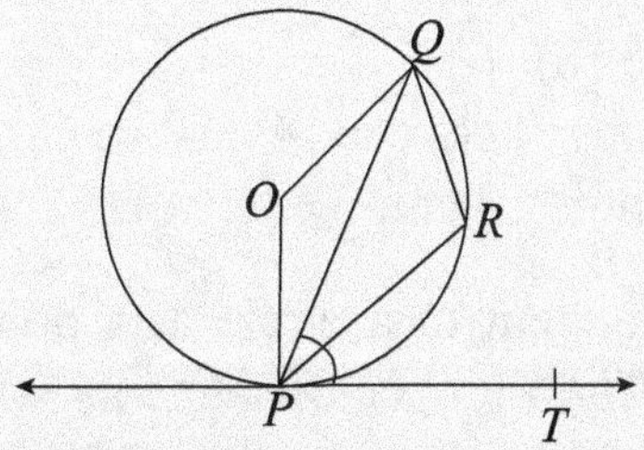

(iii) In the given figure, PQ is a chord of a circle with center O and PT is a tangent. If $\angle QPT = 70°$, find $\angle PRQ$

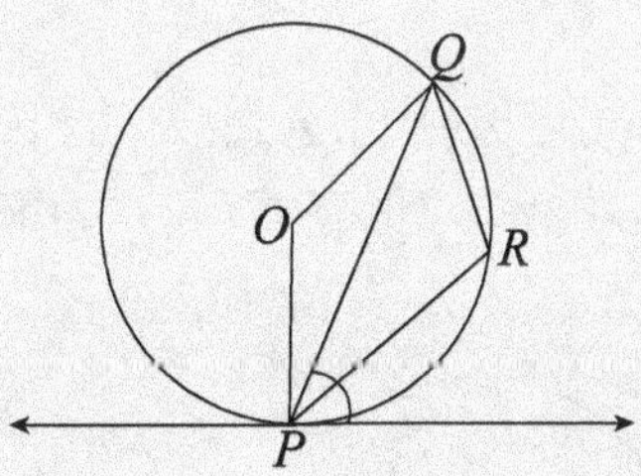

Key Point Distance between two parallel tangents of a circle is equal to diameter of the circle.

108. The distance between two parallel tangents of a circle of radius 4 cm is ________. **(2020)**

Solution:

The distance between two parallel tangents of a circle = 2 × radius = $2 \times 4 = 8$ cm

109. The distance between two parallel tangents of a circle of radius 7 cm is **(2023)**

(a) 7 cm (b) 14 cm

(c) $\frac{7}{2}$ cm (d) 28 cm

Solution: (b) 14 cm

110. A circle of radian 5.2 cm has two tangents AN and CD parallel to each other. What is the distance between two tangents ? **(2023)**

(a) 5.2 cm (b) 10.4 cm

(c) 20.8 cm (d) cant find

Solution: (b)

$2 \times 5.2 = 10.4$ cm

111. A circle is of radius 3 cm, the distance between two of its parallel tangents is **(2023)**

(a) 12 cm (b) 6 cm

(c) 3 cm (d) 4.5 cm

Solution: (b)

$2 \times 3 = 6$ cm

112. The tangents drawn at the extremities of the diameter of a circle are always: **(2025)**

(a) Parallel (b) Perpendicular

(c) Equal (d) Intersecting

Solution: (a)

Parallel

Exercise 7.16

(i) The distance between two parallel tangents of a circle of radius 8 cm is ________.

(ii) The distance between two parallel tangents of a circle of diameter 12 cm is

(a) 12 cm (b) 6 cm

(c) 24 cm (d) 30 cm

(iii) A circle of radius 5.5 cm has two tangents AB and CD parallel to each other. What is the distance between two tangents?

(a) 5.5 cm (b) 10 cm

(c) 10.5 cm (d) 11 cm

(iv) A circle is of radius 3.6 cm The distance between two of is parallel tangents is

(a) 3.6 cm (b) 1.8 cm

(c) 6 cm (d) 7.2 cm

113. In the given circle in Figure-1, number of tangents parallel to tangent PQ is **(2020)**

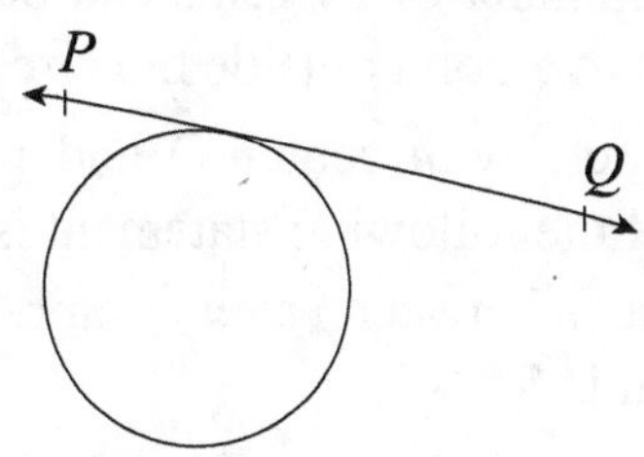

Figure-1

(a) 0 (b) many

(c) 2 (d) 1

Solution: (d) 1

114. A line intersecting a circle in two points is called a __________ **(2020)**

Solution:

Secant

115. How many tangents can be drawn to a circle from a point on it? **(2023)**

(a) one (b) two
(c) infinite (d) zero

Solution: (b) two

116. A line intersecting a circle in two distinct points is called a **(2024)**

(a) secant (b) chord
(c) diameter (d) tangent

Solution: (b) secant

117. A tangent to a circle is a line that touches the circle at **(2024)**

(a) one point only
(b) two points
(c) three points
(d) infinite number of points

Solution: (a) one point only.

118. Which of the following statements is false ? **(2025)**

(a) Infinite number of tangents can be drawn to a circle.
(b) Infinite number of tangents can be drawn to a circle from a point outside the circle.
(c) Infinite number of secants can be drawn to a circle from a point outside the circle.
(d) Angle between tangents and diameter at point of contact is 90°

Solution: (b)

Infinite number of tangents can be drawn to a circle from a point outside the circle.

119. For a circle with center O and radius 5 cm, which of the following statement is true ?

P: Distance between every pair of parallel tangent is 5 cm.

Q: Distance between every pair of parallel tangent is 10 cm.

R: Distance between every pair of parallel tangent must be between 5 cm and 10 cm.

S: There dose not exist a point outside the circle from where length of tangent is 5 cm. **(2025)**

(a) P (b) Q
(c) R (d) S

Solution: (b)

Q: Distance between every pair of parallel tangent is 10 cm.

120. PQ is tangent to a circle at a point P on the circle. The number of tangents which can be drawn to the circle parallel to PQ is: **(2026)**

(a) 2 (b) 1
(c) many (d) zero

Solution: (b) 1

Key Point In the given figure, PT is tangents to a circle centred at O. If $\angle POT = a°$, then $\angle OTP = 90° - a$

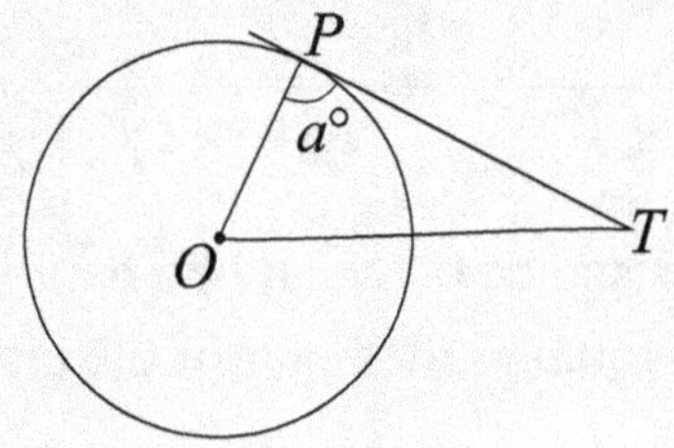

121. In figure-3,PT is tangent to a circle centred at O. Find the value of $\angle OTP$ if $\angle POT = 75°$ **(2020)**

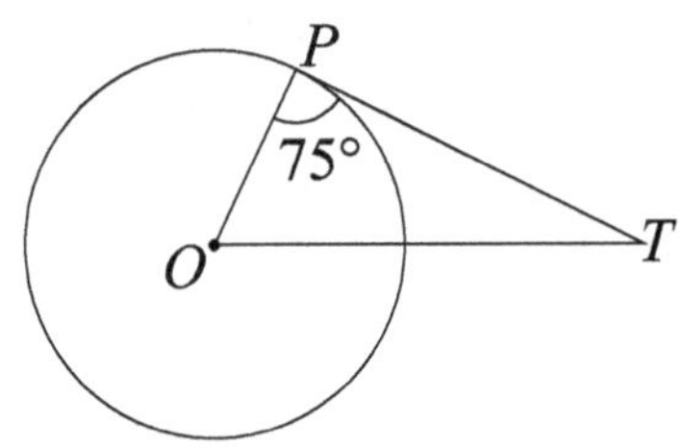

Solution:

In ΔOPT $\angle OPT = 90°$

$\therefore \angle POT + \angle OTP = 90°$

$\Rightarrow 75° + \angle OTP = 90°$

$\therefore \angle OTP = 15°$

122. In the given figure, PQ is a tangent to the circle with centre O. If $\angle OPQ = x$, $\angle POQ = y$, then $(x + y)$ is **(2023)**

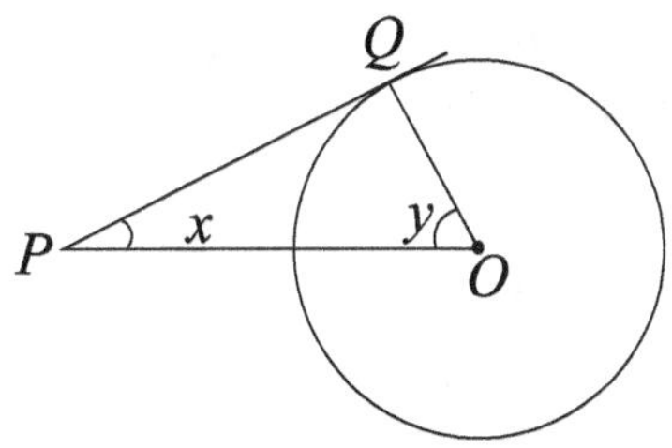

(a) 45° (b) 90°
(c) 60° (d) 180°

Solution:

In ΔPQO $\angle PQO = 90°$

$\therefore \angle OPQ + \angle POQ = 90°$

$\therefore x + y = 90°$

123. In the given figure, PA is a tangent from an external point P to a circle with centre O. If $\angle AOP = 70°$, then the measure of $\angle APO$ is **(2024)**

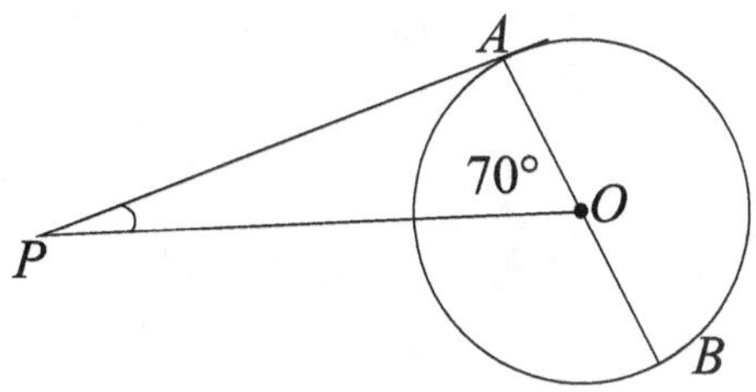

(a) 70° (b) 90°
(c) 110° (d) 20°

Solution:

In ΔPAO, $\angle PAO = 90°$

$\therefore \angle AOP + \angle APO = 90°$

$\Rightarrow 60° + \angle APO = 90°$

$\therefore \angle APO = 20°$

124. In the adjoining figure; TS is a tangent to a circle with center O. The value of $2x°$ is **(2025)**

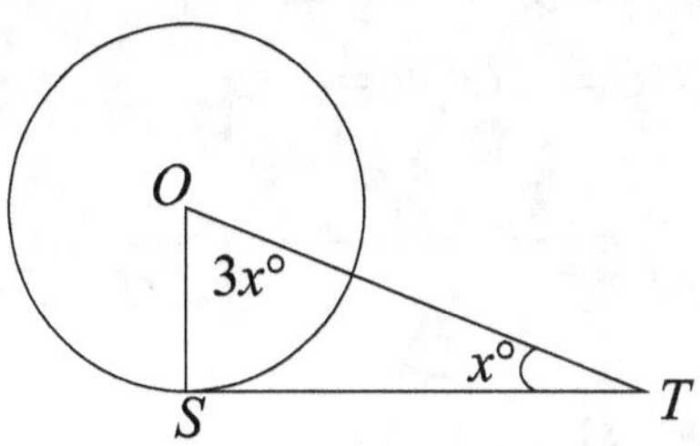

(a) 22.5 (b) 45
(c) 67.5 (d) 90

Solution:

$3x° + 2x° = 90$

$\Rightarrow 5x° = 90$

$\Rightarrow x° = 18$

$\therefore 2x° = 36$

Exercise 7.17

(i) In the given figure, PT in tangent to a circle centred at O. Find the value of $\angle OTP$ if $\angle POT = 52°$

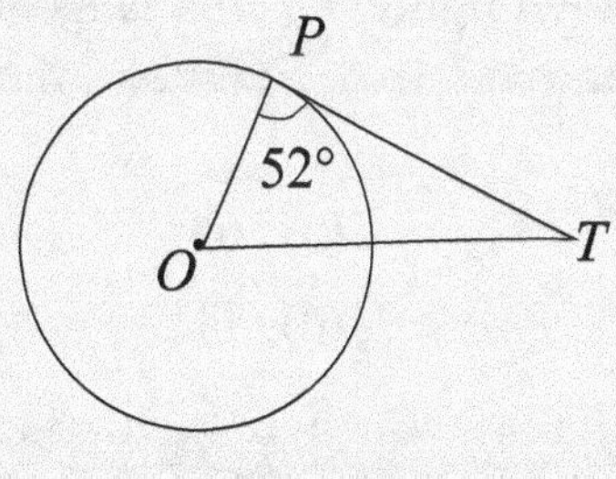

(ii) In the given figure, PQ is a tangent to the circle with centre O. If $\angle OPQ = x$, $\angle POQ = x + 20°$, then find the value of x.

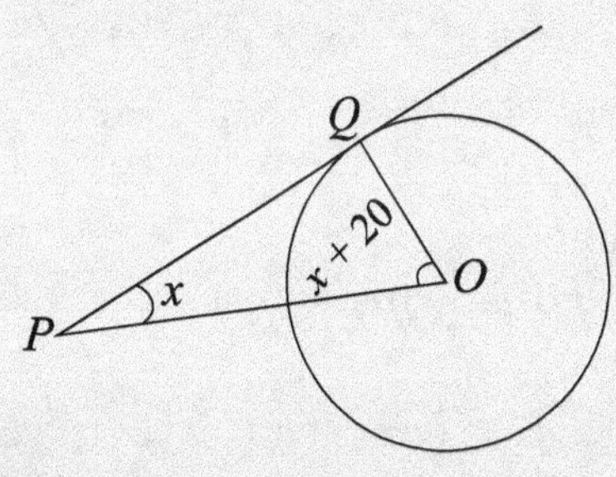

(a) 70° (b) 50°
(c) 35° (d) 40°

(iii) In the given figure, PA is a tangent from an external point P to a circle with centre O. If $\angle AOP = 44°$, then the measure of $\angle APO$ is

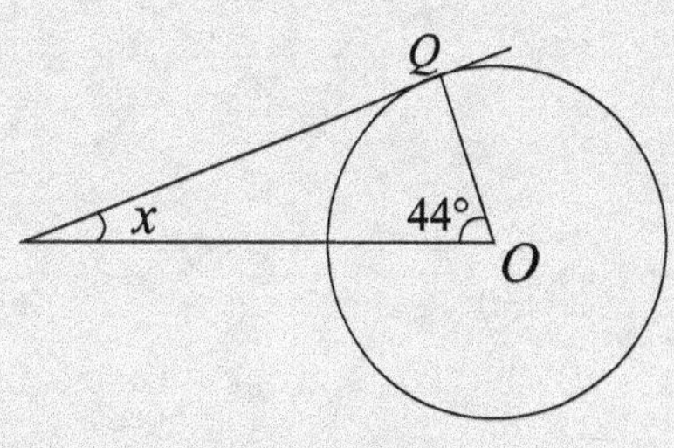

(a) 44° (b) 22°
(c) 50° (d) 46°

Key Point In the figure, PT in tangent at T to the circle with centre O. If $\angle TPO = a°$, then $\angle POL = (90 + a)°$.

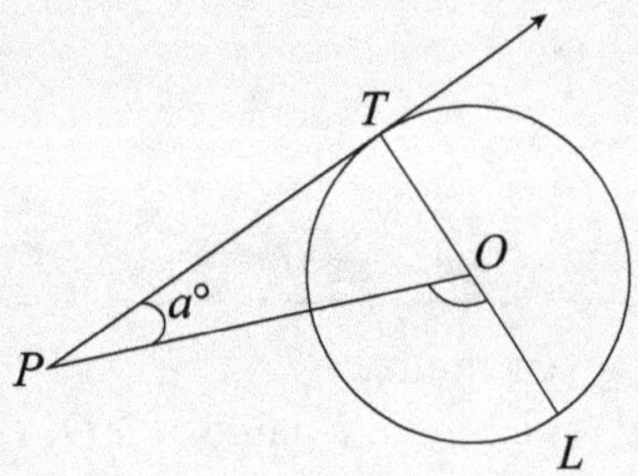

125. In the given figure, PT in a tangent at T to the circle centre O. If $\angle TPO = 25°$, then x in equal to **(2023)**

(a) 25° (b) 65°
(c) 90° (d) 115°

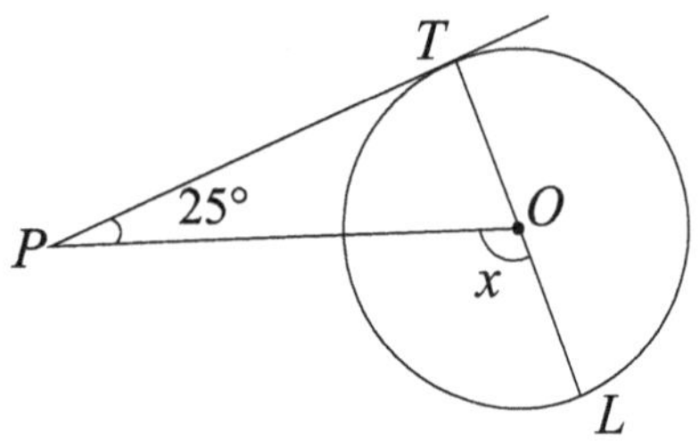

Solution:

$x = \angle OTP + \angle OPT$

$= 90° + 25°$

$= 115°$

Exercise 7.18

(i) In the given figure, PT is a tangent at T to the circle centre O. If $\angle TPO = 30°$, then x is equal to

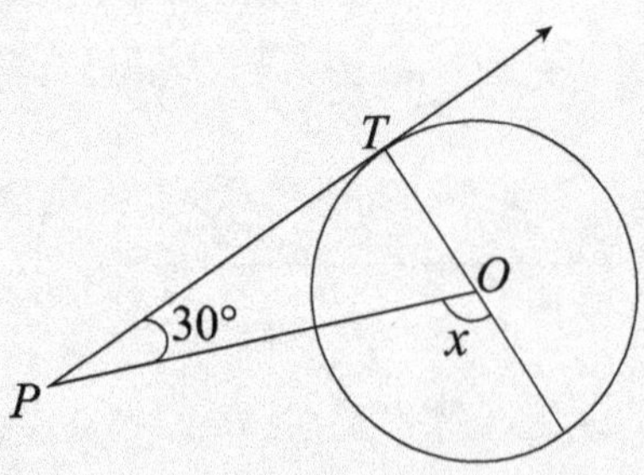

(a) 60° (b) 30°
(c) 90° (d) 120°

(ii) In the given figure, PT is a tangent at T to the centre O. If $\angle TPO = 15°$, then x is equal to

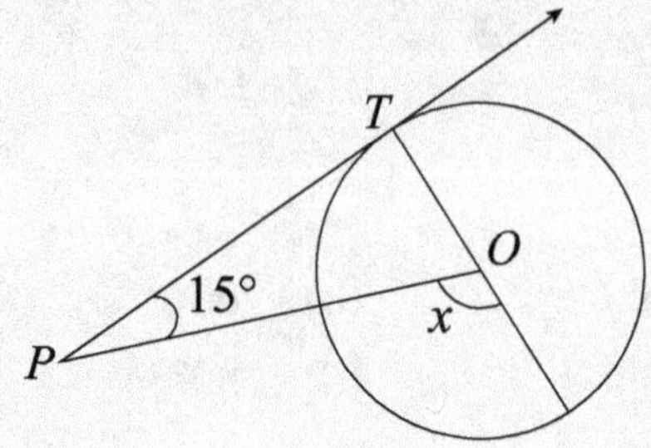

(a) 100° (b) 85°
(c) 105° (d) 120°

(iii) In the given figure, PT is a tangent at T to the centre O. If $\angle TPO = 35°$, then x is equal to

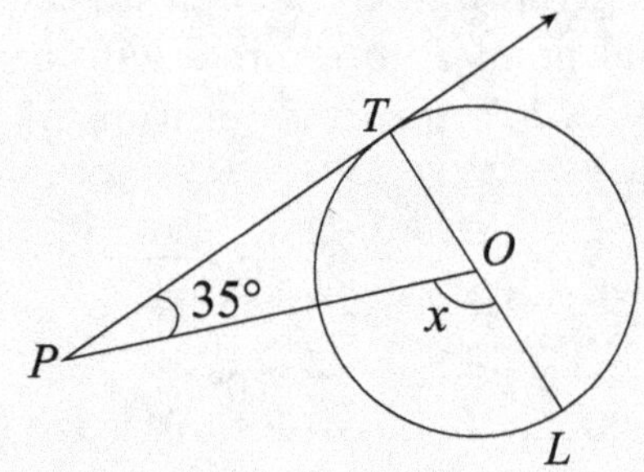

(a) 35° (b) 65°
(c) 105° (d) 125°

126. QP is a tangent to a circle with centre O at a point P on the circle. If ΔOPQ is isosceles, then $\angle OQP$ equals **(2020)**

(a) 30° (b) 45°
(c) 60° (d) 90°

Solution: (b)

In ΔPOQ $\angle POQ$ $\angle P = 90°$

and OP = PQ

$\therefore \angle POQ = \angle OQP = 45°$

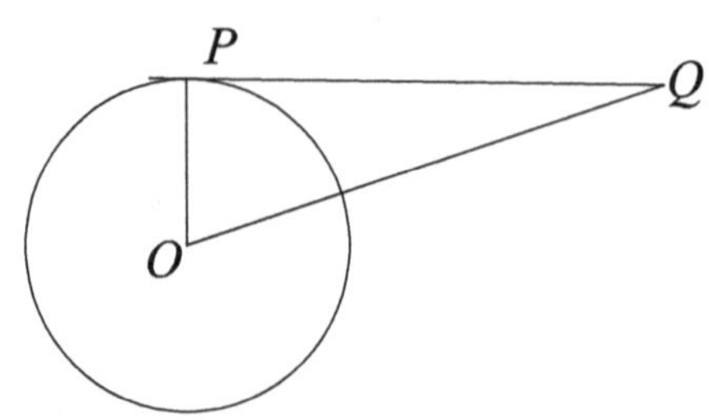

Key Point In the given figure O is the centre of the circle and PA is a tangent to the circle. If $\angle OAB = a°$, then $\angle OPA$ in equal to $\left(\frac{a}{2}\right)°$

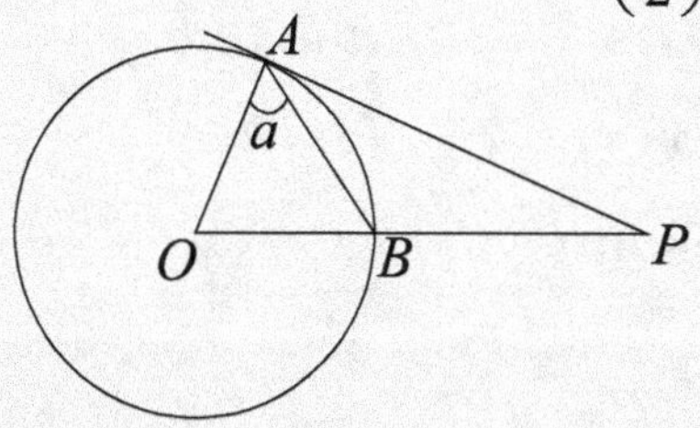

127. In the given figure, O is the centre of the circle and PA is a tangent to the circle. If $\angle OAB = 60°$, then $\angle OPA$ is equal to **(2023)**

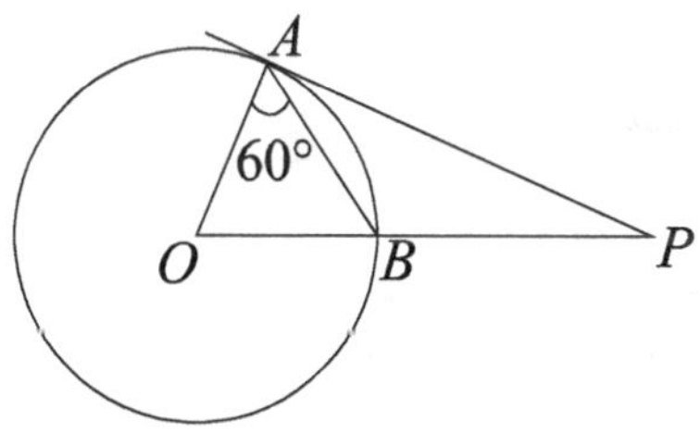

(a) 60° (b) 30°
(c) 15° (d) 20°

Solution: (b)

In ΔAOB

$OA = OB$

$\therefore \angle OAB = \angle OBA = 60°$

$\therefore \angle AOB = 60°$

In ΔOAP

$\angle OAP + \angle AOP + \angle OPA = 180°$

$\Rightarrow 90° + 60° + \angle OPA = 180°$

$\therefore \angle OPA = 30°$

128. In the given figure, PQ is tangent to the circle O. S is a point on the circle such that $\angle SQT = 55°$. The $m\angle QPS$ is **(2026)**

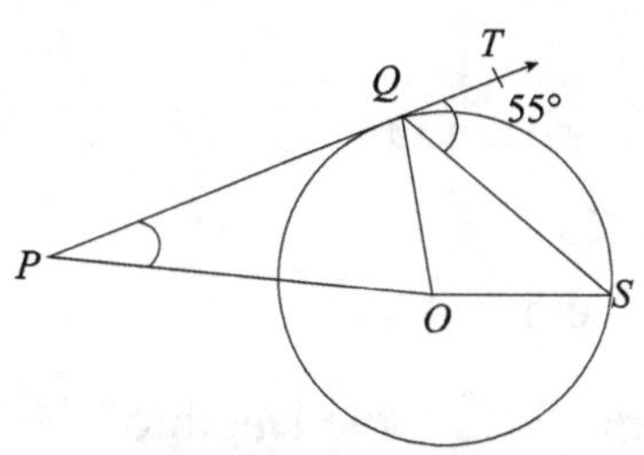

(a) 55° (b) 20°
(c) 35° (d) 70°

Solution:(b)

$\angle OQS = 90° - \angle SQT$

$= 90° - 55°$

$= 35°$

$\therefore \angle OSQ = 35°$

$\therefore \angle QOS = 180° - (\angle OSQ + \angle OQS)$

$= 180° - (35° + 35°)$

$= 110°$

Now, $\angle QOS = \angle OQP + \angle QPO$

$\Rightarrow 110° = 90° + \angle QPS$

$\therefore \angle QPS = 20°$

Exercise 7.19

(i) In the given figure, O is the centre of the circle and PA is a tangent to the circle. If $\angle OAB = 50°$ then $\angle OPA$ is equal to

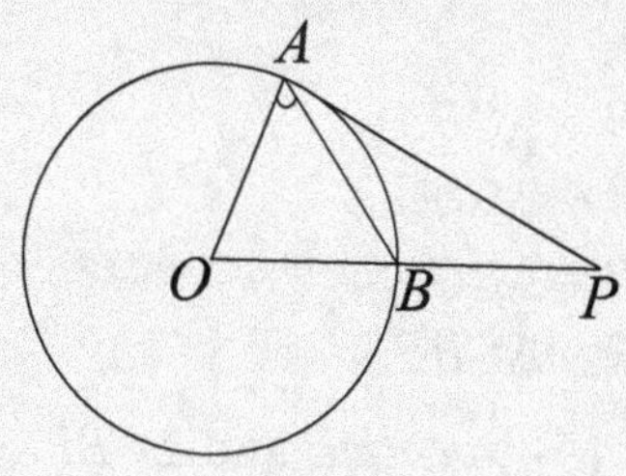

(a) 40° (b) 50°
(c) 10° (d) 80°

(ii) In the given figure, O is the centre of the circle and PA is a tangent to the circle. If $\angle OAB = 65°$ then $\angle OPA$ is equal to

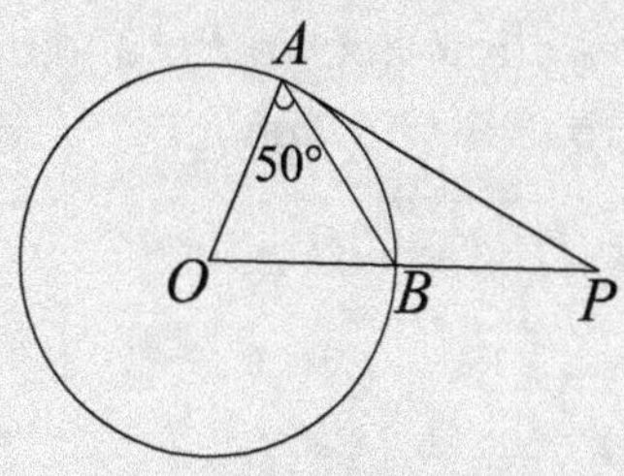

(a) 35° (b) 32°
(c) 40° (d) 33.5°

(iii) In the given figure, O is the centre of the circle and PA is a tangent to the circle. If $\angle OAB = 70°$ then $\angle OPA$ is equal to

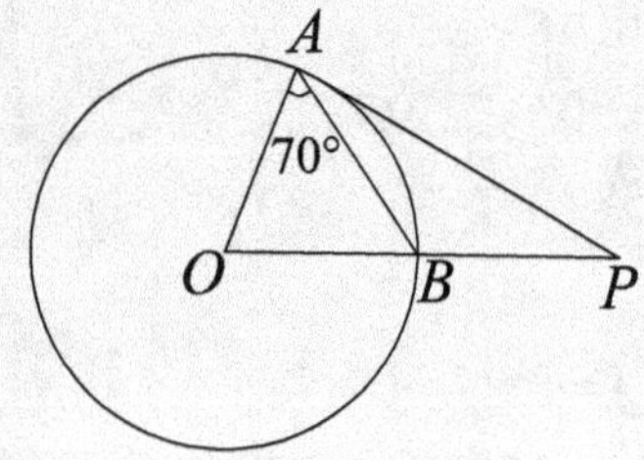

(a) 40° (b) 45°
(c) 70° (d) 80°

129. *AB* and *CD* are two chords of a circle intersecting at *P*. Choose the correct statement from the following **(2024)**

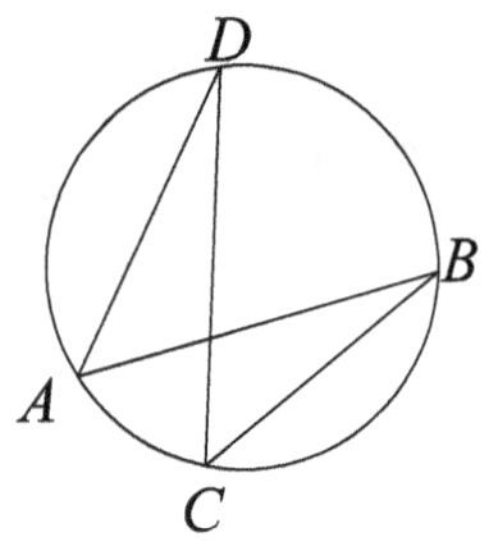

(a) $\Delta ADP \sim \Delta CBA$
(b) $\Delta ADP \sim \Delta BPC$
(c) $\Delta ADP \sim \Delta BCP$
(d) $\Delta ADP \sim \Delta CBP$

Solution: (d) In ΔADP and ΔCBP

$\angle ADP = \angle CBP$

$\angle APD = \angle CPB$

$\angle DAP = \angle BCP$

$\therefore \Delta ADP \sim \Delta CBP$

Key Point From an external point *P*, the length of the tangent to a circle at *Q* in *PQ* and distance of *P* from the centre *O* is *OP*, then

(i) $OP = \sqrt{PQ^2 + OQ^2}$

(ii) $OQ = \sqrt{OP^2 - QP^2}$

(iii) $PQ = \sqrt{OP^2 - OQ^2}$

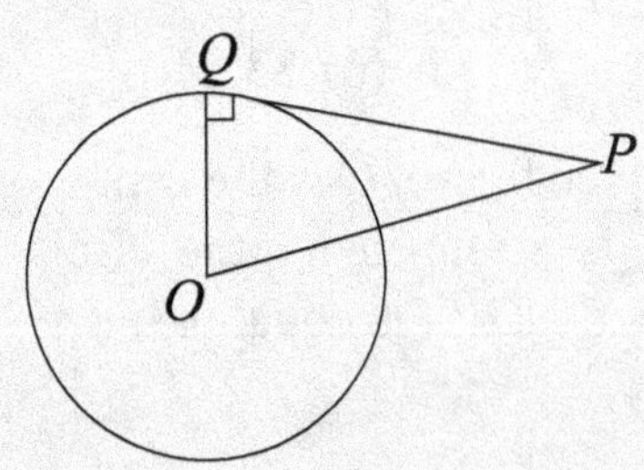

130. From an external point *Q*, the length of the tangent to a circle is 5 cm and the distance of *Q* from the centre is 8 cm. The radius of the circle is **(2020)**

(a) 39 cm (b) 3 cm
(c) $\sqrt{39}$ cm (d) 7 cm

Solution: (b)

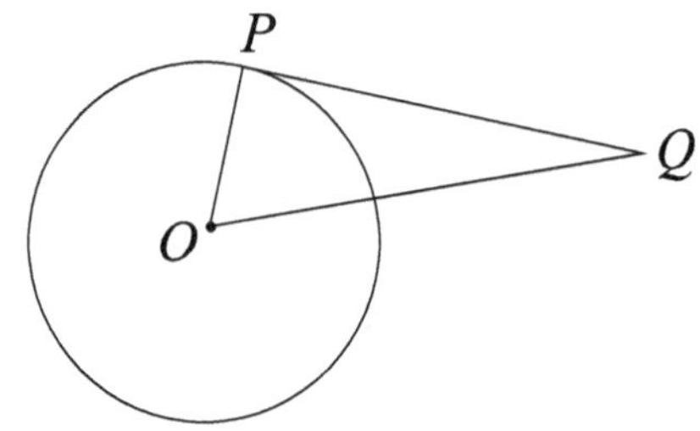

$OQ^2 = OP^2 + PQ^2$

$\Rightarrow 8^2 = OP^2 + 5^2$

$\Rightarrow OP^2 = 39$

$\therefore OP = \sqrt{39}$ cm

131. From an external point *Q*, the length of tangent to a circle is 12 cm and the distance of *Q* from the centre of circle is 13 cm. The radius of circle (in cm) is **(2020)**

Solution: (b)

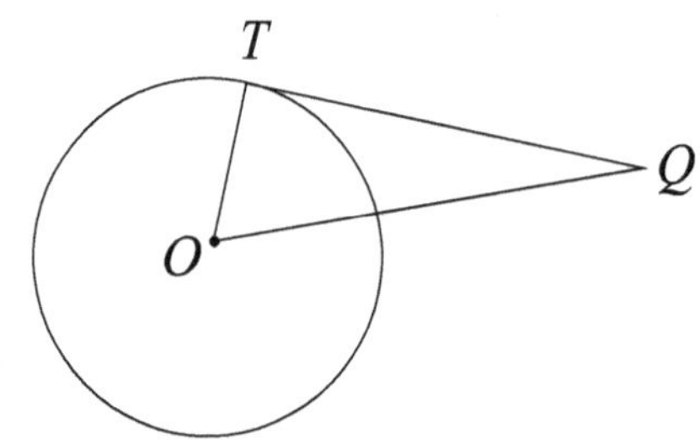

In ΔOTQ, $\angle OTQ = 90°$

$\therefore OQ^2 = OT^2 + TQ^2$

$\Rightarrow 13^2 = OT^2 + 12^2$

$\Rightarrow OT^2 = 169 - 144$

$\Rightarrow OT^2 = 25$

$\therefore OT = 5$ cm

132. In figure –3, find the length of the tangent *PQ* drawn from the point *P* to a circle with centre at *O*, given that $OP = 12$ cm and $OQ = 5$ cm

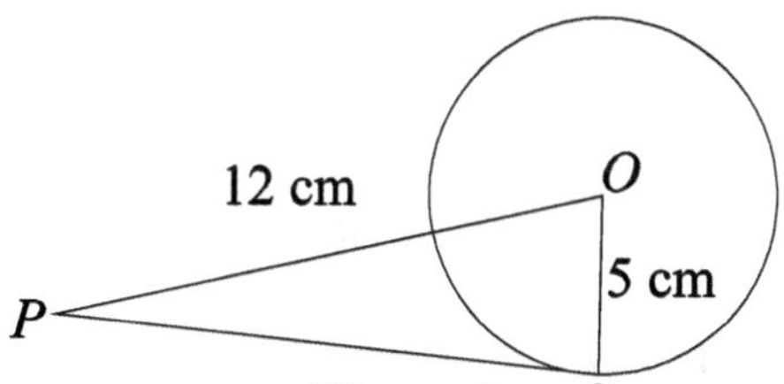

Figure-3

Solution: (b)

In ΔPQO $\angle PQO = 90°$

$\therefore OP^2 = PQ^2 + OQ^2$

$\Rightarrow 12^2 = PQ^2 + 5^2$

$\Rightarrow PQ^2 = 144 - 25$

$\Rightarrow PQ^2 = 119$

$\therefore PQ = \sqrt{119}$ cm

133. The length of tangent drawn to a circle of radius 9 cm from a point 41 cm from the centre is **(2023)**

Solution: (b)

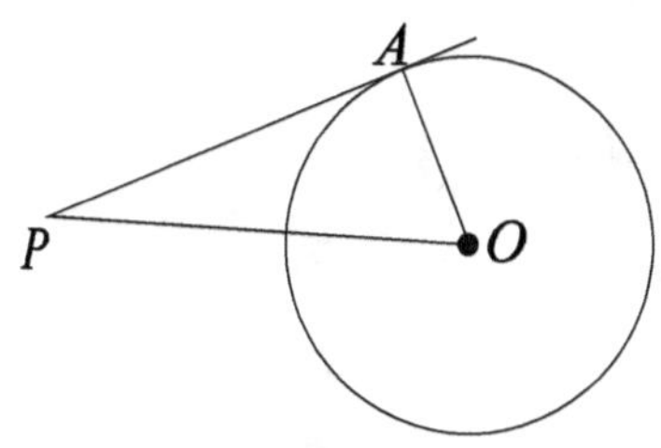

In ΔPAO, $\angle PAO = 90°$

$\therefore OP^2 = OA^2 + PA^2$

$\Rightarrow 41^2 = 9^2 + PA^2$

$\Rightarrow PA^2 = 1600$

$\therefore PA^2 = 40$ cm

134. The length of the tangent from an external point A to a circle of radian 3 cm in 4 cm. The distance of A from the centre of the circle is. **(2023)**

(a) 7 cm (b) 5 cm
(c) $\sqrt{7}$ cm (d) 25 cm

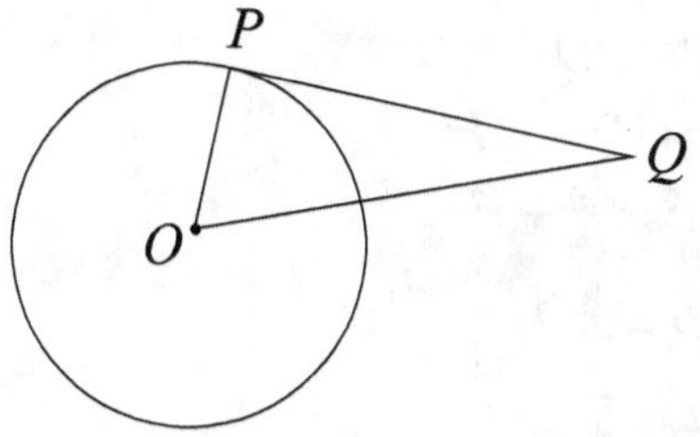

Solution: (b)

In ΔAPO, $\angle APO = 90°$

$\therefore OA^2 = OP^2 + AP^2$

$\Rightarrow OA^2 = 3^2 + 4^2$

$\Rightarrow OA^2 = 25$

$\therefore OA^2 = 5$ cm

135. In Fig 1, on a circle of radius 7 cm, tangent PT in shown from a point P such that $PT = 24$ cm. If O is the centre of the circle, then the length of PR is **(2020)**

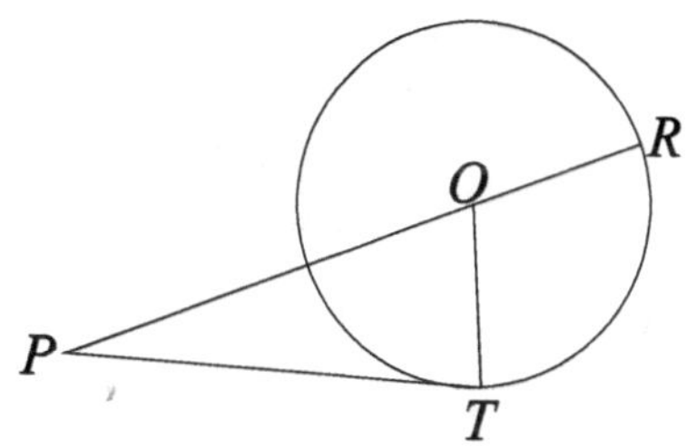

Fig. 1

(a) 30 cm (b) 28 cm
(c) 32 cm (d) 25 cm

Solution:

In ΔPTO, $\angle PTO = 90°$

$\therefore OP^2 = PT^2 + OT^2$

$\Rightarrow OP^2 = 24^2 + 7^2$

$\Rightarrow OP^2 = 625$

$\therefore OP = 25$ cm

Now, $PR = OP + OR = 25 + 7 = 32$ cm

136. In the given figure, PA is a tangent to a circle with center O. If OP = 10 cm, then the length of AP is **(2025)**

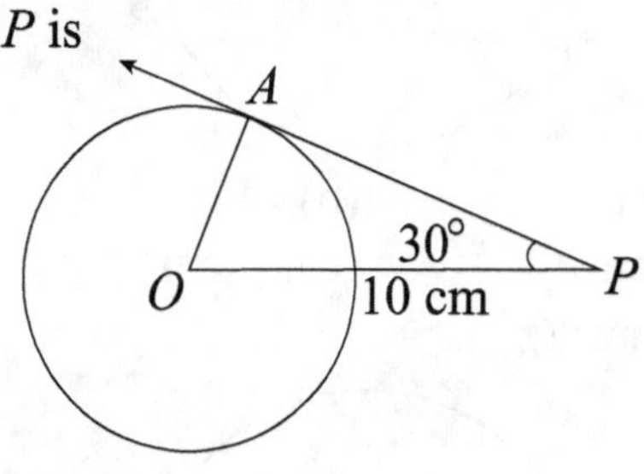

(a) $10\sqrt{3}$ cm (b) 20 cm
(c) 5 cm (d) $5\sqrt{3}$ cm

Solution:

$$\frac{AP}{OP} = \cos 30°$$

$$\Rightarrow \frac{AP}{10} = \frac{\sqrt{3}}{2}$$

$AP = 5\sqrt{3}$ cm

137. PQ is tangent to the circle centred at O. If OQ = 3 cm, PQ = 5 cm, then OP is equal to **(2025)**

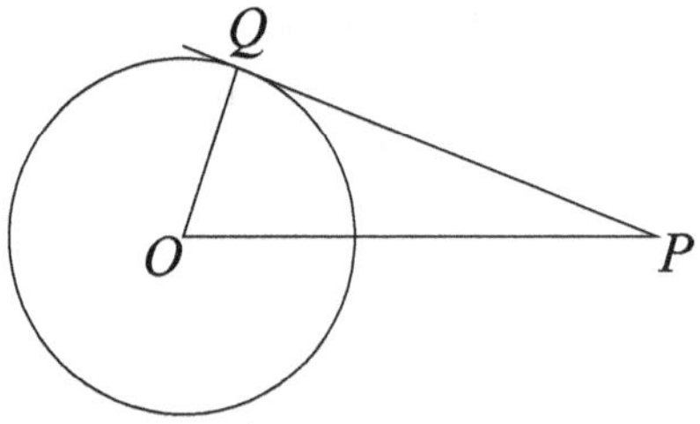

(a) 5 cm (b) 4 cm

(c) $\sqrt{15}$ cm (d) $\sqrt{34}$ cm

Solution: (d)

In ΔPQO, $\angle PQO = 90°$

$\therefore OP^2 = OQ^2 + PQ^2$

$\Rightarrow OP^2 = 3^2 + 5^2$

$\Rightarrow OP^2 = 34$

$\therefore OP = \sqrt{34}$ cm

138. In the adjoining figure, the sum of radii of two concertric circles is 16 cm. The length of chord AB which toucher the inner circle at P is 16 cm. The difference of the radii of the given circle is **(2025)**

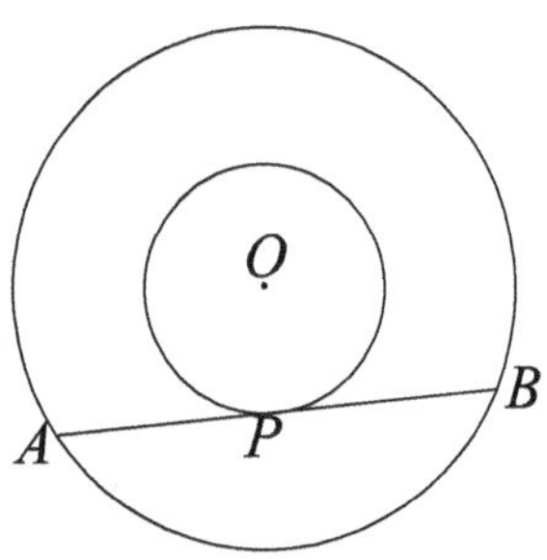

(a) 8 cm (b) 4 cm

(c) 2 cm (d) 3 cm

Solution: (b)

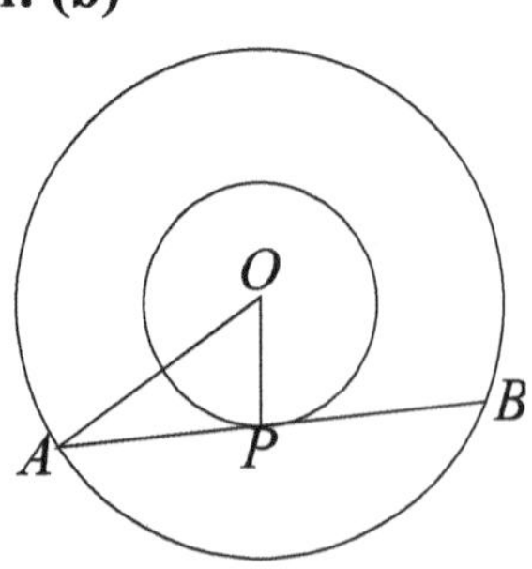

Let, $OP = r$

$\therefore OA = 16 - r$

$$AP = \frac{1}{2}AB = \frac{1}{2} \times 16 = 8$$

Now, in ΔAPO, $\angle P = 90°$

$\therefore OA^2 = AP^2 + OP^2$

$\Rightarrow (16 - r)^2 = 8^2 + r^2$

$\Rightarrow 256 - 32r + r^2 = 64 + r^2$

$\Rightarrow 32r = 192$

$\therefore r = 6$

$\therefore OP = 6$ cm and $OA = 16 - 6 = 10$ cm

$\therefore OA - OP = 10 - 6 = 4$ cm

139. In the adjoining figure, AB is the chord of the larger circle touching the smaller circle, The center of both the circles is O. If $AB = 2r$ and $OP = r$, then the radius of larger circle is: **(2025)**

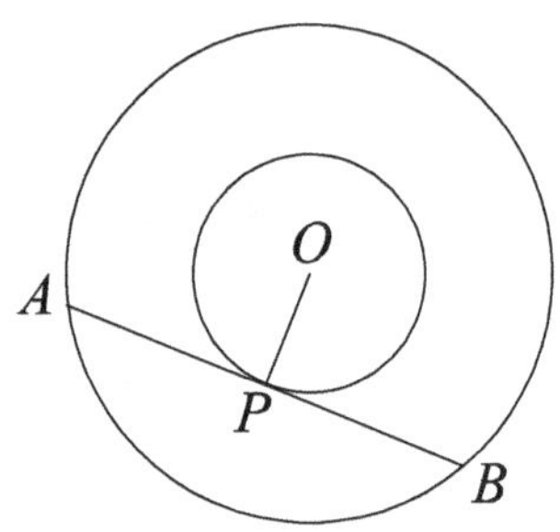

(a) $2r$ (b) $3r$

(c) $2\sqrt{2}\,r$ (d) $\sqrt{2}\,r$

Solution:

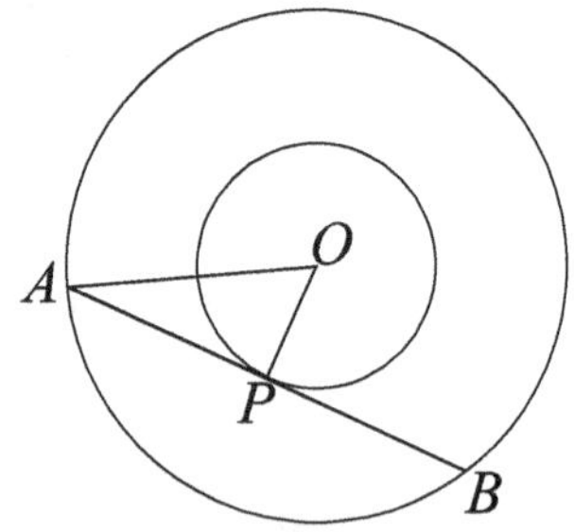

$OP \perp AB$

$\therefore \angle OPA = 90°$

$\therefore$ In ΔAPO

$OA^2 = OP^2 + AP^2$

$\Rightarrow OA^2 = r^2 + r^2$

$$\left[\because AP = \frac{1}{2}AB\right]$$

$\Rightarrow OA^2 = 2r^2$

$\therefore OA = \sqrt{2}\,r$

140. In the adjoining figure, AC is diameter of larger circle with center O. AB is tangent to smaller circle with center O. If $OD = r$, BC is equal to: **(2025)**

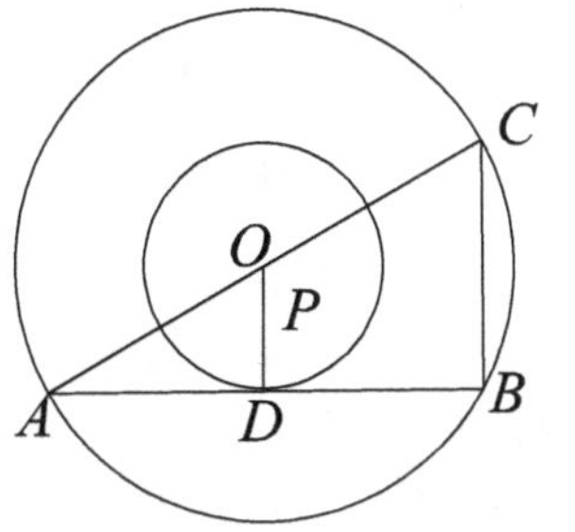

(a) r (b) $\frac{3r}{2}$

(c) $2r$ (d) $4r$

Solution: (c)

$OD \perp AB$ and $CB \perp AB$

$\therefore OD \parallel BC$

$\therefore \Delta OAD \sim \Delta CAB$

$\therefore \frac{OA}{AC} = \frac{AD}{AB} = \frac{OD}{BC}$

$\Rightarrow \frac{OA}{AC} = \frac{OD}{BC}$

$\Rightarrow \frac{OA}{2OA} = \frac{OD}{BC}$

$\Rightarrow \frac{1}{2} = \frac{r}{BC}$

$\Rightarrow BC = 2r$

141. PQ is tangent to a circle with centre O. If $OQ = a$, $OP = a + 2$ and $PQ = 2b$, then the relation between a and b is: **(2026)**

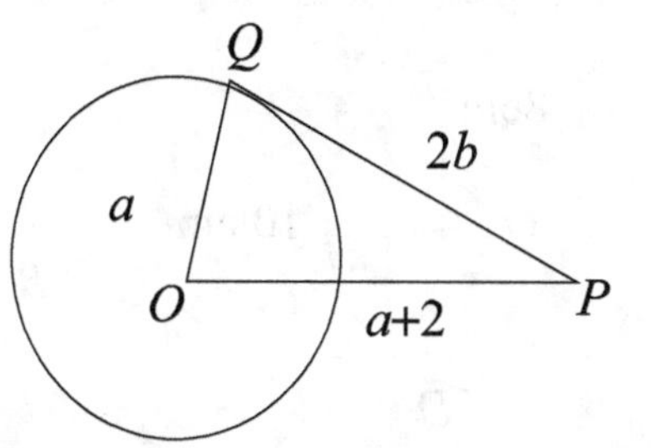

(a) $a^2 + (a+2)^2 = (2b)^2$ (b) $b^2 = a + 4$

(c) $2a^2 + 1 = b^2$ (d) $b^2 = a + 1$

Solution: (d)

In ΔPQO, $\angle PQO = 90°$

$\therefore OP^2 = OQ^2 + PQ^2$

$\Rightarrow (a+2)^2 = a^2 + (2b)^2$

$\Rightarrow a^2 + 4a + 4 = a^2 + 4b^2$

$\Rightarrow 4a + 4 = 4b^2$

$\Rightarrow a + 1 = b^2$

$\therefore b^2 = a + 1$

Exercise 7.20

(i) In the given figure, the length $PB =$ ________ cm

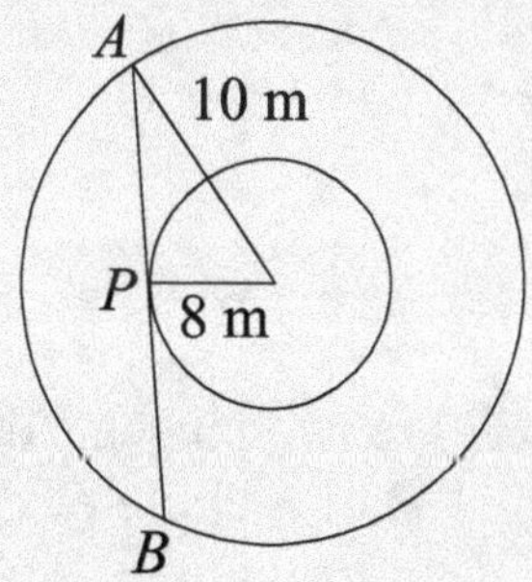

(ii) In the given figure, the length $PB =$ ________ cm

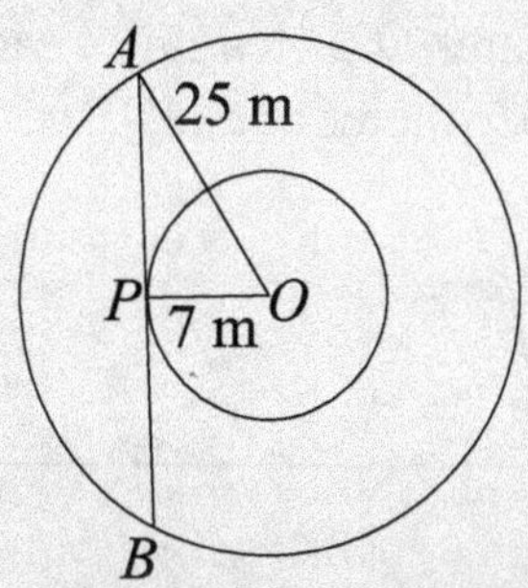

(iii) In the given figure, the length $PB =$ ________ cm

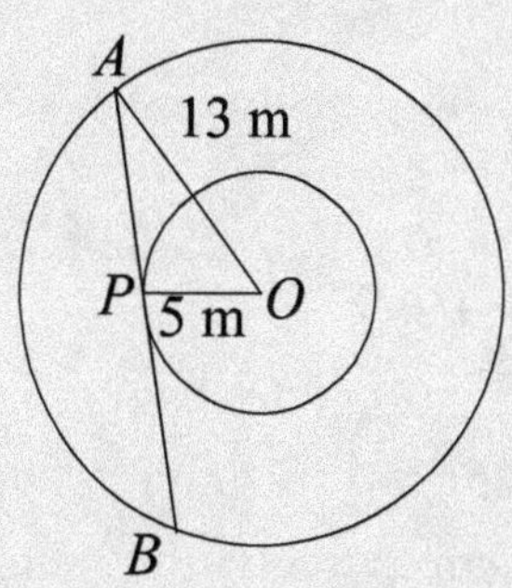

(iv) Two com-centric circle of radii 5 cm and 3 cm are given. Find the length of the chord of the larger circle which tangent the smaller circle

(v) Two com-centric circle of radii 13 cm and 12 cm are given. Find the length of the chord of the larger circle which tangent the smaller circle

(vi) Two com-centric circle of radii 25 cm and 24 cm are given. Find the length of the chord of the larger circle which tangent the smaller circle

Key Point The length of tangents drawn from an external point to a circle are equal.

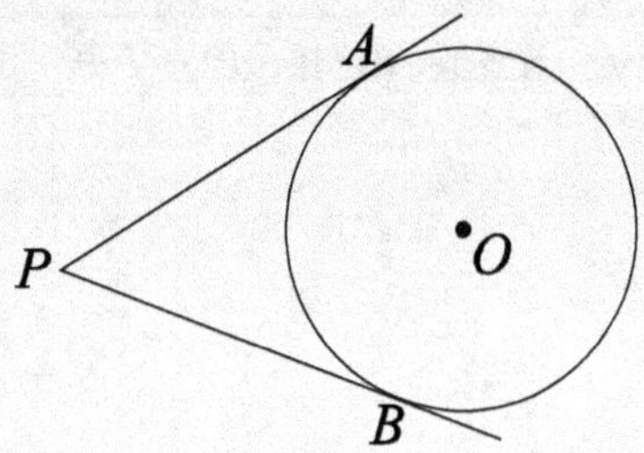

If PA and PB are two tangents drawn from an external point P of a circle of centre O, then $PA = PB$

142. PQ and PR are tangents to the circle of radius 3 cm and center O. If length of each tangent is 4 cm, then perimeter of ΔOQP is: **(2025)**

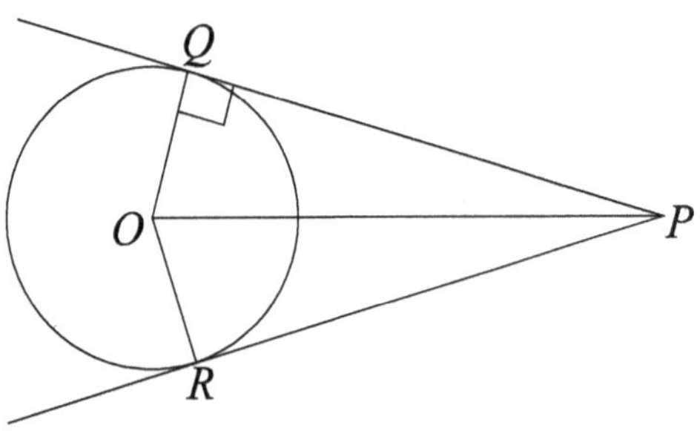

Solution:

In ΔOQP

$OP^2 = OQ^2 + PQ^2$

$\Rightarrow OP^2 = 3^2 + 4^2$

$\Rightarrow OP^2 = 25$

$\therefore OP = 5$ cm

$\therefore$ Perimeter of $\Delta OQP = 3 + 4 + 5$

$= 12$ cm

143. In the given figure, BC and BD are tangents to the circle with centre O and radius 9 cm. If OB = 15 cm, then the length $(BC + BD)$ is **(2023)**

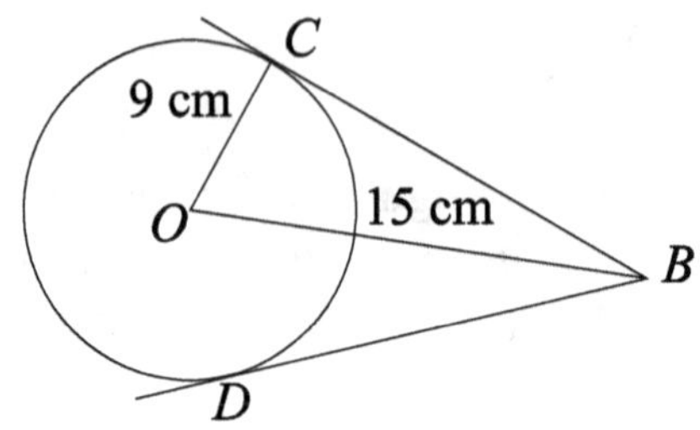

(a) 18 cm (b) 12 cm
(c) 24 cm (d) 30 cm

Solution: (b)

In ΔBCO, $\angle BCO = 90°$

$\therefore OB^2 = OC^2 + BC^2$

$\Rightarrow 15^2 = 9^2 + BC^2$

$\Rightarrow BC^2 = 144$

$\Rightarrow BC = 12$ cm

$\therefore BD = 12$ cm

$\therefore BC + BD = 12 + 12 = 24$ cm

Exercise 7.21

(i) In the given figure, BC and BD are tangents to the circle with centre O and radius 6 cm. If OB = 10 cm, then the length $(BC + BD)$ is

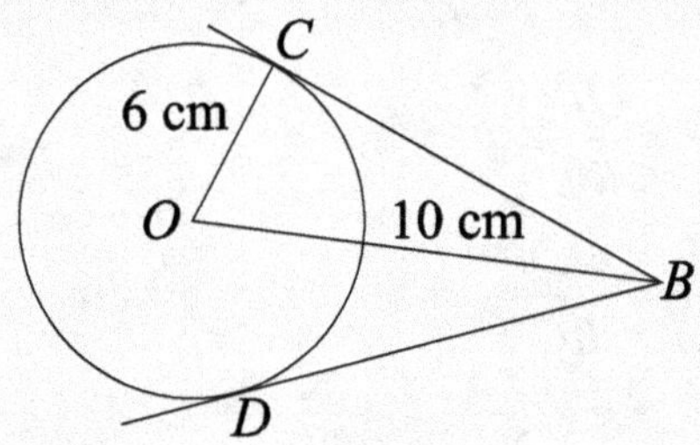

(a) 10 cm (b) 6 cm
(c) 16 cm (d) 12 cm

(ii) In the given figure, BC and BD are tangents to the circle with centre O and radius 8 cm. If OB = 10 cm, then the length $(BC + BD)$ is

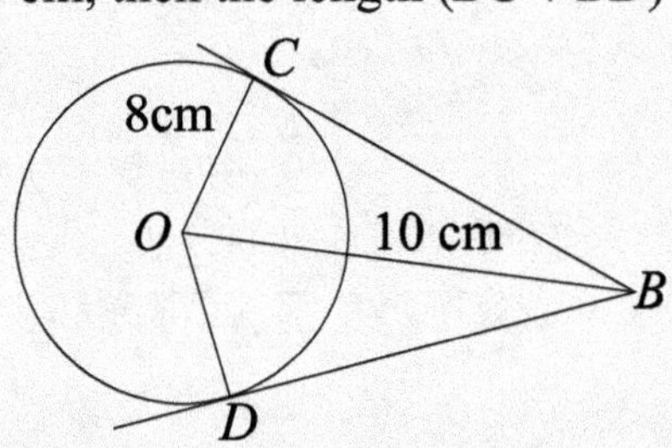

(a) 6 cm (b) 10 cm
(c) 12 cm (d) 3 cm

(iii) In the figure, BC and BD are tangents to the circle with centre O and radius 7 cm. If OB = 25 cm, then the length $(BC + BD)$ is

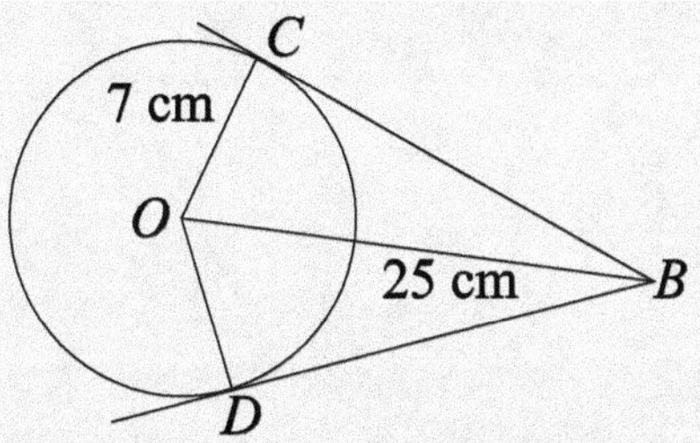

(a) 25 cm (b) 7 cm
(c) 32 cm (d) 48 cm

144. In two concertric circles, a tangent to the smaller circle will intersect the larger circle at **(2025)**

(a) Zero point (b) One point
(c) Two point (d) Three point

Solution: (c)

Two point

145. The lengths of the tangents drawn from an external point to a circle are _______ **(2020)**

Solution: Equal

146. In Figure1, ΔABC in circumscribing a circle, find the length of BC. **(2009,2019)**

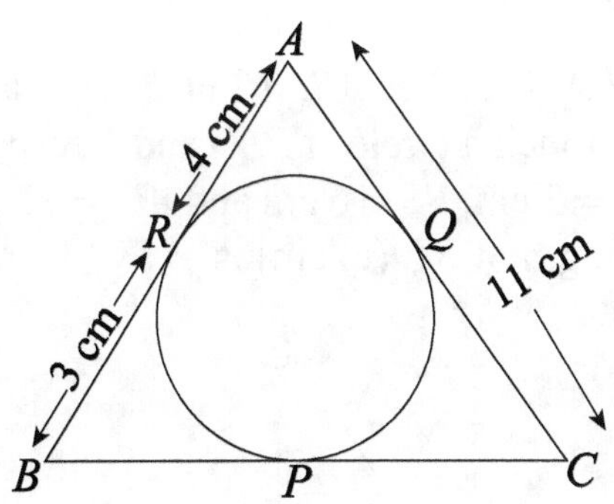

Figure 1

Solution:

$AQ = AR = 4$ cm

$\therefore QC = AC - AQ$

$= 11 - 4$

$= 7$ cm

$\therefore PC = QC = 7$ cm

$BP = BR = 3$ cm

$\therefore BC = BP + PC$

$= 3 + 7$

$= 10$ cm.

147. In Fig. 1, the sides AB, BC and CA of a triangle ABC, touch a circle at P, Q and R respectively. If $PA = 4$ cm, $BP = 3$ cm and $AC = 11$ cm, then the length of BC(in cm) is **(2012)**

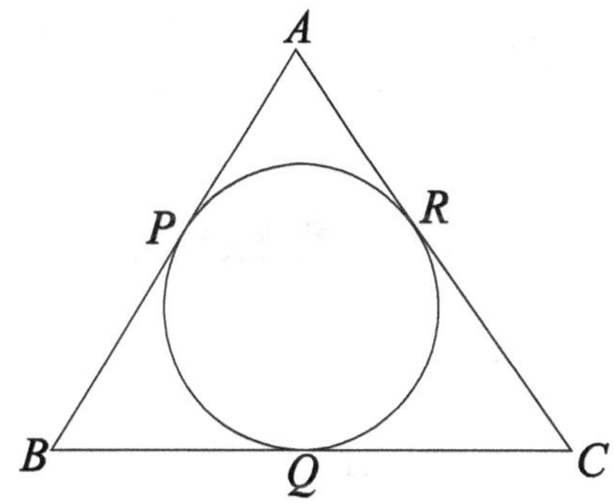

(a) 11 (b) 10
(c) 14 (d) 15

Solution: (b) 10 cm

148. In the given figure, the perimeter of ΔABC is **(2023)**

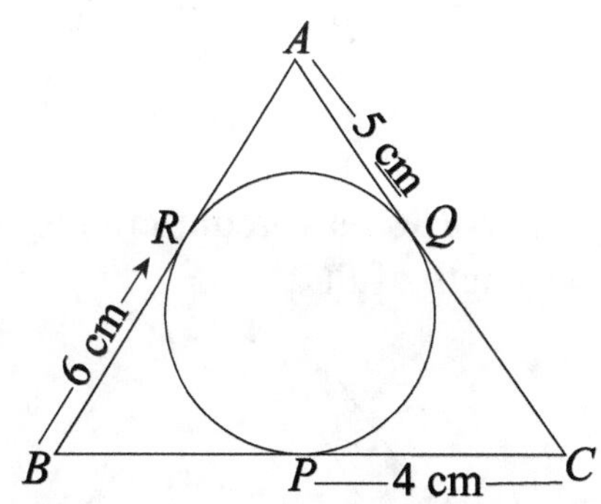

(a) 30 cm (b) 15 cm
(c) 45 cm (d) 60 cm

Solution:(a)

$BP = BR = 6$ cm

$QC = PC = 4$ cm

$AR = AQ = 5$ cm

Now, $AB = AR + BR$

$= 5 + 6$

$= 11$ cm

$BC = BP + PC$

$= 6 + 4$

$= 10$ cm

$AC = AQ + QC$

$= 5 + 4$

$= 9$ cm

$\therefore$ Perimeter of $\Delta ABC = AB + BC + AC$

$= 11 + 10 + 9$

$= 30$ cm.

Exercise 6.22

(i) In Fig. 1, ΔABC is Circumscribing a circle. Find the length of BC.

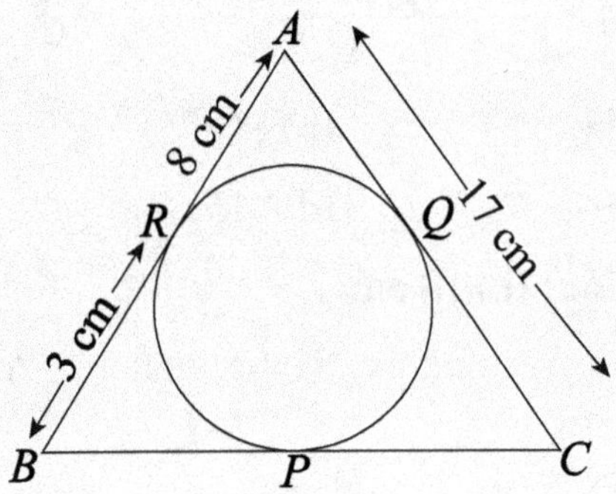

Fig -1

(ii) In Fig. 1, ΔABC is Circumscribing a circle. Find the length of BC.

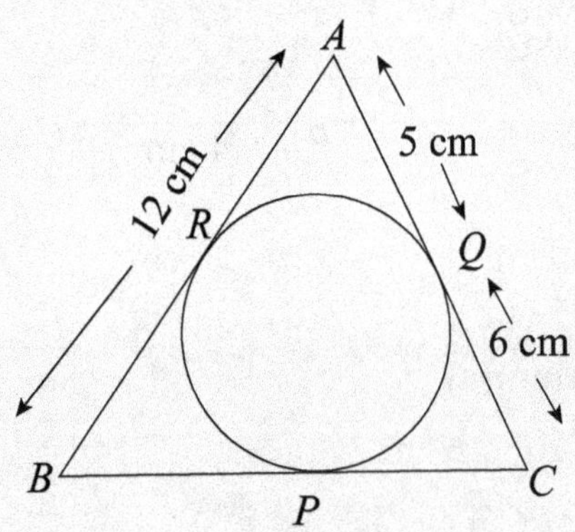

Fig -1

(iii) In Fig. 1, ΔABC is Circumscribing a circle. Find the length of BC.

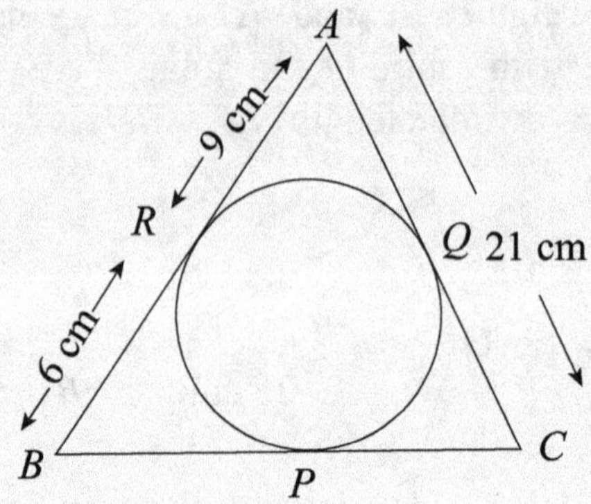

(iv) In Fig.1, the sides AB, BC and CA of a triangle ABC, touch a circle at P, Q and R respectively. If $PA = 5$ cm, $BP = 4$ cm and $AC = 12$ cm, then the length of BC (in cm) is

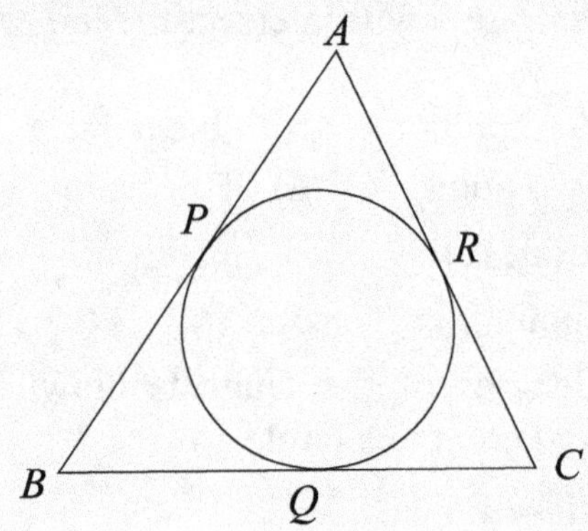

Fig. 1

(a) 10 (b) 11

(c) 12 (d) 14

(v) In Fig.1, the sides AB, BC and CA of a triangle ABC, touch a circle at P, Q and R respectively. If $PA = 3$ cm, $BP = 6$ cm and $AC = 11$ cm, then the length of BC (in cm) is

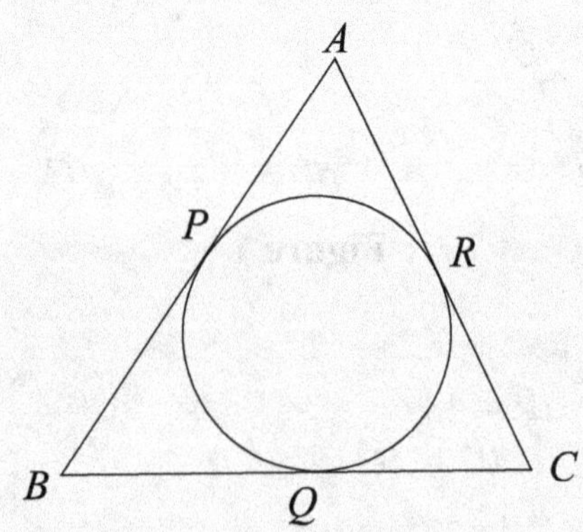

(a) 10 (b) 12

(c) 14 (d) 20

(vi) In Fig.1, the sides AB, BC and CA of a triangle ABC, touch a circle at P, Q and R respectively. If $PA = 6$ cm, $BP = 5$ cm and $AC = 13$ cm, then the length of BC (in cm) is

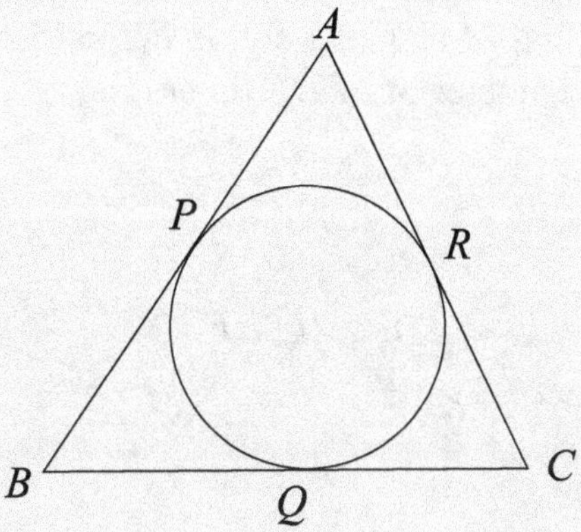

(a) 12 (b) 14
(c) 13 (d) 10

(vii) In the given figure, the perimeter of ΔABC is

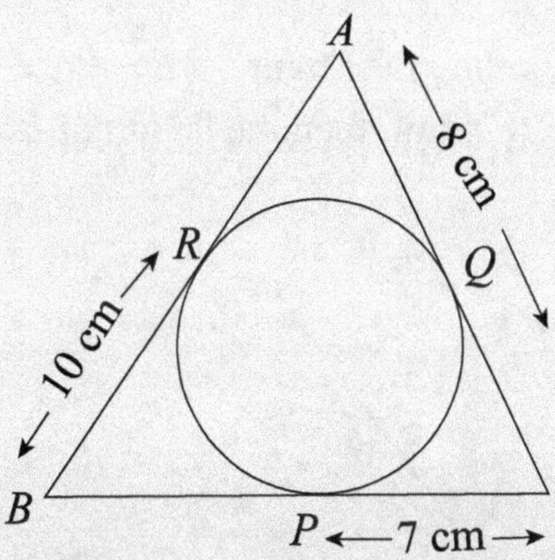

(a) 25 cm (b) 30 cm
(c) 18 cm (d) 50 cm

(viii) In the given figure, the perimeter of ΔABC is

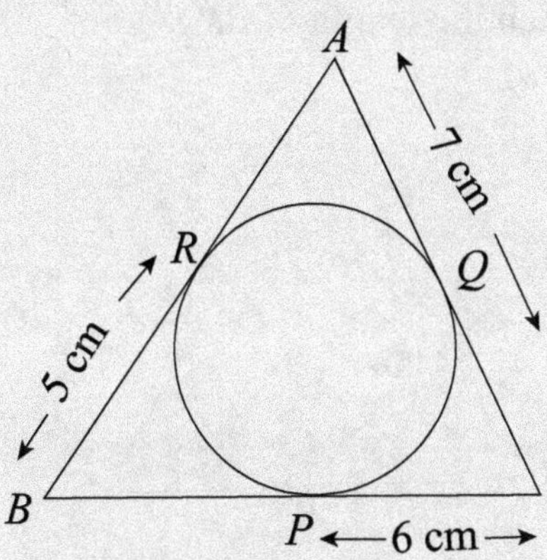

(a) 36 cm (b) 30 cm
(c) 18 cm (d) 54 cm

(ix) In the given figure, the perimeter of ΔABC is

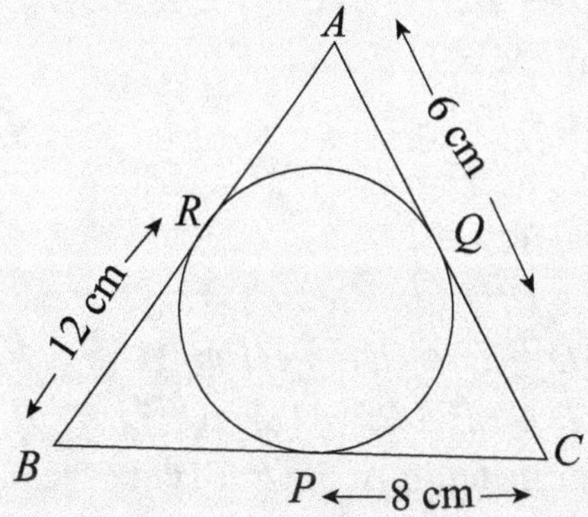

(a) 26 cm (b) 14cm
(c) 52 cm (d) 35 cm.

Key Point In the given figure, a circle touches side, DF of ΔDEF at H and touches ED and EF produced at K and M respectively. If $EK = a$ unit, Then the perimeter of ΔEDF is $2a$ unit.

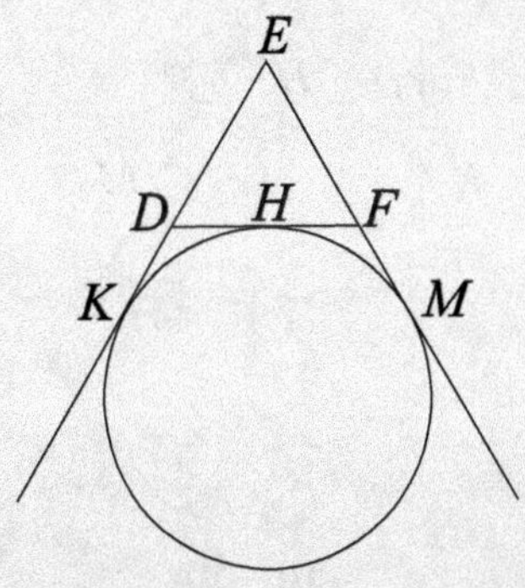

149. In Fig. 2, a circle touches the side, DF of ΔEDF at H and touches ED and EF produced at K and M respectively. If $EK = 9$ cm, then the perimeter of ΔEDF (in cm) is **(2012)**

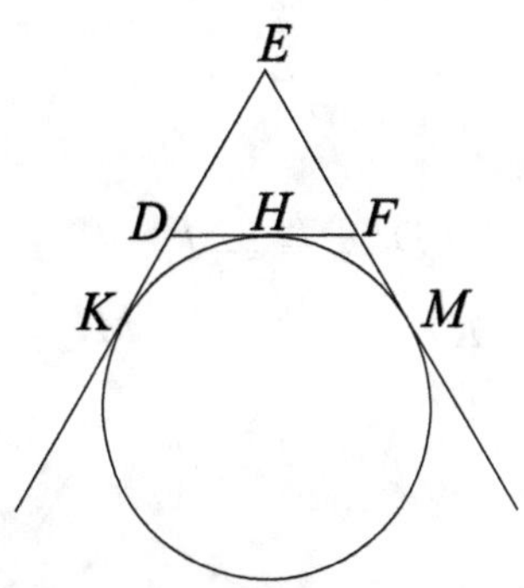

Fig. 2

Solution: (a)

$EK = 9$ cm

$\Rightarrow ED + DK = 9$

$\Rightarrow ED + DH = 9$ ...(i) $[\because DK = DH]$

again $EM = 9$ $[\because EK = EM]$

$\Rightarrow EF + FM = 9$

$\Rightarrow EF + HF = 9$ $[\because FM = HF]$ ...(ii)

(i) + (ii)

$(ED + DH) + (EF + HF) = 9 + 9$

$\Rightarrow ED + EF + (DH + HF) = 18$

$\Rightarrow ED + EF + DF = 18$ cm

$\therefore$ Perimeter of $\Delta DEF = 18$ cm.

Exercise 7.23

(i) In the given figure, a circle touches the side DF of ΔEDF at H and touches ED and EF produced at K and M respectively. If $EK = 12$ cm, then the perimeter of ΔEDF (in cm) is

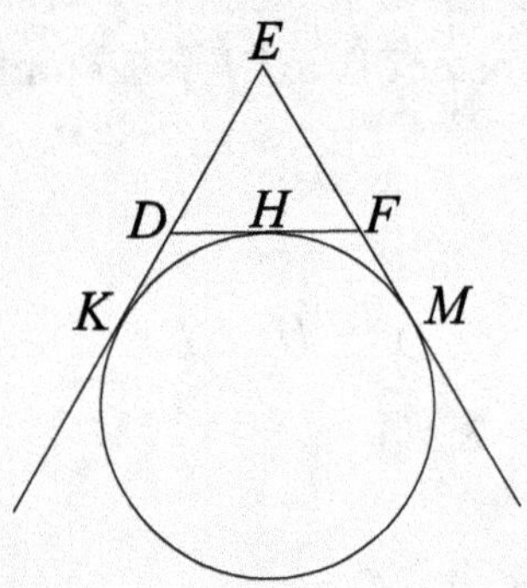

(a) 12 (b) 20
(c) 24 (d) 50

(ii) In the given figure, a circle touches the side DF of ΔEDF at H and touches ED and EF produced at K and M respectively. If $EK = 18$ cm, then the perimeter of ΔEDF (in cm) is

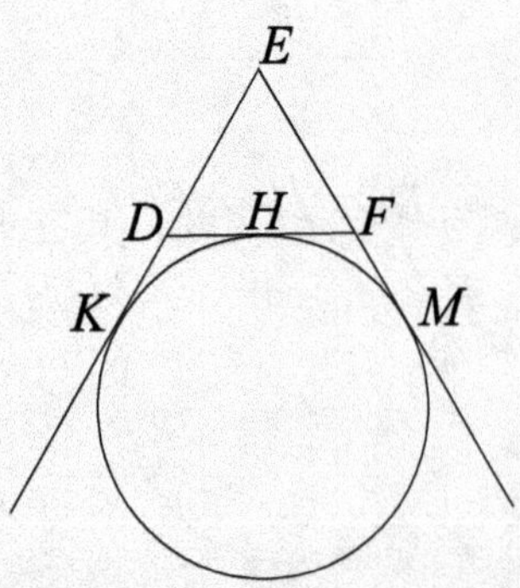

(a) 18 (b) 36
(c) 20 (d) 24

(iii) In the given figure, a circle touch the side DF of ΔEDF at H and touches ED and EF produced at K and M respectively. If $EK = 15$ cm, then the perimeter of ΔEDF (in cm) is

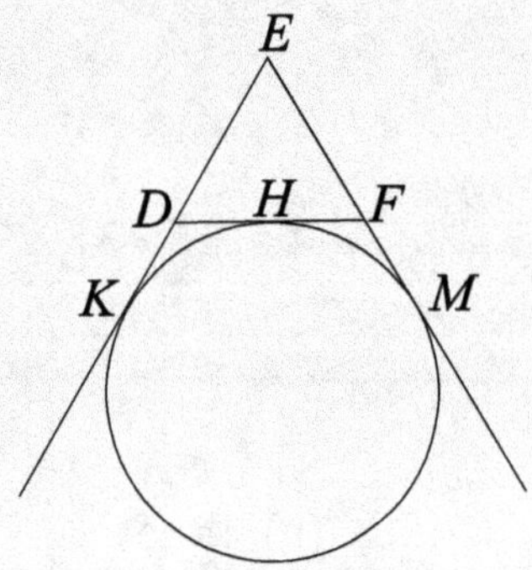

(a) 20 (b) 15
(c) 45 (d) 30

Key Point In the given figure, $AB = BC = a$ unit. If $AC = b$ unit, then the length of BP is $\left(a - \frac{b}{2}\right)$ unit.

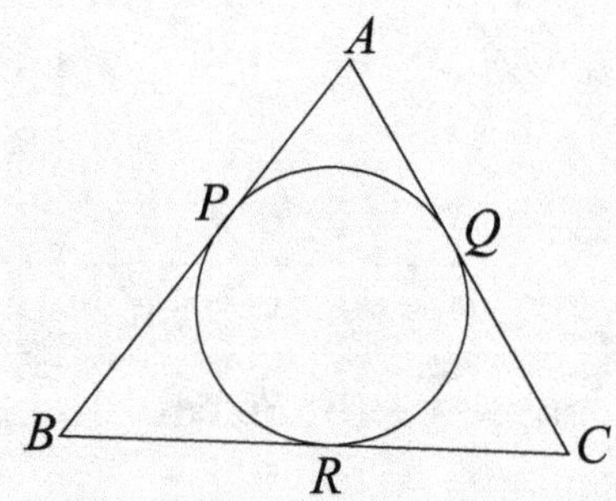

150. In the given figure, $AB = BC = 10$ cm. If $AC = 7$ cm then the length of BP is **(2023)**

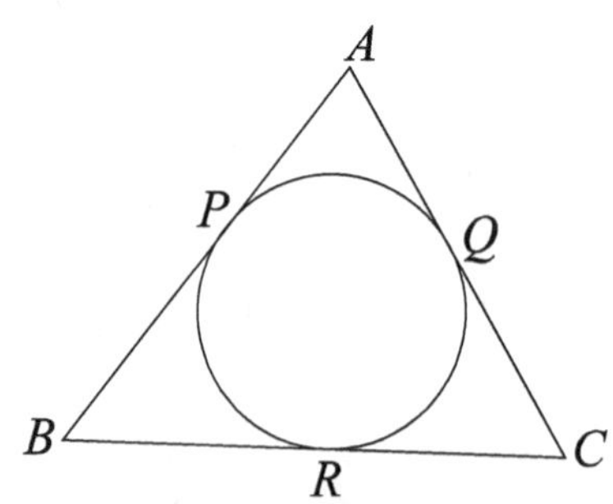

(a) 3.5 cm (b) 7 cm

(c) 6.5 cm (d) 5 cm

Solution:(c)

$AB = BC$

$\Rightarrow BP + AP = BR + RC$

$\Rightarrow AP = RC \quad [\because BP = BR]$

$\Rightarrow AQ = QC \quad [\because AP = AQ, RC = QC]$

$\Rightarrow AQ = AC - AQ$

$\Rightarrow AQ = 7 - AQ$

$\Rightarrow AQ = 3.5$ cm

$\therefore AP = 3.5$ cm

$\therefore BP = AB - AP$

$= 10 - 3.5$

$= 6.5$ cm

Exercise 7.24

(i) In the given figure, $AB = BC = 12$ cm. If $AC = 9$ cm, then the length of BP is.

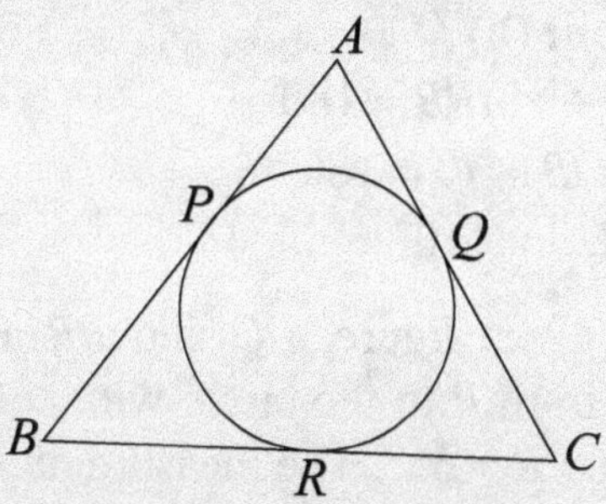

(a) 7.5 cm (b) 8 cm

(c) 6.5 cm (d) 10 cm.

(ii) In the given figure, $AB = BC = 15$ cm. If $AC = 8$ cm then the length of BP is.

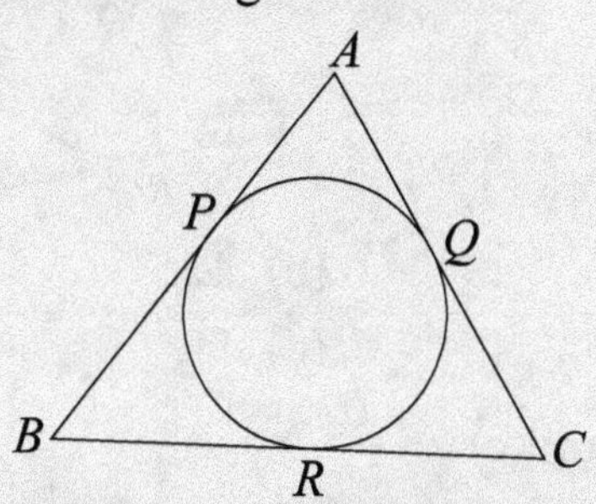

(a) 15 cm (b) 11 cm

(c) 12 cm (d) 10 cm.

(iii) In the given figure, $AB = BC = 14$ cm. If $AC = 8$ cm then the length of BP is.

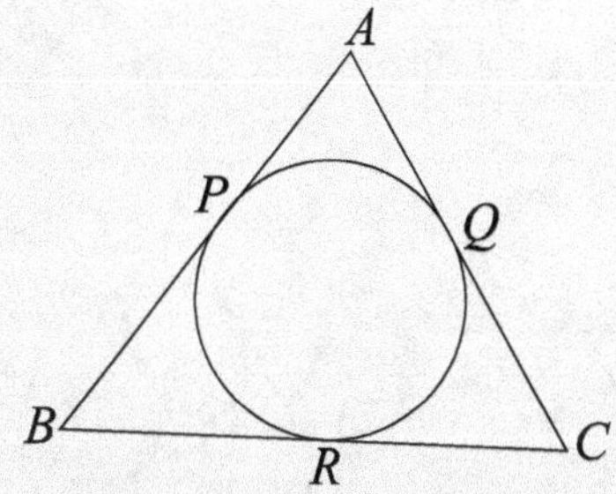

(a) 14 cm (b) 8 cm

(c) 10 cm (d) 11cm.

Key Point In the given figure, TP and TQ are tangents drawn to the circle with centre O. If $\angle POQ = a°$, then $\angle PTQ$ is equal to $(180 - a)°$

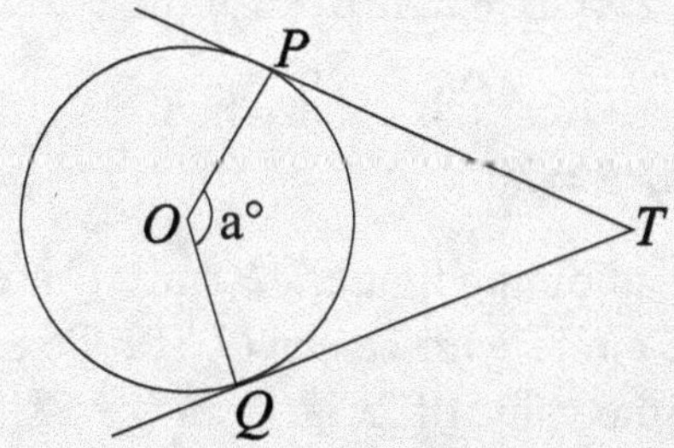

151. In Figure 2, TP and TQ are tangents drawn to the circle with centre O. If $\angle POQ = 115°$, then $\angle PTQ$ is **(2020)**

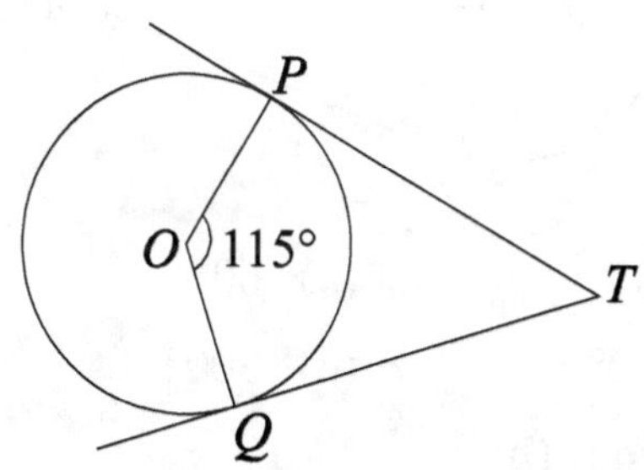

(a) 115° (b) 57.5°

(c) 55° (d) 65°

Solution: (d)

$\angle POQ + \angle PTQ = 180°$

$\Rightarrow 115° + \angle PTQ = 180°$

$\therefore \angle PTQ = 65°$

152. In the adjoining figure PA and PB are tangents to a circle with center O. The measure of angle APB is **(2025)**

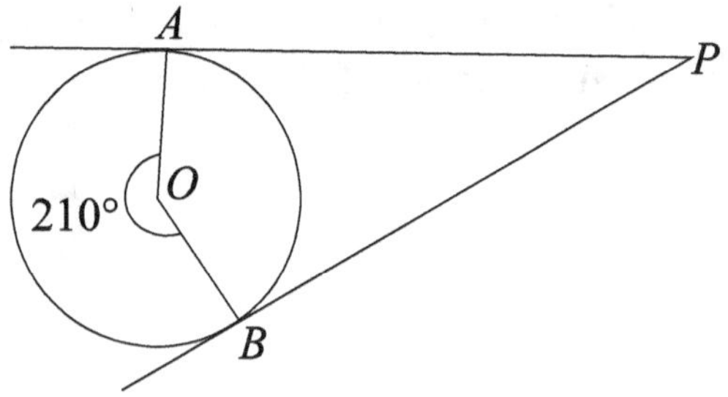

(a) 210° (b) 150°

(c) 105° (d) 30°

Solution: (d)

$\angle AOB = 360° - \text{Reflex } \angle AOB$

$= 360° - 210°$

$= 150°$

Now, $\angle AOB + \angle APB = 180°$

$\Rightarrow 150° + \angle APB = 180°$

$\therefore \angle APB = 30°$

153. In the adjoining figure, *AP* and *AQ* are tangents to the circle with center *O*. If reflex $\angle POQ$ = 210°, the value of 2*x* is.

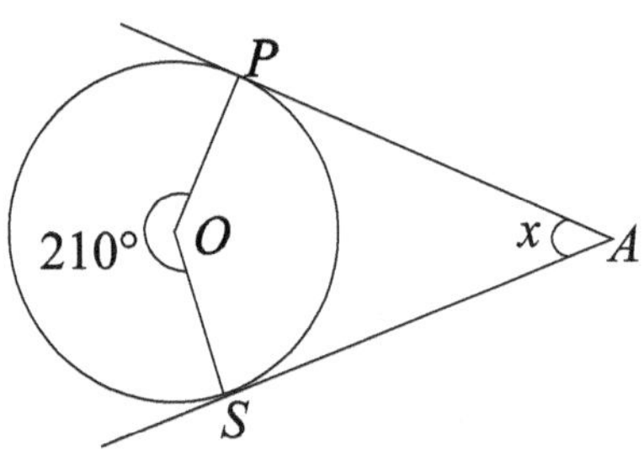

(a) 30° (b) 60°

(c) 120° (d) 300°

Solution: (b)

$\angle POQ = 360° - \text{Reflex } \angle POQ$

$= 360° - 210°$

$= 150°$

$\therefore x = 180° - 150° = 30°$

$\therefore 2x = 60°$

154. *PA* and *PB* are tangents to a circle with center with center *O*. If $\angle AOB$ = 105° then $\angle OAP$ + $\angle APB$ is equal to: **(2025)**

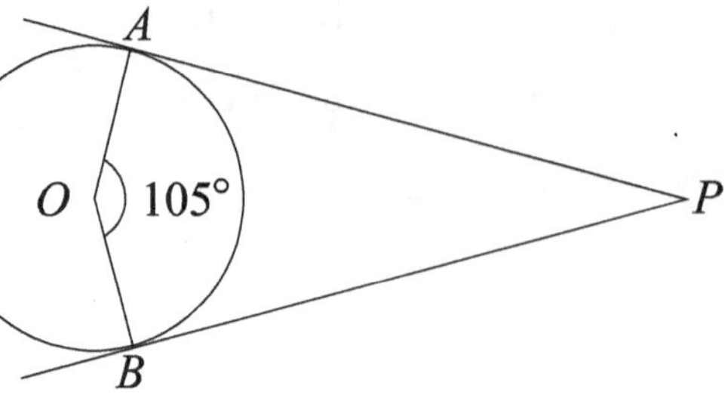

(a) 75° (b) 175°

(c) 180° (d) 165°

Solution:

$OA \perp AP$

$\therefore \angle OAP = 90°$

$\angle AOB + \angle APB = 180°$

$\Rightarrow 105° + \angle APB = 180°$

$\therefore \angle APB = 75°$

$\therefore \angle OAP + \angle APB = 90° + 75° = 165°$

155. If *PA* and *PB* are tangents to a circle with centre. *O* such that $\angle APB = 70°$, then $\angle AOB$ is **(2020)**

(a) 140° (b) 110°

(c) 35° (d) 70°

Solution: (b)

$\angle AOB + \angle APB = 180°$

$\Rightarrow \angle AOB + 70 = 180°$

$\therefore \angle AOB = 110°$

156. In the given figure, *PQ* and *PR* are tangents drawn from *P* to the circle with centre *O* such that $\angle QPR = 65°$. The measure of $\angle QOP$ is **(2023)**

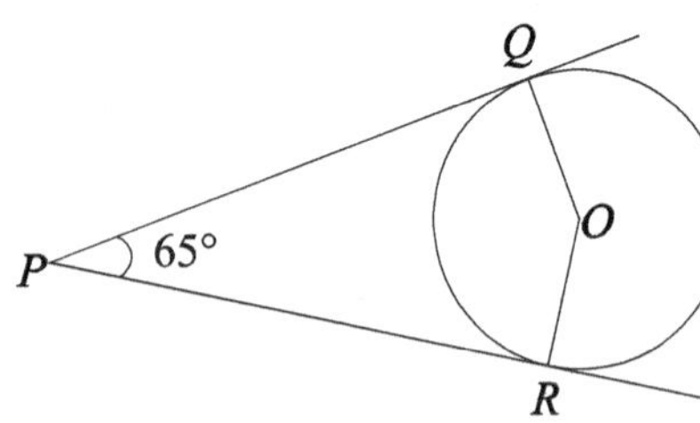

(a) 65° (b) 125°

(c) 115° (d) 90°

Solution: (c)

$\angle QPR + \angle ROQ = 180°$

$\Rightarrow 65° + \angle ROQ = 180°$

$\therefore \angle ROQ = 115°$

157. If the angle between the two radii of a circle is 130°, then the angle between the tangents at the ends of then radii is **(2024)**

(a) 50° (b) 60°

(c) 90° (d) 130°

Solution: (a)

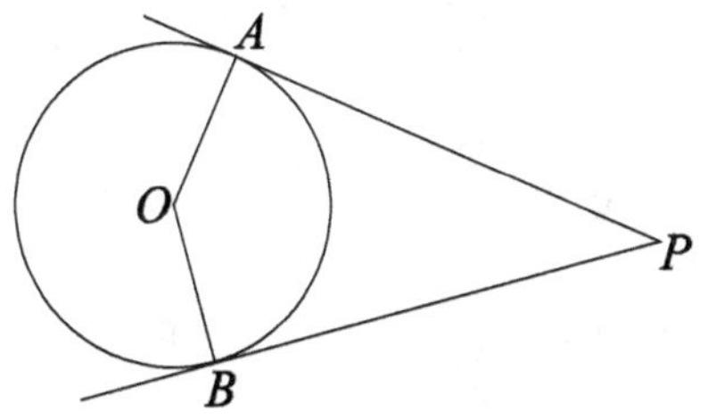

$\angle APB = 180° - \angle AOB$

$= 180° - 130°$

$= 50°$

158. In the given figure, PA and PB are tangents to a circle cented at O. if $\angle AOB = 180°$,Then $\angle APB$ is equal to : **(2026)**

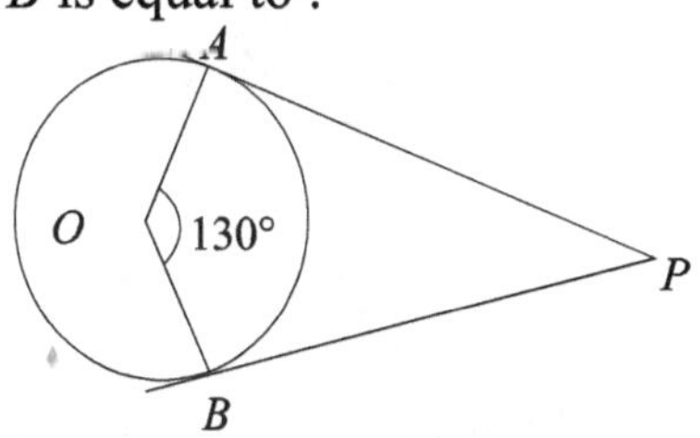

(a) 130° (b) 50°

(c) 120° (d) 90°

Solution: (b)

$\angle APB = 180° - \angle AOB$

$= 180° - 130°$

$= 50°$

159. If TP and TQ are two tangents to a circle with center O from an external point T so that $\angle POQ = 120°$, then $\angle PTQ$ is equal to : **(2026)**

(a) 60° (b) 70°

(c) 80° (d) 90°

Solution :(a)

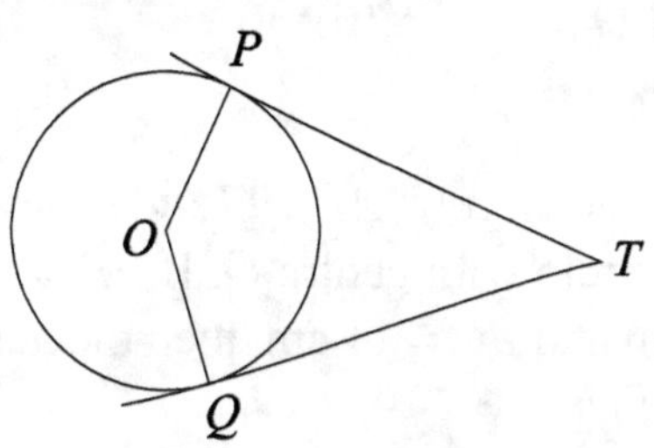

$\angle PTQ = 180° - \angle POQ$

$= 180° - 120°$

$= 60°$

Exercise 7.25

(i) In the given figure, TP and TQ are tangents drawn to the circle with centre O. If $\angle POQ = 120°$ then $\angle PTQ$ is

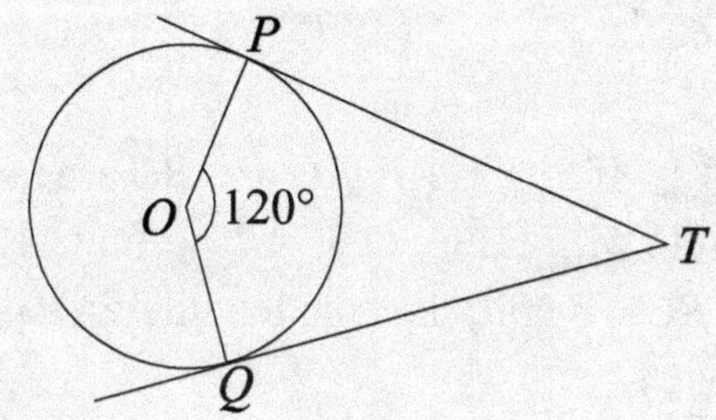

(a) 50° (b) 60°

(c) 80° (d) 55°

(ii) In the given figure, TP and TQ are tangents drawn to the circle with centre O. If $\angle POQ = 98°$ then $\angle PTQ$ is

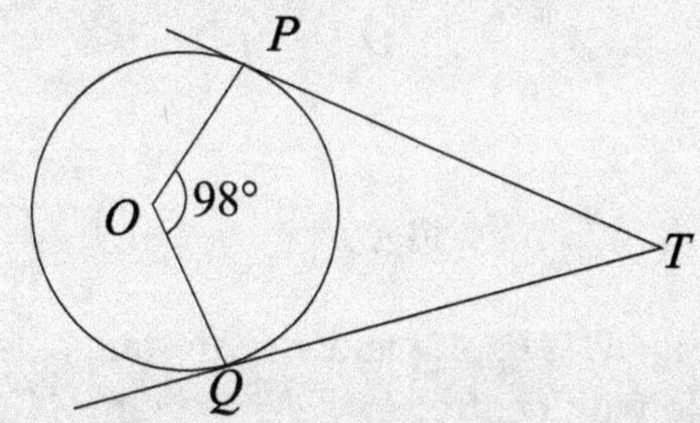

(a) 92° (b) 72°

(c) 36° (d) 82°

(iii) In the given figure, TP and TQ are tangents drawn to the circle with centre O. *If* $\angle POQ = 135°$ then $\angle PTQ$ is

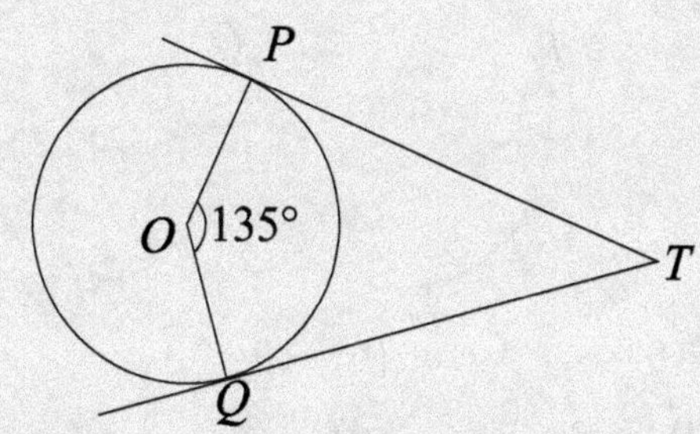

(a) 45° (b) 55°

(c) 60° (d) 65°

(iv) If PA and PB are tangent to a circle with centre O such that $\angle APB = 100°$, Then $\angle AOB$ is

(a) 100° (b) 80°

(c) 70° (d) 110°

(v) If the angle between the two radii of a circle is 105°, then the angle between the tangents at the ends of the radii is

(a) 95° (b) 70°

(c) 85° (d) 75°

Key Point If Fig. 1, AP, AQ and BC are tangents to the circle with centre O. If $AB = a$ unit, $AC = b$ unit and $BC = c$ unit, then the length of AP or AQ is $\frac{(a+b+c)}{2}$ unit.

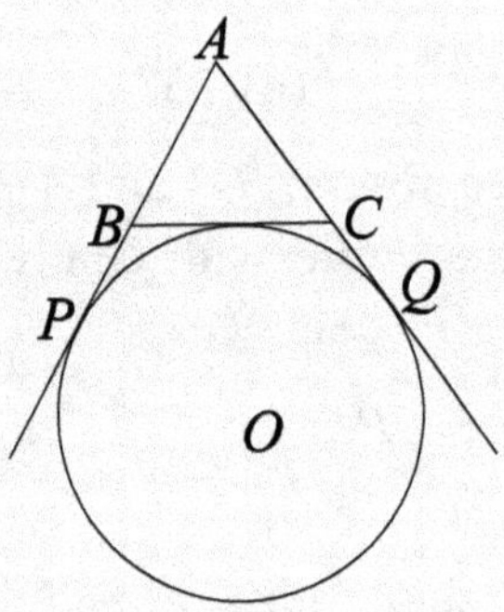

Fig. 1

160. In Fig. 1, AP, AQ and BC are tangents to the circle with centre O. If $AB = 5$ cm, $AC = 6$ cm and $BC = 4$ cm, then the length of AP (in cm) is **(2020)**

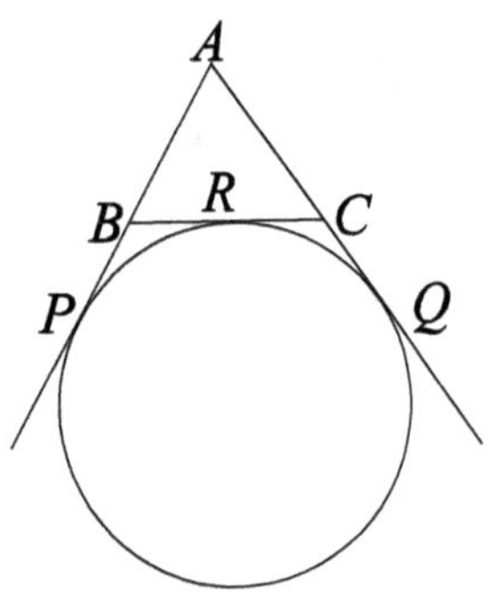

(a) 15 (b) 10

(c) 9 (d) 7.5

Solution: (d)

$AP + AQ = (AB + BP) + (AC + CQ)$

$\Rightarrow 2AP = AB + AC + BP + CQ$ [$\because AP = AQ$]

$\Rightarrow 2AP = AB + AC + BC$

$\Rightarrow 2AP = 5 + 6 + 4$

$\Rightarrow 2AP = 15$

$\therefore AP = 7.5$ cm.

Exercise 7.26

(i) In the given figure, AP, AQ and BC are tangents to the circle with centre O. If $AB = 8$ cm, $AC = 6$ cm and $BC = 7$ cm, then the length of AP(in cm) is:

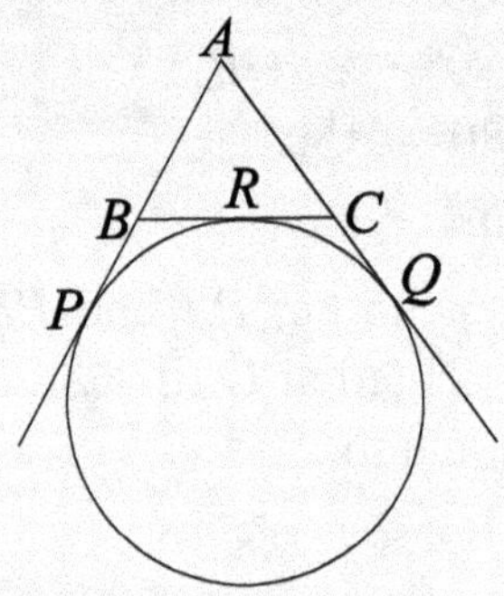

(a) 10 (b) 10.5

(c) 11 (d) 12

(ii) In the given figure, AP, AQ and BC are tangents to the circle with centre O. If AB =10 cm, AC = 8 cm and BC = 6 cm, then the length of AP (in cm) is:

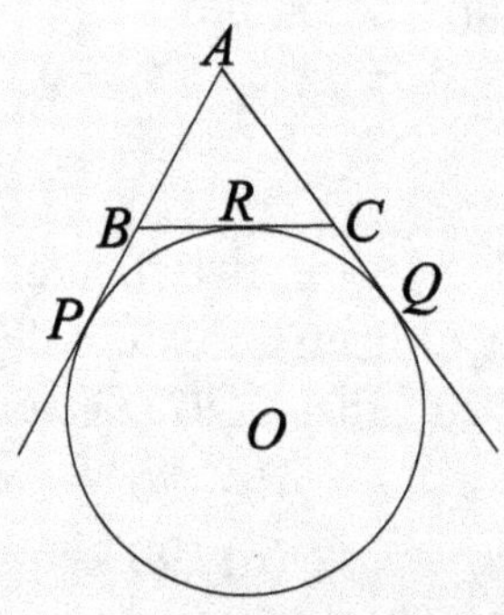

(a) 24 (b) 36

(c) 12 (d) 14

(iii) In the given figure, AP, AQ and BC are tangents to the circle with centre O. If AB = 15 cm, AC = 12 cm and BC = 11 cm, then the length of AP (in cm) is:

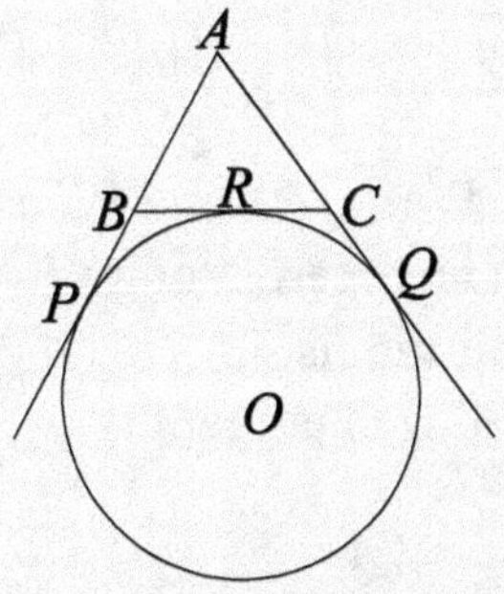

(a) 19 (b) 20
(c) 26 (d) 38

Key Point In the given figure, PA and PB are tangents drawn from an external Point P to a circle with centre C and radius a cm. If $PA \perp PB$, then the length of each tangent is 'a' unit.

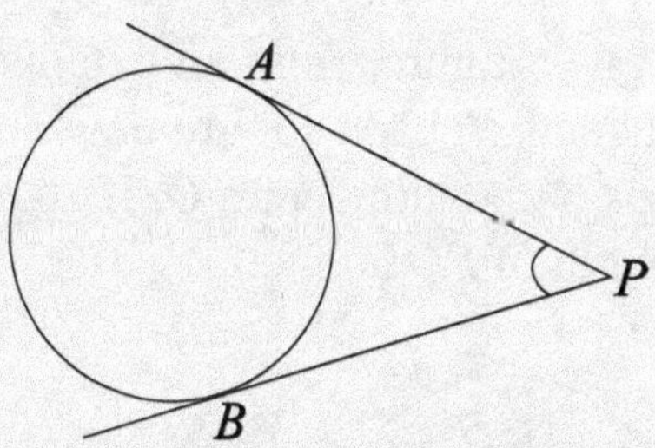

161. In Fig.1, PA and PB are two tangents drawns from an external point P to a circle with centre C and radius 4 cm. If $PA \perp PB$, then the length of each tangent is **(2013)**

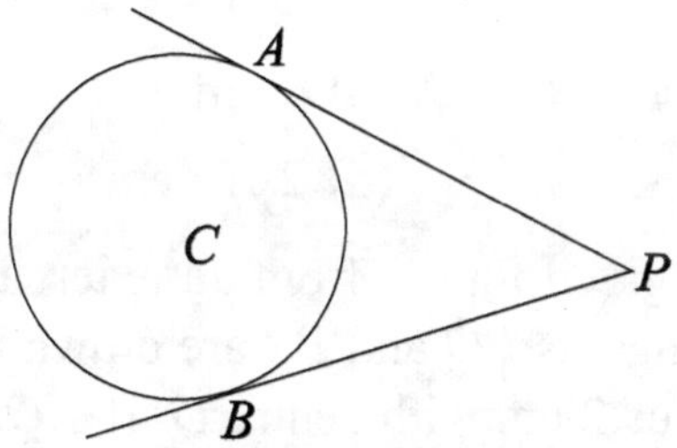

Fig. 1

(a) 3 cm (b) 4 cm
(c) 5 cm (d) 6 cm

Solution: (b)

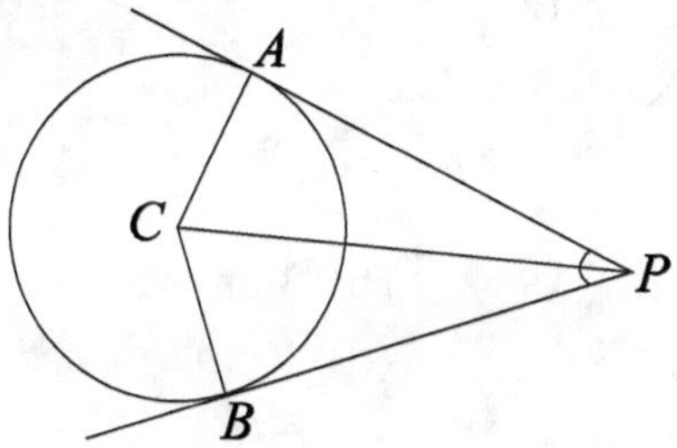

$$\angle APC = \frac{1}{2}\angle APB = \frac{1}{2} \times 90° = 45°$$

In ΔPAC, $\angle PAC = 90°$

and $\angle PCA = \angle CPA = 45°$

$\therefore AC = PA = 4$ cm.

$\therefore PA = PB = 4$ cm.

162. In figure-3, from an external point P, two tangents PQ and PR are drawn to a circle of radius 4 cm with centre O. If $\angle QPR = 90°$, then length of PQ is **(2020)**

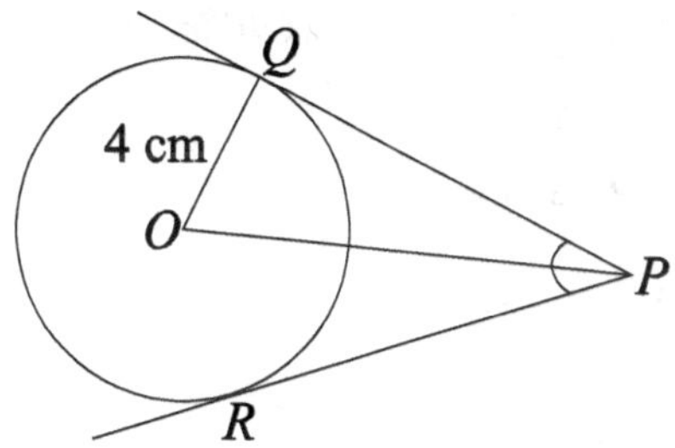

Figure 3

(a) 3 cm (b) 4 cm
(c) 2 cm (d) $2\sqrt{2}$ cm

Solution: (b)

$PQ = OQ = 4$ cm.

163. AP and AQ are tangents drawn from an external point A to a circle with centre O and inclined to each other at an angle of 90°. If the length of each tangent is 2 cm, then the radius of the circle is **(2024)**

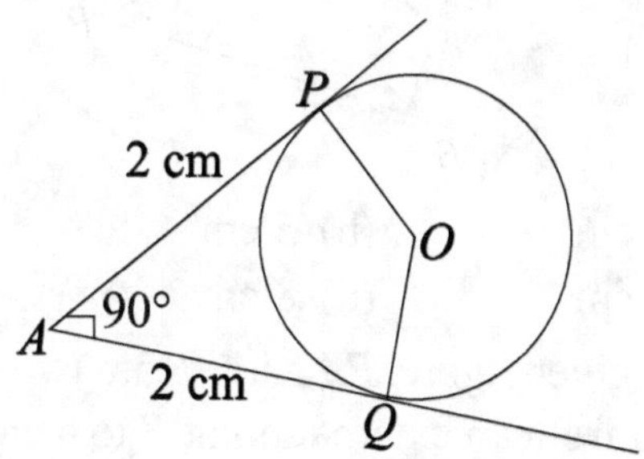

(a) 4 cm (b) 2 cm
(c) $2\sqrt{2}$ (d) 1 cm

Solution: (b)

$OP = OQ = AP = AQ = 2$ cm

164. If PQ and PR are tangents to the circle with centre O and radius 4 cm such that $\angle QPR = 90°$, then the length of OP is: **(2026)**

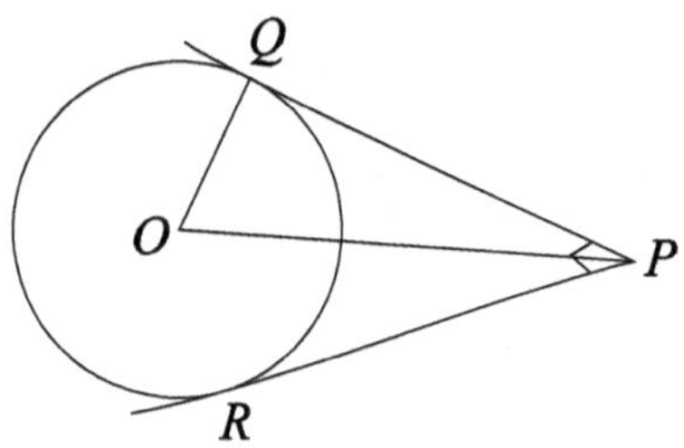

(a) 4 cm (b) $4\sqrt{2}$ cm
(c) 8 cm (d) $2\sqrt{2}$ cm

Solution:

$$\angle OPQ = \frac{1}{2}\angle QPR = \frac{1}{2}\times 90° = 45°$$

$\therefore$ In ΔPOQ

$\angle P = \angle O = 45°$

$\therefore \dfrac{OQ}{PQ} = \tan 45°$

$\Rightarrow \dfrac{OQ}{PQ} = 1$

$\Rightarrow OQ = PQ = 4$ cm

Exercise 7.27

(i) In the given figure, PA and PB are two tangents drawn from an external point P to a circle with centre C and radius 6 cm. If $PA \perp PB$, then the length of each tangent is

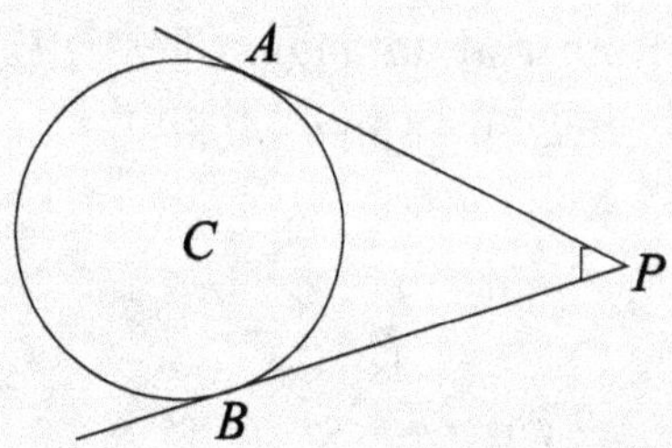

(a) 4 cm (b) 5 cm
(c) 6 cm (d) 8 cm

(ii) In the given figure, PA and PB are two tangents drawn from an external point P to a circle with centre C and radius 3.5 cm. If $PA \perp PB$, then the length of each tangent is

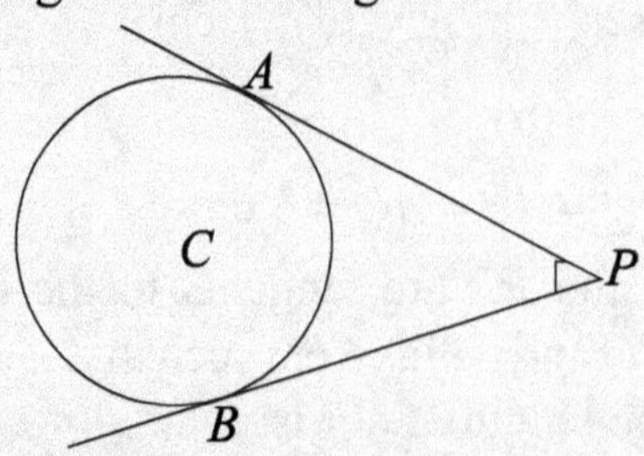

(a) 3.5 cm (b) 3 cm
(c) 4 cm (d) 7 cm

(iii) In the given figure, PA and PB are two tangents drawn from an external point P to a circle with centre C and radius 4.5 cm. If $PA \perp PB$, then the length of each tangent is

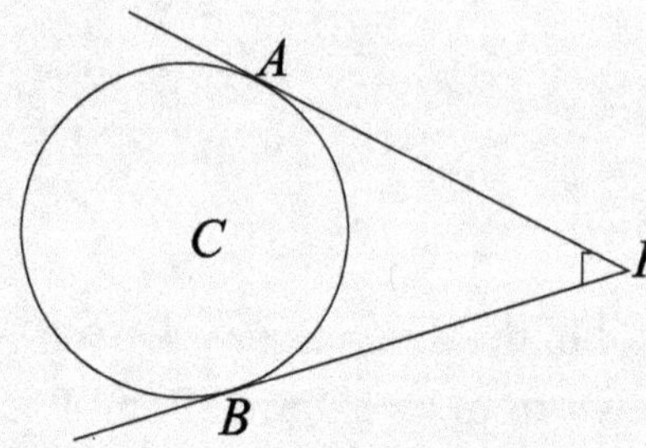

(a) 4.5 cm (b) 5 cm
(c) 9 cm (d) 10 cm

(iv) In the given figure, from an external point P, two tangents PQ and PR are drawn to a circle of radius 6 cm with centre O. If $\angle QPR = 90°$, then length of PQ is

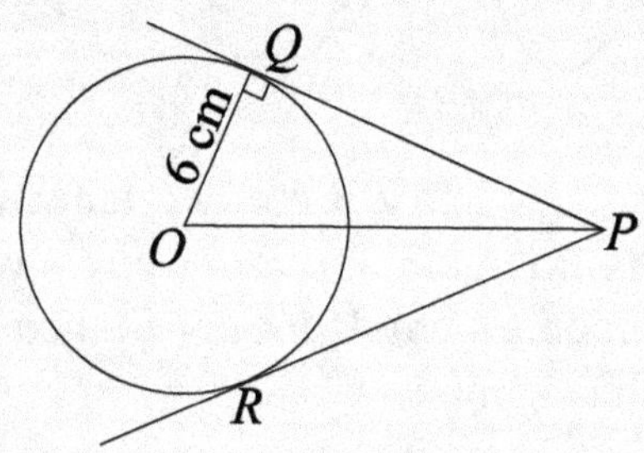

(a) 6 cm (b) 4 cm
(c) 8 cm (d) $2\sqrt{2}$ cm

(v) In the given figure, from an external point P, two tangents PQ and PR are drawn to a circle of radius 2 cm with centre O. If $\angle QPR = 90°$, then length of PQ is

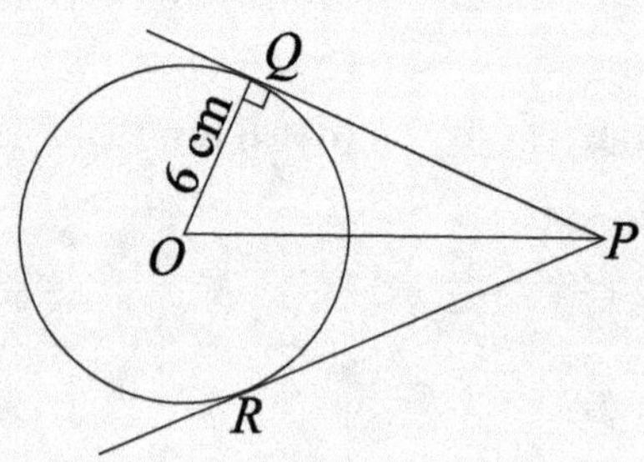

(a) 4 cm (b) 3 cm
(c) 2 cm (d) 1 cm

(vi) In the given figure, from an external point P, two tangents PQ and PR are drawn to a circle of radius 5 cm with centre O. If $\angle QPR = 90°$, then length of PQ is

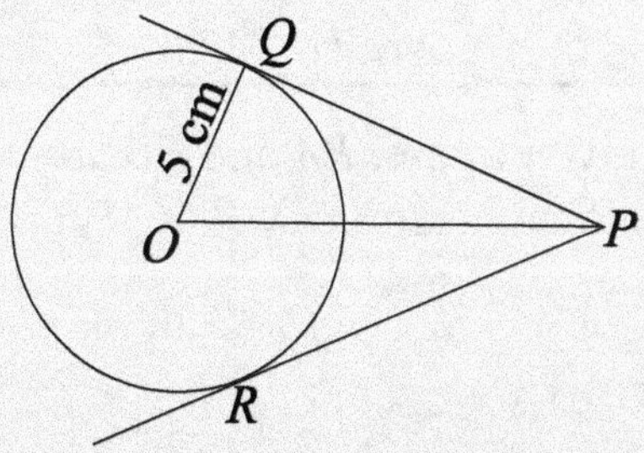

(vii) AP and 4 are tangents drawn from an external point A to a circle with centre O and inclined to each other at an angle of 90°. If the length of each tangent is 4 cm, then the radius of the circle is

(a) 2 cm (b) 4 cm
(c) 8 cm (d) $4\sqrt{2}$ cm

Key Point In the given figure, PA, and PB are tangents to the circle with centre O. If $\angle APB = a°$, then $\angle OAB = \frac{a°}{2}$

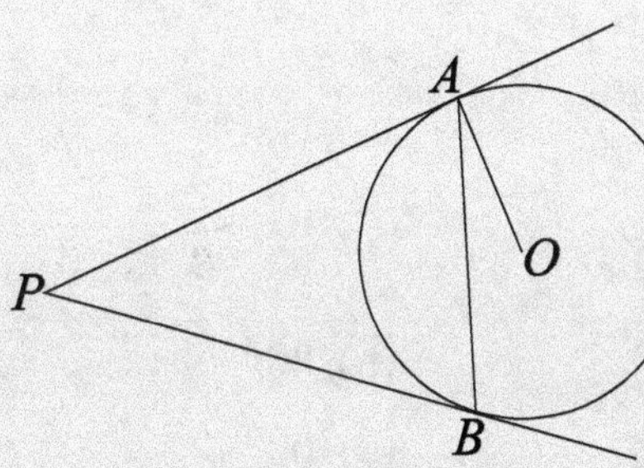

165. In figure 2, PA, and PB are tangents to the circle with centre O. If $\angle APB = 60°$, then $\angle OAB$ is **(2011)**

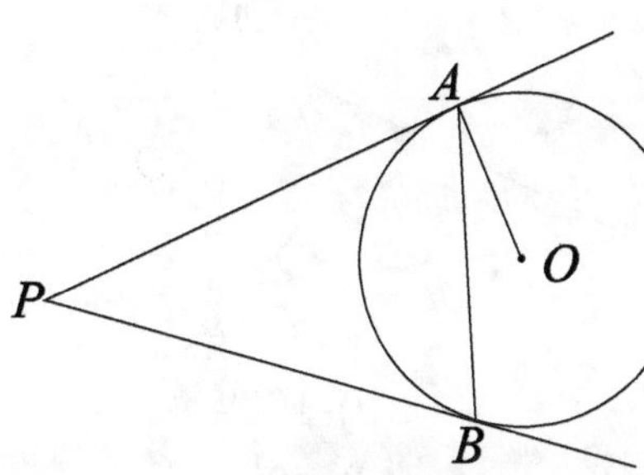

(a) 30° (b) 60°
(c) 90° (d) 15°

Solution: (a)

In ΔAPB

$AP = BP$

$\therefore \angle PAB = \angle PBA = \frac{180° - 60°}{2} = 60°$

Now $\angle OAB = \angle OAP - \angle PAB$

$= 90° - 60°$

$= 30°$

166. In Fig.1, PA and PB are tangents to the circle with centre O. Such that $\angle APB = 50°$. Write the measure of $\angle OAB$ **(2015, 2020)**

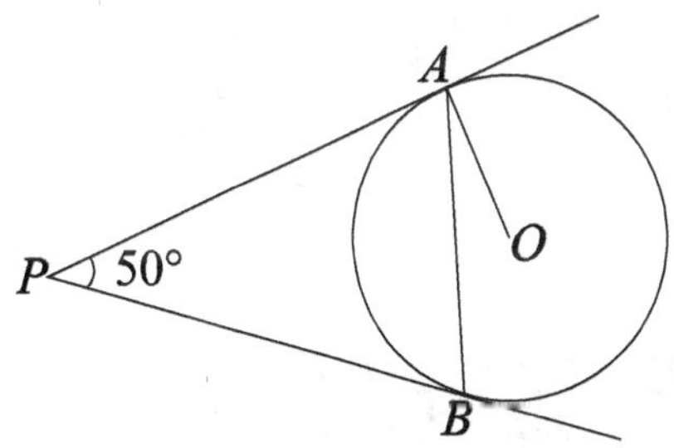

Fig. 1

Solution:

$\angle OAB = \frac{1}{2}$ of $\angle APB$

$= \frac{1}{2} \times 50°$

$= 25°$

167. In the given figure, tangents PA and PB from a point P to a circle with centre O are inclined to each other at an angle of 80°, $\angle ABO$ is equal to **(2024)**

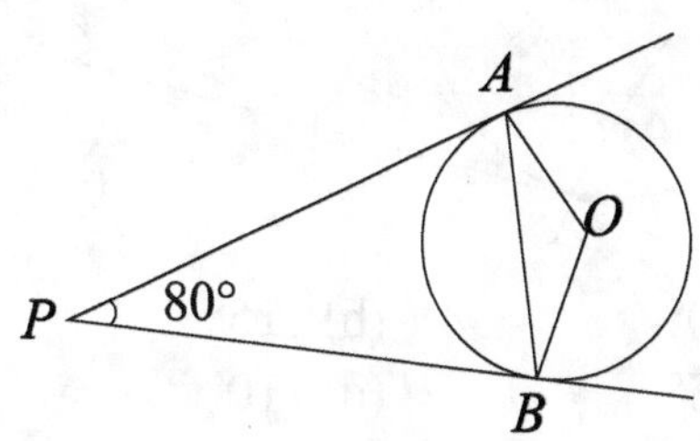

(a) 40° (b) 80°
(c) 100° (d) 50°

Solution: (a)

$\angle ABO = \frac{1}{2}$ of $\angle APB$

$$= \frac{1}{2} \times 80°$$

$$= 40°$$

168. If PA and PB are two tangents to the circle with center O such that $\angle APB = 50°$, then $\angle OAB$ is equal to **(2025)**

(a) 25° (b) 30°

(c) 40° (d) 50°

Solution: (a)

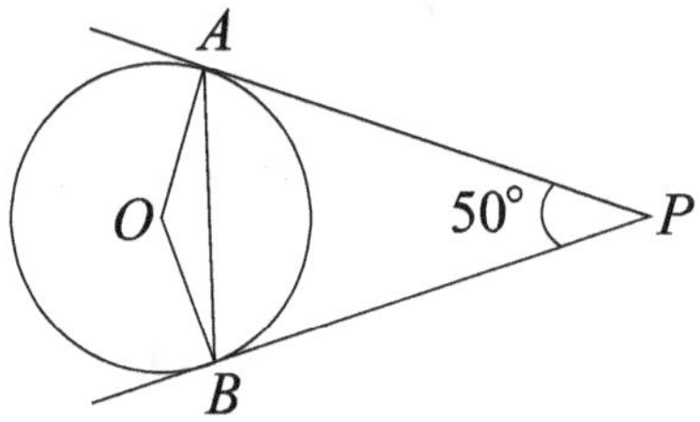

$\angle AOB = 180° - \angle APB$

$= 180° - 50°$

$= 130°$

Now, in ΔAOB

$OA = OB$

$\therefore \angle OAB = \angle OBA = \frac{180° - 130°}{2} = 25°$

169. In the given figure, PA and PB are tangents to a circle centred at O. if $\angle OAB = 15°$, then $\angle APB$ equals: **(2026)**

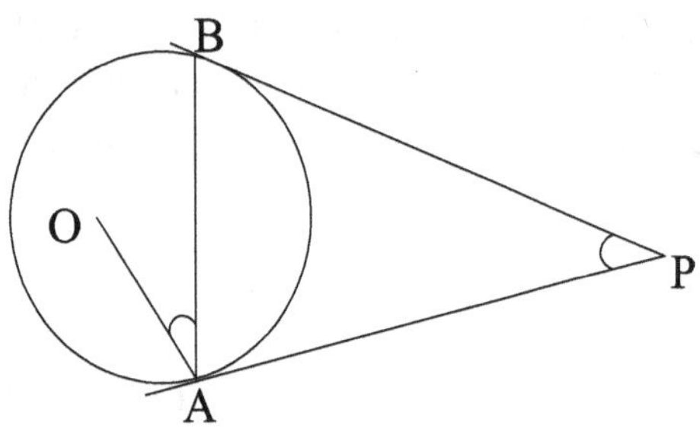

(a) 30° (b) 15°

(c) 45° (d) 10°

Solution:

$\angle PAB = 90° - \angle OAB$

$= 90° - 15°$

$= 75°$

$\therefore \angle PBA = 75°$

$\therefore \angle APB = 180° - (\angle PAB + \angle PBA)$

$= 180° - (75° + 75°)$

$= 30°$

Exercise 7.28

(i) In the given figure, PA and PB are tangents to the circle with centre O. If $\angle APB = 50°$, then $\angle OAB$ is

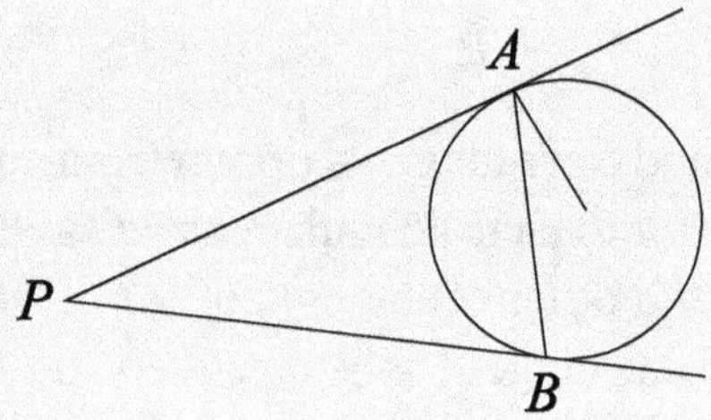

(a) 25° (b) 35°

(c) 45° (d) 50°

(ii) In the given figure, PA and PB are tangents to the circle with centre O. If $\angle APB = 80°$, then $\angle OAB$ is

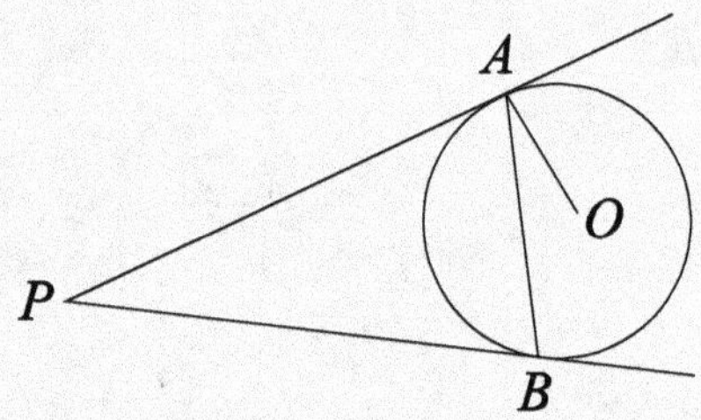

(a) 50° (b) 80°

(c) 40° (d) 60°

(iii) In the given figure, PA and PB are tangents to the circle with centre O. If $\angle APB = 100°$, then $\angle OAB$ is

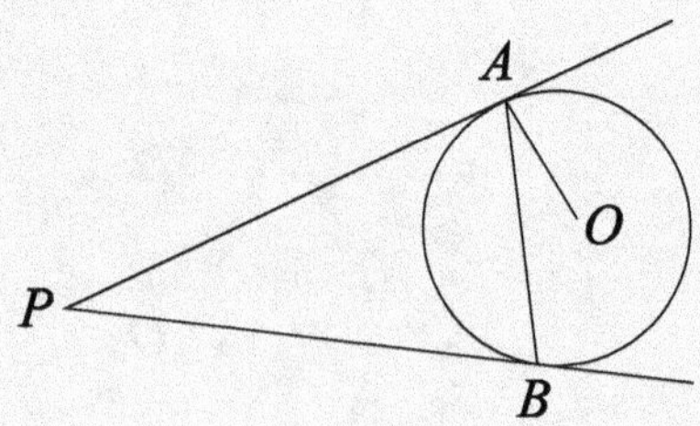

(a) 30° (b) 100°

(c) 50° (d) 40°

(iv) In the given figure, tangents PA and PB from a point P to a circle with centre O are inclined to each other at an angle of 100° $\angle ABO$ in equal to.

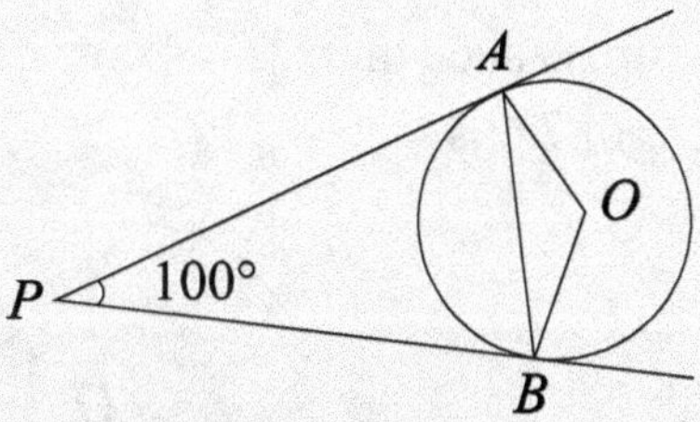

(a) 100° (b) 50°
(c) 80° (d) 110°

(v) In the given figure, tangents PA and PB from a point P to a circle with center O are inclined to each other at an angle of 70°, $\angle ABO$ is equal to.

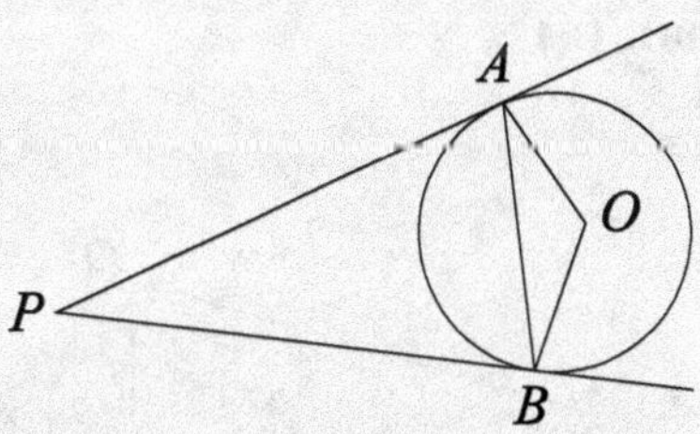

(a) 70° (b) 35°
(c) 140° (d) 110°

Key Point From an externel point P, tangents PA and PB are drawn to a circle with centre O. If $\angle PAB = a°$, then $\angle AOB = 2a°$

170. From an externel point P, tangents PA and PB are drawn to a circle with centre O. If $\angle PAB$ = 50 °, then find $\angle AOB$. **(2016)**

Solution:

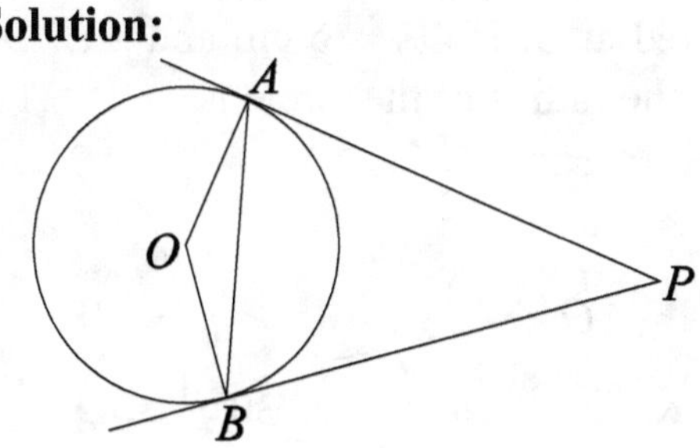

In ΔPAB

$PA = PB$

$\therefore \angle PAB = \angle PBA = 50°$

Now, in ΔPAB

$\angle PAB + \angle PBA + \angle APB = 180°$

$\Rightarrow 50° + 50° + \angle APB = 180°$

$\therefore \angle APB = 80°$

$\therefore \angle AOB = 180° - \angle APB$

$= 180° - 80°$

$= 100°$

171. PA and PB are tangents to a circle centered at O. If $\angle PBA = 65°$, then $\angle APB$ equals:

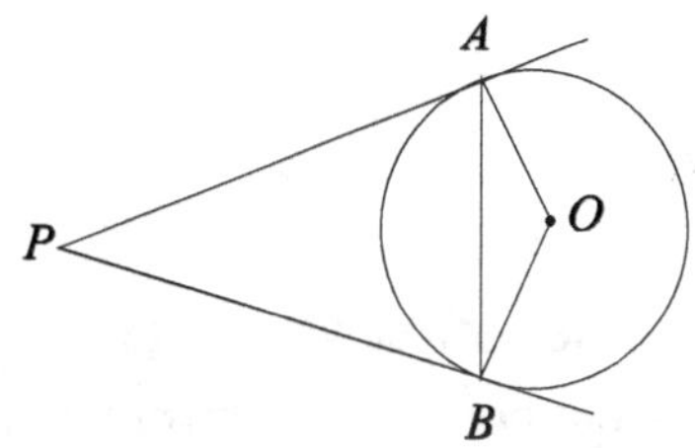

(a) 65° (b) 60°
(c) 50° (d) 35°

Solution: (c)

In ΔAPB

$PA = PB$

$\therefore \angle PAB = \angle PBA = 65°$

$\therefore \angle APB = 180° - (\angle PAB + \angle PBA)$

$= 180° - (65° + 65°)$

$= 50°$

Exercise 7.29

(i) From an externel point P, tangents PA and PB are drawn to a circle with centre O. If $\angle PAB$ = 60 °, then find $\angle AOB$.

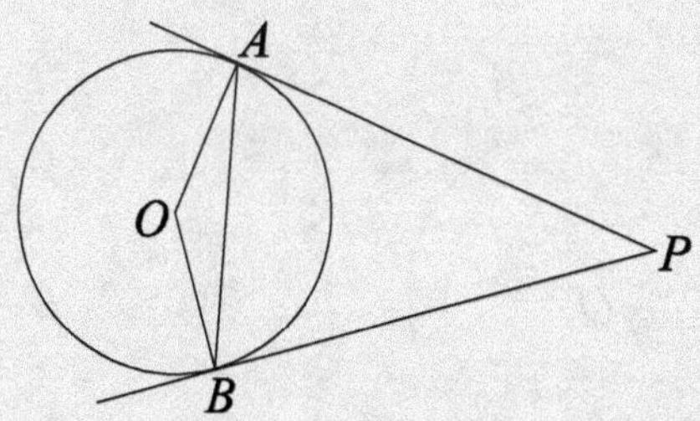

(ii) From an externel point P, tangents PA and PB are drawn to a circle with centre O. If $\angle PAB$ = 45°, then find $\angle AOB$.

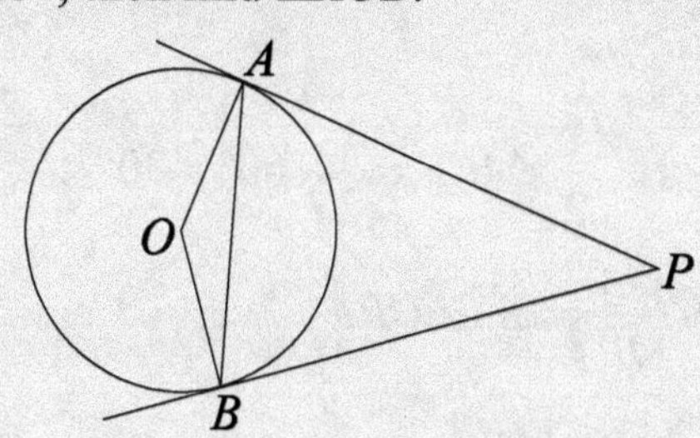

(iii) From an external point P, tangents PA and PB are drawn to a circle with centre O. If $\angle PAB = 30°$, then find $\angle AOB$.

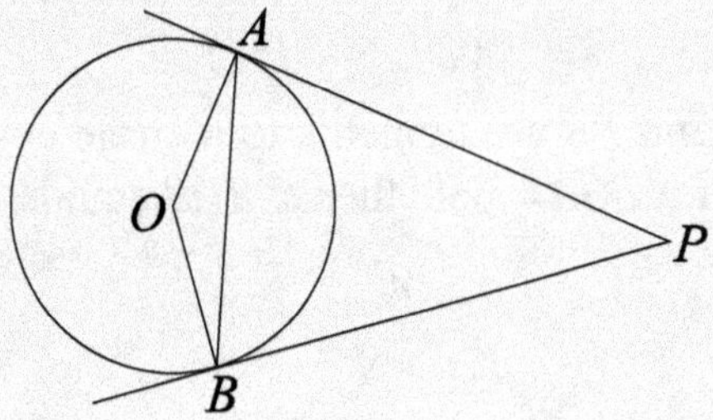

Key Point If the angle between two tangents drawn from an external point P to a circle of radius r and centre O, is θ, then the length of $OP = r\operatorname{cosec}\dfrac{\theta}{2}$

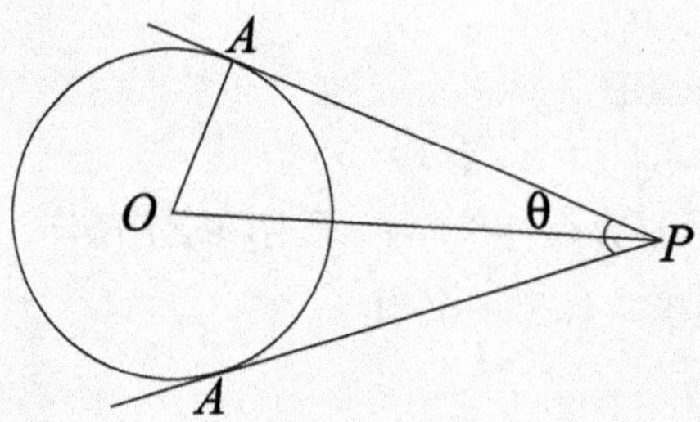

172. If the angle between two tangents drawn from an external point P to a circle of radius a and centre O, is 60°, then the length of OP. **(2017)**

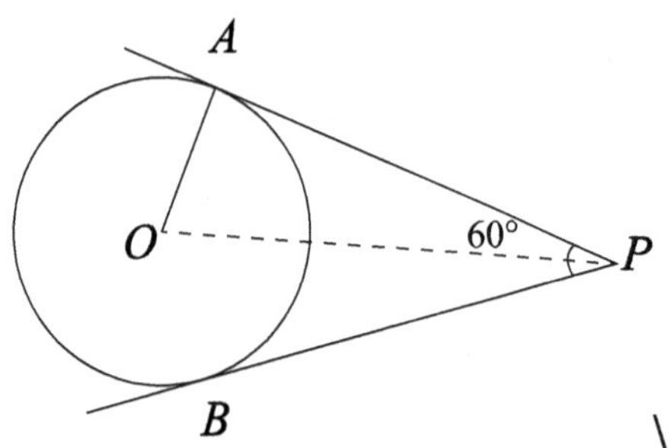

Solution:

In ΔAOP

$$\angle OPA = \frac{1}{2}\angle APB = \frac{1}{2}\times 60° = 30°$$

Now, $\dfrac{OA}{OP} = \text{Sin}\,\angle OPA$

$$\Rightarrow \frac{a}{OP} = \text{Sin}\,30°$$

$$\Rightarrow \frac{a}{OP} = \frac{1}{2}$$

$\therefore\ OP = 2a$ unit

173. In the given figure, TA is a tangent to the circle with centre O such that $OT = 4$ cm, $\angle OTA = 30°$, then length TA is **(2023)**

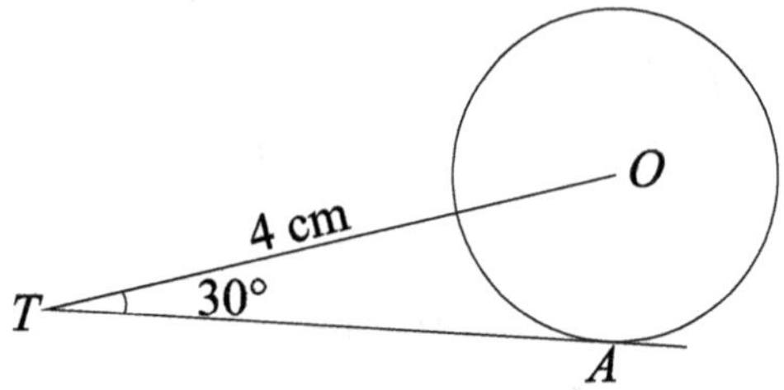

(a) $2\sqrt{3}$ cm (b) 2 cm

(c) $2\sqrt{2}$ cm (d) $\sqrt{3}$ cm.

Solution: (a)

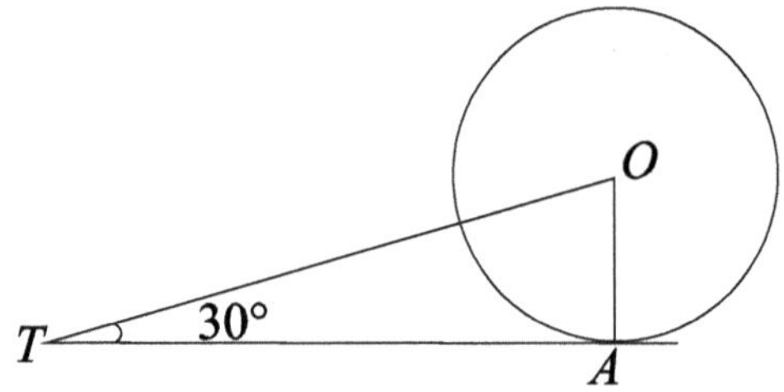

In ΔOAT, $\angle OAT = 90°$

$$\therefore \frac{TA}{OT} = \text{Cos}30°$$

$$\Rightarrow \frac{TA}{4} = \frac{\sqrt{3}}{2}$$

$$\Rightarrow TA = 2\sqrt{3}\ \text{cm}$$

174. In the given figure, AB is a tangent to the circle centred at O. If $OA = 6$ cm and $\angle OAB = 30°$, then the radius of the circle is: **(2023)**

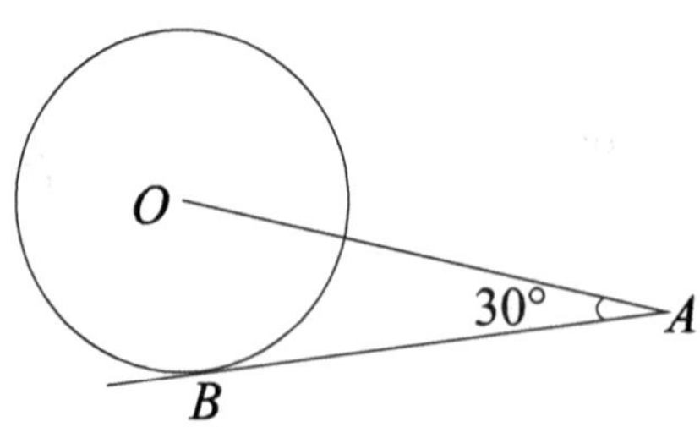

(a) 3 cm (b) $3\sqrt{3}$ cm

(c) 2 cm (d) $\sqrt{3}$ cm

Solution: (a)

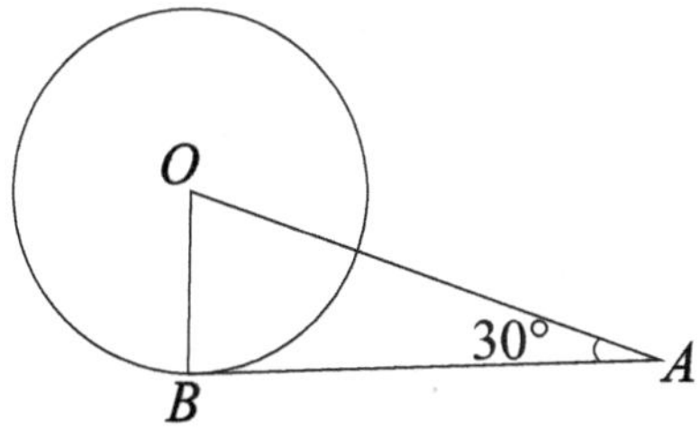

In ΔABO, $\angle ABO = 90°$

$\therefore \dfrac{OB}{OA} = \text{Sin}30°$

$\Rightarrow \dfrac{OB}{6} = \dfrac{1}{2}$

$\therefore OB = 3$ cm

175. PQ is a tangent to a circle centred at O. If the radius of the circle is 5 cm, then the length of the tangent PQ is **(2023)**

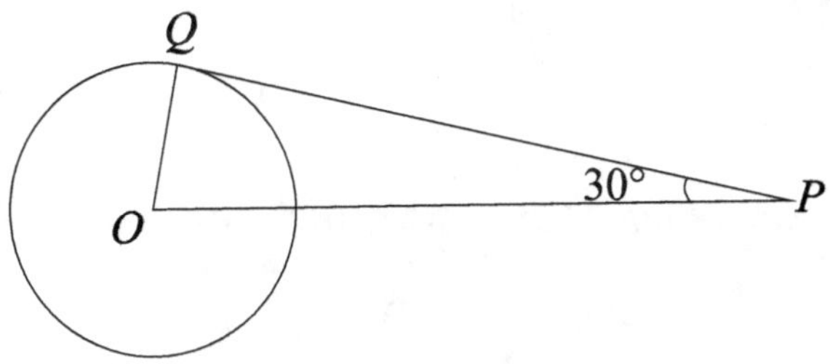

(a) $5\sqrt{3}$ cm (b) $\dfrac{10}{\sqrt{3}}$ cm

(c) 10 cm (d) $\dfrac{5}{\sqrt{3}}$ cm

Solution: (a)

In ΔPQO, $\angle PQO = 90°$

$\therefore \dfrac{OQ}{PQ} = \tan 30°$

$\Rightarrow \dfrac{5}{PQ} = \dfrac{1}{\sqrt{3}}$

$\therefore PQ = 5\sqrt{3}$ cm

176. If two tangents inclined at an angle of 60° are drawn to a circle of radius 5 cm, then the length of each tangents is **(2024)**

(a) $\dfrac{5\sqrt{3}}{2}$ cm (b) 10 cm

(c) $\dfrac{5}{\sqrt{3}}$ cm (d) $5\sqrt{3}$ cm

Solution: (d)

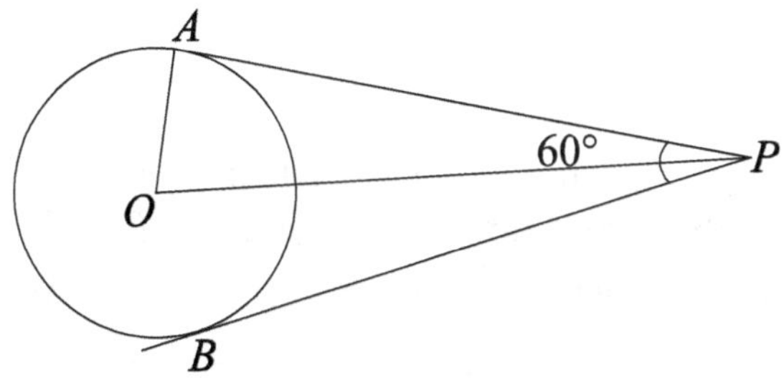

$\angle OPA = \dfrac{1}{2}\angle APB$

$= \dfrac{1}{2} \times 60° = 30°$

In ΔOAP, $\angle A = 90°$

$\therefore \dfrac{OA}{AP} = \tan 30°$

$\Rightarrow \dfrac{5}{AP} = \dfrac{1}{\sqrt{3}}$

$\therefore AP = 5\sqrt{3}$ cm

177. In the given figure, PQ and PR are tangents to a circle with centre O and radius 3 cm. If $\angle QPR = 60°$, then the length of each tangent is **(2026)**

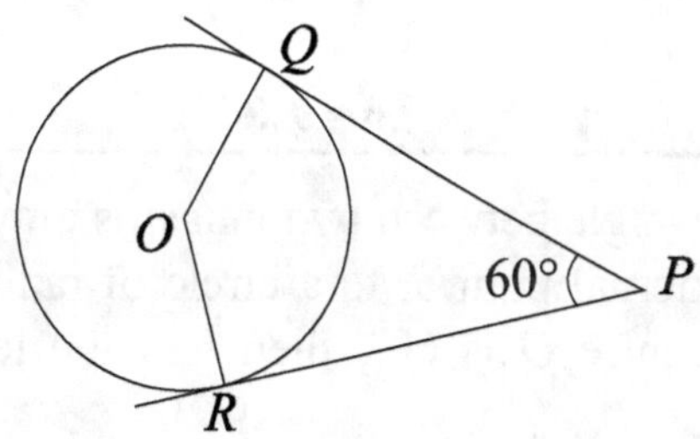

(a) $3\sqrt{3}$ cm (b) 3 cm

(c) 6 cm (d) $\sqrt{3}$ cm

Solution: (a)

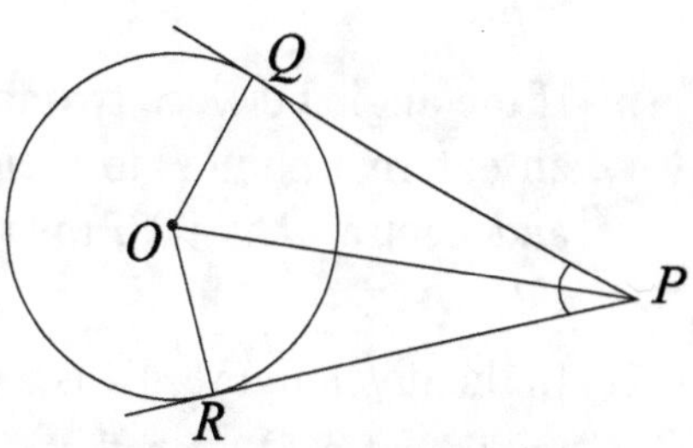

In Δ PQO, $\angle Q = 90°$

and $\angle OPQ$

$= \dfrac{1}{2}\angle QPR$

$= \dfrac{1}{2} \times 60°$

$= 30°$

$\therefore \frac{OQ}{PQ} = \tan 30°$

$\Rightarrow \frac{3}{PQ} = \frac{1}{\sqrt{3}}$

$\therefore PQ = 3\sqrt{3}$ cm

178. In the given figure, PT is a tangent to the circle with centre O and radius r. If $\angle POT = 45°$ then the length of OP is : **(2026)**

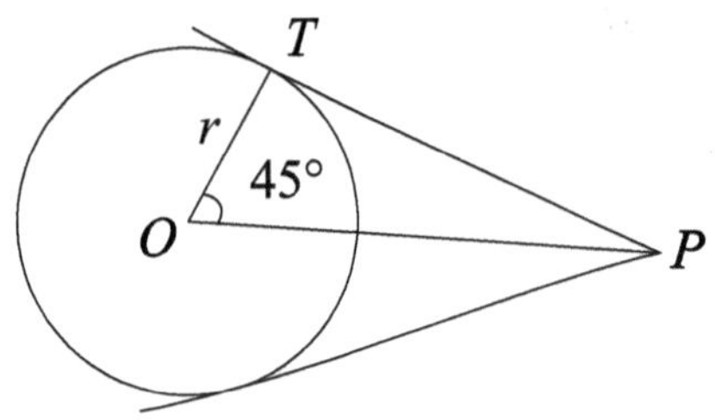

(a) $r\sqrt{2}$ (b) $\sqrt{2r}$
(c) $2r$ (d) r^2

Solution: (a)

$\frac{OT}{OP} = \sin 45°$ $[\because \angle POT = \angle OPT = 45°]$

$\Rightarrow \frac{r}{OP} = \frac{1}{\sqrt{2}}$

$\therefore OP = r\sqrt{2}$

Exercise 7.30

(i) If the angle between two tangents drawn from an external point P to a circle of radius 4 cm and centre O is 60°, then find the length of OP.

(ii) If the angle between two tangents drawn from an external point P to a circle of radius 6 cm and centre O is 120°, then find the length of OP.

(iii) If the angle between two tangents drawn from an external point P to a circle of radius 5 cm and centre O is 90°, then find the length of OP.

(iv) In the given figure, TA is a tangent to the circle with centre O such that OT = 6 cm, $\angle OTA$ = 60°, then length TA is

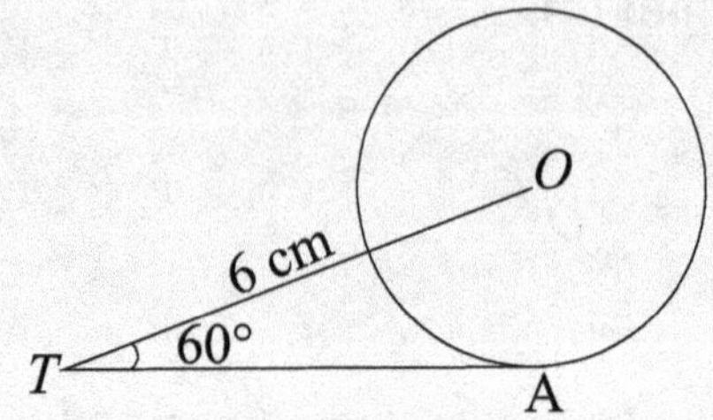

(a) 2 cm (b) 3 cm

(c) 4 cm (d) $2\sqrt{2}$ cm

(v) In the given figure, TA is a tangent to the circle with centre O such that OT = 8 cm, $\angle OTA$ = 45°, then length TA is

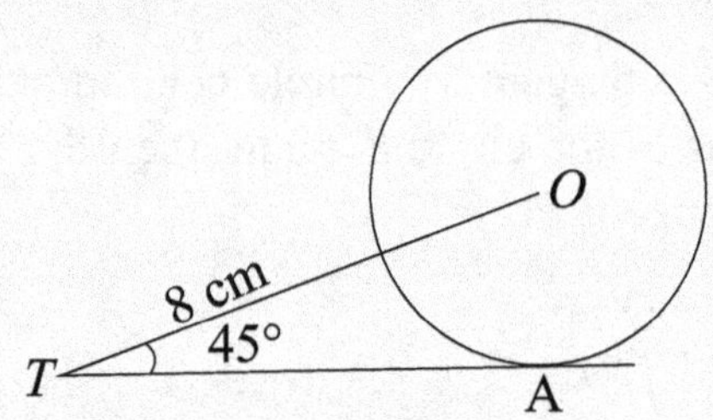

(a) 4 cm (b) $3\sqrt{3}$ cm

(c) $4\sqrt{3}$ cm (d) $4\sqrt{2}$ cm

(vi) In the given figure, TA is a tangent to the circle with centre O such that OT =12 cm, $\angle OTA$ = 30°, then length TA is

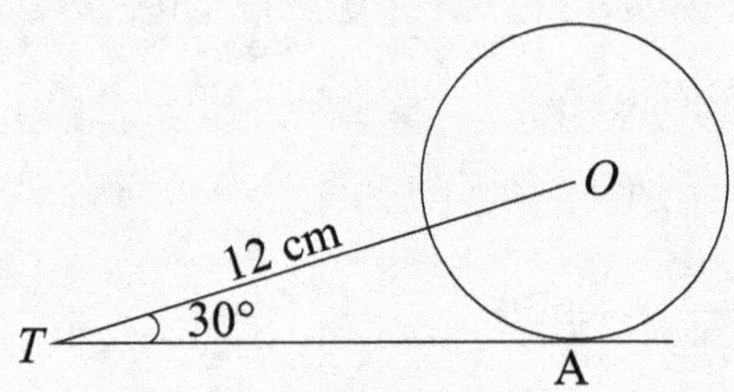

(a) $6\sqrt{3}$ (b) 6 cm

(c) $8\sqrt{2}$ cm (d) 10 cm

179. In the given figure, AC and AB are tangents to a circle centred at O. If $\angle COD$ = 120°, then $\angle BAO$ is equel to **(2023)**

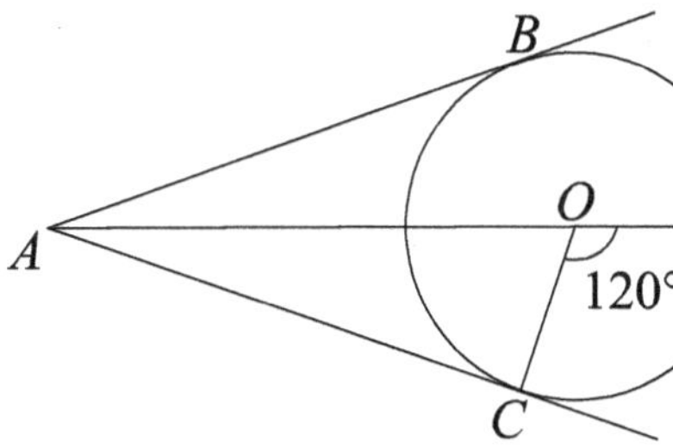

(a) 30° (b) 60°

(c) 45° (d) 90°

Solution: (a)

$\angle COD = \angle CAO + \angle OCA$

$\Rightarrow 120° = \angle CAO + 90°$

$\Rightarrow \angle CAO = 30°$

$\therefore \angle BAO = 30°$ [$\because \angle CAO = \angle BAO$]

Exercise 7.31

(i) In the given figure, *AC* and *AB* are tangents to a circle centred at *O*. If $\angle COD = 150°$, then $\angle BAO$ is equal to:

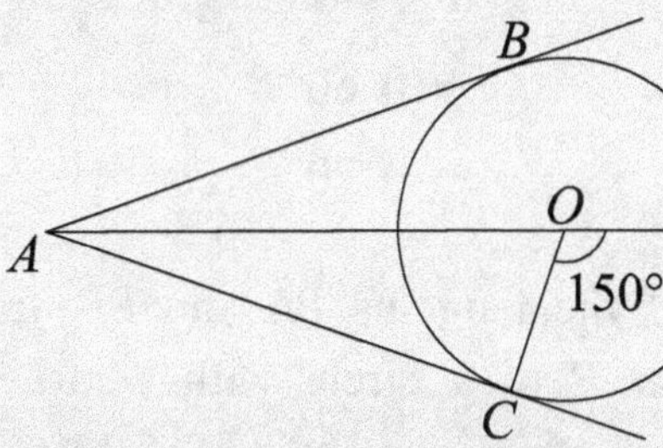

(a) 30° (b) 45°

(c) 60° (d) 90°

(ii) In the given figure, *AC* and *AB* are tangents to a circle centred at *O*. If $\angle COD = 130°$, then $\angle BAO$ is equal to:

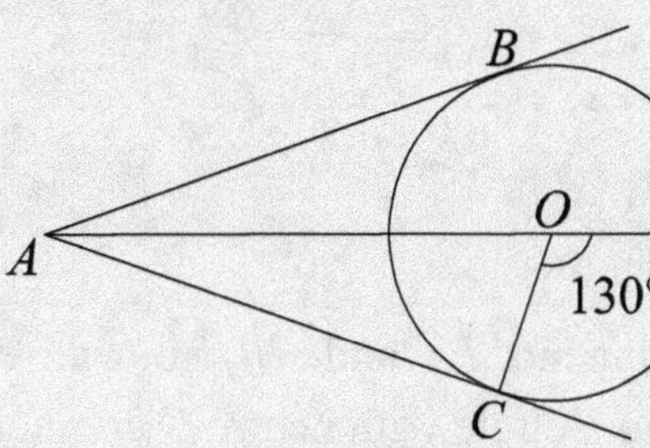

(a) 30° (b) 40°

(c) 50° (d) 60°

(iii) In the given figure, *AC* and *AB* are tangents to a circle centred at *O*. If $\angle COD = 125°$, then $\angle BAO$ is equal to:

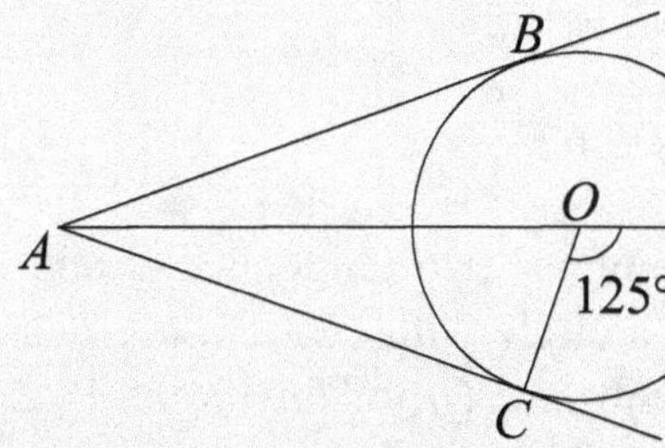

(a) 30° (b) 45°

(c) 30° (d) 35°

Key Point In the given figure, *PA* and *PB* are tangents from external point *P* to a. Circle with centre *C* and *Q* is any point on the circle. If $\angle APB = a°$, then $\angle AQB = \frac{180° - a}{2}$.

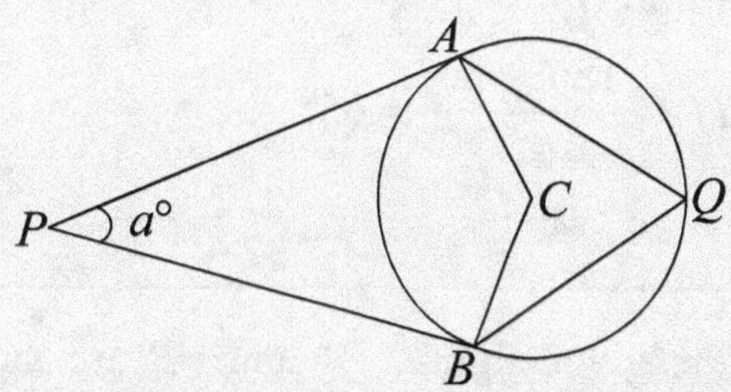

180. In the given figure, *PA* and *PB* are tangents from external point *P* to a circle with centre *C* and *Q* is any point on the circle. Then the measure of $\angle AQB$ is **(2023)**

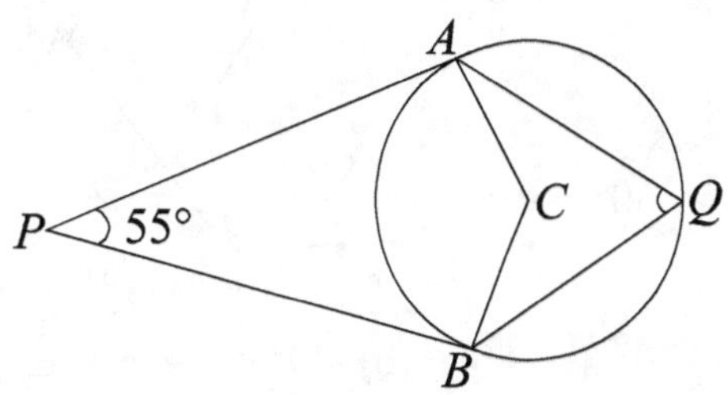

(a) $62\frac{1°}{2}$ (b) 125°

(c) 55° (d) 90°

Solution: (a)

$\angle ACB + \angle APB = 180°$

$\Rightarrow \angle ACB + 55° = 180°$

$\Rightarrow \angle ACB = 125°$

Now, $\angle ACB = 2\angle AQB$

$\Rightarrow 2\angle AQB = 125°$

$\therefore \angle AQB = 62\frac{1}{2}°$

181. From a point P two tangents PQ and PR are drawn to a circle with centre at O. T is a point on the major arc QR of the circle. If $\angle QPR = 50°$, then $\angle QTR$ equals **(2023)**

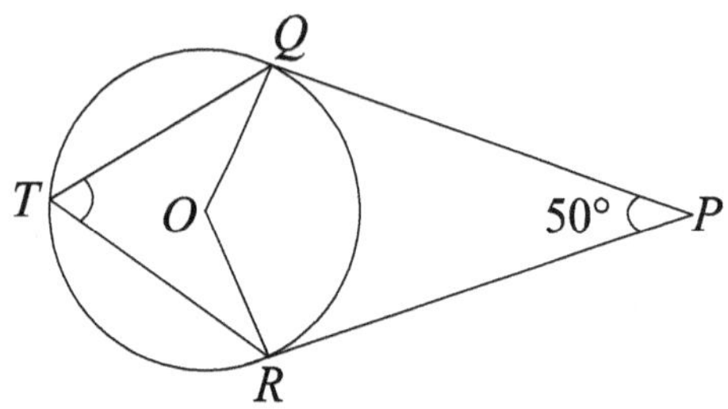

Solution: (c)

$$\angle QTR = \frac{180° - 50°}{2} = 65°$$

Exercise 7.32

(i) In the given figure, PA and PB are tangents from external point P to a circle with centre C and Q is any point on the circle. Then the measure of $\angle AQB$ is:

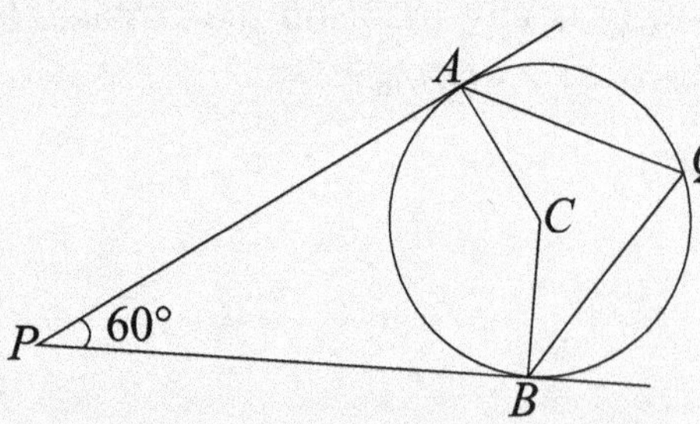

(a) 120° (b) 90°

(c) 65° (d) 60°

(ii) In the given figure, PA and PB are tangents from external point P to a circle with centre C and Q is any point on the circle. Then the measure of $\angle AQB$ is:

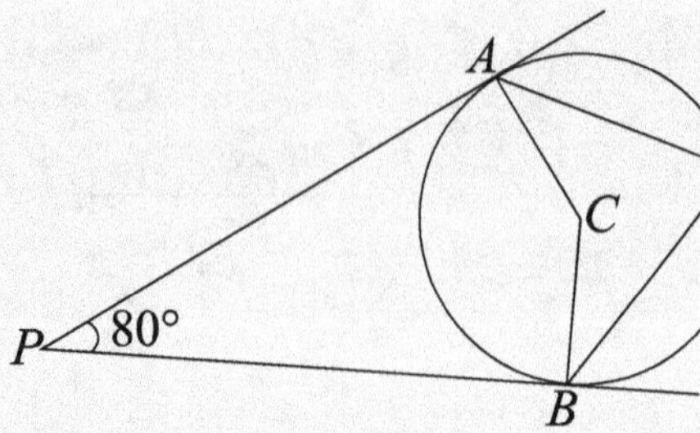

(a) 40° (b) 55°

(c) 45° (d) 50°

(iii) In the given figure, PA and PB are tangents from external point P to a circle with centre C and Q is any point on the circle. Then the measure of $\angle AQB$ is:

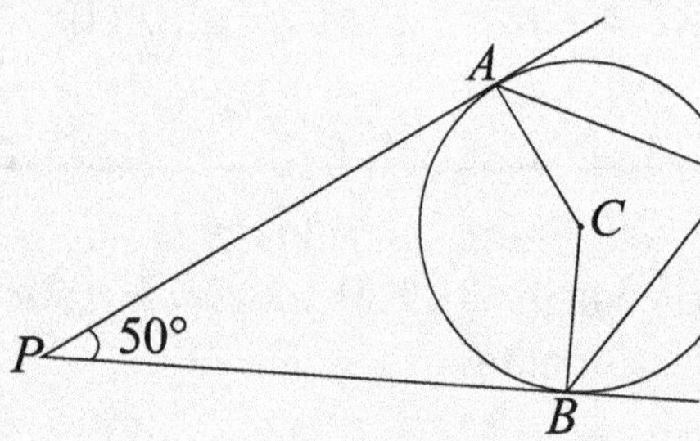

(a) 50° (b) 60°

(c) 65° (d) 70°

Key Point If the tangents PA and PB from an external point P to a circle with centre O are inclined to each other at an angle of $a°$, then $\angle POA = \frac{1}{2}(180° - a)$

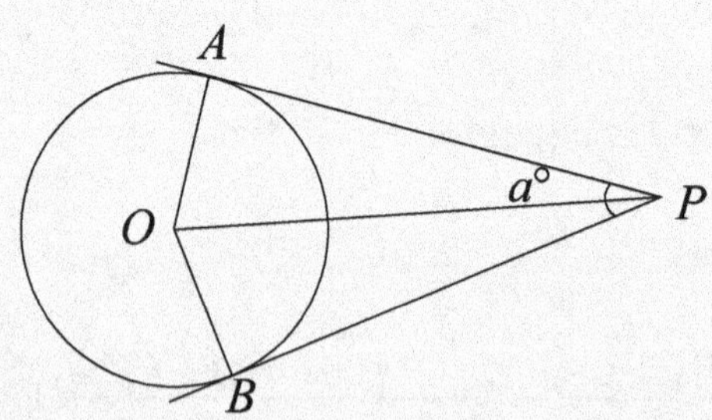

182. If the tangents PA and PB from an external point P to a circle with centre O are inclined to each other at an angle of $a°$, then $\angle POA$ equals **(2020)**

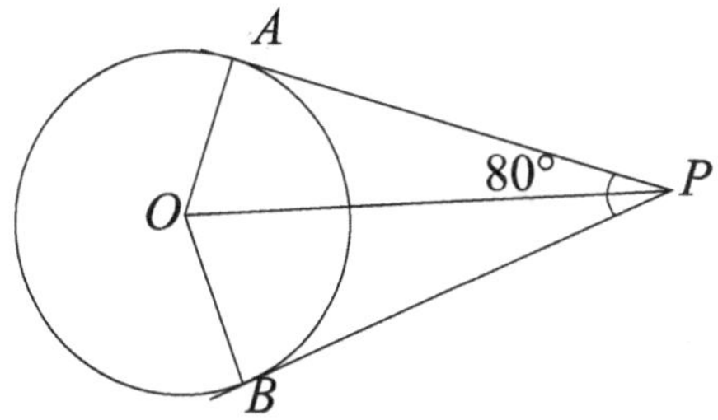

(a) 50° (b) 60°
(c) 70° (d) 80°

Solution: (a)

$\angle AOB + \angle APB = 180°$

$\Rightarrow \angle AOB + 80° = 180°$

$\Rightarrow \angle AOB = 100°$

$\Rightarrow 2\angle POA = 100°$

$\therefore \angle POA = 50°$

Exercise 7.33

(I) If the tangents *PA* and *PB* from an external point *P* to a circle with centre *O* are inclined to each other at an angle of 100°, then $\angle POA$ equals

(a) 40° (b) 50°
(c) 60° (d) 38°

(ii) If the tangents *PA* and *PB* from an external point *P* to a circle with centre *O* are inclined to each other at an angle of 60°, then $\angle POA$ equals

(a) 40° (b) 50°
(c) 60° (d) 100°

(iii) If the tangents PA and PB from an external point *P* to a circle with centre *O* are inclined to each other at an angle of 58°, then $\angle POA$ equals

(a) 61° (b) 58°
(c) 116° (d) 24°

183. Two circles touch each other externally at *P*. *AB* is a common tangent to the circles touching them at *A* and *B*. The value of $\angle APB$ is **(2014)**

(a) 30° (b) 45°
(c) 60° (d) 90°

Solution: (d)

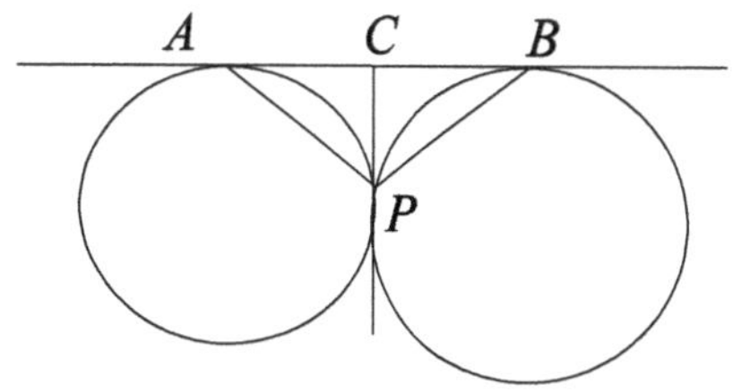

In ΔACP

$CA = CP$

$\therefore \angle CPA = \angle CAP$

$\therefore \angle CPA = \angle PAB$ ...(i)

In ΔBCP

$BC = CP$

$\therefore \angle CPB = \angle CBP$

$\therefore \angle CPB = \angle PBA$ ...(ii)

Now, In ΔAPB

$\angle APB + \angle PAB + \angle PBA = 180°$

$\Rightarrow \angle APB + \angle CPA + \angle CPB = 180°$ [from (i) and (ii)]

$\Rightarrow \angle APB + \angle APB = 180°$

$\Rightarrow 2\angle APB = 180°$

$\therefore \angle APB = 90°$

Key Point In the given figure, *AB* and *AC* are tangents to the circle. If $\angle ABC$ = a°, then the measure of $\angle BAC = 180° - 2a°$

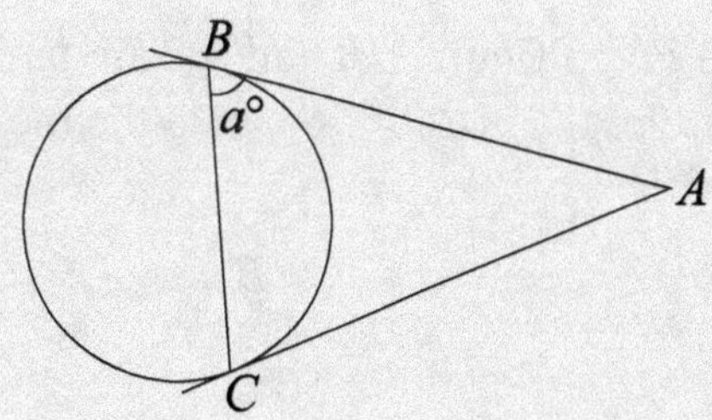

184. In the given figure, *AB* and *AC* are tangents to the circle. If $\angle ABC = 42°$, then the measure of $\angle BAC$ is **(2024)**

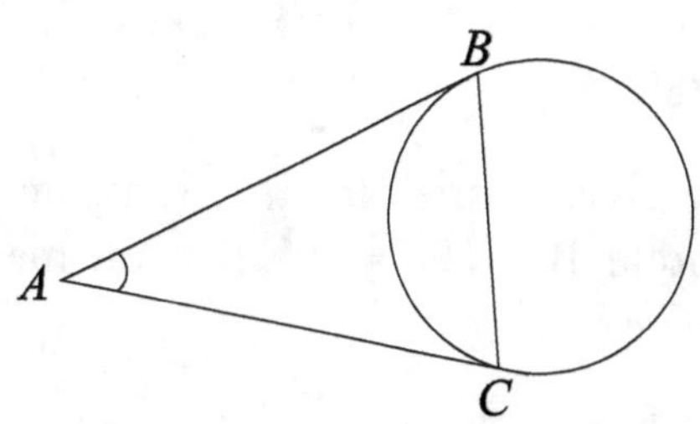

(a) 96° (b) 42°
(c) 106° (d) 86°

Solution: (a)

In ΔABC

$AB = AC$

$\therefore \angle ABC = \angle ACB = 42°$

$\therefore \angle BAC = 180° - 2\angle ABC$

$= 180° - 2 \times 42°$

$= 180° - 84°$

$= 96°$

Exercise 7.34

(i) In the given figure, AB and AC are tangents to the circle. If $\angle ABC = 60°$, then the measure of $\angle BAC$ is

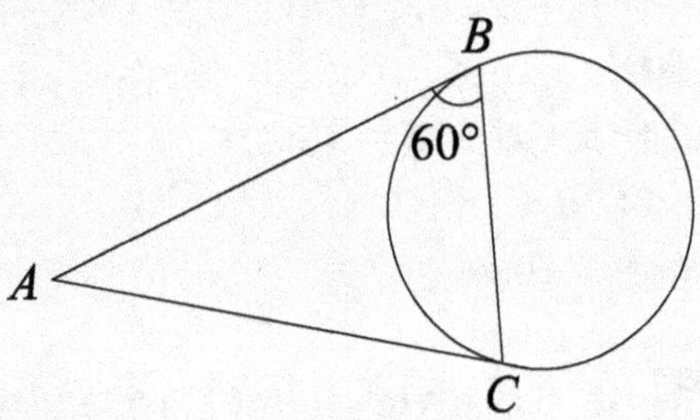

(a) 42° (b) 80°
(c) 60° (d) 120°

(ii) In the given figure, AB and AC are tangents to the circle. If $\angle ABC = 56°$, then the measure of $\angle BAC$ is

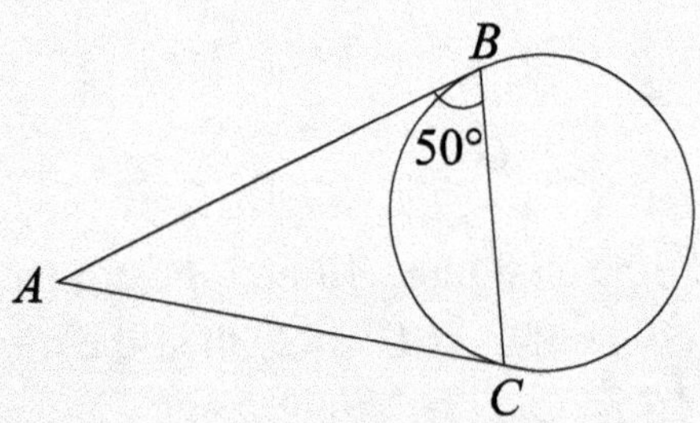

(a) 78° (b) 68°
(c) 88° (d) 98°

(iii) In the given figure, AB and AC are tangents to the circle. If $\angle ABC = 70°$, then the measure of $\angle BAC$ is

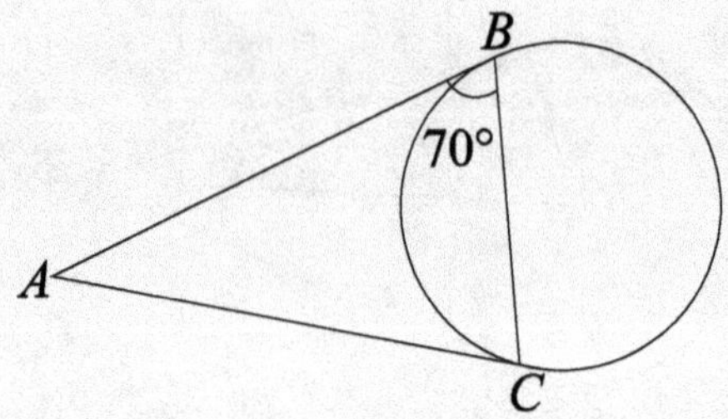

(a) 40° (b) 70°
(c) 30° (d) 80°

Key Point In the given figure tangents PA and PB to the circle centred at O, from the point P are perpendicular to each other. If $PA = a$ unit, then length of AB will be $\sqrt{2}$ a unit

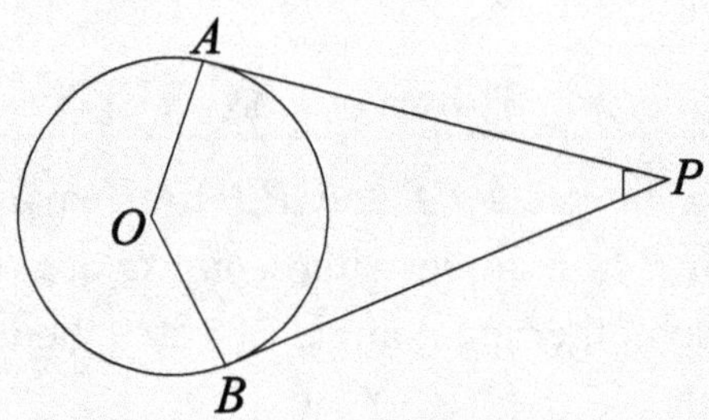

185. In the given figure tangents PA and PB to the circle centred at O, from the point P are perpendicular to each other. If $PA = 5$ cm, then length of AB is equal to **(2024)**

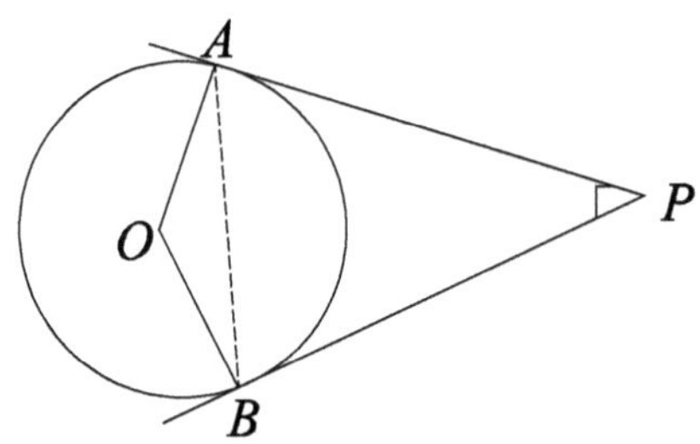

(a) 5 cm (b) $5\sqrt{2}$ cm
(c) $2\sqrt{5}$ cm (d) 10 cm

Solution: (b)

In ΔAPB, $\angle APB = 90°$

$\therefore AB^2 = PA^2 + PB^2$

$\Rightarrow AB^2 = 5^2 + 5^2$ [$\because PA = PB$]

$\Rightarrow AB^2 = 50$

$\therefore AB = 5\sqrt{2}$ cm

Exercise 7.35

(i) In the given figure tangents PA and PB to the circle centred at O, from the point P are perpendicular to each other. If $PA = 4$ cm, then length of AB is equal to

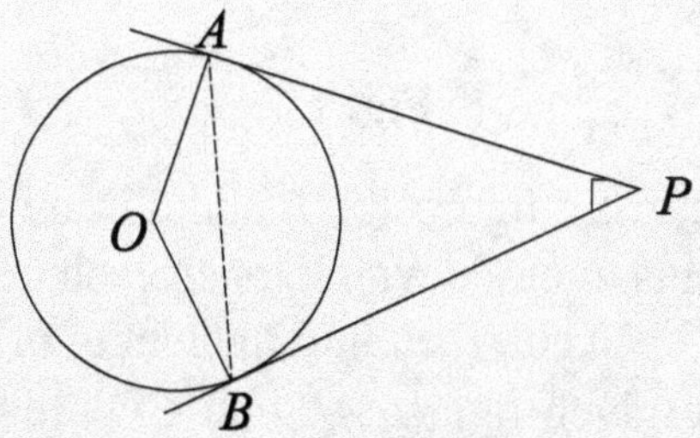

(a) 4 cm (b) $4\sqrt{2}$ cm
(c) 8 cm (d) $8\sqrt{2}$ cm

(ii) In the given figure tangents PA and PB to the circle centred at O, from the point P are perpendicular to each other. If $PA = 5\sqrt{2}$ cm, then length of AB is equal to

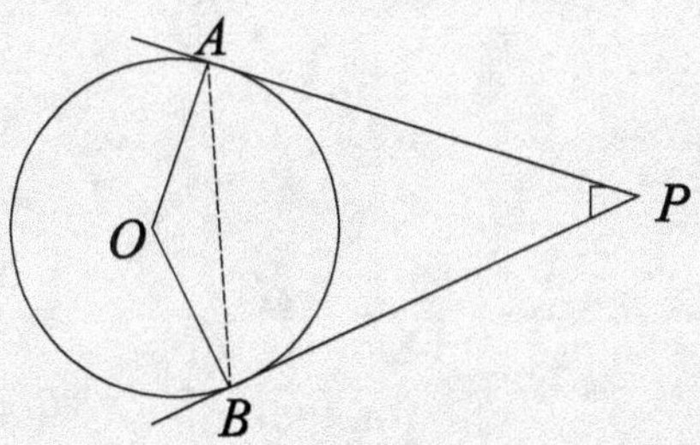

(a) $5\sqrt{2}$ cm (b) 10 cm
(c) $10\sqrt{2}$ cm (d) 20 cm

186. In Fig.2, a circle with centre O is inclined in a quadrilateral $ABCD$ such that, it touches the sides BC, AB, AD and CD at points P, Q, R and S respectively. If $AB = 29$ cm, $AD = 23$ cm, $\angle B = 90°$ and $DS = 5$ cm, then the radius of the circle (in cm) is **(2013)**

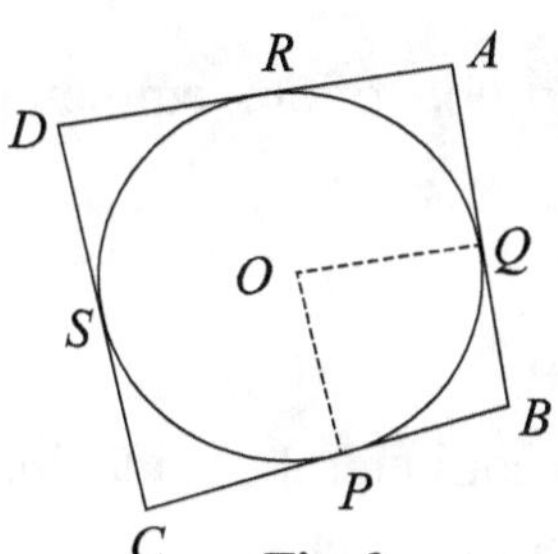

Fig. 2

(a) 11 (b) 18
(c) 6 (d) 15

Solution: (a)

$AD = 23$

$\Rightarrow AR + RD = 23$

$\Rightarrow AR + DS = 23$ [$\because RD = DS$]

$\Rightarrow AR + 5 = 23$

$\Rightarrow AR = 18$ cm

$\therefore AQ = 18$ cm

$AB = 29$

$\Rightarrow AQ + QB = 29$

$\Rightarrow 18 + QB = 29$ [$\because AQ = 18$ cm]

$\Rightarrow QB = 11$ cm

Now, in quadrilateral $BQOP$

$\angle B = \angle Q = \angle P = 90°$

$\therefore \angle O = 90°$ and $BP = BQ$

$\therefore$ $BQOP$ is a square

$\therefore OP = OQ = BQ = 11$ cm.

Exercise 7.36

(i) In fig. 1, a circle with centre O is incribed in a quadrilateral $ABCD$ such that, It touches the sides BC, AB, AD and CD at points P, Q, R and S respectively. If $AB = 30$ cm, $AD = 24$ cm, $\angle B = 90°$ and $DS = 6$ cm, then the radius of the circle (in cm) is:

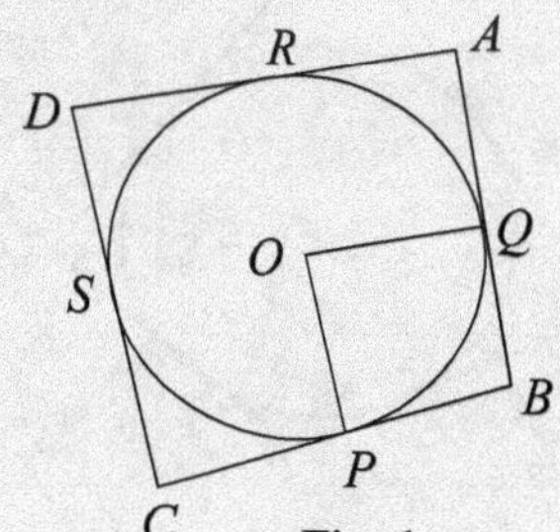

Fig. 1

(a) 10 (b) 11
(c) 12 (d) 13

(ii) In fig. 1, a circle with centre O is incribed in a quadrilateral $ABCD$ such that, It touches the sides BC, AB, AD and CD at points P, Q, R and S respectively. If $AB = 33$ cm, $AD = 22$ cm, $\angle B = 90°$ and $DS = 7$ cm, then the radius of the circle (in cm) is:

(a) 18 (b) 15

(c) 16 (d) 22

(iii) In the given figure 1, if $AB = 36$ cm, $AD = 27$ cm, $\angle B = 90°$ and $DS = 8$ cm.

(a) 10 (b) 14

(c) 15 (d) 17

187. In the given figure, QR is a common tangent to the two given circles touching external at A. The tangent at A meets QR at P. If $AP = 4.2$ cm, then the length of QR is: **(2024)**

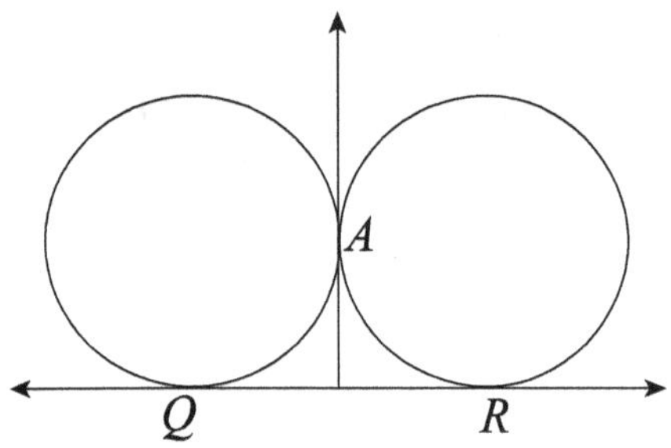

(a) 4.2 cm (b) 2.1 cm

(c) 8.4 cm (d) 6.3 cm

Solution: (c)

$QR = 2 \times AP = .2 \times 4.2 = 8.4$ cm.

188. In the given figure, the quadrilateral $PQRS$ circumscribes a circle. Here $PA + CS$ is equal to **(2023)**

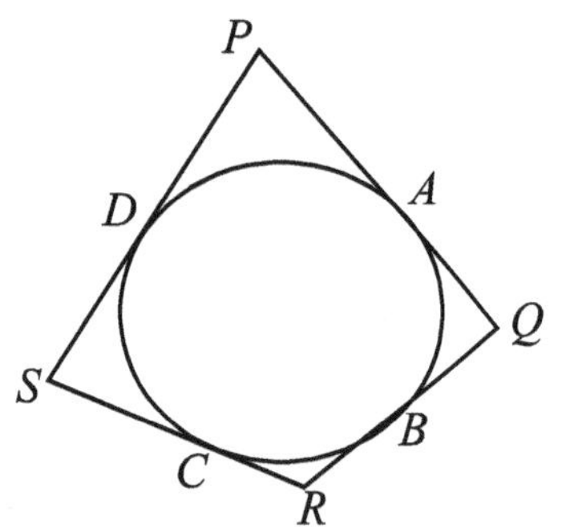

(a) QR (b) PR

(c) PS (d) PQ

Solution: (c)

$PA + CS = PD + DS = PS.$

Exercise 7.37

(i) In the given figure 1, AB and CD are common tangents to circle which touch each other at D. If $AB = 12$ cm, then find the length of CD.

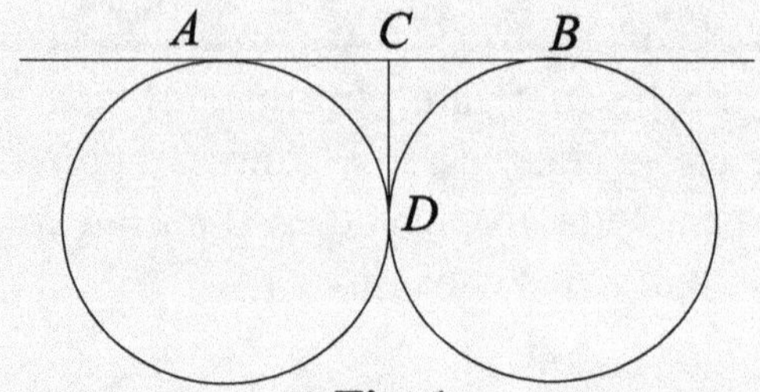

Fig. 1

(ii) In the given figure 1, AB and CD are common tangents to circle which touch each other at D. If $AB = 20$ cm, then find the length of CD.

(iii) In the given figure, AB and CD are common tangents to circle which touch each other at D. If $AB = 18$ cm, then find the length of CD.

(iv) In the given figure-2, QR is a common tangent to the two given circles touching external at A. The tangent at A meets QR at P. If $AP = 8$ cm, then the length of QR is:

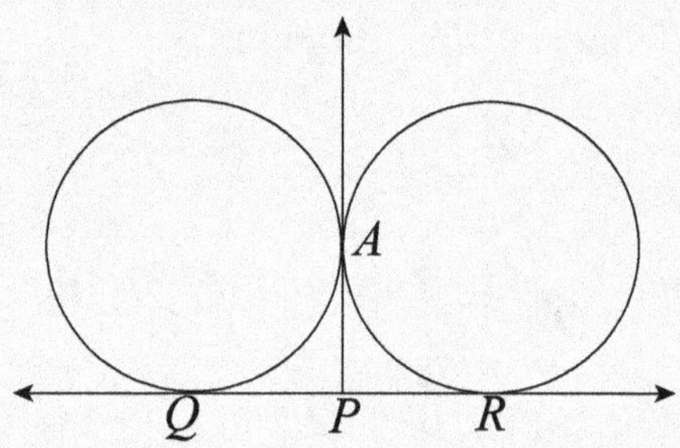

Fig. 2

(a) 8 cm (b) 4 cm

(c) 8.8 cm (d) 16 cm

(v) In the given figure, QR is a common tangent to the two given circles touching external at A. The tangent at A meets QR at P. If $AP = 7.3$ cm, then the length of QR is:

(a) 14.1 cm (b) 15 cm

(c) 14.6 cm (d) 7 cm

Key Point Altitude of an equilateral triangle $= \dfrac{\sqrt{3}}{2} \times \text{side unit}$

189. ABC is an equilateral of side $2a$, then length of one of its altitude is **(2020)**

Solution:

Altitude of the equilateral triangle $= \frac{\sqrt{3}}{2} \times \text{side}$

$= \frac{\sqrt{3}}{2} \times 2a = \sqrt{3a}$ unit.

Exercise 7.38

(i) ABC is an equilateral of side $4a$ unit, then length of one of its altitude is ________.

(ii) ABC is an equilateral of side $6\sqrt{3}$ cm, then length of one of its altitude is ________.

(iii) ABC is an equilateral of side $2\sqrt{3}$ cm, then length of one of its altitude is ________.

190. The larger of two Supplementary angle exceeds the smaller by 18 degree. What is the measure of larger angle ? **(2023)**

(a) 81° (b) 99°

(c) 36° (d) 54°

Solution: (b)

$x + (x + 18) = 180$

$\Rightarrow 2x = 162$

$\Rightarrow x = 81$

$\therefore$ The larger angle is $(81 + 18)° = 99°$

Exercise 7.39

(i) The larger of two supplementary angle exceeds the smaller by 20 degree. What is the measure of larger angle ?

(a) 100° (b) 120°

(c) 80° (d) 90°

(ii) The larger of two supplementary angle exceeds the smaller by 12 degree. What is the measure of larger angle ?

(a) 84° (b) 96°

(c) 90° (d) 92°

(iii) The larger of two supplementary angle exceeds the smaller by 30 degree. What is the measure of larger angle ?

(a) 100° (b) 102°

(c) 104° (d) 105°

191. PQ and PR are tangents to a circle with center O such that $OQ = QP$. The value of $\angle OPQ$ is equal to **(2025)**

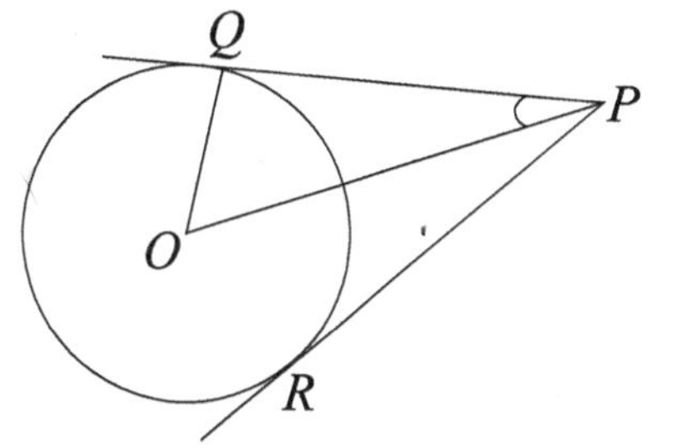

(a) 45° (b) 30°

(c) 60° (d) 90°

Solution: (a)

In ΔOPQ, $\angle OQP = 90°$

$OQ = QP$

$\therefore \angle QOP = \angle OPQ = \frac{90°}{2} = 45°$

192. In the adjoining figure, PA and PB are tangents to a circle with center O such that $\angle P = 90°$. If $AB = 3\sqrt{2}$ cm, then the diameter of the circle is **(2025)**

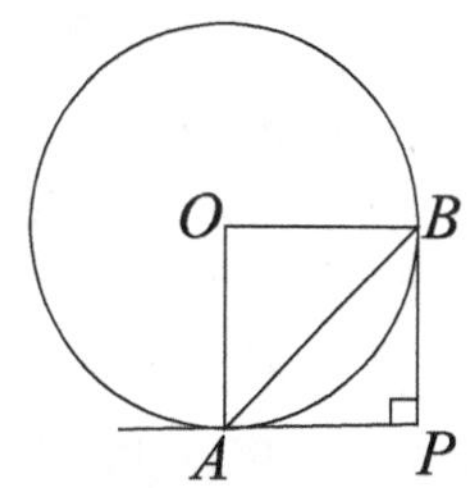

(a) $3\sqrt{2}$ cm (b) $6\sqrt{2}$ cm

(c) 3 cm (d) 6 cm

Solution: (d)

$OA^2 + OB^2 = AB^2$

$\Rightarrow r^2 + r^2 = (3\sqrt{2})^2$

$\Rightarrow 2r^2 = 18$

$\Rightarrow r^2 = 9$

$\therefore r = 3$

$\therefore d = 6$ cm

193. Two circle of radii 10 cm and 17 cm intersect at P and Q. If A and B are their centers and $PQ =$ 16 cm, then the distance AB is equal to **(2025)**

(a) 30 cm (b) 12 cm

(c) 21 cm (d) 16 cm

Solution: (c)

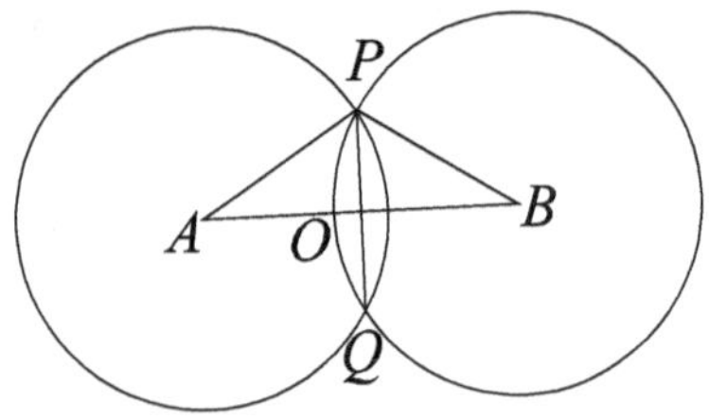

In ΔAOP, $\angle AOP = 90$ and $AP = 10$ cm

$\therefore AP^2 = OA^2 + OP^2$

$\Rightarrow 10^2 = OA^2 + 8^2$ $\quad [\because PQ = 2\,OP]$

$\Rightarrow OA^2 = 36$

$\therefore OA = 6$ cm

Now, in ΔBOP, $\angle BOP = 90$, and $BP = 17$ cm

$\therefore BP^2 = OB^2 + OP^2$

$\Rightarrow 172 = OB^2 + 8^2$

$\Rightarrow OB = 225$

$\therefore OB = 15$ cm

$\therefore AB = OA + OB$

$= 6 + 15$

$= 21$ cm

194. If the length of a chord of a circle is equal to its radius, then the angle subtended by chord at the center is: **(2025)**

(a) 60° (b) 30°

(c) 120° (d) 90°

Solution: (a) 60°

195. PQ is tangent to a circle with centre O. If $\angle POR = 65°$, then $m\angle PTR$ is **(2026)**

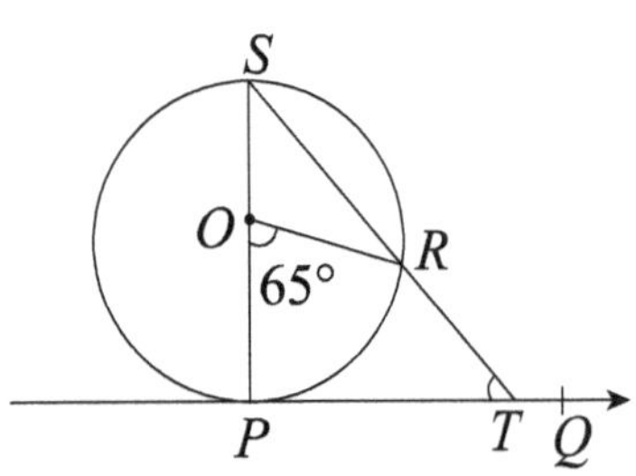

(a) 65° (b) 58.5°

(c) 57.5° (d) 45°

Solution: (b)

$\angle POR = 2\angle PSR$

$\Rightarrow 65° = 2\angle PSR$

$\Rightarrow \angle PSR = 32.5°$

$\therefore \angle PST = 32.5°$

Now, in $\Delta\, PST$, $\angle SPT = 90°$

$\therefore \angle SPT + \angle PTS + \angle PST = 180°$

$\Rightarrow 90° + \angle PTR + 32.5° = 180°$

$\Rightarrow \angle PTR + 122.5° = 180°$

$\Rightarrow \angle PTR = 58.5°$

Question number **196 to 207** are Assertion and Reason based question. Two statements are given, one labeled on Assertion (A) and the other labeled an Reason (R). Select the correct answer to these questions from the codes (A), (B), (C) and (D) an given below.

(a) Both Assertion (A) and reason (R) are true and Reason (R) is the correct explanator of Assertion (A).

(b) Both Assertion (A) and reason (R) are true and Reason (R) is the correct explanator of Assertion (A).

(c) Assertion (A) is true, but Reason (R) is false.

(d) Assertion (A) is false, but Reason (R) is trues.

196. Assertion (A): The perimeter of ΔABC is a rational number.

Reason (R): The sum of the square of two rational numbers in always rational. **(2023)**

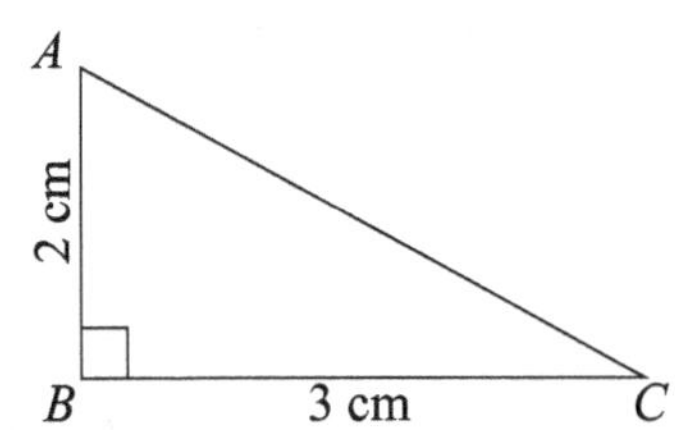

Solution: (d)

$AC = \sqrt{AB^2 + BC^2}$

$= \sqrt{2^2 + 3^2}$

$= \sqrt{13}$

$\therefore$ Perimeter of $\Delta ABC = 2 + 3 + \sqrt{13}$

$$= 5 + \sqrt{13}$$

= Irrational number

Sum of the squares of two rational numbers is always rational number.

197. Assertion (A): *ABCD* is a trapezium with DC ∥ AB. E and F are points on AD and BC respectively, such that EF ∥AB. Then $\frac{AE}{ED} = \frac{BF}{FC}$.

Reason (R): Any line parallel to parallel sides of a trapezium divides the non- Parallel sides proportionally. **(2024)**

Solution: (a)

Both Assertion (A) and Reason (R) are true and Reason is the correct explanation of Assertion (A).

198. Assertion (A): A line drawn parallel to any one side of a triangle intersects the other two sides in the same ration.

Reason (R): parallel lines cannot be drawn to any side of a triangle. **(2024)**

Solution: (c)

Assertion (A) is true, but reason (R) is false.

199. Assertion (A): A tangent to a circle is perpendicular to the radius through the point of contact.

Reason (R): the length of tangents drawn from an external point to a circle are equal. **(2023)**

Solution: (b)

Both Assertion (A) and Reason (R) are true, but Reason (R) is not the correct explanation of Assertion(A).

200. Assertion (A): If *PA* and *PB* are tangents drawn from an external point *P* to a circle with centre *O*, then the quadrilateral *AOBP* is cyclic.

Reason (R): The angle between two tangents drawn from an external point to a circle is supplementary to the angle subtended by the line segment joining the points of contact at the circle. **(2023)**

Solution: (a)

Both Assertion (A) and Reason (R) are true, but Reason (R) is the correct explanation of Assertion(A).

201. Assertion (A): The tangents drawn at the end points of a diametor of a circle, are parallel.

Reason (R): Diameter of a circle in the largest chord. **(2023)**

Solution: (b)

Both Assertion (*A*) and Reason (*R*) are true, but Reason (*R*) is not the correct explanation of Assertion (*A*).

202. Assertion (A): The tangents drawn at the end points of a diameter of a circle, are

Reason (R): Diameter of a circle in the largest chord. **(2024)**

Solution: (b)

Both Assertion (A) and Reason (R) are true, but Reason (R) is not the correct explanation of Assertion(A).

203.

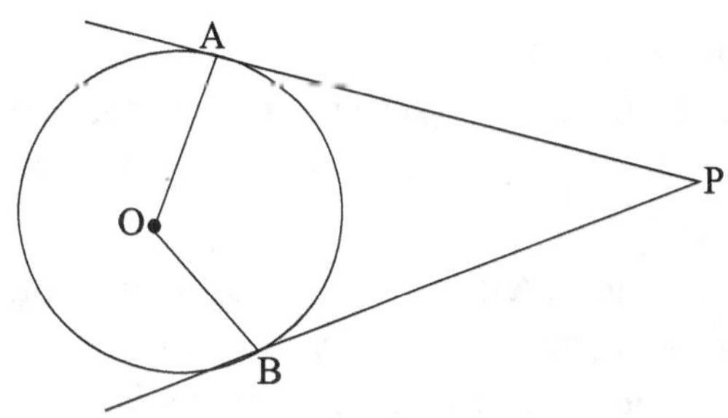

Assertion (A): If *PA* and *PB* are tangents drawn to a circle with centre *O* from an external point P, then the quadrilateral *OAPB* is a cyclic quadrilateral

Reason (R): In a cyclic quadrilateral opposite angle are equal. **(2024)**

Solution: (a)

Both Assertion (A) and reason (R) are true and Reason (R) is the correct explanation of Assertion (A).

204. Assertion(A): Tangents drawn at the end points of a diameter of a circle are always parallel to each other.

Reason(R): The lengths of tangents drawn to a circle from a point outside the circle are always equal. **(2025)**

Solution: (b)

Both Assertion(A) and Reason(R) are true but Reason(R) is not the correct explanation of Assertion(A)

205. **Assertion(A):** All congruent triangles are similar

Reason(R): In congruent triangles the ratio of corrosponding sides is 1 : 1 **(2025)**

Solution: (a)

Both Assertion (A) and reason (R) are true and Reason (R) is the correct explanation of Assertion (A).

206. **Assertion(A):** In a ΔABC, D and E are points on the sides AB and AC respectively such that $DE \parallel BC$, then $\frac{AD}{AB} = \frac{AE}{AC}$

Reason(R): If a line is drawn parallel to the one side of a triangle, intersecting the other two sides, then it divides the two sides in the same ratio. **(2025)**

Solution: (a)

Both Assertion (A) and reason (R) are true and Reason (R) is the correct explanation of Assertion (A).

207. **Assertion (A):** Radius is the smallest distance of a tangent from the centre of the circle.

Reason (R): Radius is perpendicular to the tangent **(2026)**

Solution: (a)

Both Assertion (A) and reason (R) are true and Reason (R) is the correct explanation of Assertion (A).

Exercise 7.40

Direction:

(i) **Assertion(A):** ABC and DEF are two similar triangles such that $BC = 8$ cm, $EF = 10$ cm and area of $\Delta ABC = 36$ cm², then area of $\Delta DEF = 100$ cm².

Reason(R): The areas of two similar triangles are in the ratio of the squares of the corresponding attitudes.

(ii) **Assertion(A):** The areas of two similar triangles ABC and PQR are in the ratio 9:16. If $BC = 4.5$ cm, then the length of QR is 6 cm.

Reason(R): The ratio of the areas of two similar triangles is equal to the ratio of their corresponding sides.

(iii) **Assertion(A):** If T is a point on side YZ of ΔXYZ such that $XT \perp YZ$, then $\Delta XYT \sim \Delta XZT$

Reason(R): In the given figure given bellow, if $\angle A = \angle R$, then $\Delta PAB \sim \Delta PRQ$

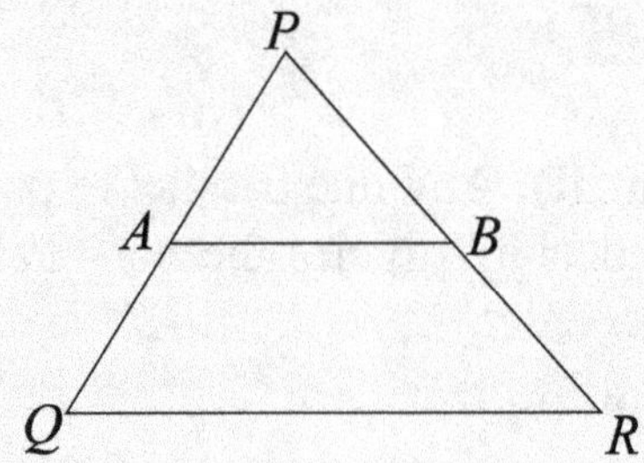

(iv) **Assertion(A):** Two similar triangles are always congruent.

Reason: If the areas of two similar triangles are equal then the triangles are congruent.

(v) **Assertion(A):** $\Delta ABC \sim \Delta DEF$ such that ar(ΔABC) = 64 cm² and ar(ΔDEF) = 81 cm², then, $AB:DE = 8:9$

Reason(R): If $\Delta ABC \sim \Delta DEF$ then

$$\frac{ar\Delta ABC}{ar\Delta DEF} = \frac{AB^2}{DE^2} = \frac{BC^2}{EF^2} = \frac{AC^2}{DF^2}$$

(vi) **Assertion(A):** In ΔABC, $DE \parallel BC$ such that $AD = (4x - 3)$ cm, $AE = (8x - 7)$ cm, $BD = (3x - 1)$ cm and $EC = (5x - 3)$ cm, then x equal to 1.

Reason(R): If a line is drawn parallel to one side of a triangle to intersect the other two sides in distinct points, then the other two sides are divided in the same ratio.

(vii) **Assertion(A):** PA and PB are two tangents to a circle with center O. Such that $\angle AOB = 102°$, then $\angle APB = 88°$

Reason(R): The length of two tangents drawn from on external point are equal.

(viii) **Assertion(A):** The two tangents are drawn to a circle from an external point, then they subtend equal angles at the center.

Reason(R): A parallelogrom circumscribing a circle is a rhombus.

(ix) Assertion(A): If in a circle, the radius of the circle is 6 cm and distance of a point from the center of a circle is 10 cm, then length of the tangent will be 8 cm.

Reason(R): $(\text{hypotenuse})^2 = (\text{Bose})^2 + (\text{Height})^2$

(x) Assertion(A): If length of a tangent from on external point to a circle is 10 cm, then length of the other tangent from the same point is 10 cm.

Reason(R): Length of the tangents drawn from on external point to a circle are equal.

(xi) Assertion(A): $ABCD$ is a trapezium with $DC \parallel AB$. E and F are points on AD and BC respectively such that $EF \parallel AB$, then $\frac{AE}{ED} = \frac{BE}{FC}$

Reason(R): Any line parallel to parallel sides of a trapezium divides the non-parallel sides proportionally.

(xii) Assertion(A): If TP and TQ are the two tangents to a circle with center O so that $\angle POQ = 131°$, then $\angle PTQ = 49°$.

Reason(R): $OPTQ$ in a cyclic quadrilateral.

(xiii) Assertion(A): At a point P of a circle with center O and radius 8 cm, a tangent PQ of length 6 cm is drawn, then $OQ = 10$ cm.

Reason(R): The tangent at any point of a circle is perpendicular to the radius through the point of contact.

(xiv) Assertion(A): The triangles ABC and DEF are similar by SSS criterion

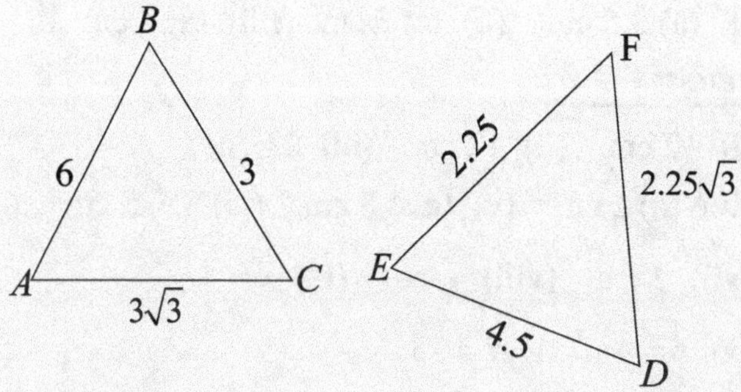

Reason(R): If the three sides of one triangle are respectively proportional to the three sides of another, then the two triangles are similar.

(xv)Assertion(A): D and E are points on the sides AB and AC respectively of a triangle ABC, if $AB = 2.56$ cm, $AC = 5.12$ cm, $AD = 0.36$ cm and $AE = 0.72$ cm, then DE $\parallel$ BC by converse of BPT.

Reason(R): Converse of basic proportionality theorem states that if a line divides any two sides of a triangle in the same ratio, then the line is parallel to the third side.w

ANSWER KEYS

Exercise – 7.1

(i) 35° (ii) 58° (iii) 30° (iv) 30°
(v) (a) 50° (vi) (d) 79°

Exercise – 7.2

(i) (a) 3.5 cm (ii) (c) 3 cm (iii) (a) 4 cm

Exercise – 7.3

(i) 12 cm (ii) 12 cm (iii) 21 cm
(iv) (a) 2 cm (v) (c) 1.8 cm (vi) (b) 5 cm
(vii) 2 cm (viii) $\frac{4}{3}$ cm (ix) $x = 4$
(x) $x = 4$ (xi) $x = 13$

Exercise – 7.4

(i) (a) 10 (ii) (d) 15 (iii) (d) 10

Exercise – 7.5

(i) 6 cm (ii) 7 cm (iii) 10 cm

Exercise – 7.6

(i) (b) $CD = 6$ cm (ii) (c) $CD = 8$ cm
(iii) (a) $CD = 14$ cm

Exercise – 7.7

(i) 16 cm (ii) 14 cm (iii) 20 cm
(iv) 36 cm (v) 27 cm (vi) 33 cm
(vii) (b) 3:2 (viii) (c) 8:15 (ix) (a) 4:9

Exercise – 7.8

(i) $30\sqrt{2}$ cm (ii) $5\sqrt{2}$ cm (iii) $\sqrt{2}$ a cm.

Exercise – 7.9

(i) 4:25 (ii) 1:16 (iii) 16:81 (iv) $\frac{1}{16}$
(v) $\frac{4}{9}$ (vi) $\frac{9}{25}$ (vii) 49:9 (viii) 25:4
(ix) 81:16 (x) 4:5 (xi) 7:10 (xii) 10:11
(xiii) 25:36 (xiv) 9:16 (xv) 1.21:2.25
(xvi) 4:9 (xvii) 9:4 (xviii) 16:25

Exercise – 7.10

(i) (b) 55° (ii) (c) 40° (iii) (b) 60°
(iv) (a)50° (v) (c) 55° (vi) 50°

Exercise – 7.11

(i) (c) 90° (ii) (b) 60° (iii) (c) 124°

Exercise – 7.12

(i) 50° (ii) 45° (iii) 38°

Exercise – 7.13

(i) (b) 50° (ii) (b) 65° (iii) (a) 52°

Exercise – 7.14

(i) 30° (ii) 20° (iii) 50°

Exercise – 7.15

(i) 130° (ii) 140° (iii) 110°

Exercise – 7.16

(i) 16 cm (ii) (a) 12 cm (iii) (d) 11cm
(iv) (d) 7.2 cm

Exercise – 7.17

(i) 48° (ii) (c) 35° (iii) (d) 46°

Exercise – 7.18

(i) (d) 120° (ii) (c) 105° (iii) (d) 125°

Exercise – 7.19

(i) (c) 10° (ii) (c) 40° (iii) (a) 40°

Exercise – 7.20

(i) 6 cm (ii) 24 cm (iii) 12 cm
(iv) 8 cm (v) 10 cm (vi) 14 cm

Exercise – 7.21

(i) (c) 16 cm (ii) (c) 12 cm (iii) (d) 48 cm

Exercise – 7.22

(i) 12 cm (ii) 13 cm (iii) 18 cm
(iv) (b) 11 cm (v) (c) 14 cm (vi) (a) 12 cm
(vii) (d) 50 cm (viii) (a) 36 cm (ix) (a) 52 cm

Exercise – 7.23

(i) (c) 24 (ii) (b) 36 (iii) (d) 30

Exercise – 7.24

(i) (a) 7.5 cm (ii) (b) 11 cm (iii) (c) 10 cm

Exercise – 7.25

(i) (b) 60° **(ii)** (d) 82° **(iii)** (a) 45°
(iv) (b) 80° **(v)** (d) 75°

Exercise – 7.26

(i) (b) 10.5 **(ii)** (c) 12 **(iii)** (a) 19

Exercise – 7.27

(i) (c) 6 cm **(ii)** (a) 3.5 cm **(iii)** (a) 4.5 cm
(iv) (a) 6 cm **(v)** (c) 2 cm **(vi)** 5 cm
(vii) (b) 4 cm

Exercise – 7.28

(i) (a) 25° **(ii)** (c) 40° **(iii)** (c) 50°
(iv) (b) 50° **(v)** (b) 35°

Exercise – 7.29

(i) 1 20° **(ii)** 90° **(iii)** 60°

Exercise – 7.30

(i) 8 cm **(ii)** $4\sqrt{3}$ cm **(iii)** $5\sqrt{2}$ cm
(iv) (b) 3 cm **(v)** (d) $4\sqrt{2}$ cm **(vi)** (a) $6\sqrt{3}$ cm

Exercise – 7.31

(i) (c) 60° **(ii)** (b) 40° **(iii)** **(d)** 35°

Exercise – 7.32

(i) (d) 60° **(ii)** (d) 50° **(iii)** **(c)** 65°

Exercise – 7.33

(i) (a) 40° **(ii)** (c) 60° **(iii)** **(a)** 61°

Exercise – 7.34

(i) (c) 60° **(ii)** (b) 68° **(iii)** **(a)** 40°

Exercise – 7.35

(i) (b) $4\sqrt{2}$ cm **(ii)** (b) 10 cm

Exercise – 7.36

(i) (c) 12 **(ii)** (a) 18 **(iii)** **(d)** 17

Exercise – 7.37

(i) 6 cm **(ii)** 10 cm **(iii)** 9 cm
(iv) (d) 16 cm **(v)** (c) 14.6 cm

Exercise – 7.38

(i) $2a\sqrt{3}$ unit **(ii)** 9 cm **(iii)** 3 cm

Exercise – 7.39

(i) (a) 100° **(ii)** (b) 96° **(iii)** **(d)** 105°

Exercise – 7.40

(i) d **(ii)** a **(iii)** d **(iv)** d
(v) a **(vi)** c **(vii)** d
(viii) b **(ix)** a **(x)** a **(xi)** a
(xii) a **(xiii)** a **(xiv)** a **(xv)** a

8 Trigonometry

Key Point **(i)** $\sin\theta = \frac{p}{h}$

(ii) $\cos\theta = \frac{b}{h}$

(iii) $\tan\theta = \frac{p}{b}$

(iv) $\cot\theta = \frac{b}{p}$

(v) $\operatorname{cosec}\theta = \frac{h}{p}$

(vi) $\sec\theta = \frac{h}{b}$

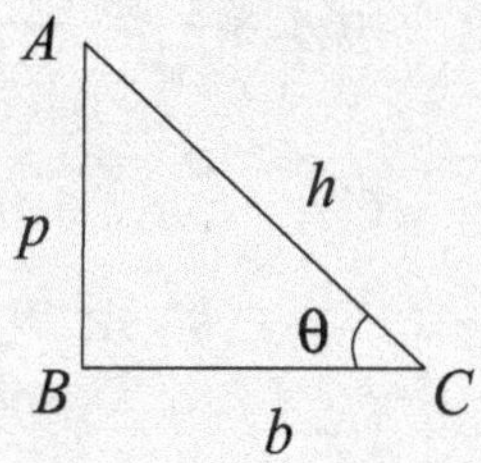

1. If $\sin A = \frac{3}{4}$, calculate $\sec A$. **(2019)**

Solution :

$\sin A = \frac{3}{4}$, Let $BC = 3k$

$AC = 4k$

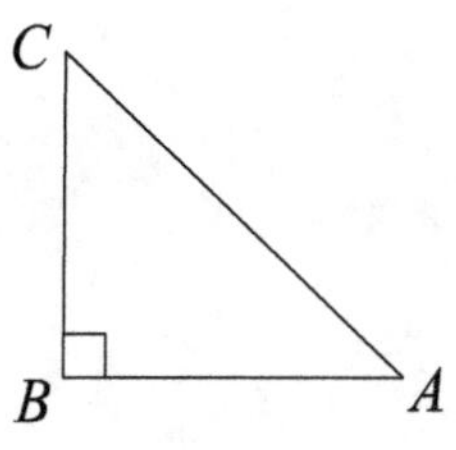

Where k is positive

By Pythagoras theorm we have

$AC^2 = AB^2 + BC^2$

$\Rightarrow (4k)^2 = AB^2 + (3k)^2$

$\Rightarrow 16k^2 = AB^2 + 9k^2$

$\Rightarrow AB^2 = 7k^2$

$\therefore AB = \sqrt{7}\ k$

$\therefore \sec A = \frac{AC}{AB}$

$= \frac{4k}{\sqrt{7}\ k}$

$= \frac{4}{\sqrt{7}}$

$= \frac{4\sqrt{7}}{7}$

2. If $\sin\theta = \frac{5}{13}$, then the value of $\tan\theta$ is **(2020)**

Solution :

$\sin\theta = \frac{5}{13} = \frac{5k}{13k}$

$\therefore AB = 5k$

$AC = 13k$

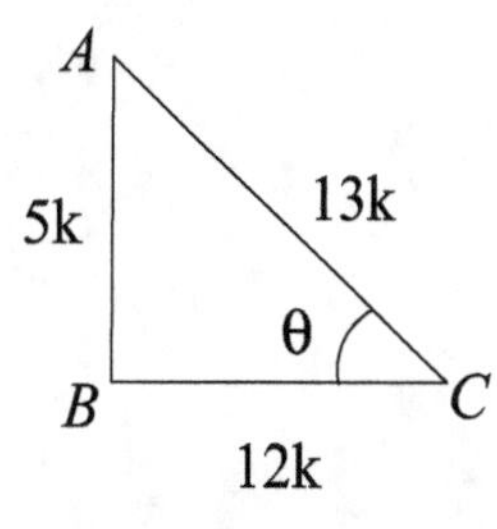

Now,

$AC^2 = AB^2 + BC^2$

$\Rightarrow (13k)^2 = (5k)^2 + BC^2$

$\Rightarrow BC^2 = 169k^2 - 25k^2$

$\Rightarrow BC^2 = 144k^2$

$\therefore BC = 12k$

$\therefore \tan\theta = \dfrac{AB}{BC}$

$= \dfrac{5k}{12k}$

$= \dfrac{5}{12}$

3. If $\sin\theta = \dfrac{a}{b}$, then $\sec\theta$ is equal to $(0° \le 0 \le 90°)$. **(2023)**

(a) $\dfrac{a}{\sqrt{b^2 - a^2}}$ (b) $\dfrac{b}{\sqrt{b^2 - a^2}}$

(c) $\dfrac{\sqrt{b^2 - a^2}}{b}$ (d) $\dfrac{\sqrt{b^2 - a^2}}{a}$

Solution :(b)

$\sin\theta = \dfrac{a}{b} = \dfrac{ak}{bk}$

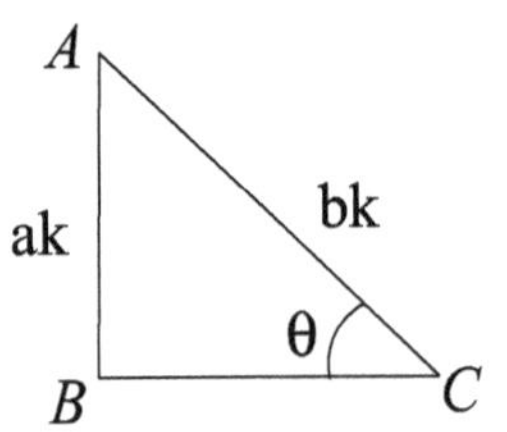

$\therefore AB = ak$

$AC = bk$

Now,

$AC^2 = AB^2 + BC^2$

$\Rightarrow (bk)^2 = (ak)^2 + BC^2$

$\Rightarrow b^2k^2 = a^2k^2 + BC^2$

$\Rightarrow BC^2 = b^2k^2 - a^2k^2$

$\Rightarrow BC^2 = k^2(b^2 - a^2)$

$\therefore BC = k\sqrt{b^2 - a^2}$

$\sec\theta = \dfrac{AC}{BC}$

$= \dfrac{bk}{k\sqrt{b^2 - a^2}}$

$= \dfrac{b}{\sqrt{b^2 - a^2}}$

4. If $\sin A = \dfrac{2}{3}$, then value of $\cot A$ is **(2024)**

(a) $\dfrac{\sqrt{5}}{2}$ (b) $\dfrac{3}{2}$

(c) $\dfrac{5}{4}$ (d) $\dfrac{2}{3}$

Solution :(a)

$\sin A = \dfrac{2}{3} = \dfrac{2k}{3k}$

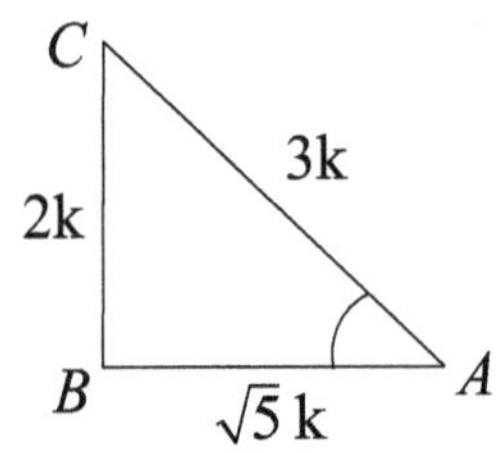

$\therefore BC = 2k$

$AC = 3k$

Now,

$AC^2 = BC^2 + AB^2$

$\Rightarrow (3k)^2 = (2k)^2 + AB^2$

$\Rightarrow AB^2 = 5k^2$

$\therefore AB = \sqrt{5}\,k$

$\cot A = \dfrac{AB}{BC}$

$= \dfrac{\sqrt{5}k}{2k}$

$= \dfrac{\sqrt{5}}{2}$

5. If $\sin\theta = \dfrac{1}{3}$, then $\sec\theta$ is equal to : **(2024)**

(a) $\dfrac{2\sqrt{2}}{3}$ (b) $\dfrac{3}{2\sqrt{2}}$

(c) 3 (d) $\dfrac{1}{\sqrt{3}}$

Solution : (b)

$\sin\theta = \frac{1}{3} = \frac{k}{3k}$

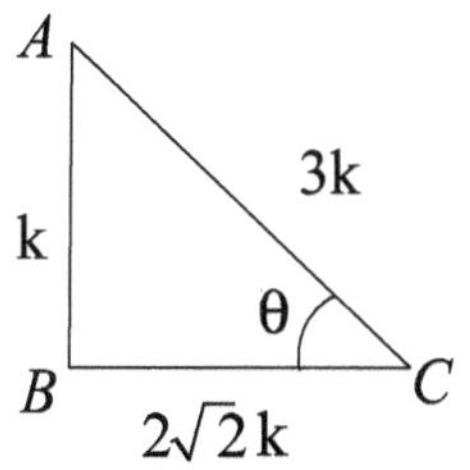

$\therefore AB = k$

$AC = 3k$

Now,

$AC^2 = AB^2 + BC^2$

$\Rightarrow (3k)^2 = (k)^2 + BC^2$

$\Rightarrow 9k^2 = k^2 + BC^2$

$\Rightarrow BC^2 = 8k^2$

$\therefore BC = 2\sqrt{2}\,k$

$\sec\theta = \frac{AC}{BC}$

$= \frac{3k}{2\sqrt{2}k}$

$= \frac{3}{2\sqrt{2}}$

6. If $\sin A = \frac{3}{5}$, then value of $\cot A$ is : **(2024)**

(a) $\frac{3}{4}$ (b) $\frac{4}{3}$

(c) $\frac{4}{5}$ (d) $\frac{5}{4}$

Solution : (b)

$\sin A = \frac{3}{5} = \frac{3k}{5k}$

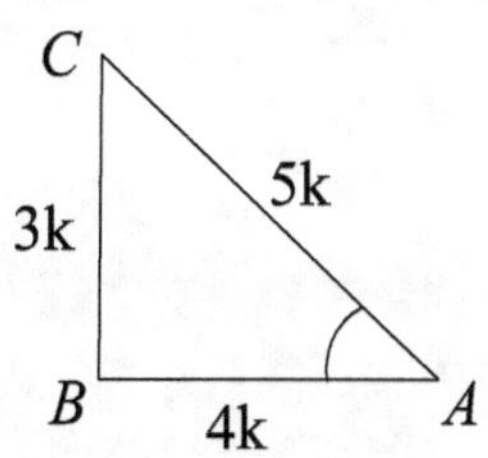

$\therefore BC = 3k$

$AC = 5k$

Now,

$AC^2 = BC^2 + AB^2$

$\Rightarrow (5k)^2 = (3k)^2 + AB^2$

$\Rightarrow 25k^2 = 9k^2 + AB^2$

$\Rightarrow AB^2 = 16k^2$

$\therefore AB = 4k$

$\cot A = \frac{AB}{BC}$

$= \frac{4k}{3k}$

$= \frac{4}{3}$

7. If $\tan\alpha = \frac{5}{12}$, find the value of $\sec\alpha$. **(2019)**

Solution :

$\tan\alpha = \frac{5}{12} = \frac{5k}{12k}$

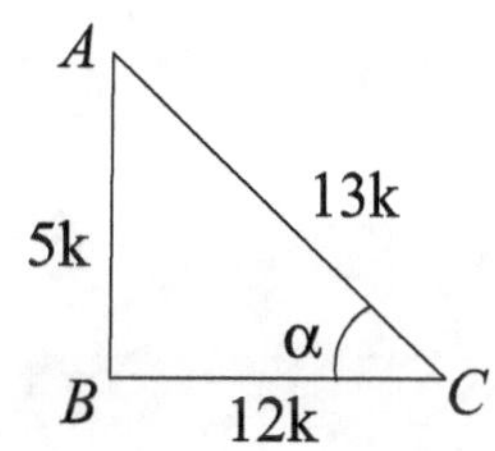

$\therefore AB = 5k$

$BC = 12k$

Now,

$AC^2 = AB^2 + BC^2$

$\Rightarrow AC^2 = (5k)^2 + (12k)^2$

$\Rightarrow AC^2 = 25k^2 + 144k^2$

$\Rightarrow AC^2 = 169k^2$

$\therefore AC = 13k$

$\sec\alpha = \frac{AC}{BC}$

$= \dfrac{13k}{12k}$

$= \dfrac{13}{12}$

8. If $\cot\theta = \dfrac{12}{5}$, then the value of $\sin\theta$ is : **(2020)**

Solution :

$\cot\theta = \dfrac{12}{5} = \dfrac{12k}{5k}$

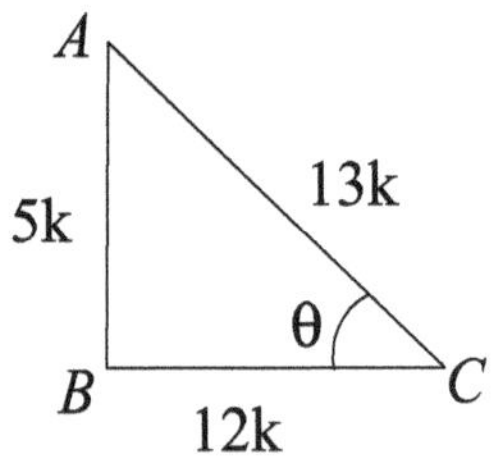

$\therefore\ BC = 5k$

$AB = 12k$

Now,

$AC^2 = AB^2 + BC^2$

$\Rightarrow AC^2 = (5k)^2 + (12k)^2$

$\Rightarrow AC^2 = 25k^2 + 144k^2$

$\Rightarrow AC^2 = 169k^2$

$\therefore\ AC = 13k$

$\sin\theta = \dfrac{AB}{AC}$

$= \dfrac{5k}{13k}$

$= \dfrac{5}{13}$

9. If $\sec\theta = \dfrac{25}{7}$, then find the value of $\cot\theta$ is : **(2020)**

Solution :

$\sec\theta = \dfrac{25}{7} = \dfrac{25k}{7k}$

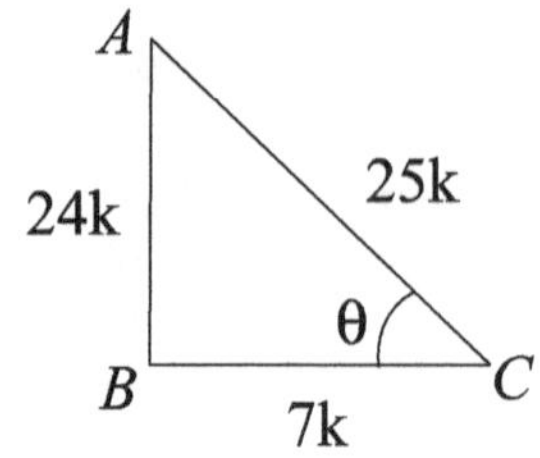

$\therefore\ AC = 25k$

$BC = 7k$

Now,

$AC^2 = AB^2 + BC^2$

$\Rightarrow (25k)^2 = AB^2 + (7k)^2$

$\Rightarrow AB^2 = 625k^2 - 49k^2$

$\Rightarrow AB^2 = 576k^2$

$\therefore\ AB = 24k$

$\cot\theta = \dfrac{BC}{AB}$

$= \dfrac{7k}{24k}$

$= \dfrac{7}{24}$

10. If $\operatorname{cosec}\theta = \dfrac{5}{4}$, then find the value of $\cot\theta$: **(2020)**

Solution :

$\operatorname{cosec}\theta = \dfrac{5}{4} = \dfrac{5k}{4k}$

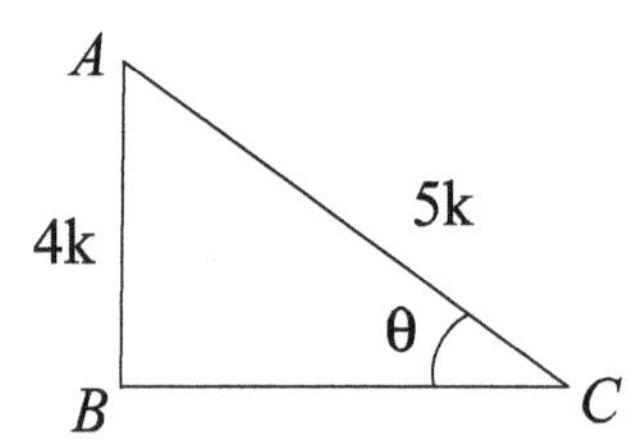

$\therefore\ AB = 4k$

$AC = 5k$

Now,

$AC^2 = AB^2 + BC^2$

$\Rightarrow (5k)^2 = (4k)^2 + BC^2$

$\Rightarrow 25k^2 = 16k^2 + BC^2$

$\Rightarrow BC^2 = 9k^2$

$\therefore\ BC = 3k$

$\cot\theta = \dfrac{BC}{AB}$

$= \dfrac{3k}{4k}$

$= \dfrac{3}{4}$

11. If $\tan\theta = \dfrac{x}{y}$, then $\cos\theta$ is equal to : **(2023)**

(a) $\dfrac{x}{\sqrt{x^2+y^2}}$ (b) $\dfrac{y}{\sqrt{x^2+y^2}}$

(c) $\dfrac{x}{\sqrt{x^2-y^2}}$ (d) $\dfrac{y}{\sqrt{x^2-y^2}}$

Solution :(b)

$\tan\theta = \dfrac{x}{y} = \dfrac{kx}{ky}$

$\therefore AB = kx$

$BC = ky$

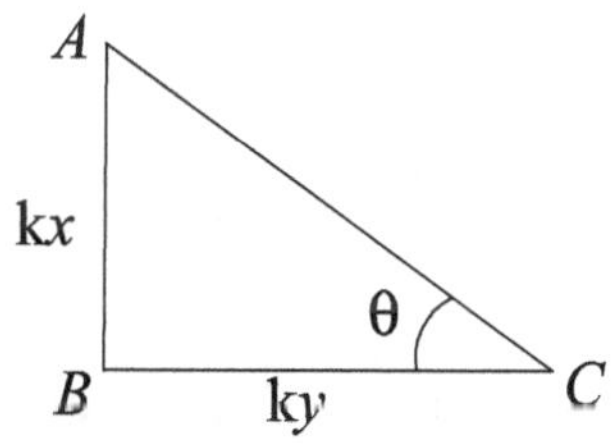

Now,

$AC^2 = AB^2 + BC^2$

$\Rightarrow AC^2 = (kx)^2 + (ky)^2$

$\Rightarrow AC^2 = k^2x^2 + k^2y^2$

$\Rightarrow AC^2 = k^2(x^2 + y^2)$

$\therefore AC = k\sqrt{x^2+y^2}$

$\cos\theta = \dfrac{BC}{AC}$

$= \dfrac{ky}{k\sqrt{x^2+y^2}}$

$= \dfrac{y}{\sqrt{x^2+y^2}}$

12. If $4\sec\theta - 5 = 0$, then the value of $\cot\theta$ is : **(2024)**

(a) $\dfrac{3}{4}$ (b) $\dfrac{4}{5}$

(c) $\dfrac{5}{3}$ (d) $\dfrac{4}{3}$

Solution :(d)

$4\sec\theta - 5 = 0$

$\Rightarrow 4\sec\theta - 5$

$\Rightarrow \sec\theta = \dfrac{5}{4} = \dfrac{5k}{4k}$

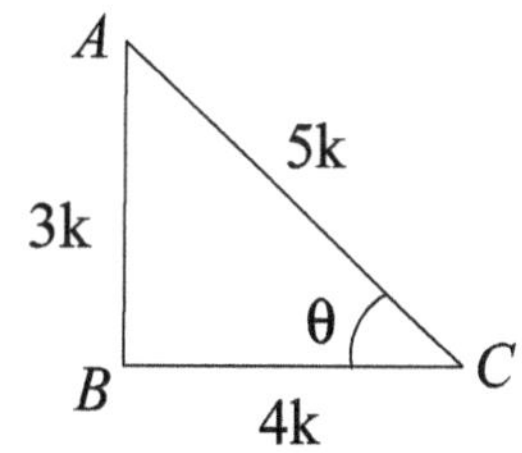

$\therefore BC = 4k$

$AC = 5k$

$AC^2 = AB^2 + BC^2$

$\Rightarrow (5k)^2 = AB^2 + 4k^2$

$\Rightarrow 25k^2 = AB^2 + 16k^2$

$\Rightarrow AB^2 = 9k^2$

$\therefore AB = 3k$

$\cot\theta = \dfrac{BC}{AB}$

$= \dfrac{4k}{3k}$

$= \dfrac{4}{3}$

13. If $5\tan\theta - 12 = 0$, then the value of $\sin\theta$ is : **(2024)**

(a) $\dfrac{5}{12}$ (b) $\dfrac{12}{13}$

(c) $\dfrac{5}{13}$ (d) $\dfrac{12}{5}$

Solution : (b)

$5\tan\theta - 12 = 0$

$\Rightarrow 5\tan\theta - 12 = 0$

$\Rightarrow \tan\theta = \dfrac{12}{5} = \dfrac{12k}{5k}$

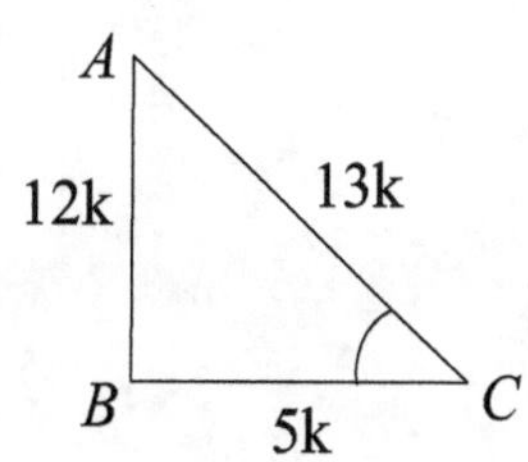

$AC^2 = AB^2 + BC^2$

$\Rightarrow AC^2 = (12k)^2 + (5k)^2$

$\Rightarrow AC^2 = 169k^2$

$\Rightarrow AC^2 = 13k$

$\Rightarrow AC = 13k$

$\therefore \sin\theta = \dfrac{AB}{AC}$

$= \dfrac{12k}{13k}$

$= \dfrac{12}{13}$

14. If $5 \cos A - 4 = 0$, then the value of $\tan A$ is : **(2024)**

(a) $\frac{3}{4}$ (b) $\frac{4}{3}$

(c) $\frac{3}{5}$ (d) $\frac{4}{5}$

Solution :(a)

$5 \cos A - 4 = 0$

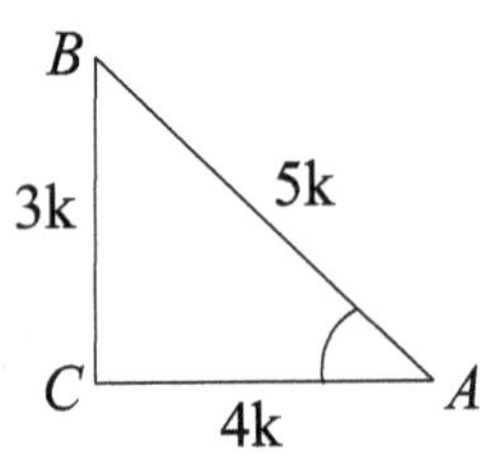

$\Rightarrow 5 \cos A - 4$

$\Rightarrow \cos A = \dfrac{4}{5} = \dfrac{4k}{5k}$

$\therefore AC = 4k$

$AB = 5k$

$AB^2 = AC^2 + BC^2$

$\Rightarrow (5k)^2 = (4k)^2 + BC^2$

$\Rightarrow 25k^2 = 16k^2 + BC^2$

$\Rightarrow BC^2 = 9k^2$

$\therefore BC = 3k$

$\tan A = \dfrac{BC}{AC}$

$= \dfrac{3k}{4k}$

$= \dfrac{3}{4}$

15. If $\cos\theta = \dfrac{p}{q}$, then $\sin\theta$ is equal to : **(2024)**

(a) $\dfrac{q}{\sqrt{q^2 - p^2}}$ (b) $\dfrac{q}{p}$

(c) $\dfrac{\sqrt{q^2 - p^2}}{q}$ (d) $\dfrac{p}{\sqrt{q^2 - p^2}}$

Solution : (c)

$\cos\theta = \dfrac{p}{q} = \dfrac{pk}{qk}$

$\therefore BC = pk$

$AC = qk$

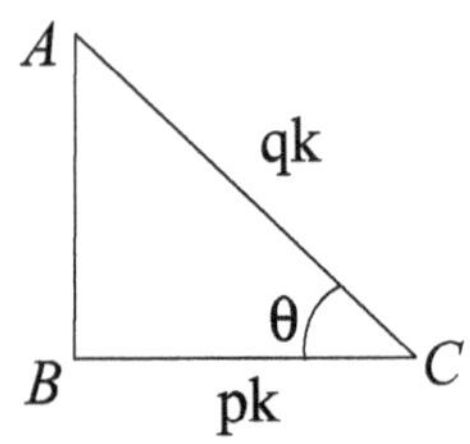

$AC^2 = AB^2 + BC^2$

$\Rightarrow (qk)^2 = AB^2 + (pk)^2$

$\Rightarrow q^2k^2 = AB^2 + p^2k^2$

$\Rightarrow AB^2 = q^2k^2 - p^2k^2$

$\therefore AB = k\sqrt{q^2 - p^2}$

$\sin\theta = \dfrac{AB}{AC}$

$= \dfrac{k\sqrt{q^2 - p^2}}{kq}$

$= \dfrac{\sqrt{q^2 - p^2}}{q}$

16. In a right triangle ABC, right – angled at A, if $\sin B = \dfrac{1}{4}$, then the value of $\sec B$ is **(2025)**

(a) 4 (b) $\dfrac{\sqrt{15}}{4}$

(c) $\sqrt{15}$ (d) $\dfrac{4}{\sqrt{15}}$

Solution: (d)

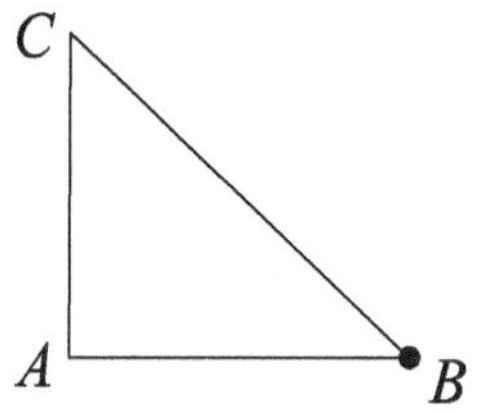

$\sin B = \frac{1}{4}$

Let $AC = k$

$BC = 4k$

$\therefore AB = \sqrt{(4k)^2 - k^2}$

$= \sqrt{15}\,k$

$\therefore \sec B = \frac{BC}{AB}$

$= \frac{4k}{\sqrt{15}\,k}$

$= \frac{4}{\sqrt{15}}$

17. If $\sin A = \frac{2}{3}$, then $\cos A$ is equal to: **(2025)**

(a) $\frac{3}{2}$ (b) $\frac{\sqrt{5}}{3}$

(c) $\frac{1}{3}$ (d) $\frac{1}{\sqrt{3}}$

Solution: (b)

$\sin A = \frac{2}{3}$

$\Rightarrow \sin^2 A = \frac{4}{9}$

$\Rightarrow 1 - \cos^2 A = \frac{4}{9}$ $[\because \sin^2 A + \cos^2 A = 1]$

$\Rightarrow -\cos^2 A = \frac{4}{9} - 1$

$\Rightarrow -\cos^2 A = \frac{-5}{9}$

$\Rightarrow \cos^2 A = \frac{5}{9}$

$\therefore \cos A = \frac{\sqrt{5}}{3}$

18. If $\tan A = \frac{1}{2}$, then $\sin A$ is equal to: **(2025)**

(a) $\frac{2}{\sqrt{5}}$ (b) $\frac{1}{\sqrt{3}}$

(c) $\frac{1}{\sqrt{5}}$ (d) 1

Solution: (c)

$\tan A = \frac{1}{2}$

$\Rightarrow \cot A = 2$

$\Rightarrow \cot^2 A = 4$

$\Rightarrow \text{cosec}^2 A - 1 = 4$

$\Rightarrow \text{cosec}^2 A = 5$

$\Rightarrow \text{cosec } A = \sqrt{5}$

$\therefore \sin A = \frac{1}{\sqrt{5}}$

19. If $\cos A = \frac{4}{5}$, then the value of tan A is: **(2026)**

(a) $\frac{3}{5}$ (b) $\frac{3}{4}$

(c) $\frac{4}{3}$ (d) $\frac{5}{3}$

Solution: (b)

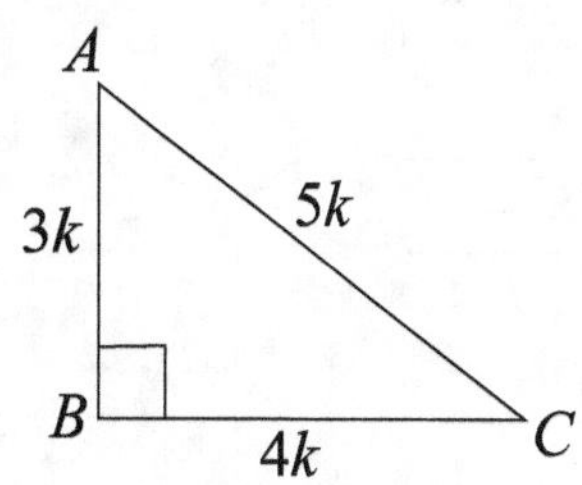

$\cos A = \frac{4}{5}$

$\therefore \tan A = \frac{3k}{4k} = \frac{3}{4}$

20. If $\sin \theta = \frac{1}{\sqrt{11}}$, then $\cot \theta$ equals: **(2026)**

(a) $\frac{\sqrt{11}}{\sqrt{10}}$ (b) $\frac{\sqrt{10}}{\sqrt{11}}$

(c) $\sqrt{10}$ (d) $\sqrt{11}$

Solution : (c)

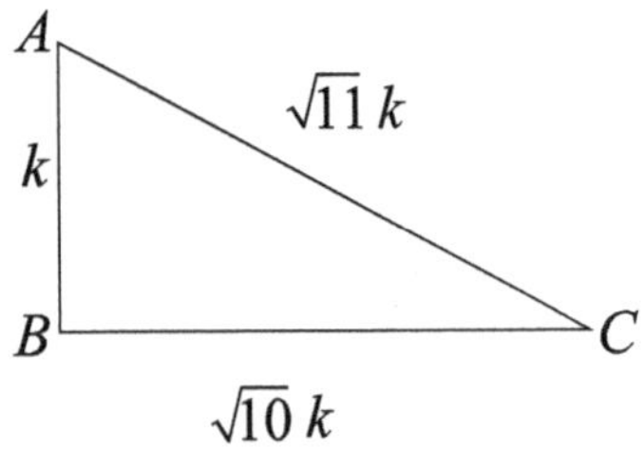

$$\cot\theta = \frac{BC}{AB}$$

$$= \frac{\sqrt{10}k}{k}$$

$$= \sqrt{10}$$

21. Given that $\sin\theta = \frac{a}{b}$, then $\cos\theta$ is equal to: **(2026)**

(a) $\frac{b}{\sqrt{b^2 - a^2}}$ (b) $\frac{b}{a}$

(c) $\frac{\sqrt{b^2 - a^2}}{b}$ (d) $\frac{a}{\sqrt{b^2 - a^2}}$

Solution: (c)

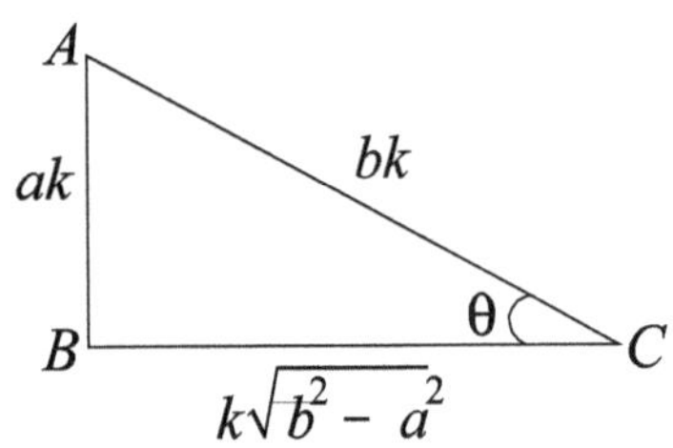

$$\cos\theta = \frac{k\sqrt{b^2 - a^2}}{bk}$$

$$= \frac{\sqrt{b^2 - a^2}}{b}$$

22. If value of $\cot\theta$ is $\sqrt{5}$, then $\sin\theta$ equals: **(2026)**

(a) $\frac{1}{\sqrt{6}}$ (b) $\sqrt{6}$

(c) $\frac{\sqrt{5}}{6}$ (d) $\frac{1}{2}$

Solution: (a)

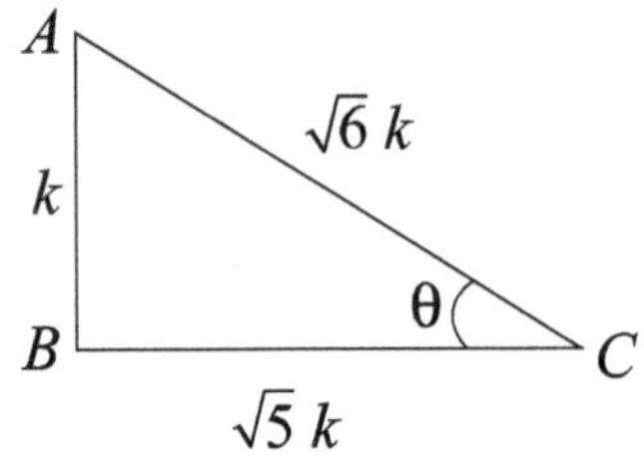

$$\sin\theta = \frac{k}{\sqrt{6}k} = \frac{1}{\sqrt{6}}$$

23. Given $\cot\theta = 3$, the value of $\cos\theta$ is: **(2026)**

(a) $\frac{1}{3}$ (b) $\frac{1}{\sqrt{10}}$

(c) $\frac{3}{\sqrt{10}}$ (d) $\frac{\sqrt{10}}{3}$

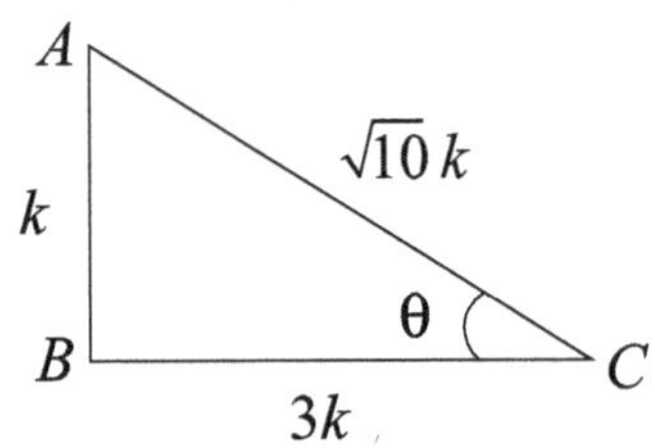

Solution: (c)

$$\cos\theta = \frac{3k}{\sqrt{10}k}$$

$$= \frac{3}{\sqrt{10}}$$

24. If $2\tan A = 3$, then of $\sec A$ equals: **(2026)**

(a) $\sqrt{\frac{13}{2}}$ (b) $\frac{\sqrt{13}}{4}$

(c) $\frac{2}{\sqrt{13}}$ (d) $\frac{\sqrt{13}}{2}$

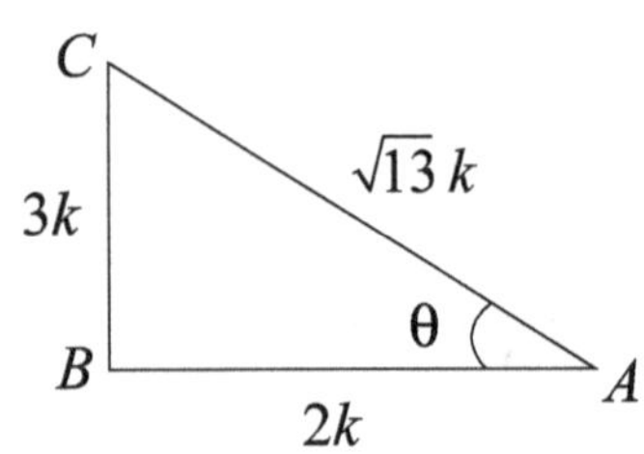

Solution: (d)

$\tan A = \frac{3}{2}$

$\therefore \sec A = \frac{\sqrt{13}k}{2k}$

$= \frac{\sqrt{13}}{2}$

Exercise 8.1

(i) If $\sin A = \frac{4}{5}$, calculate $\cos A$.

(ii) If $\sin A = \frac{5}{13}$, calculate $\sec A$.

(iii) If $\sin A = \frac{3}{5}$, calculate $\tan A$.

(iv) If $\sin A = \frac{2}{3}$, calculate $\cot A$.

(v) If $\tan \alpha = \frac{4}{5}$, find the value of $\sec \alpha$.

(vi) If $\tan \alpha = \frac{4}{7}$, find the value of $\sec \alpha$.

(vii) If $\tan \alpha = \frac{7}{24}$, find the value of $\sec \alpha$.

(viii) If $\sec \theta = \frac{5}{3}$, then find the value of $\cot \theta$

(ix) If $\sec \theta = \frac{5}{4}$, then find the value of $\cot \theta$.

(x) If $\sec \theta = \frac{24}{25}$, then find the value of $\cot \theta$

(xi) If $\text{cosec}\, \theta = \frac{25}{7}$, then find the value of $\cot \theta$.

(xii) If $\text{cosec}\, \theta = \frac{5}{3}$, then find the value of $\cot \theta$.

(xiii) If $\text{cosec}\, \theta = \frac{3}{2}$, then find the value of $\cot \theta$.

Key Point Relation between Trigonometric Ratios:

(i) $\sin \theta = \frac{1}{\cos ec\theta}$

(ii) $\text{cosec}\, \theta = \frac{1}{\sin\theta}$

(iii) $\cos \theta = \frac{1}{\sec\theta}$

(iv) $\sec \theta = \frac{1}{\cos\theta}$

(v) $\tan \theta = \frac{1}{\cot\theta}$

(vi) $\cot \theta = \frac{1}{\tan\theta}$

Trigonometric Identities

(i) $\sin^2 \theta + \cos^2 \theta = 1$

(ii) $\sin^2 \theta = 1 - \cos^2 \theta$

(iii) $\cos^2 \theta = 1 - \sin^2 \theta$

(iv) $\sec^2 - \tan^2 \theta = 1$

(v) $\sec^2 \theta = 1 + \tan^2 \theta$

(vi) $\tan^2 0 = \sec^2 0 - 1$

(vii) $\text{cosec}^2 \theta - \cot^2 \theta = 1$

(viii) $\text{cosec}^2 \theta = 1 + \cot^2 \theta$

(ix) $\cot^2 \theta = \text{cosec}^2 \theta - 1$

25. If $\tan A = \frac{5}{12}$, find the value of $(\sin A + \cos A)$ $\sec A$ **(2008)**

Solution :

$(\sin A + \cos A) \sec A$

$= \sin A. \sec A + \cos A. \sec A$

$= \frac{\sin A}{\cos A} + 1$

$= \tan A + 1$

$= \frac{5}{12} + 1$

$= \frac{17}{12}$

26. If $\sin \theta = \frac{3}{4}$, then $\frac{(\sec^2 \theta - 1) \cos^2 \theta}{\sin \theta}$ equals **(2023)**

Solution :(b)

$\frac{(\sec^2 \theta - 1) \cos^2 \theta}{\sin \theta}$

$= \frac{(\tan^2\theta.\cos^2\theta)}{\sin\theta}$

$= \frac{\sin^2\theta}{\sin\theta}$

$= \sin\theta$

$= \frac{3}{4}$

27. If $\cos\theta = \frac{3}{7}$, then $\frac{\cos\theta}{1-\sin^2\theta}$ is equal to **(2023)**

(a) $\frac{3}{\sqrt{40}}$ (b) $\frac{3}{7}$

(c) $\frac{7}{3}$ (d) $\frac{7}{\sqrt{40}}$

solution :(c)

$\frac{\cos\theta}{1-\sin^2\theta}$

$= \frac{\cos\theta}{\cos^2\theta}$

$= \frac{1}{\cos\theta}$

$= \frac{1}{\frac{3}{7}} = \frac{7}{3}$

28. If $\cot\theta = \frac{7}{8}$, then the value of $\frac{(1+\sin\theta)(1-\sin\theta)}{(1+\cos\theta)(1-\cos\theta)}$ = ... **(2020)**

Solution :

$\frac{(1+\sin\theta)(1-\sin\theta)}{(1+\cos\theta)(1-\cos\theta)}$

$= \frac{1-\sin^2\theta}{1-\cos^2\theta}$

$= \frac{\cos^2\theta}{\sin^2\theta}$

$= \cot^2\theta$

$= \left(\frac{7}{8}\right)^2$

$= \frac{49}{64}$

29. If $\tan A = \frac{2}{5}$, then the value of $\frac{1-\cos^2 A}{1-\sin^2 A}$ is **(2023)**

(a) $\frac{25}{4}$ (b) $\frac{4}{25}$

(c) $\frac{4}{5}$ (d) $\frac{5}{4}$

Solution :

$\frac{1-\cos^2 A}{1-\sin^2 A}$

$= \frac{\sin^2 A}{\cos^2 A}$

$= \tan^2 A$

$= \left(\frac{2}{5}\right)^2$

$= \frac{4}{25}$

30. If $\sin\theta = \cos\theta$ ($0° < \theta < 90°$), then the value of $\sec\theta \,.\, \sin\theta$ is: **(2025)**

(a) $\frac{1}{\sqrt{2}}$ (b) $\sqrt{2}$

(c) 0 (d) 1

Solution: (d)

$\sin\theta = \cos\theta$

$\Rightarrow \tan\theta = 1$

Now, $\sec\theta \,.\, \sin\theta$

$= \frac{1}{\cos\theta}.\sin\theta$

$= \tan\theta$

$= 1$

31. If $\sin\theta + \cos\theta = \sqrt{2}\cos\theta$, ($0 \neq 90°$), then $\tan\theta$ is equal to **(2025)**

(a) $\sqrt{2}+1$ (b) $\sqrt{2}-1$

(c) $-\sqrt{2}$ (d) $\sqrt{2}$

Solution: (b)

$\sin\theta + \cos\theta = \sqrt{2}\cos\theta$

$\Rightarrow \frac{\sin\theta+\cos\theta}{\cos\theta}=\sqrt{2}$

$\Rightarrow \frac{\sin\theta}{\cos\theta}+\frac{\cos\theta}{\cos\theta}=\sqrt{2}$

$\Rightarrow \tan\theta+1=\sqrt{2}$

$\therefore \tan\theta=\sqrt{2}-1$

32. Simplest from of $\frac{\sec A}{\sqrt{\sec^2 A-1}}$ is: **(2026)**

(a) $\sin A$ (b) $\tan A$

(c) $\operatorname{cosec} A$ (d) $\cos A$

Solution: (c)

$\frac{\sec A}{\sqrt{\sec^2 A-1}}$

$=\frac{\sec A}{\tan A}$

$=\sec A \cot A$

$=\operatorname{cosec} A$

33. If $\cos A=\frac{1}{2}$, then the value of $\sin^2 A+2\cos^2 A$ is: **(2026)**

(a) $\frac{3}{2}$ (b) $\frac{5}{4}$

(c) -1 (d) $\frac{1}{2}$

Solution: (b)

$\sin^2 A+2\cos^2 A$

$=\sin^2 A+\cos^2 A+\cot^2 \text{A}$

$=1+\cos^2 A$

$=1+\left(\frac{1}{2}\right)^2$

$=1+\frac{1}{4}$

$=\frac{5}{4}$

34. $\frac{1+\tan^2 A}{1+\cot^2 A}$ equal to: **(2026)**

(a) $\tan^2 A$ (b) -1

(c) $-\tan^2 A$ (d) $\cot^2 A$

Solution: (a)

$\frac{1+\tan^2 A}{1+\cot^2 A}$

$=\frac{\sec^2 A}{\operatorname{cosec}^2 A}$

$=\frac{\sin^2 A}{\cos^2 A}$

$=\tan^2 A$

Exercise 8.2

(i) If $\tan A=\frac{3}{4}$, find the value of $(\sin A+\cos A)$ $\sec A$.

(ii) If $\tan A=\frac{11}{15}$, find the value of $(\sin A+\cos A)$ $\sec A$.

(iii) If $\tan A=\frac{4}{5}$, find the value of $(\sin A+\cos A)$ $\sec A$.

(iv) If $\cos\theta=\frac{4}{5}$, then $\frac{(\operatorname{cosec}^2\theta-1)\sin^2\theta}{\cos\theta}$ equals.

(a) $\frac{4}{5}$ (b) $\frac{5}{4}$

(c) 4 (d) 5

(v) If $\sin\theta=\frac{5}{8}$, then $\frac{(\sec^2\theta-1)\cos^2\theta}{\sin\theta}$ equals.

(a) $\frac{1}{4}$ (b) $\frac{3}{4}$

(c) $\frac{5}{8}$ (d) $\frac{25}{64}$

(vi) If $\sin\theta=\frac{3}{7}$, then $\frac{(\sec^2\theta-1)\cos^2\theta}{\sin\theta}$ equals.

(a) $\frac{3}{4}$ (b) $\frac{7}{4}$

(c) $\frac{3}{7}$ (d) $\frac{4}{7}$

(vii) If $\cos\theta=\frac{3}{5}$, then $\frac{\cos\theta}{1-\sin^2\theta}$ is equal to

(a) $\frac{3}{5}$ (b) $\frac{5}{3}$

(c) $\frac{3}{4}$ (d) $\frac{3}{2}$

(viii) If $\cos\theta = \frac{3}{7}$, then $\frac{\cos\theta}{1-\sin^2\theta}$ is equal to

(a) $\frac{3}{4}$ (b) $\frac{4}{5}$

(c) $\frac{3}{7}$ (d) $\frac{7}{3}$

(ix) If $\sin\theta = \frac{4}{5}$, then $\frac{\sin\theta}{1-\cos^2\theta}$ is equal to

(a) $\frac{5}{4}$ (b) $\frac{4}{5}$

(c) $\frac{3}{4}$ (d) $\frac{3}{7}$

(x) If $\cot\theta = \frac{3}{4}$, then the value of $\frac{(1+\sin\theta)(1-\sin\theta)}{(1+\cos\theta)(1-\cos\theta)} =$ ....

(xi) If $\cot\theta = \frac{3}{5}$, then the value of $\frac{(1+\sin\theta)(1-\sin\theta)}{(1+\cos\theta)(1-\cos\theta)}$ = ...

(xii) If $\cot\theta = \frac{9}{10}$, then the value of $\frac{(1+\sin\theta)(1-\sin\theta)}{(1+\cos\theta)(1-\cos\theta)} =$ ....

(xiii) If $\tan A = \frac{3}{4}$, then the value of $\frac{1-\cos^2 A}{1-\sin^2 A}$ is

(xiv) If $\cot A = \frac{3}{5}$, then the value of $\frac{1-\sin^2 A}{1-\cos^2 A}$ is...........................

(xv) If $\tan A = \frac{2}{3}$, then the value of $\frac{1-\cos^2 A}{1-\sin^2 A}$ is ....

35. If $\sec^2\theta\,(1+\sin\theta)(1-\sin\theta) = k$, then find the value of k. **(2009)**

Solution :

$\sec^2\theta\,(1+\sin\theta)(1-\sin\theta) = k$

$= \sec^2\theta\,(1-\sin^2\theta) = k$

$= \sec^2\theta\cos^2\theta = k$

$\therefore\ k = 1$

Exercise 8.3

(i) If $\operatorname{cosec}^2\theta\,(1+\cos\theta)(1-\cos\theta) = k$, then find the value of k.

(ii) If $\sec^2\theta\,(1+\sin\theta)(1-\sin\theta) = k^2$, then find the value of k.

(iii) If $\tan^2\theta\,(\operatorname{cosec}\theta+1)(\operatorname{cosec}\theta-1) = k$, then find the value of k.

(iv) If $\cot^2\theta\,(\sec\theta+1)(\sec\theta-1) = k$, then find the value of k.

36. If $5\tan\theta = 3$, then what is the value of $\left(\frac{5\sin\theta-3\cos\theta}{4\sin\theta+3\cos\theta}\right)$ **(2020)**

Solution :

$5\tan\theta = 3$

$\Rightarrow \tan\theta = \frac{3}{5}$

$$\frac{5\sin\theta-3\cos\theta}{4\sin\theta+3\cos\theta}$$

$$= \frac{\frac{5\sin\theta-3\cos\theta}{\cos\theta}}{\frac{4\sin\theta+3\cos\theta}{\cos\theta}}$$

$$= \frac{5\frac{\sin\theta}{\cos\theta}-3\frac{\cos\theta}{\cos\theta}}{4\frac{\sin\theta}{\cos\theta}+3\frac{\cos\theta}{\cos\theta}}$$

$$= \frac{5\tan\theta-3}{4\tan\theta+3}$$

$$= \frac{5\times\frac{3}{5}-3}{4\times\frac{5}{5}+3}$$

$= \dfrac{3-3}{\frac{12}{5}+3}$

$= \dfrac{0}{\frac{17}{5}}$

$= 0$

37. If 3 tan θ = 4, then what is the value of $\left(\dfrac{3\sin\theta+2\cos\theta}{3\sin\theta-2\cos\theta}\right)$ **(2020)**

Solution :

$3\tan\theta = 4$

$\Rightarrow \tan\theta = \dfrac{4}{3}$

Now, $\dfrac{3\sin\theta+2\cos\theta}{3\sin\theta-2\cos\theta}$

$= \dfrac{3\tan\theta+2}{3\tan\theta-2}$

[Dividing num. and denom. By cos θ]

$= \dfrac{3\times\frac{4}{3}+2}{3\times\frac{4}{3}-2}$

$= \dfrac{6}{2}$

$= 3$

38. if 2tan A = 3, then what is the value of $\left(\dfrac{4\sin A+3\cos A}{4\sin A-3\cos A}\right)$ is **(2023)**

(a) $\dfrac{7}{\sqrt{13}}$ (b) $\dfrac{1}{\sqrt{13}}$

(c) 3 (d) does not exist

Solution: (c)

$2\tan A = 3$

$\Rightarrow \tan A = \dfrac{3}{2}$

$\dfrac{4\sin A+3\cos A}{4\sin A-3\cos A}$

$= \dfrac{4\tan A+3}{4\tan A-3}$

[Dividing num. and denom. By cos θ]

$= \dfrac{4\times\frac{3}{2}+3}{4\times\frac{3}{2}-3}$

$= \dfrac{6+3}{6-3}$

$= \dfrac{9}{3}$

$= 3$

39. If tan θ = $\dfrac{5}{12}$, then what is the value of $\left(\dfrac{\sin\theta+\cos\theta}{\sin\theta-\cos\theta}\right)$ **(2023)**

(a) $-\dfrac{17}{7}$ (b) $\dfrac{17}{7}$

(c) $\dfrac{17}{13}$ (d) $-\dfrac{7}{13}$

Solution: (a)

$\dfrac{\sin\theta+\cos\theta}{\sin\theta-\cos\theta}$

$= \dfrac{\tan\theta+1}{\tan\theta-1}$

[Dividing num. and denom. by cos θ]

$= \dfrac{\frac{5}{12}+1}{\frac{5}{12}-1}$

$= \dfrac{\frac{17}{12}}{\frac{-7}{12}}$

$= -\dfrac{17}{7}$

40. For an acute angle θ, if $\sin\theta = \frac{1}{9}$, then value of $\frac{9\operatorname{cosec}\theta + 1}{9\operatorname{cosec} 0 - 1}$ is

(a) 0 (b) $\frac{80}{81}$

(c) 1 (d) $\frac{82}{80}$

Solution: (d)

$$\frac{9\operatorname{cosec}\theta + 1}{9\operatorname{cosec}\theta - 1}$$

$$= \frac{9\times\frac{1}{\sin\theta}+1}{9\times\frac{1}{\sin\theta}-1}$$

$$= \frac{9+\sin\theta}{9-\sin\theta}$$

$$= \frac{9+\frac{1}{9}}{9-\frac{1}{9}}$$

$$= \frac{82}{80}$$

Exercise 8.4

(i) If $a\tan\theta = b$, then what is the value of $\left(\frac{a\sin\theta - b\cos\theta}{b\sin\theta + a\cos\theta}\right)$

(ii) If $b\tan\theta = a$, then what is the value of $\left(\frac{a\sin\theta - b\cos\theta}{a\sin\theta + b\cos\theta}\right)$

(iii) If $2\tan\theta = 3$, then what is the value of $\left(\frac{4\sin\theta - 3\cos\theta}{2\sin\theta + 6\cos\theta}\right)$

(iv) If $7\tan\theta = 4$, then $\frac{7\sin\theta - 3\cos\theta}{7\sin\theta + 3\cos\theta}$

41. In figure 1, $PS = 3$ cm, $QS = 4$ cm, $\angle PSQ = 90^0$, $PQ \perp RQ$ and $RQ = 9$ cm, Evaluate tan θ. **(2019)**

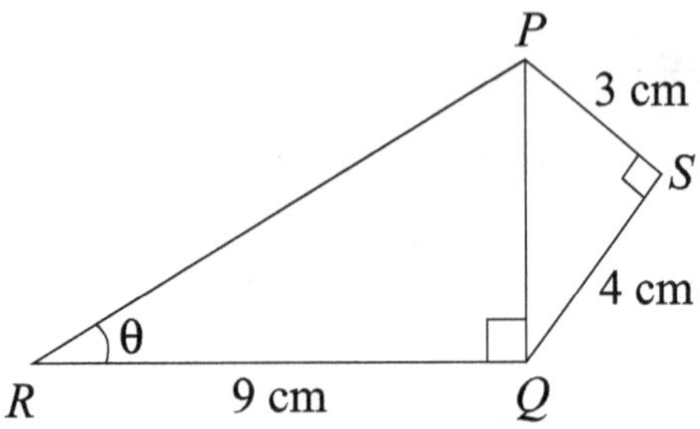

Figure - 1

Solution :

In Δ PSQ, $\angle S = 90°$

∴ By Pythagoras theorem

$PQ^2 = PS^2 + QS^2$

$\Rightarrow PQ^2 = 3^2 + 4^2$

$\Rightarrow PQ^2 = 9 + 16$

$\Rightarrow PQ^2 = 25$

$\therefore PQ = 5$ cm

Now, in Δ PQR, $\angle Q = 90°$

$$\therefore \tan\theta = \frac{PQ}{RQ}$$

$$= \frac{5}{9}$$

Exercise 8.5

(i) In figure 1, $PS = 6$ cm, $QS = 8$ cm, $\angle PSQ = 90°$, $PQ \perp RQ$ and $RQ = 12$ cm, Evaluate tan θ.

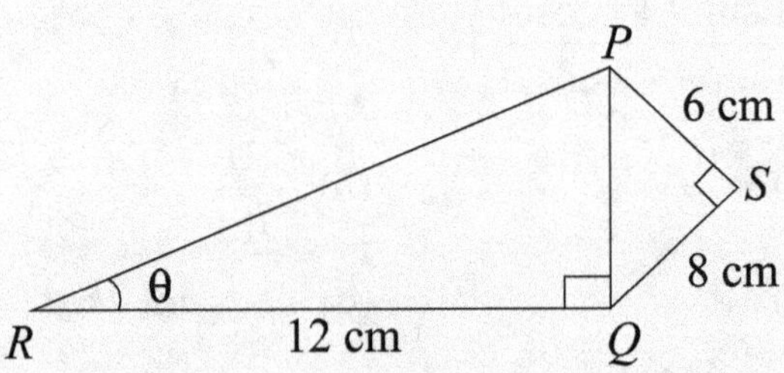

Figure - 1

(ii) In figure 1, $PS = 3$ cm, $QS = 12$ cm, $\angle PSQ = 90°$, $PQ \perp RQ$ and $RQ = 20$ cm, Evaluate tan θ.

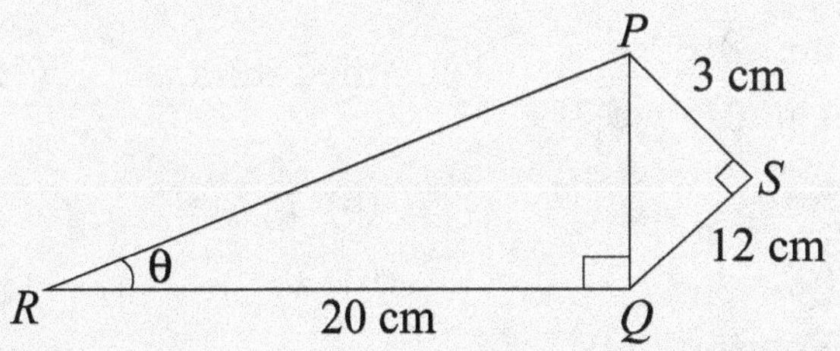

Figure - 1

(iii) In figure 1, $PS = 7$ cm, $QS = 24$ cm, $\angle PSQ = 90°$, $PQ \perp RQ$ and $RQ = 32$ cm, Evaluate $\tan\theta$.

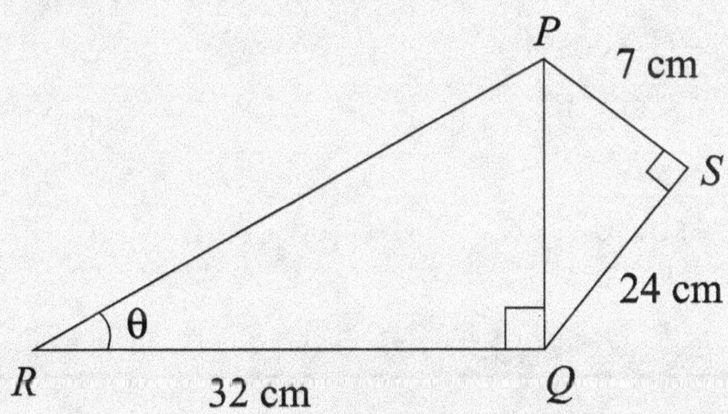

Figure - 1

42. The value of $\left(\sin^2\theta + \frac{1}{1+\tan^2\theta}\right) =$ ______ **(2020)**

Solution :

$\sin^2\theta + \frac{1}{1+\tan^2\theta}$

$= \sin^2\theta + \frac{1}{\sec^2\theta}$

$= \sin^2\theta + \cos^2\theta$

$= 1$

43. The value of $\left(\frac{1}{\sec^2 A} + \frac{1}{\text{cosec}^2 A}\right)$ is: **(2025)**

(a) More than 1 (b) 1

(c) 0 (d) –1

Solution: (b)

$\left(\frac{1}{\sec^2 A} + \frac{1}{\text{cosec}^2 A}\right)$

$= \cos^2 A + \sin^2 A$

$= 1$

44. $(\sin\theta + \cos\theta)^2 + (\sin\theta - \cos\theta)^2 =$ **(2025)**

(a) 1 (b) 2

(c) $2 + 2\sin\theta\cos\theta$ (d) $2 + 4\sin\theta\cos\theta$

Solution:

$(\sin\theta + \cos\theta)^2 + (\sin\theta - \cos\theta)^2$

$= 2(\sin^2\theta + \cos^2\theta)$

$= 2 \times 1$

$= 2$

Exercise 8.6

(i) The value of $\left(\cos^2\theta + \frac{1}{1+\cot^2\theta}\right) =$ ______

(ii) The value of $2\sin^2\theta + \frac{2}{1+\tan^2\theta} =$ ______

45. The value $(1+\tan^2\theta)(1+\sin\theta)(1-\sin\theta) =$ ______ **(2020)**

Solution :

$(1+\tan^2\theta)(1+\sin\theta)(1-\sin\theta)$

$= \sec^2\theta\,(1-\sin^2\theta)$

$= \sec^2\theta.\cos^2\theta$

$= 1$

Exercise 8.7

(i) The value of $(1+\cot^2\theta)(1+\cos\theta)(1-\cos\theta)$ is ______.

(ii) The value of $(1+\tan^2\theta)(1+\sin\theta)(\sin\theta-1)$ is ______

iii. The value of $(\cot^2\theta + 1)(\cos\theta + 1)(\cos\theta - 1)$ is ______

46. What is the value of $\left(\frac{1}{1+\cot^2\theta} + \frac{1}{1+\tan^2\theta}\right)$? **(2020)**

Solution :

$\frac{1}{1+\cot^2\theta} + \frac{1}{1+\tan^2\theta}$

$= \frac{1}{\text{cosec}^2\theta} + \frac{1}{\text{Sec}^2\theta}$

$= \sin^2\theta + \cos^2\theta$

$= 1$

Exercise 8.8

(i) What is the value of $\left(\frac{2}{1+\cot^2\theta}+\frac{2}{1+\tan^2\theta}\right)$

(ii) What is the value of $3\left(\frac{1}{1+\tan^2\theta}+\frac{1}{1+\cot^2\theta}\right)$

47. Simplest from of $\frac{1+\tan^2 A}{1+\cot^2 A}$ is ______ **(2020)**

48. $\frac{1+\tan^2 A}{1+\cot^2 A}$ is equal to **(2024)**

(a) $\sec^2 A$ (b) -1

(c) $\cot^2 A$ (d) $\tan^2 A$

Solution : (d)

$\frac{1+\tan^2 A}{1+\cot^2 A}$

$=\frac{\sec^2 A}{\operatorname{cosec}^2 A}$

$=\frac{\sin^2 A}{\cos^2 A}$

$=\tan^2 A$

Exercise 8.9

(i) Simplest from of $\frac{1+\cot^2 A}{1+\tan^2 A}$ is______

(ii) Simplest from of $\frac{\sec^2 A-1}{\operatorname{cosec}^2 A-1}$ is________

(iii) Simplest from of $\frac{1}{1+\tan^2 A}$ is________

(iv) Simplest from of $\frac{1}{1+\cot^2 A}$ is ________.

49. Simplest form of $(1-\cos^2 A)(1+\cot^2 A)$ is____

Solution :

$(1-\cos^2 A)(1+\cot^2 A)$

$=\sin^2 A.\operatorname{cosec}^2 A$

$=1$

Exercise 8.10

(i) Simplest form of $(1-\sin^2 A)(1+\tan^2 A)$ is-

(ii) Simplest form of $(1-\cos^2 A)(1+\tan^2 A)$ is-

(iii) Simplest form of $(1-\sin^2 A)(1+\cot^2 A)$ is-

50. $\frac{\cos^2\theta}{\sin^2\theta}-\frac{1}{\sin^2\theta}$, in simplified from is **(2023)**

(a) 2 (b) 1

(c) -1 (d) 0

Solution: (c)

$\frac{\cos^2\theta}{\sin^2\theta}-\frac{1}{\sin^2\theta}$

$=\frac{\cos^2\theta-1}{\sin^2\theta}$

$=\frac{-(1-\cos^2\theta)}{\sin^2\theta}$

$=\frac{-\sin^2\theta}{\sin^2\theta}$

$=-1$

Exercise 8.11

(i) $\frac{1}{\sin^2\theta}-\frac{\cos^2\theta}{\sin^2\theta}$ in simplified from is.

(a) 1 (b) -1

(c) 0 (d) 2

(ii) $\frac{1}{\cos^2\theta}-\frac{\sin^2\theta}{\cos^2\theta}$ in simplified from is.

(a) 0 (b) -1

(c) 1 (d) 4

(iii) $\frac{\sin^2\theta}{\cos^2\theta}-\frac{1}{\cos^2\theta}$ in simplified from is.

(a) -1 (b) 0

(c) 1 (d) 2

51. $(\sec\theta+\tan\theta)(1-\sin\theta)$ is equal to **(2024)**

(a) $\sec\theta$ (b) $\sin\theta$

(c) $\operatorname{cosec}\theta$ (d) $\cos\theta$

Solution :(d)

$(\sec\theta+\tan\theta)(1-\sin\theta)$

$=\left(\frac{1}{\cos\theta}+\frac{\sin\theta}{\cos\theta}\right)(1-\sin\theta)$

$= \frac{(1+\sin\theta)}{\cos\theta}(1-\sin\theta)$

$= \frac{1-\sin^2\theta}{\cos\theta}$

$= \frac{\cos^2\theta}{\cos\theta}$

$= \cos\theta$

Exercise 8.12

(i) $(\text{cosec }\theta + \cot\theta)(1-\cos\theta)$ is equal to ______

(a) $\sin\theta$ (b) $\cos\theta$

(c) $\text{cosec }\theta$ (d) $\cot\theta$

(ii) $(\sec\theta - \tan\theta)(1+\sin\theta)$ is equal to______

(a) $\sec\theta$ (b) $\cot\theta$

(c) $\cos\theta$ (d) $\text{cosec }\theta$

(iii) $(\text{cosec }\theta - \cot\theta)(1+\cos\theta)$ is equal to______

(a) $\sin\theta$ (b) $\cos\theta$

(c) $\cot\theta$ (d) $\text{cosec }\theta$

52. $\frac{\text{cosec}^2 A - \cot^2 A}{1-\sin^2 A}$ is equal to : **(2024)**

(a) $\sin^2 A$ (b) $\cos^2 A$

(c) $\sec^2 A$ (d) $\tan^2 A$

Solution : (C)

$\frac{\text{cosec}^2 A - \cot^2 A}{1-\sin^2 A}$

$= \frac{1}{\cos^2 A}$

$= \sec^2 A$

53. The value of $\tan^2\theta - \left(\frac{1}{\cos\theta}\times\sec\theta\right)$ is, **(2025)**

(a) 1 (b) 0

(c) –1 (d) 2

Solution: (c)

$\tan^2\theta - \left(\frac{1}{\cos\theta}\times\sec\theta\right)$

$= \tan^2\theta - \sec\theta \,.\, \sec\theta$

$= \tan^2\theta - \sec^2\theta$

$= -1$

54. The value of $(\tan A \text{ cosec } A)^2 - (\sin A \sec A)^2$ is: **(2025)**

(a) 0 (b) 1

(c) –1 (d) 2

Solution: (b)

$(\tan A \text{ cosec } A)^2 - (\sin A \,.\, \sec A)^2$

$= \left(\frac{\sin A}{\cos A}.\frac{1}{\sin A}\right)^2 - \left(\sin A.\frac{1}{\cos A}\right)^2$

$= \sec^2 A - \tan^2 A$

$= 1$

55. Which the following is a trigonometric identify ? **(2025)**

(a) $\sin^2\theta = 1 + \cos^2\theta$

(b) $\text{cosec}^2\theta + \cot^2\theta = 1$

(c) $\sec^2 = 1 + \tan^2\theta$

(d) $\sin^2\theta = 2\sin\theta$

Solution: (c)

$\sec^2 = 1 + \tan^2\theta$

56. The value of $\left(\tan^2 A - \frac{1}{\cos^2 A}\right)$ is: **(2025)**

(a) More than 1 (b) 1

(c) 0 (d) –1

Solution: (d)

$\tan^2 A - \frac{1}{\cos^2 A}$

$= \tan^2 A - \sec^2 A$

$= -(\sec^2 A - \tan^2 A)$

$= -1$

57. The value of $\left(\cot^2 A - \frac{1}{\sin^2 A}\right)$ is: **(2025)**

(a) More than 1 (b) 1

(c) 0 (d) –1

Solution: (d)

$\left(\cot^2 A - \frac{1}{\sin^2 A}\right)$

$= \cot^2 A - \text{cosec}^2 A$

$= -(\text{cosec}^2 A - \cot^2 A)$

$= -1$

58. The value of $(\sin^2 A + \cos^2 A) + (\sec^2 A - \tan^2 A) - (\cot^2 A - \text{cosec}^2 A)$ is **(2025)**

(a) 1 (b) –1

(c) 3 (d) –2

Solution: (c)

$(\sin^2 A + \cos^2 A) + (\sec^2 A - \tan^2 A) - (\cot^2 A - \text{cosec}^2 A)$

$= 1 + 1 - (-1)$

$= 1 + 1 + 1$

$= 3$

59. $(\sec\theta - \cos\theta)^2 + \sin^2\theta - \tan^2\theta = ?$

(a) 0 (b) 1

(c) 2 (d) 4

Solution: (a)

$(\sec\theta - \cos\theta)^2 + \sin^2\theta - \tan^2\theta$

$= \sec^2\theta - 2\sec\theta\cos\theta + \cos^2\theta + \sin^2\theta - \tan^2\theta$

$= \sec^2\theta - 2\times 1 + \cos^2\theta + \sin^2\theta - \tan^2\theta$

$= (\sec^2\theta - \tan^2\theta) + (\sin^2\theta + \cos^2\theta) - 2$

$= 1 + 1 - 2$

$= 0$

60. If $\sec\theta - \tan\theta = 2$, then $\sec\theta + \tan\theta$ is equal to: **(2025)**

(a) $\frac{1}{2}$ (b) $\sqrt{2}$

(c) $\frac{1}{\sqrt{2}}$ (d) 2

Solution:

$\sec^2\theta - \tan^2\theta = 1$

$\Rightarrow (\sec\theta + \tan\theta)(\sec\theta - \tan\theta) = 1$

$\Rightarrow (\sec\theta + \tan\theta).\ 2 = 1$

$\Rightarrow \sec\theta + \tan\theta = \frac{1}{2}$

Exercise 8.13

(i) $\frac{\sec^2 A - \tan^2 A}{1-\cos^2 A}$ is equal to________

(a) $\sin^2 A$ (b) $\cos^2 A$

(c) $\sec^2 A$ (d) $\text{cosec}^2 A$

(ii) $\frac{\text{cosec}^2 A - \cot^2 A}{1-\cos^2 A}$ is equal to________

(a) $\sin^2 A$ (b) $\sec^2 A$

(c) $\tan^2 A$ (d) $\text{cosec}^2 A$

(iii) $\frac{\sec^2 A - \tan^2 A}{1-\sin^2 A}$ is equal to________

(a) $\sec^2 A$ (b) $\text{cosec}^2 A$

(c) $\tan^2 A$ (d) $\cot^2 A$

61. $\left(\frac{1}{\tan^2\theta} - \frac{1}{\sin^2\theta}\right)$ is equal to **(2024)**

(a) 1 (b) –1

(c) $\sec^2\theta$ (d) $\sin^2\theta$

Solution :(b)

$\frac{1}{\tan^2\theta} - \frac{1}{\sin^2\theta}$

$= \cot^2\theta - \text{cosec}^2\theta$

$= -\left(\text{cosec}^2\theta - \cos^2\theta\right)$

$= -1$

Exercise 8.14

(i) $\left(\frac{1}{\cot^2\theta} - \frac{1}{\cos^2\theta}\right)$ is equal to ________

(a) 1 (b) –1

(c) 0 (d) $\sec^2\theta$

(ii) $\left(\frac{4}{\tan^2\theta} - \frac{4}{\sin^2\theta}\right)$ is equal to ________

(a) 1 (b) 2

(c) –4 (d) 0

(iii) $\left(\frac{3}{\cos^2\theta} - \frac{3}{\cot^2\theta}\right)$ is equal to ________

(a) –3 (b) 3

(c) 0 (d) 1

(iv) $\left(\frac{5}{\cot^2\theta} - \frac{5}{\cos^2\theta}\right)$ is equal to -________

(a) 1 (b) 5

(c) –5 (d) 0

62. If $\sin A + \sin^2 A = 1$, then find the value of the expression $(\cos^2 A + \cos^4 A)$ **(2020)**

Solution :

$\sin A + \sin^2 A = 1$

$\Rightarrow \sin A = 1 - \sin^2 A$

$\Rightarrow \sin A = \cos^2 A$

Now, $\cos^2 A + \cos^4 A$

$= \cos^2 A + (\cos^2 A)^2$

$= \cos^2 A + (\sin A)^2$ $[\because \cos^2 A = \sin A]$

$= \cos^2 A + \sin^2 A$

$= 1$

Exercise 8.15

(i) If $\cos A + \cos^2 A = 1$, then find the value of the expression $(\sin^2 A + \sin^4 A)$ ______is___

63. $8\cot^2 A - 8\,\mathrm{cosec}^2 A$ is equal to : **(2020)**

(a) 8 (b) $\frac{1}{8}$

(c) –8 (d) $-\frac{1}{8}$

Solution :

$8\cot^2 A - 8\,\mathrm{cosec}^2 A$

$= -8(\mathrm{cosec}^2 A - \cot^2 A)$

$= -8 \times 1$

$= -8$

Exercise 8.16

(i) $2\cot^2 A - 2\,\mathrm{cosec}^2 A$ is equal to :

(a) 2 (b) –2

(c) 0 (d) 1

(ii) $a\cot^2 A - a\,\mathrm{cosec}^2 A$ is equal to :

(a) a (b) 0

(c) 1 (d) $-a$

(iii) $\frac{3}{2}\mathrm{cose}^2\theta - \frac{3}{2}\cot^2\theta$ is equal to :

(a) 3 (b) 2

(c) $\frac{2}{3}$ (d) $\frac{3}{2}$

64. $\sec\theta$ when expressed in the terms of $\cot\theta$ is equal to : **(2023)**

(a) $\frac{1+\cot^2\theta}{\cot\theta}$ (b) $\sqrt{1+\cot^2\theta}$

(c) $\frac{\sqrt{1+\cot^2\theta}}{\cot\theta}$ (d) $\frac{\sqrt{1-\cot^2\theta}}{\cot\theta}$

Solution: (c)

$\sec\theta = \sqrt{1+\tan^2\theta}$

$= \sqrt{1+\frac{1}{\cot^2\theta}}$

$= \sqrt{\frac{\cot^2\theta+1}{\cot^2 0}}$

$= \frac{\sqrt{1+\cot^2\theta}}{\cot\theta}$

Exercise 8.17

(i) $\tan\theta$ when expressed in the terms of $\cos\theta$ is equal to :

(a) $\sqrt{1+\cos^2\theta}$ (b) $\frac{\sqrt{\cot^2\theta-1}}{\cos\theta}$

(c) $\frac{\sqrt{1-\cot^2\theta}}{\cos\theta}$ (d) $\frac{\sqrt{1-\sin^2\theta}}{\cos\theta}$

(ii) $\mathrm{cosec}\,\theta$ when expressed in the terms of $\tan\theta$ is equal to :

(a) $\frac{1+\tan^2\theta}{\tan\theta}$ (b) $\frac{\sqrt{\tan^2\theta+1}}{\tan\theta}$

(c) $\frac{\sqrt{1-\tan^2\theta}}{\tan\theta}$ (d) $\frac{\sqrt{\tan^2\theta-1}}{\tan\theta}$

65. $(\sec^2\theta - 1)(\mathrm{cosec}^2\theta - 1)$ is equal to **(2023)**

(a) –1 (b) 1

(c) 0 (d) 2

Solution : (b)

$(\sec^2\theta - 1)(\mathrm{cosec}^2\theta - 1)$

$= \tan^2\theta . \cot^2\theta$

$= 1$

66. If $\sec\theta - \tan\theta = \frac{1}{3}$, then the value of $(\sec\theta + \tan\theta)$ is : **(2023)**

(a) $\frac{4}{3}$ (b) $\frac{2}{3}$

(c) $\frac{1}{3}$ (d) 3

Solution : (d)

$\sec^2\theta - \tan^2\theta = 1$

$\Rightarrow (\sec\theta + \tan\theta)(\sec\theta - \tan\theta) = 1$

$\Rightarrow (\sec\theta + \tan\theta)\ \frac{1}{3} = 1$

$\therefore\ \sec\theta + \tan\theta = 3$

67. If $\sec\theta - \tan\theta = m$, then the value of $(\sec\theta + \tan\theta)$ is : **(2024)**

(a) $1 - \frac{1}{m}$ (b) $m^2 - 1$

(c) $\frac{1}{m}$ (d) $-m$

Solution : (c)

$\sec^2\theta - \tan^2\theta = 1$

$\Rightarrow (\sec\theta + \tan\theta)(\sec\theta - \tan\theta) = 1$

$\Rightarrow (\sec\theta + \tan\theta)\ m = 1$

$\therefore\ \sec\theta + \tan\theta = \frac{1}{m}$

Exercise 8.18

(i) If $\sec\theta - \tan\theta = \frac{1}{4}$, then the value of $(\sec\theta + \tan\theta)$ is :

(a) 1 (b) 4

(c) $\frac{2}{3}$ (d) 0

(ii) If $\sec\theta + \tan\theta = \frac{2}{3}$, then the value of $(\sec\theta - \tan\theta)$ is :

(a) 2 (b) 3

(c) $\frac{3}{2}$ (d) 1

(iii) If $\sec\theta - \tan\theta = \frac{4}{5}$, then the value of $(\sec\theta + \tan\theta)$ is :

(a) 4 (b) 5

(c) $\frac{4}{5}$ (d) $\frac{5}{4}$

68. If θ is an acute angle of a right angled triangle then which of the following equation is not true. **(2023)**

(a) $\sin\theta . \cot\theta = \cos\theta$

(b) $\text{cosec}\,\theta . \tan\theta = \sin\theta$

(c) $\text{cosec}^2\theta - \cot^2\theta = 1$

(d) $\tan^2\theta - \sec^2\theta = 1$

Solution : (d)

L.H.S. $= \tan^2\theta - \sec^2\theta$

$= \frac{\sin^2\theta}{\cos^2\theta} - \frac{1}{\cos^2\theta}$

$= \frac{\sin^2\theta - 1}{\cos^2\theta}$

$= \frac{-(1-\sin^2\theta)}{\cos^2\theta}$

$= \frac{-\cos^2\theta}{\cos^2\theta}$

$= -1$

$\neq$ R.H.S.

Exercise 8.19

(i) If θ is an acute angle of a right angled triangle, then which of the following equation is not true ? **(2023)**

(a) $\cot\theta . \tan\theta = \sin\theta$

(b) $\sin^2\theta + \cos^2\theta = 1$

(c) $\cos^2\theta - \sin^2\theta = 2\cos^2\theta - 1$

(d) $\cot^2\theta - \text{cosec}^2\theta = 1$

(ii) If θ is an acute angle of a right angled triangle, then which of the following equation is not true ?

(a) $2\sin^2\theta\text{-}1 = \cos^2\theta\text{-}\sin^2\theta$

(b) $\cos\theta.\sec\theta = 1$

(c) $\cot\theta.\tan\theta = 1$

(d) $\sec^2\theta - \tan^2\theta = 1$

(iii) If θ is an acute angle of a right angled triangle, then which of the following equation is not true ?

(a) $2\cos^2\theta - 1 = 1 - 2\sin^2\theta$

(b) $\sec^2\theta + \operatorname{cosec}^2\theta = \sec^2\theta.\operatorname{cosec}^2\theta$

(c) $(1 - \sin^2\theta)\sec^2\theta = 2$

(d) $(1 + \tan^2\theta)\cos^2\theta = 1$

69. Which of the following is true for all values of θ $(0° \le \theta \le 90°)$? **(2023)**

(a) $\sin^2\theta - \cos^2\theta = 1$

(b) $\operatorname{cosec}^2\theta - \sec^2\theta = 1$

(c) $\sec^2\theta - \tan^2\theta = 1$

(d) $\cot^2\theta - \tan^2\theta = 1$

Solution : (c)

$\sec^2\theta - \tan^2\theta = 1$

70. $9\sec^2 A - 9\tan^2 A$ is equal to **(2023)**

(a) 9 (b) 0

(c) 8 (d) $\frac{1}{9}$

Solution : (a)

$9\sec^2 A - 9\tan^2 A$

$= 9(\sec^2 A - \tan^2 A)$

$= 9 \times 1$

$= 9$

71. $8(\cos^2 A + \sin^2 A)$ is equal to **(2023)**

(a) 1 (b) 0

(c) 9 (d) 8

Solution : (d)

$8(\cos^2 A + \sin^2 A)$

$= 8 \times 1$

$= 8$

72. $(\cos^2 A - \sin^2 A)$ on simplification, gives **(2023)**

(a) $2\sin^2 A - 1$ (b) $2\sin^2 A + 1$

(c) $2\cos^2 A + 1$ (d) $2\cos^2 A - 1$

Solution : (d)

$\cos^2 A - \sin^2 A$

$= \cos^2 A - (1 - \cos^2 A)$

$= \cos^2 A - 1 + \cos^2 A$

$= 2\cos^2 A - 1$

73. If $3x = \operatorname{cosec}\theta$ and $\frac{3}{x} = \cot\theta$, find the value of $3\left(x^2 - \frac{1}{x^2}\right)$ **(2010)**

Solution :

$3x = \operatorname{cosec}\theta$ ....(i)

$\frac{3}{x} = \cot\theta$ ....(ii)

(i)? – (ii)?

$\Rightarrow (3x)^2 - \left(\frac{3}{x}\right)^2 = \operatorname{cosec}^2\theta - \cot^2\theta$

$\Rightarrow 9x^2 - \frac{9}{x^2} = 1$

$\Rightarrow 9\left(x^2 - \frac{1}{x^2}\right) = 1$

$\therefore 3\left(x^2 - \frac{1}{x^2}\right) = \frac{1}{3}$

Exercise 8.20

(i) If $x = \operatorname{cosec}\theta$ and $\frac{1}{x} = \cot\theta$, then find the value of $2\left(x^2 - \frac{1}{x^2}\right)$.

(ii) If $ax = \operatorname{cosec}\theta$ and $\frac{a}{x} = \cot\theta$, then find the value of $\left(x^2 - \frac{1}{x^2}\right)$.

(iii) If $x = \sec\theta$ and $\frac{1}{x} = \tan\theta$, then find the value of $\left(x^2 - \frac{1}{x^2}\right)$.

(iv) If $2x = \sec\theta$ and $\frac{2}{x} = \tan\theta$, then find the value of $\left(x^2 - \frac{1}{x^2}\right)$.

(v) If $3x = \sec\theta$ and $\frac{3}{x} = \tan\theta$, then find the value of $3\left(x^2 - \frac{1}{x^2}\right)$.

74. If $x = a\sin\theta$ and $y = b\cos\theta$, write the value of $(b^2x^2 + a^2y^2)$: **(2020)**

Solution :

$b^2x^2 + a^2y^2$

$= b^2(a\sin\theta)^2 + a^2(b\cos\theta)^2$

$= a^2b^2\sin^2\theta + a^2b^2\cos^2\theta$

$= a^2b^2(\sin^2\theta + \cos^2\theta)$

$= a^2b^2.\ 1$

$= a^2b^2$

Exercise 8.21

(i) If $x = 2\sin\theta$ and $y = 3\cos\theta$, write the value of $9x^2 + 4y^2$.

(ii) If $x = \sin\theta$ and $y = 2\cos\theta$, write the value of $4x^2 + y^2$.

(iii) If $x = 4\sin\theta$ and $y = 3\cos\theta$, write the value of $9x^3 + 16y^2$.

75. If $\frac{x}{3} = 2\sin A$, $\frac{y}{3} = 2\cos A$, then the value of $x^2 + y^2$ is **(2024)**

(a) 36 (b) 9

(c) 6 (d) 18

Solution: (a)

$\frac{x}{3} = 2\sin A$

$\Rightarrow x = 6\sin A$ ....(i)

$\frac{y}{3} = 2\cos A$

$\Rightarrow y = 6\cos A$ ....(ii)

$(i)^2 + (ii)^2$

$x^2 + y^2 = (6\sin A)^2 + (6\cos A)^2$

$= 36\sin^2 A + 36\cos^2 A$

$= 36(\sin^2 A + \cos^2 A)$

$= 36 \times 1$

$= 36$

76. If $x = 4\sin\theta$, $y = 4\cos\theta$, then the value of $(x^2 + y^2)$ is **(2024)**

(a) 4 (b) $\frac{1}{4}$

(c) $\frac{1}{16}$ (d) 16

Solution: (d)

$x^2 + y^2$

$= (4\sin\theta)^2 + (4\cos\theta)^2$

$\Rightarrow x^2 + y^2 = 16\sin^2\theta + 16\cos^2\theta$

$\Rightarrow x^2 + y^2 = 16(\sin^2\theta + \cos^2\theta)$

$\Rightarrow x^2 + y^2 = 16 \times 1$

$\Rightarrow x^2 + y^2 = 16$

Exercise 8.22

(i) If $x = a\sin\theta$, $y = a\cos\theta$, then the value of $(x^2 + y^2)$ is.

(a) a (b) a^2

(c) $\frac{1}{a^2}$ (d) $\frac{4}{a^2}$

(ii) If $\frac{x}{3} = 2\sin\theta$, $\frac{y}{3} = 2\cos\theta$, then the value of $(x^2 + y^2)$ is.

(a) 10 (b) 8

(c) 36 (d) 100

(iii) If $ax = \sin\theta$, $ay = \cos\theta$, then the value of $(x^2 + y^2)$ is.

(a) $\frac{1}{a}$ (b) $\frac{1}{a^2}$

(c) a (d) $2a$

77. If $x = a\cos\theta$ and $y = b\sin\theta$, write the value of $(b^2x^2 + a^2y^2)$ is **(2024)**

(a) a^2b^2 (b) ab

(c) a^4b^4 (d) $a^2 + b^2$

Solution: (a)

$x = a\cos\theta$

$\Rightarrow \frac{x}{a} = \cos\theta$ ....(i)

$y = b \sin\theta$

$\Rightarrow \frac{y}{b} = \sin\theta$ ....(ii)

$(i)^2 + (ii)^2$

$\Rightarrow \left(\frac{x}{a}\right)^2 + \left(\frac{y}{b}\right)^2 = \cos^2\theta + \sin^2\theta$

$\Rightarrow \frac{x^2}{a^2} + \frac{y^2}{b^2} = 1$

$\Rightarrow \frac{b^2x^2 + a^2y^2}{a^2b^2} = 1$

$\Rightarrow b^2x^2 + a^2y^2 = a^2b^2$

Exercise 8.23

(i) If $x = 2\cos\theta$, and $y = 3\sin\theta$, then the value of $(9x^2 + 4y^2)$ is.

(a) 9 (b) 4

(c) 36 (d) 12

(ii) If $x = 3\cos\theta$, and $y = 4\sin\theta$, then the value of $(16x^2 + 9y^2)$ is.

(a) 144 (b) 12

(c) 100 (d) 1

(iii) If $x = a\sin\theta$, and $y = b\cos\theta$, then the value of $\frac{x^2}{a^2} + \frac{y^2}{b^2}$ is.

(a) a (b) b

(c) 1 (d) ab

78. If $\sin\theta = \cos\theta$ ($0° < \theta < 90°$), then the value of ($\sec\theta.\ \sin\theta$) is : **(2024)**

(a) $\frac{1}{\sqrt{2}}$ (b) $\sqrt{2}$

(c) 1 (d) 0

Solution : (c)

$\sin\theta = \cos\theta$

$\Rightarrow \frac{\sin\theta}{\cos\theta} = 1$

$\Rightarrow \tan\theta = 1$

Now, $\sec\theta.\ \sin\theta$

$= \frac{1}{\cos\theta}.\ \sin\theta$

$= \frac{\sin\theta}{\cos\theta}$

$= \tan\theta$

$= 1$

Exercise 8.24

(i) If $\sin\theta = \cos\theta$ ($0° < \theta < 90°$), then the value of ($\text{cosec}\,\theta.\ \cos\theta$) is :

(a) $\frac{1}{\sqrt{2}}$ (b) $\sqrt{2}$

(c) 1 (d) 2

(ii) If $\sin\theta = \cos\theta$, ($0° < \theta < 90°$), then the value of $\frac{1}{\cos\theta\,\text{cosec}\,\theta}$ is

(a) 1 (b) 0

(c) $\sqrt{2}$ (d) $\sqrt{3}$

(iii) If $\sin\theta = \cos\theta$, ($0° < \theta < 90°$), then the value of $\frac{1}{\sin\theta\sec\theta}$ is

(a) 0 (b) $\sqrt{2}$

(c) 1 (d) $\frac{1}{\sqrt{3}}$

Key Point

$\angle A$	0°	30°	45°	60°	90°
sin A	0	$\frac{1}{2}$	$\frac{1}{\sqrt{2}}$	$\frac{\sqrt{3}}{2}$	1
cos A	1	$\frac{\sqrt{3}}{2}$	$\frac{1}{\sqrt{2}}$	$\frac{1}{2}$	0
tan A	0	$\frac{1}{\sqrt{3}}$	1	$\sqrt{3}$	not defined
cot A	not defined	$\sqrt{3}$	1	$\frac{1}{\sqrt{3}}$	0
sec A	1	$\frac{2}{\sqrt{3}}$	$\sqrt{2}$	2	not defined
cosec A	not defined	2	$\sqrt{2}$	$\frac{2}{\sqrt{3}}$	1

79. Evaluate: $\sin^2 60° + 2\tan 45° - \cos^2 30°$ **(2019)**

Solution:

$\sin^2 60° + 2\tan 45° - \cos^2 30°$

$$= \left(\frac{\sqrt{3}}{2}\right)^2 + 2\times 1 - \left(\frac{\sqrt{3}}{2}\right)^2$$

$$= 2$$

Exercise 8.25

(i) Evaluate: $2\sin^2 60° - \tan 45° - 2\cos^2 30$

(ii) Evaluate: $\frac{3}{2}\cos^2 30° + 2\tan 45° - \frac{3}{2}\sin^2 30°$

(iii) Evaluate: $\sin^2 60° - 3\tan 45° - \cos^2 30°$

80. Fill in the blank: $\frac{2\tan^2 60°}{1+\tan^2 30°}$ is ________ **(2020)**

Solution:

$$\frac{2\tan^2 60°}{1+\tan^2 30°}$$

$$= \frac{2\left(\sqrt{3}\right)^2}{1+\left(\frac{1}{\sqrt{3}}\right)^2}$$

$$= \frac{2.3}{1+\frac{1}{3}}$$

$$= \frac{6}{\frac{4}{3}}$$

$$= 6\times\frac{3}{4}$$

$$= \frac{9}{2}$$

Exercise 8.26

(i) Fill in the blank: value of $\frac{2\tan^2 45°}{1+\tan^2 60°}$ is ________

(ii) Value of $\frac{2\tan^2 30°}{1+\tan^2 60°}$ is ________

(iii) Value of $\frac{2\tan^2 60°}{1-\tan^2 30°}$ is ________

81. Evaluate: $2\sec 30° \times \tan 60°$ **(2020)**

Solution:

$2\sec 30° \times \tan 60°$

$$= 2\times\frac{2}{\sqrt{3}}\times\sqrt{3}$$

$$= 4$$

Exercise 8.27

(i) Evaluate: $2\text{cosec } 60° \times \cot 30°$

(ii) Evaluate: $2\sec 45° \times \cot 45°$

(iii) Evaluate: $2\text{cosec } 60° \times \tan 60°$

82. Write the value of $\sin^2 30° + \cos^2 60°$ **(2020)**

Solution:

$\sin^2 30° + \cos^2 60°$

$$= \left(\frac{1}{2}\right)^2 + \left(\frac{1}{2}\right)^2$$

$$= \frac{1}{4}+\frac{1}{4}$$

$$= \frac{2}{4}$$

$$= \frac{1}{2}$$

Exercise 8.28

(i) Write the value of $\sin^2 45° + \cos^2 45°$

(ii) Write the value of $\sin^2 60° + \cos^2 30°$

(iii) Write the value of $\sin^2 30° + \cos^2 30°$

83. $\left(\frac{2\tan 30°}{1+\tan^2 30°}\right)$ is equal to **(2023)**

(a) $\sin 60°$ (b) $\cos 60°$

(c) $\tan 60°$ (d) $\sin 30°$

Solution: (a)

$$\left(\frac{2\tan 30°}{1+\tan^2 30°}\right)$$

$$= \frac{2 \times \frac{1}{\sqrt{3}}}{1+\left(\frac{1}{\sqrt{3}}\right)^2}$$

$$= \frac{\frac{2}{\sqrt{3}}}{1+\frac{1}{3}}$$

$$= \frac{\frac{2}{\sqrt{3}}}{\frac{4}{3}}$$

$$= \frac{2}{\sqrt{3}} \times \frac{3}{4}$$

$$= \frac{\sqrt{3}}{2}$$

$= \sin 60°$

84. $\frac{2\tan 30°}{1-\tan^2 30°}$ is equal to **(2024)**

(a) $\cos 60°$ (b) $\sin 60°$

(c) $\tan 60°$ (d) $\sin 30°$

Solution: (c)

$$\frac{2\tan 30°}{1-\tan^2 30°}$$

$$= \frac{2 \times \frac{1}{\sqrt{3}}}{1-\left(\frac{1}{\sqrt{3}}\right)^2}$$

$$= \frac{\frac{2}{\sqrt{3}}}{1-\frac{1}{3}}$$

$$= \frac{\frac{2}{\sqrt{3}}}{\frac{2}{3}}$$

$$= \frac{2}{\sqrt{3}} \times \frac{3}{2}$$

$= \sqrt{3}$

$= \tan 60°$

Exercise 8.29

(i) $\frac{2\tan 60°}{1+\tan^2 60°}$ is equal to

(a) $\sqrt{3}$ (b) $\frac{2}{\sqrt{3}}$

(c) 1 (d) $\frac{\sqrt{3}}{2}$

(ii) $\frac{1-\tan^2 30°}{1+\tan^2 30°}$ is equal to **(2023)**

(a) $\sin 60°$ (b) $\cos 60°$

(c) $\tan 60°$ (d) $\cos 30°$

(iii) $\frac{1-\tan^2 60°}{1+\tan^2 60°}$ is equal to

(a) $\frac{1}{2}$ (b) $-\frac{1}{2}$

(c) $\frac{\sqrt{3}}{2}$ (d) $\sqrt{3}$

85. The value of $2\sin^2 30° + 3\tan^2 60° - \cos^2 45°$ is **(2023)**

(a) $3\sqrt{3}$ (b) $\frac{19}{2}$

(c) $\frac{9}{4}$ (d) 9

Solution: (d)

$2\sin^2 30° + 3\tan^2 60° - \cos^2 45°$

$$= 2\left(\frac{1}{2}\right)^2 + 3\left(\sqrt{3}\right)^2 - \left(\frac{1}{\sqrt{2}}\right)^2$$

$$= 2 \times \frac{1}{4} + 3 \times 3 - \frac{1}{2}$$

$$= \frac{1}{2} + 9 - \frac{1}{2}$$

$= 9$

Exercise 8.30

(i) The value of $4\sin^2 30° + \tan^2 60° - 2\cos^2 45°$

(a) 4 (b) 0

(c) 3 (d) 2

(ii) The value of 4 sin² 60° – tan² 30° + 4 cos² 45°

(a) $\frac{14}{3}$ (b) 3

(c) –3 (d) 0

(iii) The value of 2 sin² 30° – tan² 60° + 2 cos² 45°

(a) 0 (b) 2

(c) 3 (d) 1

86. The value of $\frac{\sin 90° + \cos 60°}{\sec 45° + \tan 45°}$ is **(2023)**

(a) 1 (b) $\frac{3}{2}(\sqrt{2}+1)$

(c) $\frac{3}{2}(\sqrt{2}-1)$ (d) $\frac{1+\sqrt{3}}{\sqrt{2}+1}$

Solution: (c)

$$\frac{\sin 90° + \cos 60°}{\sec 45° + \tan 45°}$$

$$= \frac{1+\frac{1}{2}}{\sqrt{2}+1}$$

$$= \frac{\frac{3}{2}}{(\sqrt{2}+1)}$$

$$= \frac{3}{2(\sqrt{2}+1)}$$

$$= \frac{3(\sqrt{2}-1)}{2(\sqrt{2}+1)(\sqrt{2}-1)}$$

$$= \frac{3(\sqrt{2}-1)}{2(2-1)}$$

$$= \frac{3(\sqrt{2}-1)}{2}$$

Exercise 8.31

(i) The value of $\frac{\sin 90° - \cos 60°}{\sec 45° + \tan 45°}$ is

(a) $\sqrt{2}-1$ (b) $\frac{\sqrt{2}-1}{2}$

(c) $\frac{\sqrt{2}+1}{2}$ (d) 1

(ii) The value of $\frac{\sin 90° - \cos 60°}{\sec 45° - \tan 45°}$ is

(a) $\sqrt{2}-1$ (b) $\frac{\sqrt{2}-1}{2}$

(c) $\frac{1}{\sqrt{2}+1}$ (d) $\frac{\sqrt{2}+1}{2}$

(iii) The value of $\frac{\sin 90° + \cos 60°}{\sec 45° - \tan 45°}$ is

(a) $\frac{3(\sqrt{2}-1)}{2}$ (b) $\frac{3(\sqrt{2}+1)}{2}$

(c) $\sqrt{2}$ (d) 1

87. The value of 5 sin² 90° – 2 cos² 0° is **(2023)**

(a) –2 (b) 5

(c) 3 (d) –3

Solution:

5 sin² 90° – 2 cos² 0°

= 5.1² – 2.1²

= 5 – 2

= 3

Exercise 8.32

(i) The value of 5 sin² 90° + 2 cos² 0° is

(a) 5 (b) 2

(c) 3 (d) 7

(ii) The value of 3 sin² 90° – 4 cos² 0° is

(a) 0 (b) 1

(c) –1 (d) 7

(iii) The value of sin² 90° – 3 cos² 0° is

(a) 3 (b) 1

(c) 0 (d) –2

88. $\left(\frac{2}{3}\sin 0° - \frac{4}{5}\cos 0°\right)$ is equal to **(2023)**

(a) $\frac{2}{3}$ (b) $-\frac{4}{5}$

(c) 0 (d) $-\frac{2}{5}$

Solution: (b)

$\frac{2}{3}\sin 0° - \frac{4}{5}\cos 0°$

$= \frac{2}{3} \times 0 - \frac{4}{5} \times 1$

$= -\frac{4}{5}$

Exercise 8.33

(i) $\left(\frac{\sqrt{3}}{4}\sin 0° - \cos 0°\right)$ is equal to

(a) $\frac{\sqrt{3}}{4}$ (b) 0

(c) –1 (d) 1

(ii) $\left(3\sin 0° - \frac{1}{2}\cos 0°\right)$ is equal to

(a) 3 (b) $\frac{1}{2}$

(c) 0 (d) $-\frac{1}{2}$

(iii) (4 sin 0° – 5 cos 0°) is equal to

(a) 4 (b) –1

(c) 1 (d) –5

89. ($3\sin^2 30° - 4\cos^2 60°$) is equal to **(2023)**

(a) $\frac{5}{4}$ (b) $-\frac{3}{4}$

(c) $-\frac{1}{4}$ (d) $-\frac{9}{4}$

Solution: (c)

$3\sin^2 30° - 4\cos^2 60°$

$= 3\times\left(\frac{1}{2}\right)^2 - 4\times\left(\frac{1}{2}\right)^2$

$= 3\times\frac{1}{4} - 4\times\frac{1}{4}$

$= \frac{3-4}{4}$

$= -\frac{1}{4}$

Exercise 8.34

(i) ($\sin^2 30° - \cos^2 60°$) is equal to

(a) 1 (b) –1

(c) 0 (d) 3

(ii) ($5\sin^2 30° + 3\cos^2 60°$) is equal to

(a) 5 (b) 3

(c) 2 (d) $\frac{1}{2}$

(iii) ($9\sin^2 30° - \cos^2 60°$) is equal to

(a) 1 (b) 2

(c) 3 (d) 4

90. Evaluate: $\frac{2\tan 45° \times \cos 60°}{\sin 30°}$ **(2020)**

Solution:

$\frac{2\tan 45° \times \cos 60°}{\sin 30°}$

$= \frac{2\times 1\times\frac{1}{2}}{\frac{1}{2}}$

$= 2$

Exercise 8.35

(i) Evaluate: $\frac{4\tan 45° \times \cos 60°}{\sin 30°}$

(ii) Evaluate: $\frac{\tan 45° \times \cos 60°}{\sin 30°}$

(iii) Evaluate: $\frac{2\tan 45° \times \cos 60°}{2\sin 30°}$

91. The value of $\tan^2 60° + \sin^2 45°$ is ________ **(2020)**

Solution:

$\tan^2 60° + \sin^2 45°$

$= (\sqrt{3})^2 + \left(\frac{1}{\sqrt{2}}\right)^2$

$= 3 + \frac{1}{2}$

$= 3\frac{1}{2}$

Exercise 8.36

(i) The value of $2 \tan^2 60° + 4\sin^2 45°$ is

(ii) The value of $3 \tan^2 60° - 2\sin^2 45°$ is

(iii) The value of $\tan^2 60° - 6\sin^2 45°$ is

92. The value of $\cos^2 45° + \cot^2 45°$ is ________ **(2020)**

Solution:

$\cos^2 45° + \cot^2 45°$

$= \left(\frac{1}{\sqrt{2}}\right)^2 + 1^2$

$= \frac{1}{2} + 1$

$= \frac{3}{2}$

Exercise 8.37

(i) The value of $2 \cos^2 45° + \cot^2 45°$ is ________

(ii) The value of $4 \cos^2 45° + \cot^2 45°$ is ________

(iii) The value of $3\cos^2 45° - \frac{1}{2} \cot^2 45°$ is ________

93. $\left[\frac{3}{4}\tan^2 30° - \sec^2 45° + \sin^2 60°\right]$ is equal to **(2023)**

(a) -1 (b) $\frac{5}{6}$

(c) $-\frac{3}{2}$ (d) $\frac{1}{6}$

Solution:

$\left[\frac{3}{4}\tan^2 30° - \sec^2 45° + \sin^2 60°\right]$

$= \frac{3}{4}\left(\frac{1}{\sqrt{3}}\right)^2 - (\sqrt{2})^2 + \left(\frac{\sqrt{3}}{2}\right)^2$

$= \frac{3}{4} \times \frac{1}{3} - 2 + \frac{3}{4}$

$= \frac{1}{4} + \frac{3}{4} - 2$

$= 1 - 2$

$= -1$

94. The value of $\left(\frac{1}{2}\tan^2 45° - \cos^2 60°\right)$ is **(2026)**

(a) 0 (b) $-\frac{1}{2}$

(c) $\frac{1}{4}$ (d) $-\frac{1}{4}$

Solution: (c)

$\frac{1}{2}\tan^2 45° - \cos^2 60°$

$= \frac{1}{2} \times (1)^2 - \left(\frac{1}{2}\right)^2$

$= \frac{1}{2} - \frac{1}{4}$

$= \frac{1}{4}$

95. The value of $\left(\frac{1}{3}\cot^2 30° - \frac{1}{2}\sec^2 60°\right)$ is: **(2026)**

(a) -1 (b) -2

(c) $\frac{5}{8}$ (d) $\frac{7}{8}$

Solution: (a)

$\frac{1}{3}\cot^2 30° - \frac{1}{2}\sec^2 60°$

$= \frac{1}{3}(\sqrt{3})^2 - \frac{1}{2}.(2)^2$

$= \frac{1}{3} \times 3 - \frac{1}{2} \times 4$

$= 1 - 2$

$= -1$

96. The value of $\frac{\tan^2 60°-1}{\tan^2 60°+1}$ is equal to the value of which of the following? **(2026)**

(a) tan 60° (b) cos 60°

(c) sec 60° (d) sin 60°

Solution: (b)

$\frac{\tan^2 60°-1}{\tan^2 60°+1}$

$=\frac{(\sqrt{3})^2-1}{(\sqrt{3})^2+1}$

$=\frac{3-1}{3+1}$

$=\frac{2}{4}$

$=\frac{1}{2}$

$= \cos 60°$

97. The value of 2 sin 45° cos 45° is same as the value of: **(2026)**

(a) sin 0° (b) cos 0°

(c) tan 90° (d) cot 90°

Solution: (b)

2 sin 45° cos 45°

$=2\times\frac{1}{\sqrt{2}}\times\frac{1}{\sqrt{2}}$

$=2\times\frac{1}{2}$

$=1$

$=\cos 0°$

98. The value of 3 cos 30° – 4 cos³ 30° is same as the value of: **(2026)**

(a) sin 90° (b) cos 90°

(c) tan 90° (d) sec 90°

Solution: (b)

3 cos 30° – 4 cos³ 30°

$=3\times\frac{\sqrt{3}}{2}-4\times\left(\frac{\sqrt{3}}{2}\right)^3$

$=\frac{3\sqrt{3}}{2}-4\times\frac{3\sqrt{3}}{8}$

$=\frac{3\sqrt{3}}{2}-\frac{3\sqrt{3}}{2}$

$=0$

$=\cos 90°$

Exercise 8.38

(i) $\left[\frac{3}{4}\tan^2 30°-3\sec^2 45°+\sin^2 60°\right]$ is equal to.

(a) 2 (b) –2

(c) 0 (d) 1

(ii) $\left[\frac{3}{4}\tan^2 30°-3\sec^2 45°+\sin^2 60°\right]$ is equal to

(a) –2 (b) 0

(c) 1 (d) 4

(iii) [3 tan² 30° – 4 sec² 45° + 4 sin² 60°] is equal to

(a) 1 (b) –1

(c) 3 (d) 0

99 $\left[\frac{5}{8}\sec^2 60°-\tan^2 60°+\cos^2 45°\right]$ is equal to **(2023)**

(a) $-\frac{5}{3}$ (b) $-\frac{1}{2}$

(c) 0 (d) $-\frac{1}{4}$

Solution: (c)

$\frac{5}{8}\sec^2 60°-\tan^2 60°+\cos^2 45°$

$=\frac{5}{8}\times 2^2-(\sqrt{3})^2+\left(\frac{1}{\sqrt{2}}\right)^2$

$=\frac{5}{8}\times 4-3+\frac{1}{2}$

$=\frac{5}{2}-3+\frac{1}{2}$

$= \frac{5}{2} + \frac{1}{2} - 3$

$= 3 - 3$

$= 0$

Exercise 8.39

(i) $\left[\frac{3}{8}\sec^2 60° - 2\tan^2 60° + \cos^2 45°\right]$ is equal to

(a) 0 (b) 2

(c) –2 (d) –4

(ii) $\left[\frac{7}{8}\sec^2 60° - 3\tan^2 60° + 4\cos^2 45°\right]$ is equal to

(a) $-\frac{7}{2}$ (b) 5

(c) –5 (d) 10

(iii) $\left[\frac{3}{2}\sec^2 60° - \tan^2 60° + 2\cos^2 45°\right]$ is equal to

(a) 4 (b) 7

(c) 0 (d) 2

100. If $\sin x + \cos y = 1$; $x = 30°$ and y is an acute angle, find the value of y. **(2019)**

Solution:

$\sin x + \cos y = 1$

$\Rightarrow \sin 30° + \cos y = 1$

$\Rightarrow \frac{1}{2} + \cos y = 1$

$\Rightarrow \cos y = \frac{1}{2}$

$\Rightarrow \cos y = \cos 60°$

$\therefore y = 60°$

101. If θ is an acute angle and $7 + 4\sin\theta = 9$, then the value of θ is **(2025)**

(a) 90° (b) 30°

(c) 45° (d) 60°

Solution:

$7 + 4\sin\theta = 9$

$\Rightarrow 4\sin\theta = 2$

$\Rightarrow \sin\theta = \frac{1}{2}$

$\Rightarrow \sin\theta = \sin 30°$

$\therefore \theta = 30°$

102. If $\alpha + \beta = 90°$ and $\alpha = 2\beta$, then $\cos^2\alpha + \sin^2\beta$ is equal to: **(2025)**

(a) 0 (b) $\frac{1}{2}$

(c) 1 (d) 2

Solution: (b)

$\alpha + \beta = 90°$

$\Rightarrow 2\beta + \beta = 90°$

$\Rightarrow 3\beta = 90°$

$\Rightarrow \beta = 30°$

$\therefore \alpha = 2\beta = 2 \times 30 = 60°$

$\therefore \cos^2\alpha + \sin^2\beta$

$= \cos^2 60° + \sin^2 30°$

$= \left(\frac{1}{2}\right)^2 + \left(\frac{1}{2}\right)^2$

$= \frac{1}{4} + \frac{1}{4} = \frac{1}{2}$

103. If $7\cos^2\theta + 3\sin^2\theta = 4$, then the value of θ is: **(2025)**

(a) 30° (b) 45°

(c) 60° (d) 90°

Solution: (c)

$7\cos^2\theta + 3\sin^2\theta = 4$

$\Rightarrow 7(1 - \sin^2\theta) + 3\sin^2\theta = 4$

$\Rightarrow 7 - 7\sin^2\theta + 3\sin^2\theta = 4$

$\Rightarrow -4\sin^2\theta = -3$

$\Rightarrow \sin^2\theta = \frac{3}{4}$

$\Rightarrow \sin\theta = \frac{\sqrt{3}}{2}$

$\Rightarrow \sin\theta = \sin 60°$

$\therefore \theta = 60°$

104. If $\sin 30° \tan 45° = \frac{\sec 60°}{k}$, then the value of k is: **(2025)**

(a) 4 (b) 3
(c) 2 (d) 1

Solution: (a)

$\sin 30° \tan 45° = \frac{\sec 60°}{k}$

$\Rightarrow \frac{1}{2} \times 1 = \frac{2}{k}$

$\Rightarrow k = 4$

105. tan 2*A* = 3 tan *A* is true, when the measure of ∠*A* is **(2025)**

(a) 90° (b) 60°
(c) 45° (d) 30°

Solution: (d)

LHS = tan 2*A*

$= \tan (2 \times 30°)$

$= \tan 60°$

$= \sqrt{3}$

RHS = 3 tan *A*

$= 3 \tan 30°$

$= 3 \times \frac{1}{\sqrt{3}} = \sqrt{3}$

106. Sec *A* = 2 cos *A* is true for *A* = **(2025)**

(a) 0° (b) 30°
(c) 45° (d) 60°

Solution: (c)

$\sec A = 2\cos A$

$\Rightarrow \frac{1}{\cos A} = 2\cos A$

$\Rightarrow 2\cos^2 A = 1$

$\Rightarrow \cos^2 A = \frac{1}{2}$

$\Rightarrow \cos A = \frac{1}{\sqrt{2}}$

$\Rightarrow \cos A = \cos 45°$

$\therefore A = 45°$

107. $\frac{1-\tan^2 30°}{1+\tan^2 30°}$ is equal to **(2025)**

(a) sin 60° (b) cos 60°
(c) tan 60° (d) sec 60°

Solution: (b)

$\frac{1-\tan^2 30°}{1+\tan^2 60°}$

$= \frac{1-\left(\frac{1}{\sqrt{3}}\right)^2}{1+\left(\frac{1}{\sqrt{3}}\right)^2}$

$= \frac{1-\frac{1}{3}}{1+\frac{1}{3}}$

$= \frac{\frac{2}{3}}{\frac{4}{3}}$

$= \frac{1}{2}$

$= \cos 60°$

108. The value of $\frac{2\tan 60°}{1-\tan^2 60°}$ is same as the value of **(2025)**

(a) –tan 30° (b) –tan 60°
(c) 2 sin 60° (d) 2 cos 60°

Solution: (b)

$\frac{2\tan 60°}{1-\tan^2 60°}$

$= \frac{2\sqrt{3}}{1-\left(\sqrt{3}\right)^2}$

$= \frac{2\sqrt{3}}{1-3}$

$= \frac{2\sqrt{3}}{-2}$

$= -\sqrt{3}$

$= -\tan 60°$

109. The value of $(1 - 2\sin^2 60°)$ is same as that of **(2025)**

(a) $\sin 30°$ (b) $-\sin 30°$
(c) $\cos 60°$ (d) $-\cos 30°$

Solution: (b)

$1 - 2\sin^2 60°$

$$= 1 - 2 \times \left(\frac{\sqrt{3}}{2}\right)^2$$

$$= 1 - 2 \times \frac{3}{4}$$

$$= 1 - \frac{3}{2}$$

$$= -\frac{1}{2}$$

$= -\sin 30°$

110. If $x\left(\frac{2\tan 30°}{1+\tan^2 30°}\right) = y\left(\frac{2\tan 30°}{1-\tan^2 30°}\right)$, then $x : y =$ **(2025)**

(a) 1 : 1 (b) 1 : 2
(c) 2 : 1 (d) 4 : 1

Solution:

$$x\left(\frac{2\tan 30°}{1+\tan^2 30°}\right) = y\left(\frac{2\tan 30°}{1-\tan^2 30°}\right)$$

$$\Rightarrow x.\left\{\frac{2.\frac{1}{\sqrt{3}}}{1+\left(\frac{1}{\sqrt{3}}\right)^2}\right\} = y\left\{\frac{2.\frac{1}{\sqrt{3}}}{1-\left(\frac{1}{\sqrt{3}}\right)^2}\right\}$$

$$\Rightarrow x.\left(\frac{\frac{2}{\sqrt{3}}}{1+\frac{1}{3}}\right) = y\left(\frac{\frac{2}{\sqrt{3}}}{1-\frac{1}{3}}\right)$$

$$\Rightarrow x.\left(\frac{2}{\sqrt{3}} \times \frac{3}{4}\right) = y\left(\frac{2}{\sqrt{3}} \times \frac{3}{2}\right)$$

$$\Rightarrow x.\frac{\sqrt{3}}{2} = y\sqrt{3}$$

$$\Rightarrow \frac{x}{y} = 2$$

$\therefore x : y = 2 : 1$

111. If $x = \cos 30° - \sin 30°$ and $y = \tan 60° - \cot 60°$, then **(2025)**

(a) $x = y$ (b) $x > y$
(c) $x < y$ (d) $x > 1, y < 2$

Solution: (c)

$x = \cos 30° - \sin 30°$

$$= \frac{\sqrt{3}}{2} - \frac{1}{2}$$

$$= \frac{1}{2}\left(\sqrt{3} - 1\right) < 1$$

$y = \tan 60° - \cot 60°$

$$= \sqrt{3} - \frac{1}{\sqrt{3}}$$

$$= \frac{3-1}{\sqrt{3}}$$

$$= \frac{2}{\sqrt{3}} > 1$$

$\therefore x < y$

112. If $x = 2\sin 60° \cos 60°$ and $y = \sin^2 30° - \cos^2 30°$ and $x^2 = ky^2$, the value of k is **(2025)**

(a) $\sqrt{3}$ (b) $-\sqrt{3}$
(c) 3 (d) -3

Solution: (c)

$x = 2\sin 60° \cos 60°$

$$= 2 \times \frac{\sqrt{3}}{2} \times \frac{1}{2}$$

$$= \frac{\sqrt{3}}{2}$$

$y = \sin^2 30° - \cos^2 30°$

$$= \left(\frac{1}{2}\right)^2 - \left(\frac{\sqrt{3}}{2}\right)^2$$

$$= \frac{1}{4} - \frac{3}{4}$$

$= \frac{-2}{4}$

$= -\frac{1}{2}$

Now, $x^2 = ky^2$

$\Rightarrow \left(\frac{\sqrt{3}}{2}\right)^2 = k.\left(-\frac{1}{2}\right)^2$

$\Rightarrow \frac{3}{4} = k \times \frac{1}{4}$

$\Rightarrow k = 3$

113. If $\sin\theta - \cos\theta = 0$, then the value of $\sin^6\theta + \cos^6\theta$ is **(2025)**

(a) 1 (b) $\frac{1}{8}$

(c) $\frac{3}{4}$ (d) $\frac{1}{4}$

Solution: (d)

$\sin\theta - \cos\theta = 0$

$\Rightarrow \sin\theta = \cos\theta$

$\Rightarrow \tan\theta = 1$

$\Rightarrow \tan\theta = \tan 45°$

$\therefore \theta = 45°$

Now, $\sin^6\theta + \cos^6\theta$

$= (\sin 45°)^6 + (\cos 45°)^6$

$= \left(\frac{1}{\sqrt{2}}\right)^6 + \left(\frac{1}{\sqrt{2}}\right)^6$

$= \frac{1}{8} + \frac{1}{8}$

$= \frac{2}{8}$

$= \frac{1}{4}$

114. Which of the following statements is false ? **(2025)**

(a) $\tan 45° = \cot 45°$

(b) $\sin 90° = \tan 45°$

(c) $\sin 30° = \cos 30°$

(d) $\sin 45° = \cos 45°$

Solution: (c)

$\sin 30° = \frac{1}{2}$

$\cos 30° = \frac{\sqrt{3}}{2}$

115. The value of $\frac{2\tan 60°}{1-\tan^2 60°}$ is: **(2025)**

(a) -3 (b) $\sqrt{3}$

(c) $-\frac{1}{\sqrt{3}}$ (d) $-\sqrt{3}$

Solution: (d)

$\frac{2\tan 60°}{1-\tan 60°}$

$= \frac{2\times\sqrt{3}}{1-\left(\sqrt{3}\right)^2}$

$= \frac{2\sqrt{3}}{1-3}$

$= \frac{2\sqrt{3}}{-2}$

$= -\sqrt{3}$

116. If $\sqrt{2}\sin\theta = 1$, then $\cot\theta \times \operatorname{cosec}\theta$ is equal to **(2025)**

(a) $\frac{1}{\sqrt{2}}$ (b) $\frac{1}{2\sqrt{2}}$

(c) $\sqrt{2}$ (d) $\frac{1}{2}$

Solution: (c)

$\sqrt{2}\sin\theta = 1$

$\Rightarrow \sin\theta = \frac{1}{\sqrt{2}}$

$\Rightarrow \sin\theta = \sin 45°$

$\therefore \theta = 45°$

Now, $\cot\theta \times \operatorname{cosec}\theta$

$= \cot 45° \times \text{cosec } 45°$

$= 1 \times \sqrt{2}$

$= \sqrt{2}$

117. If $\tan A = 1$, then $3 \sin A + \cos A$ is equal to: **(2025)**

(a) $4\sqrt{2}$ (b) 4

(c) $2\sqrt{2}$ (d) $4 \times 45°$

Solution: (c)

$\tan A = 1$

$\Rightarrow \tan A = \tan 45°$

$\therefore A = 45°$

$3 \sin A + \cos A$

$= 3 \sin 45° + \cos 45°$

$= 3 \times \frac{1}{\sqrt{2}} + \frac{1}{\sqrt{2}}$

$= \frac{4}{\sqrt{2}}$

$= 2\sqrt{2}$

118. If $\sin A = \cos A$, then $\frac{1-\tan A}{1+\tan A}$ is equal to **(2025)**

(a) 1 (b) –1

(c) 0 (d) Not defined

Solution: (c)

$\sin A = \cos A$

$\Rightarrow \frac{\sin A}{\cos A} = 1$

$\Rightarrow \tan A = \tan 45°$

$\therefore A = 45°$

Now, $\frac{1-\tan A}{1+\tan A}$

$= \frac{1-\tan 45°}{1+\tan 45°}$

$= \frac{1-1}{1+1} = 0$

119. If $2 \sin A = 1$, then the value of $\tan A + \cot A$ is: **(2026)**

(a) $\sqrt{3}$ (b) $\frac{4}{\sqrt{3}}$

(c) $\frac{\sqrt{3}}{2}$ (d) 1

Solution: (b)

$2 \sin A = 1$

$\Rightarrow \sin A = \frac{1}{2}$

$\Rightarrow \sin A = \sin 30°$

$\therefore A = 30°$

Now, $\tan A + \cot A$

$= \tan 30° + \cot 30°$

$= \frac{1}{\sqrt{3}} + \sqrt{3}$

$= \frac{1+3}{\sqrt{3}}$

$= \frac{4}{\sqrt{3}}$

120. If the length of the shadow of a tower is $\sqrt{3}$ times that of its height then attitude of the sum is: **(2026)**

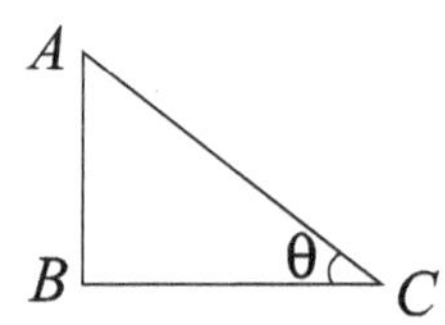

(a) 45° (b) 30°

(c) 60° (d) 15°

Solution :figure

$\tan \theta = \frac{AB}{BC}$

$\Rightarrow \tan \theta = \frac{AB}{\sqrt{3}AB}$

$\Rightarrow \tan \theta = \frac{1}{\sqrt{3}}$

$\Rightarrow \tan \theta = \tan 30°$

$\therefore \theta = 30°$

121. Given that $\sin 2\alpha = \frac{\sqrt{3}}{2}$, the value of $\sin 3\alpha$ is: **(2026)**

(a) $\frac{3\sqrt{3}}{4}$ (b) $\frac{1}{2}$

(c) 1 (d) $\frac{\sqrt{3}}{4}$

Solution: (c)

$\sin 2\alpha = \frac{\sqrt{3}}{2}$

$\Rightarrow \sin 2\alpha = \sin 60°$

$\therefore 2\alpha = 60°$

$\therefore \alpha = 30°$

$\therefore \sin 3\alpha$

$= \sin(3\times 30°)$

$= \sin 90°$

$= 1$

Exercise 8.40

(i) If $\sin x + \cos y = 1$, $y = 60°$ and x is an acute angle, find the value of x.

(ii) If $\sin x + \cos y = \sqrt{3}$, $x = 60°$ and y is an acute angle, find the value of y.

(iii) If $\sin x + \cos y = \sqrt{2}$, $x = 45°$ and y is an acute angle, find the value of y.

122. Evaluate: $(\sec A + \tan A)(1 - \sin A)$ of $A = 60°$ **(2020)**

Solution:

$(\sec A + \tan A)(1 - \sin A)$

$= (\sec 60° + \tan 60°)(1 - \sin 60°)$

$= \left(2+\sqrt{3}\right)\left(1-\frac{\sqrt{3}}{2}\right)$

$= \left(2+\sqrt{3}\right)\left(\frac{2-\sqrt{3}}{2}\right)$

$= \frac{1}{2}\left(2^2-3\right)$

$= \frac{1}{2}\times 1$

$= \frac{1}{2}$

Exercise 8.41

(i) Evaluate: $(\sec A + \tan A)(1 - \sin A)$ of $A = 45°$

(ii) Evaluate: $(\sec A + \tan A)(1 - \sin A)$ of $A = 30°$

(iii) Evaluate: $(\sec A - \tan A)(1 + \sin A)$ of $A = 60°$

123. If $\tan A = 1$, then $2 \sin A \cos A =$ ________ **(2020)**

Solution:

$\tan A = 1$

$\Rightarrow \tan A = \tan 45°$

$\therefore A = 45°$

Now, $2 \sin A \cos A$

$= 2 \sin 45° \cos 45°$

$= 2.\frac{1}{\sqrt{2}}.\frac{1}{\sqrt{2}}$

$= 1$

Exercise 8.42

(i) If $\tan A = \frac{1}{\sqrt{3}}$, then $2 \sin A \cos A =$ ________.

(ii) If $\tan A = \sqrt{3}$, then $2 \sin A \cos A =$ ________.

(iii) If $\tan A = 1$, then $3 \sin A \cos A =$ ________.

124. If $\tan(A + B) = \sqrt{3}$ and $\tan(A - B) = \frac{1}{\sqrt{3}}$, $A > B$, then the value of A is ________ **(2020)**

Solution:

$\tan(A + B) = \sqrt{3}$

$\Rightarrow \tan(A + B) = \tan 60°$

$\therefore A + B = 60°$ ....(i)

$\tan(A - B) = \frac{1}{\sqrt{3}}$

$\Rightarrow \tan(A - B) = \tan 30°$

$\therefore A - B = 30°$ ....(ii)

Solving equation (i) and (ii), we get $A = 45°$

Exercise 8.43

(i) If $\tan(A + B) = \sqrt{3}$ and $\tan(A - B) = 1$, $A > B$, then the value of A is ________.

(ii) If $\tan(A + B) = 1$ and $\tan(A - B) = \frac{1}{\sqrt{3}}$, $A > B$, then the value of A is ________.

125. If $\cos A = \frac{\sqrt{3}}{2}$, $0° < A < 90°$, then A is equal to **(2020)**

(a) $\frac{\sqrt{3}}{2}$ (b) 30°

(c) 60° (d) 1

Solution: (b)

$\cos A = \frac{\sqrt{3}}{2}$

$\Rightarrow \cos A = \cos 30°$

$\therefore A = 30°$

126. If $2\cos\theta = 1$, then the value of θ is **(2023)**

(a) 45° (b) 60°

(c) 30° (d) 90°

Solution: (b)

$2\cos\theta = 1$

$\Rightarrow \cos\theta = \frac{1}{2}$

$\Rightarrow \cos\theta = \cos 60°$

$\therefore \theta = 60°$

127. If $\sqrt{3}\tan\theta = 1$, then the value of θ is **(2023)**

(a) 30° (b) 45°

(c) 60° (d) 90°

Solution:

$\sqrt{3}\tan\theta = 1$

$\Rightarrow \tan\theta = \frac{1}{\sqrt{3}}$

$\Rightarrow \tan\theta = \tan 30°$

$\therefore \theta = 30°$

128. The value of θ for which $2\sin^2\theta = \frac{1}{2}$, $0° \le \theta \le 90°$ is **(2024)**

(a) 30° (b) 60°

(c) 45° (d) 90°

Solution: (a)

$2\sin^2\theta = \frac{1}{2}$

$\Rightarrow \sin^2\theta = \frac{1}{4}$

$\Rightarrow \sin\theta = \frac{1}{2}$ $[\because 0° \le \theta \le 90°]$

$\Rightarrow \sin\theta = \sin 30°$

$\therefore \theta = 30°$

129. If $\sin^2\theta = \frac{3}{4}$, then θ is **(2024)**

(a) 30° (b) 45°

(c) 60° (d) 90°

Solution: (c)

$\sin^2\theta = \frac{3}{4}$

$\Rightarrow \sin\theta = \frac{\sqrt{3}}{2}$

$\Rightarrow \sin\theta = \sin 60°$

$\therefore \theta = 60°$

130. If $\tan^2\theta = 3$,where θ is an acute angle, then the value of θ is **(2024)**

(a) 30° (b) 60°

(c) 0° (d) 45°

Solution: (b)

$\tan^2\theta = 3$

$\Rightarrow \tan\theta = \sqrt{3}$ $[\because 0° \le \theta \le 90°]$

$\Rightarrow \tan\theta = \tan 60°$

$\therefore \theta = 60°$

Exercise 8.44

(i) If $\cos\theta = \frac{1}{\sqrt{2}}$, $0° < \theta < 90°$, then θ is equal to

(a) 30° (b) 45°

(c) 60° (d) 90°

(ii) If $2\sqrt{3}\cos\theta = 3$, $0° < \theta < 90°$, then θ is equal to

(a) 30° (b) 45°

(c) 60° (d) 90°

(iii) If $\tan\theta = 1$, $0° < \theta < 90°$, then θ is equal to

(a) 15° (b) 30°

(c) 45° (d) 60°

(iv) The value of θ for which $4 \sin^2 \theta = 3$, $0° < \theta < 90°$ is

(a) 30° (b) 45°
(c) 60° (d) 90°

(v) The value of θ for which $2 \sin^2 \theta = 1$, $0° < \theta < 90°$ is

(a) 30° (b) 45°
(c) 60° (d) 90°

(vi) If $\tan^2 \theta - 1 = 0$ and θ is an acute angle, then the value of θ is

(a) 30° (b) 15°
(c) 75° (d) 45°

(vii) If $3 \tan^2 A - 9 = 0$ and θ is an acute angle, then the value of θ is

(a) 30° (b) 45°
(c) 60° (d) 90°

(viii) If $2 \tan^2 \theta = 6$ and θ is an acute angle, then the value of θ is

(a) 30° (b) 45°
(c) 60° (d) 90°

131. For $\theta = 30°$, the value of $(2 \sin \theta \, . \, \cos \theta)$ is **(2024)**

(a) 1 (b) $\frac{\sqrt{3}}{2}$

(c) $\frac{\sqrt{3}}{4}$ (d) $\frac{3}{2}$

Solution: (b)

$2 \sin \theta \cos \theta$

$= 2 \sin 30° \cos 30°$

$= 2.\frac{1}{2}.\frac{\sqrt{3}}{2}$

$= \frac{\sqrt{3}}{2}$

Exercise 8.45

(i) For $\theta = 60°$, the value of $(2 \sin \theta \cos \theta)$ is

(a) $\frac{1}{2}$ (b) $\frac{\sqrt{3}}{2}$

(c) $\frac{\sqrt{3}}{4}$ (d) $\sqrt{3}$

(ii) For $\theta = 45°$, the value of $(2 \sin \theta \cos \theta)$ is

(a) 1 (b) $\frac{1}{\sqrt{2}}$

(c) $\frac{1}{2}$ (d) $\frac{\sqrt{3}}{4}$

(iii) For $\theta = 30°$, the value of $\cos^2 \theta - \sin^2 \theta$ is

(a) $\frac{\sqrt{3}}{2}$ (b) $\frac{\sqrt{3}}{4}$

(c) $\frac{3}{4}$ (d) $\frac{1}{2}$

132. If $\sin \theta = 1$, then the value of $\frac{1}{2} \sin\left(\frac{\theta}{2}\right)$ is **(2024)**

(a) $\frac{1}{2\sqrt{2}}$ (b) $\frac{1}{\sqrt{2}}$

(c) $\frac{1}{2}$ (d) 0

Solution: (a)

$\sin \theta = 1$

$\Rightarrow \sin \theta = \sin 90°$

$\therefore \theta = 90°$

Now, $\frac{1}{2} \sin\left(\frac{\theta}{2}\right)$

$= \frac{1}{2} \sin 45°$

$= \frac{1}{2}.\frac{1}{\sqrt{2}} = \frac{1}{2\sqrt{2}}$

Exercise 8.46

(i) If $\sin \theta = \frac{\sqrt{3}}{2}$, then the value of $\frac{1}{2}\sin\left(\frac{\theta}{2}\right)$ is

(a) $\frac{1}{2}$ (b) $\frac{\sqrt{3}}{2}$

(c) $\frac{1}{4}$ (d) 1

(ii) If $\sin \theta = \frac{1}{2}$, then the value of $3 \sin (2\theta)$ is

(a) 3 (b) $\frac{3}{2}$

(c) $3\sqrt{3}$ (d) $\frac{3\sqrt{3}}{2}$

(iii) If $\sin\theta = 1$, then the value of $2\sin\frac{\theta}{2}$ is

(a) $\sqrt{2}$ (b) 2

(c) 1 (d) $\frac{1}{2}$

133. If $\tan A = 3\cot A$, then the measure of the angle A is: **(2024)**

(a) 15° (b) 30°
(c) 45° (d) 60°

Solution: (d)

$\tan A = 3\cot A$

$\Rightarrow \tan A = 3\frac{1}{\tan A}$

$\Rightarrow \tan^2 A = 3$

$\Rightarrow \tan A = \sqrt{3}$

$\Rightarrow \tan A = \tan 60°$

$\therefore A = 60°$

Exercise 8.47

(i) If $\cot A = 3\tan A$, then the measure of the angle A is.

(a) 15° (b) 30°
(c) 45° (d) 60°

(ii) If $\cot A = \tan A$, then the measure of the angle A is

(a) 15° (b) 30°
(c) 45° (d) 60°

134. If $2\sin\theta = \text{cosec}\,\theta$, then the measure of the angle is **(2024)**

(a) 15° (b) 30°

(c) 45° (d) 60°

135. If $\sin^2\theta = \frac{1}{2}$, then the value of $\tan^2\theta$ is: **(2024)**

(a) $\frac{1}{\sqrt{3}}$ (b) $\sqrt{3}$

(c) 0 (d) 1

Solution: (d)

$\sin^2\theta = \frac{1}{2}$

$\Rightarrow \sin\theta = \frac{1}{\sqrt{2}}$

$\Rightarrow \sin\theta = \sin 45°$

$\therefore \theta = 45°$

Now, $\tan^2\theta$

$= \tan^2 45°$

$= 1^2$

$= 1$

Exercise 8.48

(i) If $4\sin^2\theta = 3$, then the value of $\tan^2\theta$ is

(a) 3 (b) $\sqrt{3}$

(c) 1 (d) $\frac{1}{3}$

(ii) If $4\sin^2\theta = 1$, then the value of $\tan^2\theta$ is

(a) 1 (b) 3

(c) $\frac{1}{\sqrt{3}}$ (d) $\frac{1}{3}$

(iii) If $2\sin^2\theta = 1$, then the value of $3\tan^2\theta$ is

(a) 1 (b) 3
(c) 2 (d) 0

136. The value of $\cos 60° - \text{cosec}\,30° + \tan 45°$ is **(2024)**

(a) 1 (b) $-\frac{1}{2}$

(c) $\frac{1}{2}$ (d) -1

Solution: (b)

$\cos 60° - \text{cosec}\,30° + \tan 45°$

$= \frac{1}{2} - 2 + 1$

$= \frac{1-4+2}{2}$

$= -\frac{1}{2}$

Exercise 8.49

(i) The value of 2 cos 60° – 3 cosec 30° + 2 tan 45° is

(a) 0 (b) 1
(c) 6 (d) 3

(ii) The value of 4 cos 60° – cosec 30° + 3 tan 45° is

(a) 1 (b) 2
(c) 3 (d) 4

(iii) The value of 2 cos 60° – cosec 30° + tan 45° is

(a) 0 (b) 1
(c) 2 (d) –1

137. For what value of θ, $\sin^2\theta + \sin\theta + \cos^2\theta$ is equal to 2 ? **(2024)**

(a) 45° (b) 0°
(c) 90° (d) 30°

Solution: (c)

$\sin^2\theta + \sin\theta + \cos^2 = 2$

$\Rightarrow \sin^2\theta + \cos^2\theta + \cos\theta = 2$

$\Rightarrow 1 + \cos\theta = 2$

$\Rightarrow \cos\theta = 1$

$\Rightarrow \cos\theta = \cos 0°$

$\therefore \theta = 0°$

Exercise 8.50

(i) For what value of θ, $\sin^2\theta + \cos\theta + \cos^2\theta$ is equal to $\frac{3}{2}$?

(a) 0° (b) 30°
(c) 60° (d) 90°

(ii) For what value of θ, $\sin^2\theta + \sin\theta + \cos^2\theta$ is equal to $\frac{3}{2}$?

(a) 15° (b) 30°
(c) 60° (d) 45°

(iii) For what value of θ, $\sec^2\theta + \sin\theta - \tan^2\theta$ is equal to 2 ?

(a) 30° (b) 45°
(c) 60° (d) 90°

138. If $\cos(\alpha+\beta) = 0$, then the value of $\cos\left(\frac{\alpha+\beta}{2}\right)$ is equal to: **(2024)**

(a) $\frac{1}{\sqrt{2}}$ (b) $\frac{1}{2}$
(c) 0 (d) $\sqrt{2}$

Solution: (a)

$\cos(\alpha+\beta) = 0$

$\Rightarrow \cos(\alpha+\beta) = \cos 90°$

$\therefore \alpha+\beta = 90°$

$\therefore \cos\left(\frac{\alpha+\beta}{2}\right)$

$= \cos\left(\frac{90°}{2}\right)$

$= \cos 45°$

$= \frac{1}{\sqrt{2}}$

Exercise 8.51

(i) It $\sin(\alpha+\beta) = 1$, then the value of $\sin\left(\frac{\alpha+\beta}{2}\right)$ is equal to

(a) $\frac{1}{\sqrt{2}}$ (b) $\sqrt{2}$
(c) 1 (d) $\frac{1}{2}$

(ii) If $\sin(\alpha+\beta) = \frac{1}{2}$, then value of $\tan(2\alpha+2\beta)$ is equal to

(a) $\frac{1}{\sqrt{3}}$ (b) $\frac{1}{2}$
(c) $\sqrt{3}$ (d) 1

(iii) If $\tan(\alpha+\beta) = 1$, then value of $\sin(2\alpha+2\beta)$ is equal to

(a) 1 (b) 0
(c) $\frac{1}{\sqrt{2}}$ (d) $\frac{\sqrt{3}}{2}$

Key Point (i) $\sin(90-\theta)=\cos\theta$

$\cos(90-\theta)=\sin\theta$

$\tan(90-\theta)=\cot\theta$

$\sec(90-\theta)=\operatorname{cosec}\theta$

$\cot(90-\theta)=\tan\theta$

$\operatorname{cosec}(90-\theta)=\sec\theta$

139. What is the value of $(\cos^2 67° - \sin^2 23°)$ **(2018)**

Solution:

$\cos^2 67° - \sin^2 23°$

$= \cos^2 67° - \sin^2 (90° - 67°)$

$= \cos^2 67° - \cos^2 67°$

$= 0$

Exercise 8.52

(i) What is the value of $\cos^2\theta - \sin^2(90° - \theta)$

(ii) What is the value of $\sin^2 46° - \cos^2 44°$

(iii) What is the value of $\cos^2 52° - \sin^2 48°$

140. Find the value of $(\cos 48° - \sin 42°)$ **(2019, 2020)**

Solution:

$\cos 48° - \sin 42°$

$= \cos 48° - \sin(90° - 48°)$

$= \cos 48° - \cos 48°$

$= 0$

141. The value of $(\sin 20° - \cos 70°)$ is _______ **(2020)**

Solution:

$\sin 20° - \cos 70°$

$= \sin(90° - 20°) - \cos 70°$

$= \cos 70° - \cos 70° = 0$

142. The value of $(\tan 27° - \cot 33°)$ is _______

Solution:

$\tan 27° - \cot 63°$

$= \tan 27° - \cot(90° - 27°)$

$= \tan 27° - \tan 27°$

$= 0$

Exercise 8.53

(i) Find the value of $(\cos 67° - \sin 23°)$

(ii) Find the value of $(\cos 71° - \sin 19°)$

(iii) Find the value of $(\cos 75° - \sin 15°)$

143. Evaluate: $\dfrac{\tan 65°}{\cot 25°}$ **(2019)**

Solution:

$\dfrac{\tan 65°}{\cot 25°}$

$= \dfrac{\tan(90° - 25°)}{\cot 25°}$

$= \dfrac{\cot 25°}{\cot 25°}$

$= 1$

Exercise 8.54

(i) Evaluate: $\dfrac{\tan 20°}{\cot 70°}$

(ii) Evaluate: $\dfrac{\tan 23°}{\cot 67°}$

(iii) Evaluate: $\dfrac{\tan 22°}{\cot 68°}$

144. Express $(\sin 67° + \cos 75°)$ in terms of trigonometric ratios of the angles between 0° and 45°. **(2019)**

Solution:

$\sin 67° + \cos 75°$

$= \sin(90° - 23°) + \cos(90° - 15°)$

$= \cos 23° + \sin 15°$

Exercise 8.55

(i) Express $(\sin 72° + \cos 67°)$ in terms of trigonometric ratios of the angles between 0° and 45°

(ii) Express $(\sin 87° + \cos 61°)$ in terms of trigonometric ratios of the angles between 0° and 45°

(iii) Express $(\sin 77° + \cos 79°)$ in terms of trigonometric ratios of the angles between 0° and 45°.

145. $\frac{\cos 80^\circ}{\sin 10^\circ}+\cos 59^\circ \operatorname{cosec} 31^\circ =$ ________ **(2020)**

Solution:

$$\frac{\cos 80^\circ}{\sin 10^\circ}+\cos 59^\circ \operatorname{cosec} 31^\circ$$

$$=\frac{\cos 80^\circ}{\sin(90^\circ-80^\circ)}+\cos 59^\circ \operatorname{cosec}(90^\circ-59^\circ)$$

$$=\frac{\cos 80^\circ}{\cos 80^\circ}+\cos 59^\circ \sec 59^\circ$$

$= 1 + 1$

$= 2$

Exercise 8.56

(i) $\frac{\cos 70^\circ}{\sin 20^\circ}+\cos 69^\circ \operatorname{cosec} 21^\circ =$ ________

(ii) $\frac{\cos 61^\circ}{\sin 29^\circ}+\cos 61^\circ \operatorname{cosec} 29^\circ =$ ________

(iii) $\frac{\cos 75^\circ}{\sin 15^\circ}+\cos 77^\circ \operatorname{cosec} 13^\circ =$ ________

146. $\frac{2\cos 67^\circ}{\sin 23^\circ}-\frac{\tan 40^\circ}{\cot 50^\circ}-\cos 0^\circ =$ ________ **(2020)**

Solution:

$$\frac{2\cos 67^\circ}{\sin 23^\circ}-\frac{\tan 40^\circ}{\cot 50^\circ}-\cos 0^\circ$$

$$=\frac{2\cos(90^\circ-23^\circ)}{\sin 23^\circ}-\frac{\tan(90^\circ-50^\circ)}{\cot 50^\circ}-1$$

$$=\frac{2\sin 23^\circ}{\sin 23^\circ}-\frac{\cot 50^\circ}{\cot 50^\circ}-1$$

$= 2 - 1 - 1$

$= 0$

147. $\frac{3\cot 40^\circ}{\tan 50^\circ}-\frac{1}{2}\left(\frac{\cos 35^\circ}{\sin 55^\circ}\right) =$ ________ **(2020)**

Solution:

$$\frac{3\cot 40^\circ}{\tan 50^\circ}-\frac{1}{2}\left(\frac{\cos 35^\circ}{\sin 55^\circ}\right)$$

$$=\frac{3\cot(90^\circ-50^\circ)}{\tan 50^\circ}-\frac{1}{2}\frac{\cos(90^\circ-55^\circ)}{\sin 55^\circ}$$

$$=\frac{3\tan 50^\circ}{\tan 50^\circ}-\frac{1}{2}\frac{\sin 55^\circ}{\sin 55^\circ}$$

$$=3\times 1-\frac{1}{2}\times 1$$

$$=\frac{5}{2}$$

148. Evaluate: $(\tan 23^\circ)\times(\tan 67^\circ)$ **(2020)**

Solution:

$(\tan 23^\circ)\times(\tan 67^\circ)$

$= \tan(90^\circ - 67^\circ)\times \tan 67^\circ$

$= \cot 67^\circ \times \tan 67^\circ$

$= 1$

149. Evaluate: $(\tan 40^\circ)\times(\tan 50^\circ)$ **(2020)**

Solution:

$(\tan 40^\circ)\times(\tan 50^\circ)$

$= \tan(90^\circ - 50^\circ)\times \tan 50^\circ$

$= \cot 50^\circ \times \tan 50^\circ$

$= 1$

Exercise 8.57

(i) $\frac{3\cos 49^\circ}{\sin 41^\circ}-\frac{\tan 42^\circ}{\cot 48^\circ}-\cos 60^\circ = ?$

(ii) $\frac{\cos 72^\circ}{\sin 18^\circ}-\frac{\tan 47^\circ}{\cot 43^\circ}-\cos 0^\circ = ?$

(iii) $\frac{4\cos 51^\circ}{\sin 39^\circ}-\frac{\tan 70^\circ}{\cot 20^\circ}-2\cos 0^\circ = ?$

(iv) $\frac{2\cos 78^\circ}{\sin 12^\circ}-\frac{\tan 37^\circ}{\cot 53^\circ}-\cos 0^\circ = ?$

(v) $\frac{3\cos 23^\circ}{\sin 67^\circ}-\frac{2\tan 41^\circ}{\cot 49^\circ}-\cos 60^\circ = ?$

(vi) $\frac{2\cot 32^\circ}{\tan 58^\circ}-\frac{1}{3}\left(\frac{\cos 60^\circ}{\sin 30^\circ}\right) = ?$

(vii) $\frac{4\cos 68^\circ}{\sin 22^\circ}-3\frac{\cos 61^\circ}{\sin 29^\circ} = ?$

(viii) $\tan 23^\circ \times \cot 67^\circ = ?$

(ix) $\tan 62^\circ \times \cot 28^\circ = ?$

(x) $\tan 30^\circ \times \cot 60^\circ = ?$

150. The value of $\frac{\tan 35^\circ}{\cot 55^\circ}+\frac{\cot 78^\circ}{\tan 12^\circ}$ is ______ **(2020)**

Solution:

$$\frac{\tan 35^\circ}{\cot 55^\circ}+\frac{\cot 78^\circ}{\tan 12^\circ}$$
$$=\frac{\tan(90^\circ-55^\circ)}{\cot 55^\circ}+\frac{\cot(90^\circ-12^\circ)}{\tan 12^\circ}$$
$$=\frac{\cot 55^\circ}{\cot 55^\circ}+\frac{\tan 12^\circ}{\tan 12^\circ}$$
$$= 1 + 1$$
$$= 2$$

Exercise 8.58

(i) The value of $\frac{\tan 70^\circ}{\cot 20^\circ}+\frac{\cot 58^\circ}{\tan 32^\circ}$ is______.

(ii) The value of $\frac{\tan 63^\circ}{\cot 27^\circ}+\frac{\cot 60^\circ}{\tan 30^\circ}$ is______.

(iii) The value of $\frac{\tan 50^\circ}{\cot 40^\circ}+\frac{\cot 50^\circ}{\tan 40^\circ}$ is______.

(iv) The value of $\frac{\tan 62^\circ}{\cot 28^\circ}+\frac{\cot 62^\circ}{\tan 28^\circ}$

(v) The value of $\frac{\tan 71^\circ}{\cot 19^\circ}+\frac{\cot 73^\circ}{\tan 17^\circ}$ is________.

151. The value of $\frac{\sin\theta}{\cos(90^\circ-\theta)}+\frac{\cos 43^\circ}{\sin 47^\circ}$ is ______ **(2020)**

Solution:

$$\frac{\sin\theta}{\cos(90^\circ-\theta)}+\frac{\cos 43^\circ}{\sin 47^\circ}$$
$$=\frac{\sin\theta}{\sin\theta}+\frac{\cos(90^\circ-47^\circ)}{\sin 47^\circ}$$
$$=1+\frac{\sin 47^\circ}{\sin 47^\circ}$$
$$= 1 + 1$$
$$= 2$$

Exercise 8.59

(i) The value of $\frac{\sin(90^\circ-\theta)}{\cos\theta}+\frac{\sin 33^\circ}{\cos 57^\circ}$ is ______

(ii) The value of $\frac{\sin\theta}{\cos(90^\circ-\theta)}+\frac{\sin 20^\circ}{\cos 70^\circ}$ is ______

(iii) The value of $\frac{\sin 27^\circ}{\cos 53^\circ}-\frac{\sin 10^\circ}{\cos 80^\circ}$ is ______

152. Find A it $\tan 2A = \cot(A - 24^\circ)$ **(2019)**

Solution:

$\tan 2A = \cot(A - 24^\circ)$

$\Rightarrow \cot(90^\circ - 2A) = \cot(A - 24^\circ)$

$\therefore 90 - 2A = A - 24^\circ$

$\Rightarrow -2A - A = -24^\circ - 90^\circ$

$\Rightarrow -3A = -114^\circ$

$\therefore A = 38^\circ$

Exercise 8.60

(i) Find A if $\tan 3A = \cot 3A$

(ii) Find A if $\tan 4A = \cot(A - 10^\circ)$

(iii) Find A if $\tan 2A = \cot(A - 30^\circ)$

153. The value of θ for which $\cos(10^\circ + \theta) = \sin 30^\circ$ is **(2020)**

(a) 50° (b) 40°
(c) 80° (d) 20°

Solution :

$\cos(10^\circ + \theta) = \sin 30^\circ$

$\Rightarrow \cos(10^\circ + \theta) = \cos 60^\circ$

$\therefore 10^0 + \theta = 60^\circ$

$\Rightarrow \theta = 50^\circ$

Exercise 8.61

(i) The value of θ for which $\cos(20^\circ + \theta) = \sin 40^\circ$ is

(a) 20° (b) 30°
(c) 50° (d) 40°

(ii) The value of θ for which $\cos(30^\circ + \theta) = \sin 10^\circ$ is

(a) 40° (b) 10°
(c) 30° (d) 50°

(iii) The value of θ for which $\cos(40^\circ + \theta) = \sin 25^\circ$ is

(a) 25° (b) 15°
(c) 40° (d) 80°

154. The value of θ for which $\sin(44^\circ + \theta) = \cos 30^\circ$ is **(2020)**

(a) 46° (b) 60°
(c) 16° (d) 90°

Solution: (c)

$\sin(44° + \theta) = \cos 30°$

$\Rightarrow \sin(44° + \theta) = \sin 60°$

$\therefore 44° + \theta = 60°$

$\Rightarrow \theta = 16°$

Exercise 8.62

(i) The value of θ for which $\sin(32° + \theta) = \cos 12°$ is

(a) 68° (b) 46°
(c) 60° (d) 72°

(ii) The value of θ for which $\sin(42° + \theta) = \cos 20°$ is

(a) 28° (b) 38°
(c) 70° (d) 42°

(iii) The value of θ for which $\sin(31° + \theta) = \cos 10°$ is **(2020)**

(a) 20° (b) 31°
(c) 69° (d) 49°

155. If $\cos A = \sin 42°$, then the value of A. **(2020)**

Solution :

$\cos A = \sin 42°$

$\Rightarrow \cos A = \sin(90° - 48°)$

$\Rightarrow \cos A = \cos 48°$

$\therefore A = 48°$

Exercise 8.63

(i) If $\cos A = \sin 20°$, then the value of A.

(ii) If $\cos A = \sin 67°$, then the value of A.

(iii) If $\cos A = \sin 28°$, then the value of A.

156. If $\sin A = \cos A$, $(0° \le A \le 90°)$, then the angle equal to **(2020)**

(a) 30° (b) 60°
(c) 0° (d) 45°

Solution :(d)

$\sin A = \cos A$

$\Rightarrow \sin A = \sin(90° - A)$

$\therefore A = 90° - A$

$\Rightarrow 2A = 90°$

$\Rightarrow A = 45°$

157. If $\tan A = \cot B$, then find the value of $A + B$ **(2020)**

Solution :

$\tan A = \cot B$

$\Rightarrow \tan A = \tan(90° - B)$

$\therefore A = 90° - B$

$\Rightarrow A + B = 90°$

Exercise 8.64

(i) If $\sec A = \operatorname{cosec} A$, then find the value of A is

(a) 15° (b) 45°
(c) 60° (d) 20°

(ii) If $\sin(90° - \theta) = \cos(90° - \theta)$, $(0° \le A \le 90°)$, then the value of θ is

(a) 45° (b) 60°
(c) 30° (d) 15°

(iii) If $\tan(90° - \theta) = \cot(90° - \theta)$, $(0° \le A \le 90°)$, then the value of θ is

(a) 15° (b) 30°
(c) 45° (d) 90°

158. The value of ($\tan 1° \tan 2° \tan 3° \ldots\ldots \tan 89°$) is equal to...... **(2020)**

Solution :

$\tan 1° \tan 2° \ldots\ldots \tan 89°$

$= (\tan 1° \tan 89°)(\tan 2° \tan 88°)\ldots(\tan 44° \tan 46°) \tan 45°$

$= (\tan 1° \cot 1°)(\tan 2° \cot 2°)\ldots(\tan 44° \cot 44°)\ 1$

$= 1.1\ldots1.1$

$= 1$

Exercise 8.65

(i) The value of ($\cot 1° \cot 2° \ldots\ldots \cot 89°$) is equal to.......

(ii) The value of ($\tan 1° \tan 3° \tan 5° \ldots\ldots \tan 89°$) is equal to.......

(iii) The value of ($\cot 1° \cot 3° \cot 5° \ldots\ldots \cot 89°$) is equal to.......

(iv) The value of ($\cot 2° \cot 4° \cot 6° \ldots\ldots \cot 88°$) is equal to.......

(v) The value of ($\tan 2° \tan 4° \tan 6° \ldots\ldots \tan 88°$) is equal to.......

159. Find the value of $(\sin^2 33° + \sin^2 57°)$ **(2019)**

Solution :

$\sin^2 33° + \sin^2 57°$

$= \sin^2 33° + \sin^2 (90° - 33°)$

$= \sin^2 33° + \cos^2 33° = 1$

160. The value of $\sin^2 65° + \sin^2 25°$ is______ **(2020)**

Solution :

$\sin^2 65° + \sin^2 25°$

$= \sin^2 65° + \sin^2 (90° - 65°)$

$= \sin^2 65° + \cos^2 65°$

$= 1$

Exercise 8.66

(i) Find the value of $(\sin^2 68° + \sin^2 22°)$

(ii) Find the value of $(\sin^2 46° + \sin^2 44°)$

(iii) Find the value of $(\sin^2 52° + \sin^2 38°)$

161. The value of $\sin 23° \cos 67° + \cos 23° \sin 67°$ is______ **(2020)**

Solution :

$\sin 23° \cos 67° + \cos 23° \sin 67°$

$= \sin 23° \cos (90° - 23°) + \cos 23° \sin (90° - 23°)$

$= \sin 23° \sin 23° + \cos 23° \cos 23°$

$= \sin^2 23^0 + \cos^2 23°$

$=1$

162. The value of $\sin 32° \cos 58° + \cos 32° \sin 58°$ is _______ **(2020)**

Solution :

$\sin 32° \cos 58° + \cos 32° \sin 58°$

$= \sin 32^0 \cos (90° - 32°) + \cos 32° \sin (90° - 32°)$

$= \sin 32° \sin 32° + \cos 32° \cos 32°$

$= \sin^2 32° + \cos^2 32° = 1$

163. The value of $\sin 20° \cos 70° + \sin 70° \cos 20°$ is______ **(2020)**

Solution :

$\sin 20° \cos 70° + \sin 70° \cos 20°$

$= \sin 20° \cos (90° - 20°) + \sin (90° - 20°) \cos 20°$

$= \sin 2\,0° \sin 2\,0° + \cos 20° \cos 20°$

$= \sin^2 20° + \cos^2 20°$

$= 1$

164. The value of $\sin 43° \cos 47° + \sin 47° \cos 43°$ is_______ **(2020)**

Solution :

$\sin 43° \cos 47° + \sin 47° \cos 43°$

$= \sin 43° \cos (90° - 43°) + \sin (90° - 43°) \cos 43°$

$= \sin 43° \sin 43° + \cos 43° \cos 43°$

$= \sin^2 43^0 + \cos^2 43°$

$= 1$

Exercise 8.67

(i) The value of $\sin 63° \cos 27° + \cos 63° \sin 27°$ is_____

(ii) The value of $\sin 52° \cos 38° + \cos 52° \sin 38°$ is_____

(iii) The value of $\sin 67° \cos 23° + \cos 67° \sin 23°$ is______

(iv) The value of $\sin 41° \cos 49° + \cos 41° \sin 49°$ is______

(v) The value of $\sin 70° \cos 20° + \cos 70° \sin 20°$ is______

(vi) The value of $\sin 30° \cos 60° + \cos 30° \sin 60°$ is______

(vii) The value of $\sin 25° \cos 65° + \cos 25° \sin 65°$ is______

(viii) The value of $\sin 35° \cos 55° + \cos 35° \sin 55°$ is______

165. The value of $\sec^2 20° - \cot^2 70°$ is______ **(2020)**

Solution :

$\sec^2 20° - \cot^2 70^0$

$= \sec^2 20° - \cot^2 (90° - 20°)$

$= \sec^2 20° - \tan^2 20°$

$= 1$

166. Which of the following statements is true? **(2025)**

(a) $\sin 20° > \sin 70°$
(b) $\sin 20° > \cos 20°$
(c) $\cos 20° > \cos 70°$
(d) $\tan 20° > \tan 70°$

Solution: (c)

$\cos 20° > \cos 70°$

167. Which of the following is not defined for $x = 90°$? **(2026)**

(a) $\cot x$ (b) $\operatorname{cosec} x$

(c) $\tan\left(\frac{x}{2}\right)$ (d) $\sec x$

Solution: (d)

$\sec x$

168. Which of the following statements is not true? **(2026)**

(a) $\sin\theta = \cos\theta$ (b) $\tan 30° = \cot 60°$

(c) $\sin 30° = \cos 60°$ (d) $\sin 45° = \frac{1}{\sec 45°}$

Solution: (a)

$\sin\theta = \cos\theta$

169. In the adjoining figure the angle of elevation of the point C from the point B is **(2026)**

(a) 30° (b) 45°

(c) 22.5° (d) 67.5°

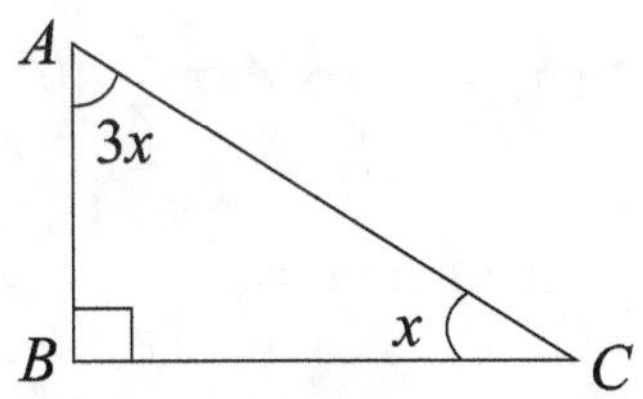

Solution: (c)

$x + 3x = 90°$

$4x = 90°$

$x = 22.5°$

Exercise 8.68

(i) The value of $\sec^2 30° - \cot^2 60°$ is______

(ii) The value of $\sec^2 67^0 - \cot^2 23°$ is______

(iii) The value of $\sec^2\theta - \cot^2(90° - \theta)$is____

170 to **177** a statement of Assertion (A) is followed by a statement of Reason (R). choose the correct option.

(a) Both, Assertion (A) and Reason (R) are true and Reason (R) is correct explanation of Assertion (A).
(b) Both, Assertion (A) and Reason (R) are true but Reason (R) is not correct explanation of Assertion (A).
(c) Assertion (A) is true but Reason (R) is false.
(d) Assertion (A) is false but Reason (R) is true.

170. Statement A (Assertion) : For $0° \le A \le 90°$, $\operatorname{cosec}\theta - \cot\theta$ and $\operatorname{cosec}\theta + \cot\theta$ are reciprocal to each other.

Statement R (Reason) : $\operatorname{cosec}^2\theta - \cot^2\theta = 1$ **(2023)**

Solution :(a)

$(\operatorname{cosec}\theta - \cot\theta)(\operatorname{cosec}\theta + \cot\theta)$

$= \operatorname{cosec}^2\theta - \cot^2\theta$

$= 1$

171.Assertion (A) : If $\sin A = \frac{1}{3}$ $(0° < A < 90°)$, then the value of $\cos A$ is $\frac{2\sqrt{2}}{3}$

Reason (R) : For every angle θ, $\sin^2\theta + \cos^2\theta = 1$ **(2024)**

Solution : (a)

$\sin^2 A + \cos^2 A = 1$

$\Rightarrow \left(\frac{1}{3}\right)^2 + \cos^2 A = 1$

$\Rightarrow \frac{1}{9} + \cos^2 A = 1$

$\Rightarrow \cos^2 A = \frac{8}{9}$

$\therefore \cos A = \frac{2\sqrt{2}}{3}$

172. Assertion(A): For an acute angle θ, $\sin\theta = \frac{3}{5}$

$\Rightarrow$ **cos** $\theta = -\frac{4}{5}$,

Reason(R): For any value of θ, $(0° \leq \theta \leq 90°)$ $\sin^2\theta + \cos^2\theta = 1$ **(2025)**

Solution: (d)

Assertion(A) is false, but Reason (R) is true.

173. Assertion(A): For an acute angle θ, $\sec\theta = 3 \Rightarrow \tan\theta = 2\sqrt{2}$

Reason(R): $\sec^2\theta = 1 - \tan^2\theta$ for all values of θ **(2025)**

Solution: (a)

$\sec\theta = 3$

$\Rightarrow \sec^2 = 9$

$\Rightarrow 1 + \tan^2\theta = 9$

$\Rightarrow \tan^2\theta = 8$

$\therefore \tan\theta = 2\sqrt{2}$

∴ Both Assertion(A) and Reason(R) are true and Reason(R) is the correct explanation of Assertion(A).

174. Assertion(A): For an acute angle θ, value of cosec θ can not be $\frac{1}{\sqrt{2}}$

Reason(R): $\text{cosec}\,\theta \geq 1$ for $0° \leq \theta \leq 90°$ **(2025)**

Solution: (a)

∴ Both Assertion(A) and Reason(R) are true and Reason(R) is the correct explanation of Assertion(A).

175. Assertion(A): In a right angle triangle *ABC*, $\angle B = 90°$. There for the value of $\cos(A + C)$ is equal to 0.

Reason(R): $A + B + C = 180°$ and $\cos 90° = 0$ **(2025)**

Solution: (a)

$\angle B = 90°$

$\therefore \angle(A + C) = 90°$

$\therefore \cos(A + C)$

$= \cos 90°$

$= 0$

176. Assertion (A): For any acute angle θ, cos θ is always less than 1.

Reason (R): In a right angled, hypotenuse is the longest side and $\cos = \frac{\text{Base}}{\text{Hypotenuses}}$ **(2026)**

Solution: (a)

∴ Both Assertion(A) and Reason(R) are true and Reason(R) is the correct explanation of Assertion(A).

177. Assertion (A): tan 2θ is not defined at θ = 45°

Reason (R): $\sin 90° \neq \cos 90°$

Solution: (b)

∴ Both Assertion(A) and Reason(R) are true and Reason(R) is the correct explanation of Assertion(A).

Exercise 8.69

(i) Statement A (Assertion) : $\sec^2 20° - \cot^2 80° = 1$

Statement R (Reason) : $\sec^2\theta - \tan^2\theta = 1$

(ii) Statement A (Assertion) : cos 1 cos 2 cos 3.... cos 180 = 0

Statement R (Reason) : $\cos 90° = 0$

(iii) Assertion (A) : If $\sin A = \frac{3}{5}$ $(0° < A < 90°)$, then the value of cos *A* is $\frac{4}{5}$

Reason (R) : For every angle θ, $\sin^2\theta + \cos^2\theta = 1$

(iv) Assertion (A) : If $\cos A = \frac{3}{4}$ $(0° < A < 90°)$, then the value of $\sin A = \frac{\sqrt{7}}{4}$

Reason (R) : For every angle θ, $\sin^2\theta + \cos^2\theta = 1$

(v) Assertion (A) : If $\sin A = \frac{3}{10}$ $(0° < A < 90°)$, then the value of $\cos A = \frac{4}{10}$

Reason (R) : For every angle θ, $\sin^2\theta + \cos^2\theta = 1$

Key Point Length of an arc of a circle is $\ell = r\theta$ where ℓ is the length of arc *r* is the radius of the circle and θ is the central angle in radians.

178. What is the length of the are of the sector of a circle with radius 14 cm and of central angle 90° ?

(a) 22 cm (b) 44 cm

(c) 88cm (d) 11 cm

Solution : (a)

$s = r\theta$

$\Rightarrow s = 14.\frac{\pi}{2}\left[\begin{array}{l}\because 180^0 = \pi \\ \therefore 90^0 = \frac{\pi}{2}\end{array}\right]$

$\Rightarrow s = 7\pi$

$\Rightarrow s = 7 \times \frac{22}{7}$

$\therefore s = 22$ cm

179. The angle of the are of a circle of radius 14 cm which subtends an angle of 60° and the center of the circle is **(2023)**

(a) $\frac{44}{3}$ cm (b) $\frac{88}{3}$ cm

(c) $\frac{308}{3}$ cm (d) $\frac{616}{3}$ cm

Solution : (a)

$s = r\theta$

$\Rightarrow s = 14.\frac{\pi}{3}\left[\begin{array}{l}\because 180^0 = \pi \\ \therefore 60^0 = \frac{\pi}{3}\end{array}\right]$

$\Rightarrow s = 14.\frac{1}{3}.\frac{22}{7}$

$\therefore s = \frac{44}{3}$ cm

180. What is length of are of a circle of radius 7 m which subtends an angle of 90° at the center of the circle is **(2023)**

(a) 22 cm (b) 11 cm

(c) $\frac{77}{2}$ cm (d) $\frac{11}{2}$ cm

Solution : (b)

$s = r\theta$

$\Rightarrow s = 7.\frac{\pi}{2}\left[\begin{array}{l}\because 180° = \pi \\ \therefore 90° = \frac{\pi}{2}\end{array}\right]$

$\Rightarrow s = \frac{7}{2} \times \frac{22}{7}$

$\therefore s = 11$ cm

181. In a circle of radius 21 cm, if an are subtends an angle of 60° at the center of the circle, then the length of the arc is **(2024)**

(a) 11 cm (b) 44 cm

(c) $\frac{22}{7}$ cm (d) 22 cm

Solution : (d)

$s = r\theta$

$\Rightarrow s = 21.\frac{\pi}{3}\left[\begin{array}{l}\because 180° = \pi \\ \therefore 60° = \frac{\pi}{3}\end{array}\right]$

$\Rightarrow s = 7 \times \frac{22}{7}.$

$\therefore s = 22$ cm

Exercise 8.70

(i) What is the length of the arc of the sector of a circle with radius 42 cm and of a central angle 60^0 ?

(a) 22cm (b) 30 cm

(c) 28 cm (d) 44cm

(ii) What is the length of the arc of the sector of a circle with radius 35 cm and of a central angle 90° ?

(a) 55 cm (b) 42 cm

(c) 110 cm (d) 21 cm

(iii) What is the length of the arc of the sector of a circle with radius 7 cm and of a central angle 30° ?

(a) $\frac{8}{3}$ cm (b) $\frac{11}{3}$ cm

(c) $\frac{10}{3}$ cm (d) 12 cm

182. The hour – hand of a clock 6 cm long. The angle swept by it between 7 : 20 am and 755 am is. **(2023)**

(a) $\left(\frac{35}{4}\right)^0$ (b) $\left(\frac{35}{2}\right)^0$

(c) 35° (d) 70°

Solution : (b)

Required angle $=\left(\frac{30}{60}\times 35\right)^0=\left(\frac{35}{2}\right)^0$

Exercise 8.71

(i) The hour – hand of a clock 8 cm long. The angle swept by it between 8 pm and 9 : 00 pm is.

(a) 20° (b) 40°

(c) $\left(\frac{35}{2}\right)^0$ (d) 30°

(ii) The hour – hand of a clock 7 cm long. The angle swept by it between 5 : 15 am and 5 : 40 am is.

(a) 25° (b) 40°

(c) $\left(\frac{25}{2}\right)^0$ (d) $\left(\frac{15}{4}\right)^0$

(iii) The hour – hand of a clock 10 cm long. The angle swept by it between 6 : 05 am and 6 : 45 am is.

(a) 20° (b) 40°

(c) $\left(\frac{35}{4}\right)^0$ (d) 15°

183. The angle of elevation of the top of a tower from a point on the ground, which is 30 m away from the foot of the tower is 45°. The length of the tower (in meter) is **(2011)**

(a) 15 (b) 30

(c) $30\sqrt{3}$ (d) $10\sqrt{3}$

Solution :(b)

Let AB be the tower and C be the point of observation,

Then $BC = 30$ m and $\angle ACB = 45°$

Let $AB = h$ meter, then

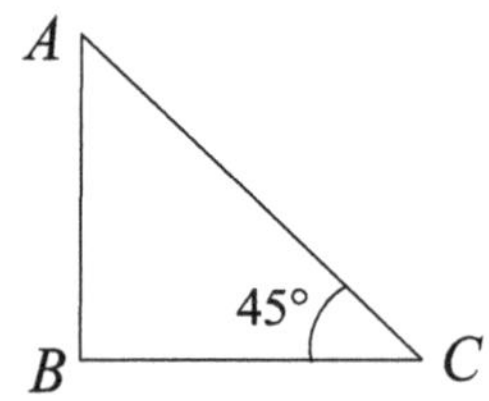

$$\frac{AB}{BC} = \tan 45° = 1$$

$$\Rightarrow \frac{h}{30} = 1$$

$$\Rightarrow h = 30 \text{ cm}$$

Exercise 8.72

(i) The angle of elevation of the top of a tower from a point on the ground, which is 30 m away from the foot of the tower is 30°. The length of the tower (in meter) is

(a) $10\sqrt{3}$ (b) 10

(c) 30 (d) $15\sqrt{2}$

(ii) The angle of elevation of the top of a tower from a point on the ground, which is 30 m away from the foot of the tower is 60°. The length of the tower (in meter) is

(a) $20\sqrt{3}$ (b) 100

(c) $60\sqrt{3}$ (d) 20

(iii) The angle of elevation of the top of a tower from a point on the ground, which is 100 m away from the foot of the tower is 45°. The length of the tower (in meter) is

(a) 50 (b) $45\sqrt{3}$

(c) 20 (d) 100

184. The length of shadow of a tower on the plane ground is $\sqrt{3}$ times the height of the tower. The angle of elevation of sun is. **(2012)**

(a) 45° (b) 30°

(c) 60° (d) 90°

Solution : (b)

Let AB be the tower and BC be its shadow.

Let $\angle ACB = \theta$

Then

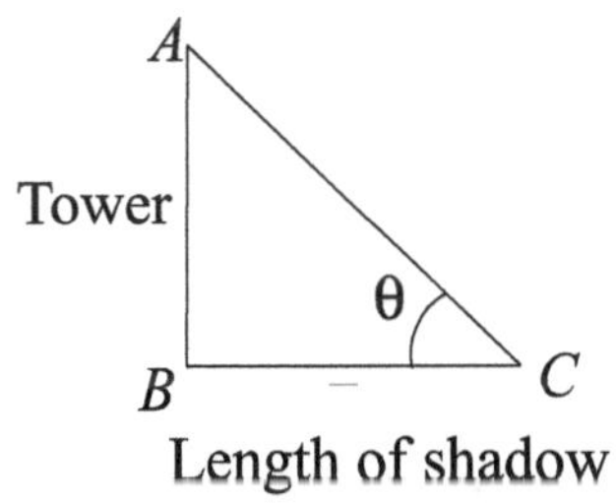

$$\frac{AB}{BC} = \frac{1}{\sqrt{3}}$$

$\Rightarrow \tan\theta = \tan 30°$

$\therefore \theta = 30°$

Exercise 8.73

(i) The length of shadow of a tower on the plane ground is $\frac{1}{\sqrt{3}}$ times the height of the tower. The angle of elevation of sun is.

(a) 45° (b) 30°

(c) 60° (d) 90°

(ii) The length of shadow of a tower on the plane ground is equal to the height of the tower. The angle of elevation of sun is.

(a) 45° (b) 30°

(c) 60° (d) 75°

185. The angle of depression of a car, standing on the ground, from the top of a 75 m high tower is 30°. The distance of the car from the base of the tower (in m.) is **(2013)**

(a) $25\sqrt{3}$ (b) $50\sqrt{3}$

(c) $75\sqrt{3}$ (d) 150

Solution : (c)

Let AB be the tower and BC be the horizontal line such that

$AB = 75$ m, $\angle ACB = 30°$

Then

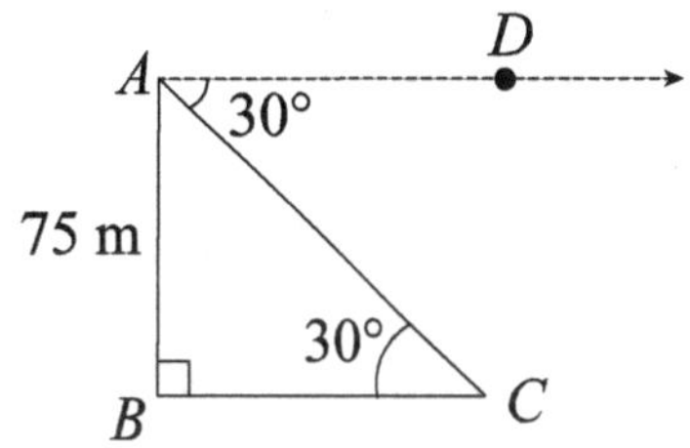

$$\frac{AB}{BC} = \tan 30°$$

$$\Rightarrow \frac{75}{BC} = \frac{1}{\sqrt{3}}$$

$\therefore BC = 75\sqrt{3}$ m

186. The angle of depression of a car parked on the road from the top of a 150 m high tower is 30°. The distance of the car from the tower (in meter) is **(2014)**

(a) $50\sqrt{3}$ (b) $150\sqrt{3}$

(c) $150\sqrt{2}$ (d) 75

Solution : (b)

Let AB be the tower and BC be the horizontal line such that

$AB = 150$ m, $\angle ACB = 30°$

Then

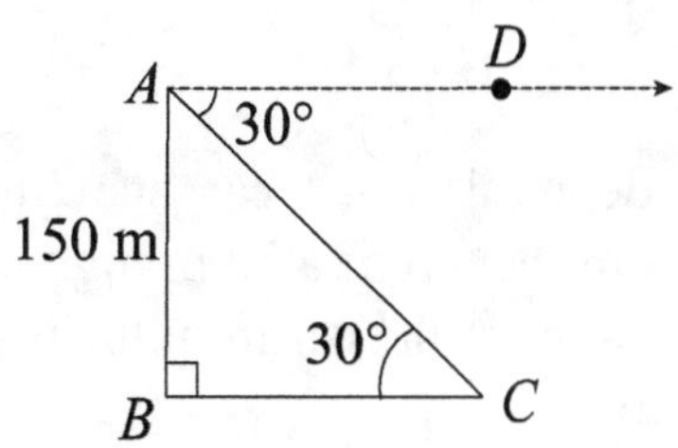

$$\frac{AB}{BC} = \tan 30°$$

$$\Rightarrow \frac{150}{BC} = \frac{1}{\sqrt{3}}$$

$\therefore BC = 150\sqrt{3}$ m

187. In the given figure, the angle of depression of the foot of building CD frem the top of tower AB is **(2026)**

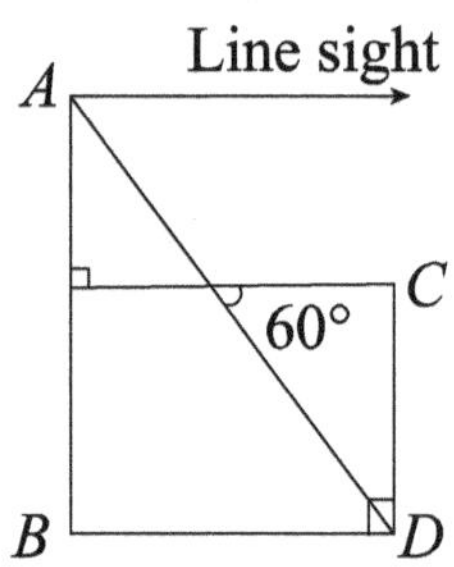

(a) 60° (b) 30°

(c) 45° (d) 90°

Solution: (a)

60°

Exercise 8.74

(i) The angle of depression of a car, standing on the ground, from the top of a 50 m high tower is 30°. The distance of the car from the base of the tower (in m.) is **(2013)**

(a) $25\sqrt{3}$ (b) $50\sqrt{3}$

(c) $75\sqrt{3}$ (d) 150

(ii) The angle of depression of a car, standing on the ground, from the top of a 60 m high tower is 30°. The distance of the car from the base of the tower (in m.) is.

(a) 60 (b) $30\sqrt{3}$

(c) $60\sqrt{3}$ (d) 75

(iii) The angle of depression of a car, standing on the ground, from the top of a 30 m high tower is 60°. The distance of the car from the base of the tower (in m.) is.

(a) 10 (b) $10\sqrt{3}$

(c) 20 (d) $30\sqrt{3}$

188. In Figure 1, a tower AB is 20 m high and BC, its shadow on the ground, is $20\sqrt{3}$ m long. Find the sun's altitude. **(2015)**

Solution :

Let AB be the tower and BC be the shadow,

Let $\angle ACB = \theta$

$$\tan\theta = \frac{AB}{BC}$$

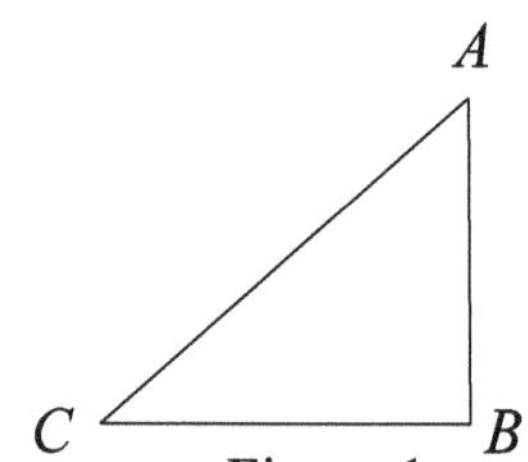

Figure 1

$$\Rightarrow \tan\theta = \frac{20}{20\sqrt{3}} = \frac{1}{\sqrt{3}}$$

$\therefore \tan\theta = \tan 30°$

$\therefore \theta = 30°$

189. If a tower 30 m high, casts a shadow $10\sqrt{3}$ m long on the ground, then what is the angle of elevation of the sun ? **(2017)**

Solution :

Let AB be the tower and BC be the shadow,

Let $\angle ACB = \theta$

$$\tan\theta = \frac{AB}{BC}$$

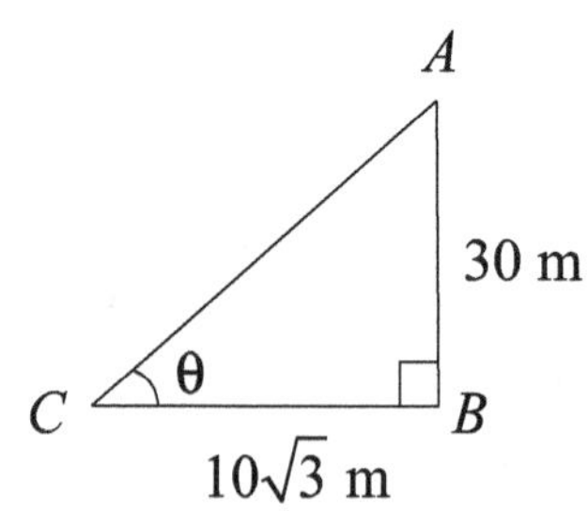

$$\Rightarrow \tan\theta = \frac{30}{10\sqrt{3}} = \sqrt{3}$$

$\therefore \tan\theta = \tan 60°$

$\therefore \theta = 60^0$

190. If a pole 6 m high casts a shadow $2\sqrt{3}$ m long on the ground, then sun elevation is **(2023)**

(a) 60° (b) 45°

(c) 30° (d) 90°

191. If a pole 6 m high casts a shadow $2\sqrt{3}$ m long on the ground, then sun's elevation is **(2024)**

(a) 30° (b) 45°

(c) 60° (d) 90°

Solution : (a)

Let AB be the tower and BC be the horizontal line such that

$BC = 2\sqrt{3}$ m,

Let $\angle ACB = \theta$

$$\tan\theta = \frac{AB}{BC}$$

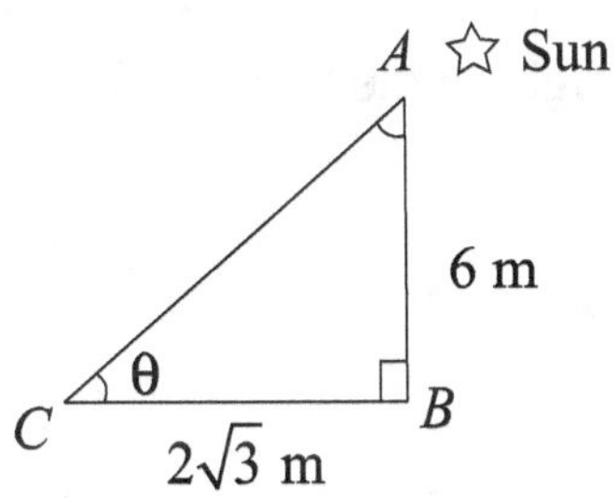

$$\Rightarrow \tan\theta = \frac{6}{2\sqrt{3}} = \sqrt{3}$$

$\therefore \tan\theta = \tan 60°$

$\therefore \theta = 60°$

192. The angle subtended by a vertical pole of height 100 m at a point on the ground $100\sqrt{3}$ m from the base is, has measure of **(2023)**

(a) 90° (b) 60°

(c) 45° (d) 30°

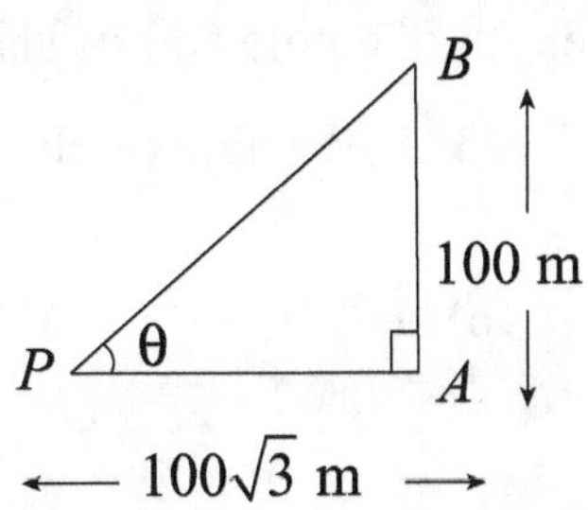

Solution : (d)

$$\tan\theta = \frac{AB}{AP}$$

$$\Rightarrow \tan\theta = \frac{100}{100\sqrt{3}} = \frac{1}{\sqrt{3}}$$

$\therefore \tan\theta = \tan 30°$

$\therefore \theta = 30°$

193. The angle subtended by a tower of height 200 meter at a point 200 meter from the base is **(2023)**

(a) 45° (b) 60°

(c) 30° (d) 90°

Solution : (a)

Let AB be the height as the tower and BP be the horizontal line such that

$BP = 200$ meters

$$\tan\theta = \frac{AB}{AP}$$

$$\Rightarrow \tan\theta = \frac{200}{200} = 1$$

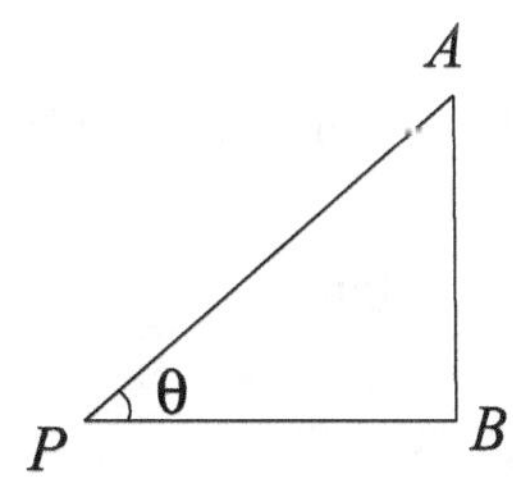

$\therefore \tan\theta = \tan 45°$

$\therefore \theta = 45°$

194. The angle of elevation of the top of a 30 m high tower at a point 30 m away from the base of the tower is **(2023)**

(a) 60° (b) 45°

(c) 30° (d) 90°

Solution : (b)

Let AB be the height of the tower and BC be the horizontal line such that

$BC = 30$ m,

Let $\angle ACB = \theta$

$$\tan\theta = \frac{AB}{BC}$$

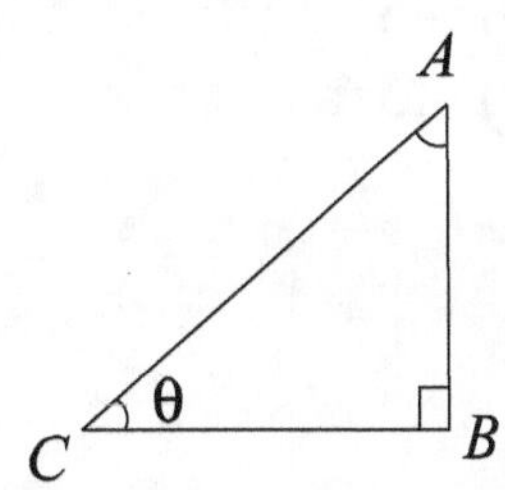

$\Rightarrow \tan\theta = \frac{30}{30} = 1$

$\therefore \tan\theta = \tan 45°$

$\therefore \theta = 45°$

195. The angle of elevation of the top of a 15m high tower at a point $15\sqrt{3}$ m away from the base of the tower is. **(2023)**

(a) 30° (b) 45°

(c) 60° (d) 90°

Solution : (b)

Let AB be the height of the tower and BC be the horizontal line such that

$BC = 15\sqrt{3}$ m

Let $\angle ACB = \theta$

$\tan\theta = \frac{AB}{BC}$

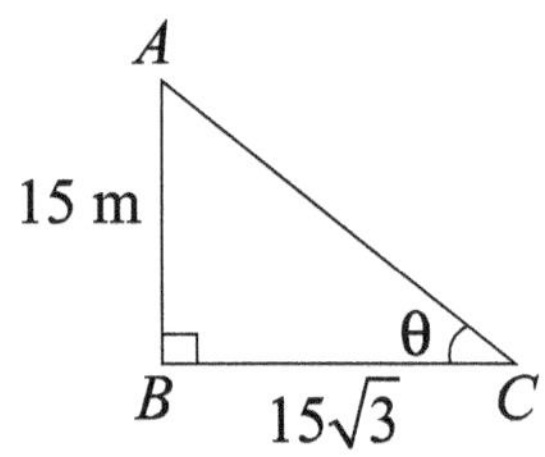

$\Rightarrow \tan\theta = \frac{15}{15\sqrt{3}} = \frac{1}{\sqrt{3}}$

$\therefore \tan\theta = \tan 30°$

$\theta = 30°$

196. In the given figure, ΔPQR is a right triangle right angled at Q. If $PQ = 4$ cm and $PR = 8$ cm, then $\angle P$ is **(2023)**

(a) 60° (b) 45°

(c) 30° (d) 15°

Solution :

$\frac{PQ}{PR} = \cos\angle P$

$\Rightarrow \cos\angle P = \frac{4}{8} = \frac{1}{2}$

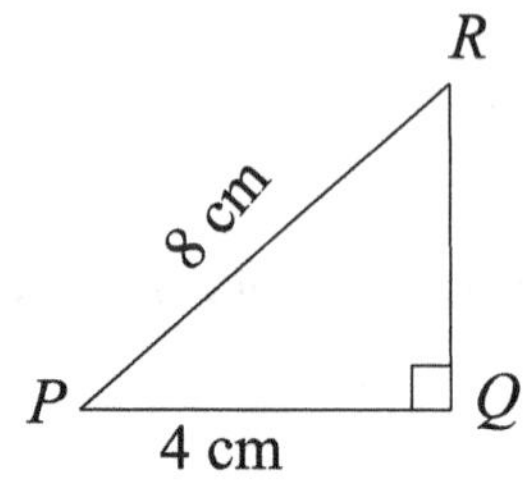

$\Rightarrow \cos P = \cos 60°$

$\therefore \angle P = 60°$

197. At some time of the day, the length of the shadow of a tower is equal to its height. Then the sun's attitude at that time is: **(2024)**

(a) 30° (b) 45°

(c) 60° (d) 90°

Solution : (b)

Let AB be the height of the tower and BC be the shadow,

Let $\angle ACB = \theta$

$\frac{AB}{BC} = \tan\theta$

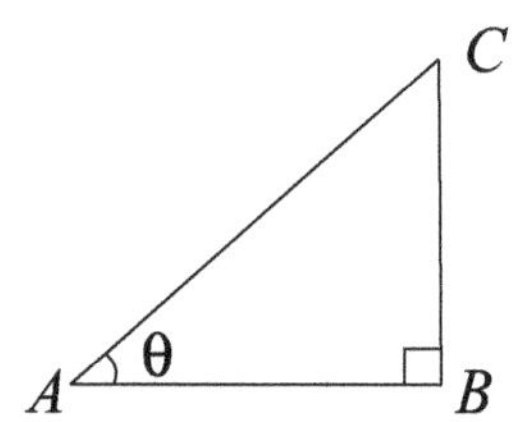

$\tan\theta = \frac{AB}{BC}$

$\Rightarrow \tan\theta = 1$ $[\because AB = BC]$

$\therefore \tan\theta = \tan 45°$

$\therefore \theta = 45°$

198. The ratio of the length of a pole and its shadow on the ground is $1 : \sqrt{3}$. The angle of elevation of the sun is : **(2024)**

(a) 90° (b) 60°

(c) 45° (d) 30°

Solution : (d)

Let AB be the length of the pole and BC be the shadow and $AB = x$, $BC = \sqrt{3}x$, $\angle ACB = \theta$

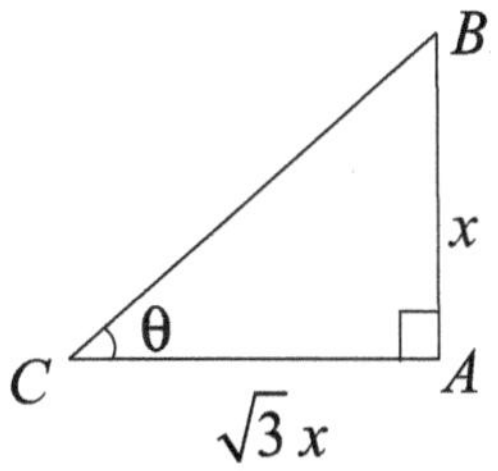

$$\frac{AB}{BC} = \tan\theta$$

$$\Rightarrow \tan\theta = \frac{x}{\sqrt{3}x} = \frac{1}{\sqrt{3}}$$

$\therefore\ \tan\theta = \tan 30°$

$\therefore \theta = 30°$

199. The length of the shadow of a tower on the plane ground is $\sqrt{3}$ times the height of the tower. The angle of elevation of the sun is : **(2024)**

(a) 30° (b) 45°

(c) 60° (d) 90°

Solution :

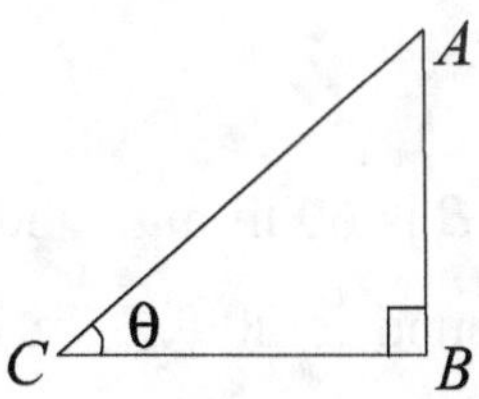

Let AB be the height of the tower and BC be the length of shadow such the $BC = \sqrt{3}\ AB$, $\angle ACB = \theta$

$$\frac{AB}{BC} = \tan\theta$$

$$\Rightarrow \tan\theta = \frac{AB}{\sqrt{3}\,AB} = \frac{1}{\sqrt{3}}$$

$\therefore\ \tan\theta. = \tan 30°$

$\therefore\ \theta = 30°$

200. A pole $7\sqrt{3}$ high casts a shadow 21 m long on the ground, then the sun's elevation is. **(2024)**

(a) 30° (b) 45°

(c) 60° (d) 90°

Solution :

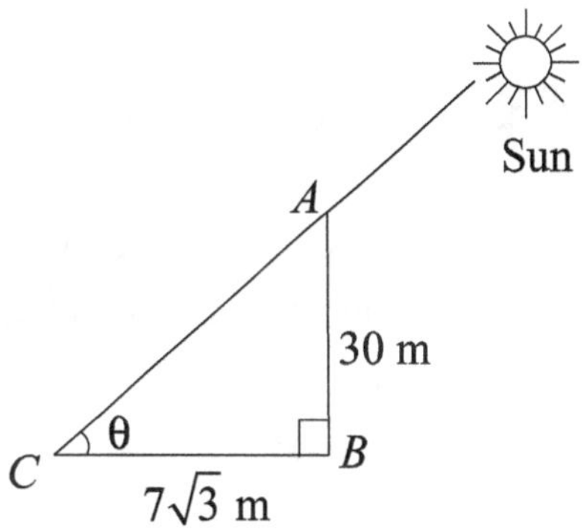

Let AB be the height of the pole and BC be the length of shadow $\angle ACB = \theta$

$$\frac{AB}{BC} = \tan\theta$$

$$\Rightarrow \frac{7\sqrt{3}}{21} = \tan\theta$$

$$\Rightarrow \tan\theta = \frac{1}{\sqrt{3}}$$

$\therefore\ \tan\theta = \tan 30°$

$\therefore\ \theta = 30°$

201. A lamp post 9 m high casts a shadow $3\sqrt{3}$ m long on the ground. The sun's elevation at this moment is. **(2024)**

(a) 60° (b) 90°

(c) 45° (d) 30°

Solution : (a)

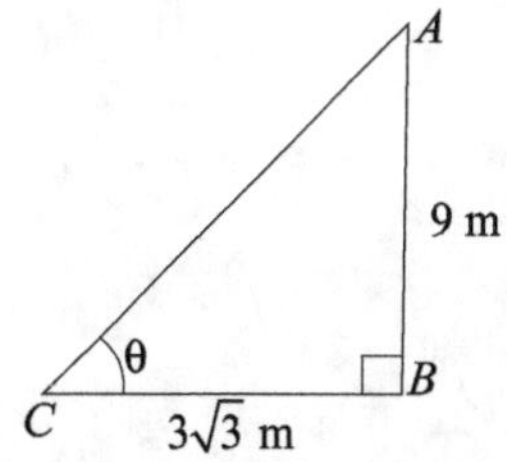

Let AB be the height of the lamp post and BC be the length of the shadow, $\angle ACB = \theta$

$$\frac{AB}{BC} = \tan\theta$$

$$\Rightarrow \frac{9}{3\sqrt{3}} = \tan\theta$$

$$\Rightarrow \tan\theta = \sqrt{3}$$

$$\therefore \tan\theta = \tan 60°$$

$$\therefore \theta = 60°$$

202. A lamp post 15 m high casts a shadow 5 $\sqrt{3}$ m long on the ground. The sun's elevation at this moment is. **(2024)**

(a) 90° (b) 60°

(c) 45° (d) 30°

Solution : (b)

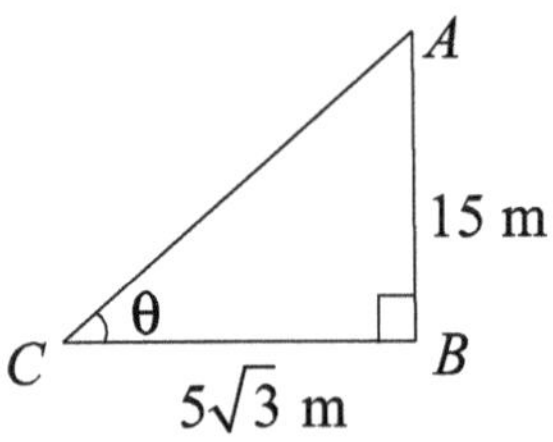

Let AB be the height of the lamp post and BC be the length of the shadow, $\angle ACB = \theta$

$$\frac{AB}{BC} = \tan\theta$$

$$\Rightarrow \tan\theta = \frac{15}{5\sqrt{3}}$$

$$\Rightarrow \tan = \sqrt{3}$$

$$\Rightarrow \tan\theta = \tan 60°$$

$$\therefore \theta = 60°$$

203. The ratio of the height of a tower and the length of its shadow on the ground is $\sqrt{3}$: 1. What is the angle of elevation of the sun. **(2017)**

Solution :

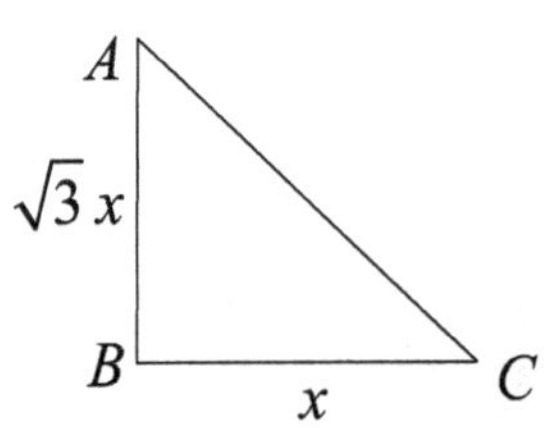

Let AB be the height of the lamp post and BC be the length of the shadow, $\angle ACB = \theta$

$$\frac{AB}{BC} = \tan\theta$$

$$\Rightarrow \tan\theta = \frac{\sqrt{3}x}{x}$$

$$\Rightarrow \tan = \sqrt{3}$$

$$\Rightarrow \tan\theta = \tan 60°$$

$$\therefore \theta = 60°$$

Exercise 8.75

(i) In Figure 1, a tower AB is 30 m high and BC, its shadow on the ground is 30 m long. Find the sun's Attitude.

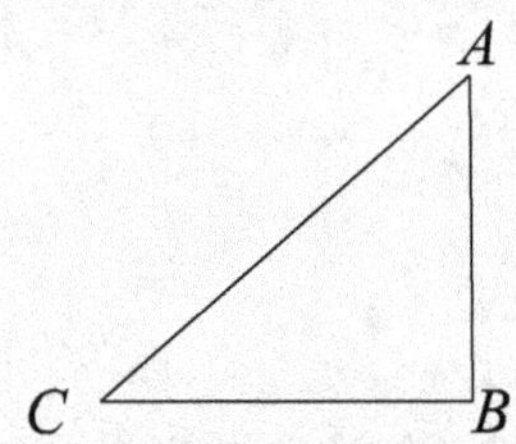

(ii) In Figure 1, a tower AB is 45 m high and BC, its shadow on the ground is 45 $\sqrt{3}$ m long. Find the sun's Attitude.

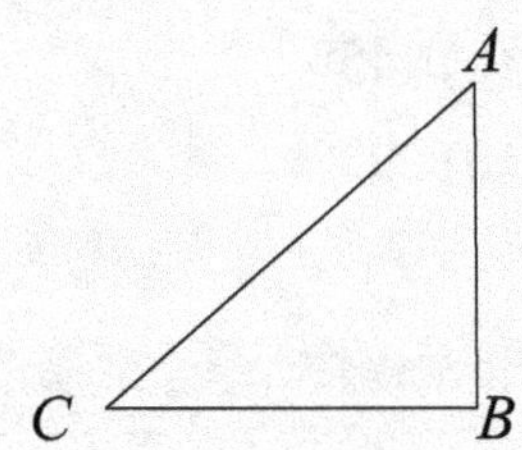

(iii) In Figure 1, a tower AB is 60 m high and BC, its shadow on the ground is 20 $\sqrt{3}$ m long. Find the sun's Attitude.

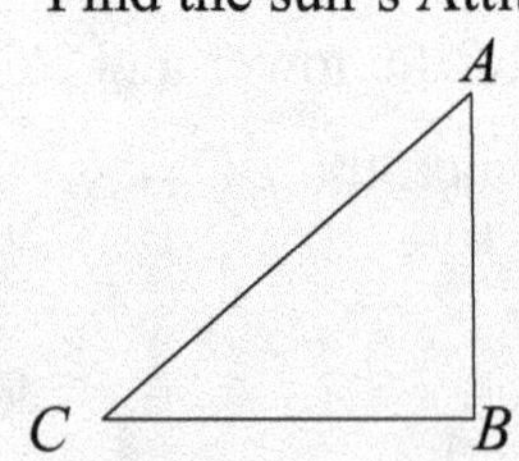

(iv) In a pole 9 m high casts a shadow 9 $\sqrt{3}$ m on the ground, then sun's elevation is

(a) 30° (b) 45°

(c) 75° (d) 90°

(v) In a pole 8 m high casts a shadow 8 m on the ground, then sun's elevation is

(a) 15° (b) 30°
(c) 45° (d) 60°

(vi) In a pole 9 m high casts a shadow $3\sqrt{3}$ m on the ground, then sun's elevation is

(a) 30° (b) 60°
(c) 75° (d) 90°

(vii) The ratio of the length of a pole and its shadow on the ground is 1 : 1. The angle of elevation of the sun is.

(a) 30° (b) 45°
(c) 60° (d) 90°

(viii) The ratio of the length of a pole and its shadow on the ground is $\sqrt{3}$: 1. The angle of elevation of the sun is.

(a) 30° (b) 60°
(c) 45° (d) 90°

(ix) The ratio of the length of a pole and its shadow on the ground is 3: $\sqrt{3}$. The angle of elevation of the sun is.

(a) 30° (b) 45°
(c) 60° (d) 90°

(x) The ratio of the height of a tower and the length of its shadow on the ground is 1 : $\sqrt{3}$. What is the angle of elevation of the sun ?

204. The tops of two towers of height x and y, standing on level ground, subtend angles of 30° and 60° respectively at the center of the line joining their feet, then find $x : y$ **(2015)**

Solution :

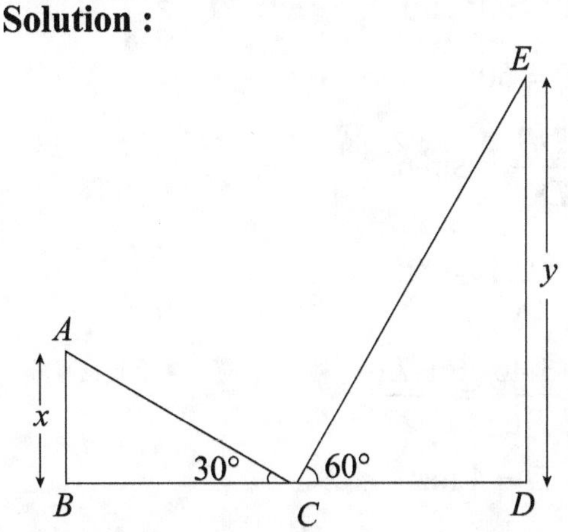

$$\frac{AB}{BC} = \tan 30°$$

$$\Rightarrow \frac{x}{BC} = \frac{1}{\sqrt{3}}$$

$$\Rightarrow . BC = \sqrt{3}x \quad \text{.........(i)}$$

$$\frac{ED}{CD} = \tan 60°$$

$$\Rightarrow \frac{y}{CD} = \sqrt{3}$$

$$\Rightarrow CD = \frac{y}{\sqrt{3}} \quad \text{............(ii)}$$

Since C is the midpoint of BD

$\therefore BC = DC$

$$\Rightarrow \sqrt{3}x = \frac{y}{\sqrt{3}}$$

$$\Rightarrow 3x = y$$

$$\Rightarrow \frac{x}{y} = \frac{1}{3}$$

$\therefore x : y = 1 : 3$

Exercise 8.76

(i) The tops of two towers of height x and y, standing on level ground, subtend angles of 60° and 30° respectively at the center of the line joining their feet, then find $x : y$.

(ii) The tops of two towers of height x and y, standing on level ground, subtend angles of 45° and 45° respectively at the center of the line joining their feet, then find $x : y$

205. A pole casts a shadow of length $2\sqrt{3}$ m on the ground, when the suns elevation in 60°. Find the height of the pole. **(2011)**

Solution :

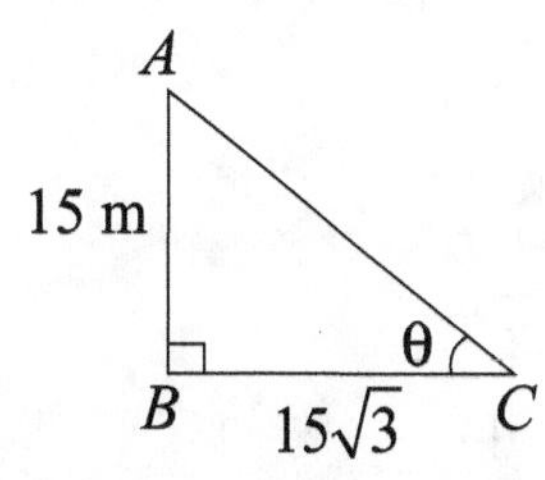

Let AB be the height of the pole and BC be the length of the shadow, $\angle ACB = 60°$

$$\frac{AB}{BC} = \tan 60^0$$

$$\Rightarrow \frac{AB}{2\sqrt{3}} = \sqrt{3}$$

$\Rightarrow AB = 6$ m.

Exercise 8.77

(i) A pole casts a shadow of length 3 $\sqrt{3}$ m on the ground, when the sun's elevation in 30°. Find the height of the pole.

(ii) A pole casts a shadow of length 27 m on the ground, when the sun's elevation in 45°. Find the height of the pole.

(iii) A pole casts a shadow of length 9 $\sqrt{3}$ m on the ground, when the sun's elevation in 60°. Find the height of the pole.

206. A ladder, lening against a wall, make an angle 60° with the horizontal. If the foot of the ladder is 2.5 m away from the wall, find the length of the ladder. **(2016)**

Solution :

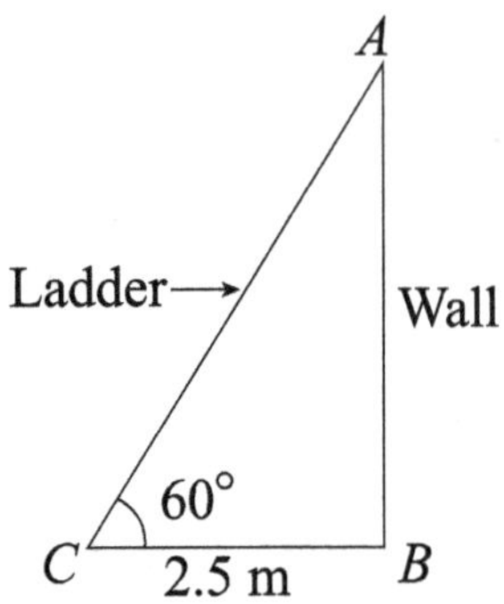

Let AC be the height of the ladder and BC be horizontal line such that

$BC = 2.5$m, $\angle ACB = 60°$

$$\frac{BC}{AC} = \cos 60°$$

$$\Rightarrow \frac{2.5}{AC} = \frac{1}{2}$$

$\Rightarrow AC = 5$ m.

Exercise 8.78

(i) A ladder, lening against a wall, make an angle 60° with the horizontal. If the foot of the ladder is 3.5 m away from the wall, find the length of the ladder.

(ii) A ladder, lening against a wall, make an angle 60° with the horizontal. If the foot of the ladder is 4.5 m away from the wall, find the length of the ladder.

(iii) A ladder, lening against a wall, make an angle 45° with the horizontal. If the foot of the ladder is 4 m away from the wall, find the length of the ladder.

207. In fig.1, AB is a 6 m high pole and CD is a ladder inclined at an angle of 60° to the horizontal and reaches up to a point D of pole. If $AD = 2.54$ m, find the length of the ladder. (use $\sqrt{3}$ = 1.73) **(2016)**

Solution :

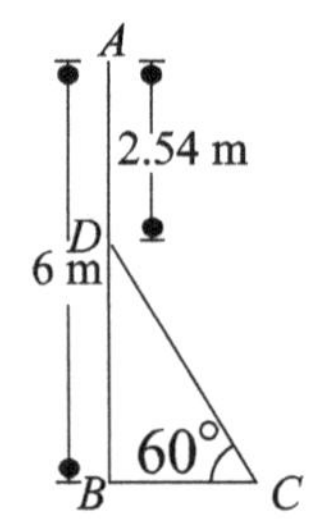

$DB = AB - AD$

$= 6 - 2.54$

$= 3.46$ m

Now, in ΔDBC

$\angle B = 90°$ and

$\angle C = 60°$

$$\therefore \frac{DB}{DC} = \sin 60°$$

$$\Rightarrow \frac{3.46}{DC} = \frac{\sqrt{3}}{2}$$

$$\Rightarrow \frac{3.46}{DC} = \frac{1.73}{2}$$

$\Rightarrow D = 4$ m.

Exercise 8.79

(i) In fig.1, AB is a 8 m high pole and CD is a ladder inclined at an angle of 60° to the horizontal and reaches up to a point D of pole. If $AD = 3$ m, find the length of the ladder.

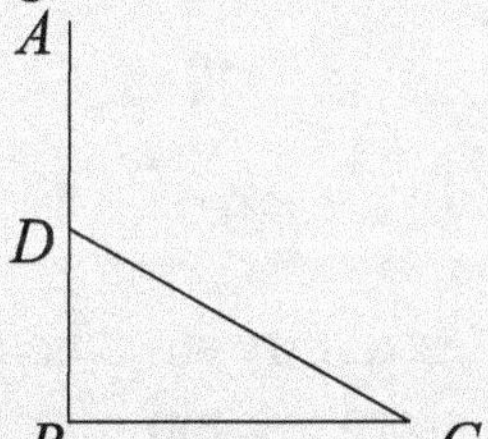

Fig.1

(ii) In fig.1, AB is a 9 m high pole and CD is a ladder inclined at an angle of 60° to the horizontal and reaches up to a point D of pole. If $AD = 4.81$ m, find the length of the ladder.

(iii) In fig.1, AB is a 10 m high pole and CD is a ladder inclined at an angle of 60° to the horizontal and reaches up to a point D of pole. If $AD = 6$ m, find the length of the ladder.

208. An observes, 1.7 m tall, is $20\sqrt{3}$ m away from a tower. The angle of elevation from the eye of observer to the top of tower is 30°. Find the height of tower. **(2016)**

Solution :

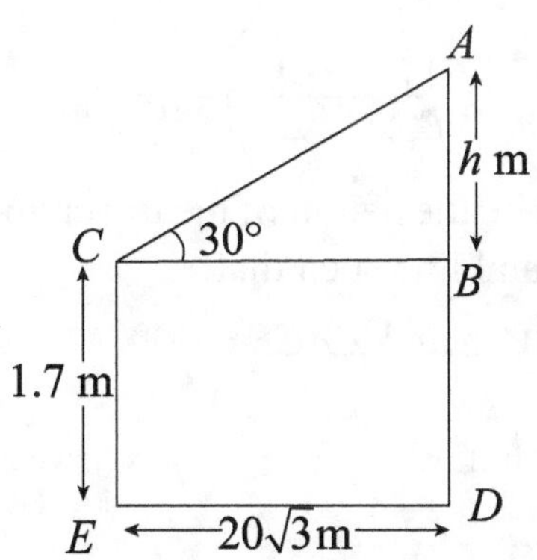

Let AD be the height of the tower, CE be the height of the observer and DE be the horizontal line such that

$DE = 20\sqrt{3}$ m

$\angle ACB = 30°$

$$\frac{AB}{BC} = \tan 30°$$

$$\Rightarrow \frac{AB}{20\sqrt{3}} = \frac{1}{\sqrt{3}}$$

$[\because BC = DE]$

$\Rightarrow AB = 20$ m.

Now,

$AD = AB + BD$

$= AB + EC\ [\because BD = EC]$

$= 20 + 1.7$

$= 21.70$ m

Exercise 8.80

(i) An observer, 1.5 m tall, is 28.5 m away from a tower. The angle of elevation from the eye of observer to the top of tower is 45°. Find the height of tower.

(ii) An observer, 2.1 m tall, is $40\sqrt{3}$ m away from a tower. The angle of elevation from the eye of observer to the top of tower is 30°. Find the height of tower.

(iii) An observer, 1.6 m tall, is 30 m away from a tower. The angle of elevation from the eye of observer to the top of tower is 30°. Find the height of tower.

209. In figure – 3, a tightly stretched rope of length 20 m is tied from the top of a vertical pole to the ground. Find the height of the pole if the angle made by the rope with the ground is 30°. **(2020)**

Solution :

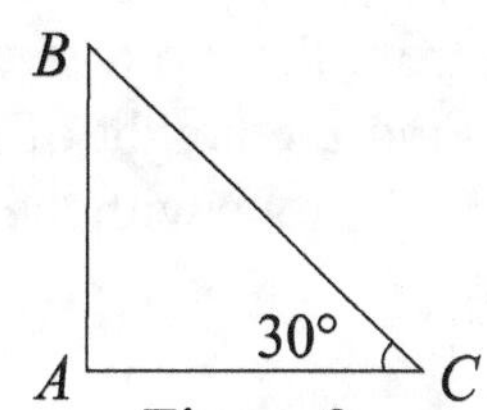

Figure-3

Let AB be the height of the vertical pole and BC be the length of rope,

$BC = 20$ m and $\angle ACB = 30^0$

$$\frac{AB}{BC} = \sin 30°$$

$$\Rightarrow \frac{AB}{20} = \frac{1}{2}$$

$\therefore AB = 10$ m.

210. A ladder 14 m long just reaches the top of a vertical wall. If the ladder makes an angle of 60° with the wall, then the height of the wall is: **(2024)**

(a) $14\sqrt{3}$ m (b) 7 m

(c) 14 m (d) $7\sqrt{3}$ m.

Solution : (d)

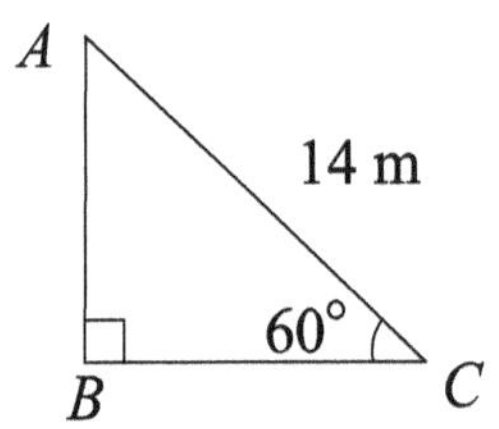

Let AB be the height of the vertical wall.

AC be the length of the ladder.

$AC = 14$ m

and $\angle ACB = 60°$

$$\frac{AB}{AC} = \sin 60°$$

$$\Rightarrow \frac{AB}{14} = \frac{\sqrt{3}}{2}$$

$\Rightarrow AB = 7\sqrt{3}$ m.

Exercise 8.81

(i) In figure – 1, a tightly stretched rope of length 30 m is tied from the top of a vertical pole to the ground. Find the height of the pole if the angle made by the rope with the ground is 30°.

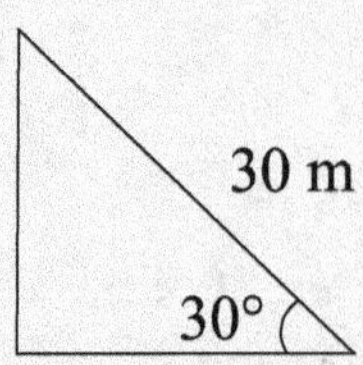

(ii) In figure – 1, a tightly stretched rope of length 40 m is tied from the top of a vertical pole to the ground. Find the height of the pole if the angle made by the rope with the ground is 30°.

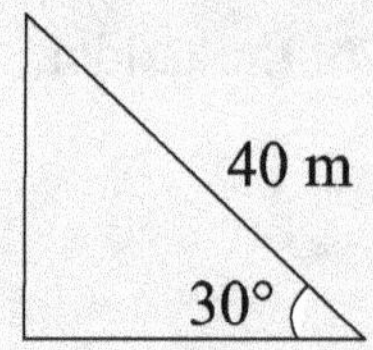

(iii) In figure – 1, a tightly stretched rope of length 60 m is tied from the top of a vertical pole to the ground. Find the height of the pole if the angle made by the rope with the ground is 30°.

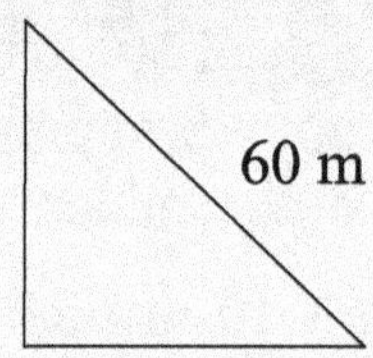

211. In figure-4, the angle of elevation of the top of a tower from a point C on the ground, which is 30 m away from the foot of the tower is 30°. Find the height of the tower. **(2020)**

Solution :

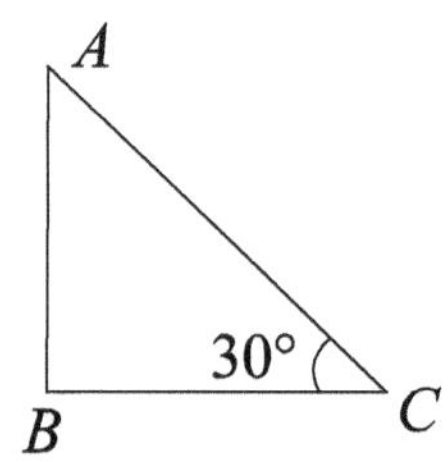

Let AB be the height of the tower and BC be the horizontal line such that

$BC = 30$ m and $\angle ACB = 30°$

$$\frac{AB}{AC} = \tan 30°$$

$$\Rightarrow \frac{AB}{30} = \frac{1}{\sqrt{3}}$$

$\Rightarrow AB = 10\sqrt{3}$ m.

212. The angle of elevation of the top of the tower AB from a point C on the ground, which is 60 m away from the foot of the tower, is 30°, as shown in figure 2. Find the height of the tower.

Solution :

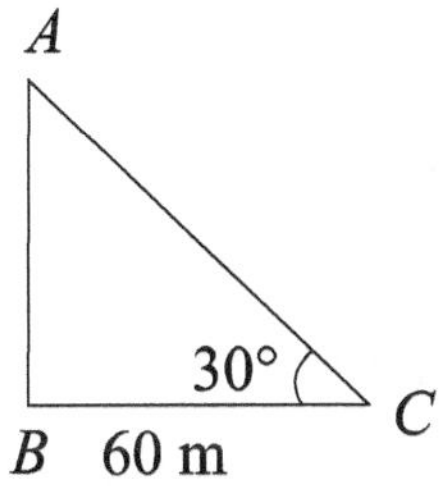

Let AB be the height of the tower and BC be the horizontal line such that

$BC = 60$ m and $\angle ACB = 30°$

$$\frac{AB}{BC} = \tan 30°$$

$$\Rightarrow \frac{AB}{60} = \frac{1}{\sqrt{3}}$$

$\therefore AB = 20\sqrt{3}$ m.

213. In Figure-2, the angle of elevation of the top of a tower AC from a point B on the ground is 60°. If the height of the tower is 20 m, find the distance of the point from the foot of the tower. **(2020)**

Solution :

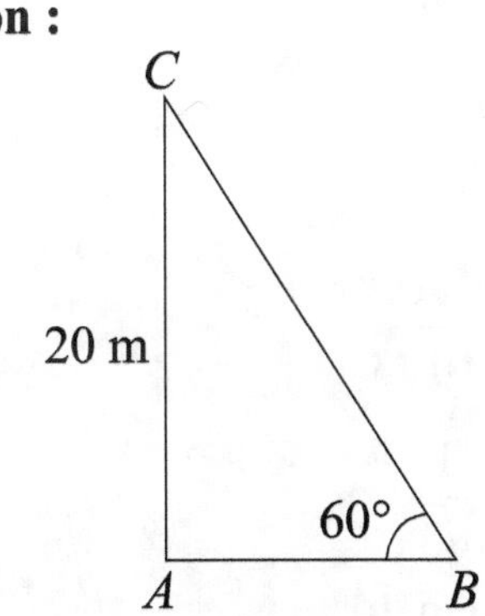

Figure-2

Let AC be the height of the tower and BC be the horizontal line such that

$\angle ABC = 60°$

$$\frac{AC}{AB} = \tan 60°$$

$$\Rightarrow \frac{20}{AB} = \sqrt{3}$$

$$\Rightarrow AB = \frac{20}{\sqrt{3}} = \frac{20\sqrt{3}}{3} \text{ m.}$$

214. A tree casts a shadow. 7 m long on the ground, when the angle of elevation of the sun is 45°. The height of the tree is **(2024)**

(a) $7\sqrt{3}$ m (b) $\frac{7}{3}\sqrt{3}$ m

(c) 7 m (d) 3.5 m.

Solution :

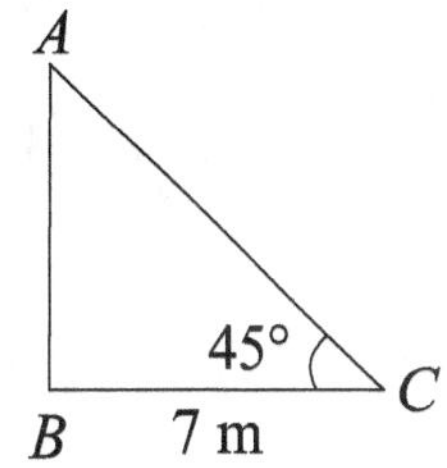

Let AB be the height of the tree and BC be the horizontal line,

$\angle ACB = 45°$

$$\frac{AB}{BC} = \tan 45°$$

$$\Rightarrow \frac{AB}{7} = 1$$

$\Rightarrow AB = 7$ m.

215. In ΔABC, right angled at C, if $\tan A = \frac{8}{7}$, then the value of $\cot B$ is **(2023)**

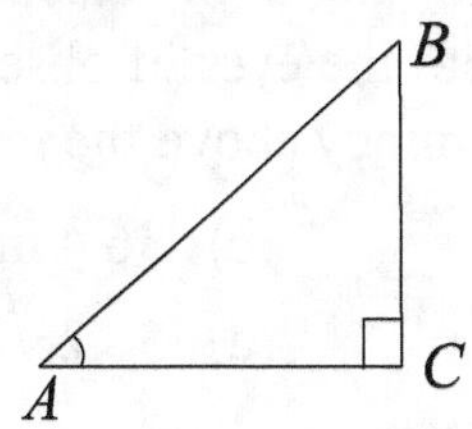

(a) $\frac{7}{8}$ (b) $\frac{8}{7}$

(c) $\frac{7}{\sqrt{13}}$ (d) $\frac{8}{\sqrt{13}}$

Solution : (b)

$$\tan A = \frac{8}{7}$$

$\Rightarrow \tan(90° - B) = \frac{8}{7}$ $[\because A + B = 90°]$

$\Rightarrow \cot B = \frac{8}{7}$

216. 30 m long rope is tightly stretched and tied from the top of pole to the ground. If the rope makes an angle of 60° with the ground, the height of the pole is: **(2025)**

(a) $10\sqrt{3}$ m (b) $30\sqrt{3}$ m

(c) 15 m (d) $15\sqrt{3}$ m

Solution: (d)

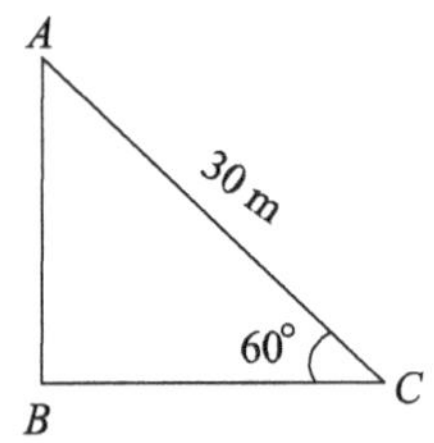

$\frac{AB}{AC} = \sin 60°$

$\Rightarrow \frac{AB}{30} = \frac{\sqrt{3}}{2}$

$\Rightarrow AB = 15\sqrt{3}$ cm

217. A observer 1.8 m tall stands away from a chimney at a distance of 38.2 m along the ground. The angle of elevation of top of chimney from the eyes of observer is 45 : The height of chimney above the ground is **(2025)**

(a) 38.2 m (b) 36.4 m

(c) 40 m (d) $38.2\sqrt{2}$ m

Solution: (c)

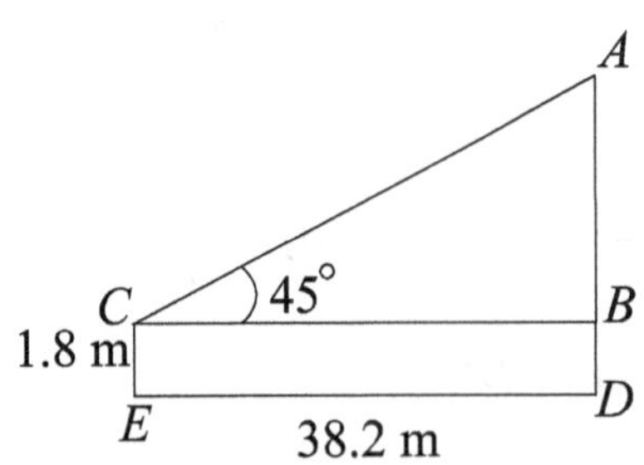

$BC = DE = 38.2$ m

$\frac{AB}{BC} = \tan 45°$

$\Rightarrow \frac{AB}{38.2} = 1$

$\Rightarrow AB = 38.2$ m

$\therefore AD = AB + DB$

$= AB + EC$

$= 38.2 + 1.8$

$= 40$ m

218. A peacock sitting on the top of a tree of height 10 m observes a snake moving on the ground. If the snake is $10\sqrt{3}$ m away from the base of the tree, then angle of depression of the snake from the eye of the peacock is **(2025)**

(a) 30°(b) 45°

(c) 60°(d) 90°

Solution:

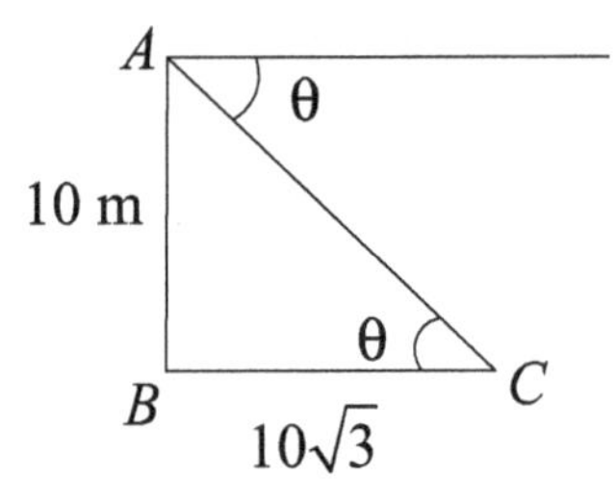

$\tan\theta = \frac{AB}{BC}$

$\Rightarrow \tan\theta = \frac{10}{10\sqrt{3}}$

$\Rightarrow \tan\theta = \frac{1}{\sqrt{3}}$

$\Rightarrow \tan\theta = \tan 30°$

$\therefore \theta = 30°$

219. In the given figure, which of the following angles represents the angle of depression **(2025)**

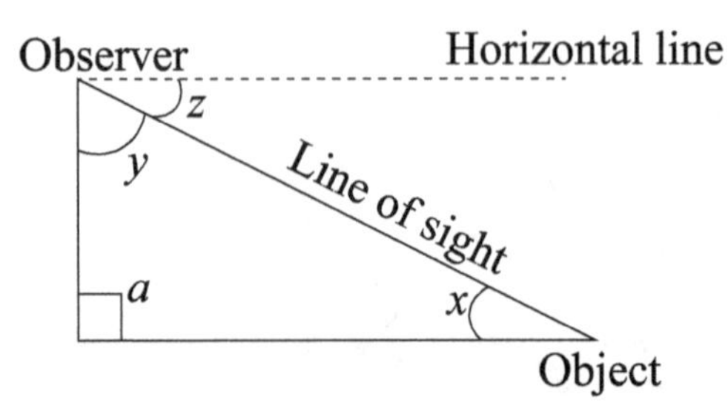

(a) x (b) y
(c) z (d) a

Solution: (c) Z

z

220. The length of the shadow of a tower when the sun's altitude changes from 30° to 60° will **(2025)**

(a) Become shorter (b) Become larger
(c) Remain same (d) Be doubled

Solution: (a)

At 30°

$\frac{h}{b_1} = \tan 30°$

$\Rightarrow \frac{h}{b_1} = \frac{1}{\sqrt{3}}$

$\Rightarrow b_1 = \sqrt{3}h$

At 60°

$\frac{h}{b_2} = \tan 60°$

$\Rightarrow \frac{h}{b_2} = \sqrt{3}$

$\Rightarrow b_2 = \frac{h}{\sqrt{3}}$

$\therefore b_1 > b_2$

221. A 8 m high tree casts a 6 m long shadow on the ground. At the same time, a flag pole caste a shadow 30 m long on the ground. The height of the flag pole is: **(2025)**

(a) 40 m (b) 22.5 m
(c) 44 m (d) 22 m

Solution: (a)

$\tan\theta = \frac{8}{6} = \frac{4}{3}$

Now, $\frac{h}{30} = \tan\theta$

$\Rightarrow \frac{h}{30} = \frac{4}{3}$

$\Rightarrow h = 40$ m.

222. From a point on the ground, which is 60 m away from the foot of a vertical tower, the angle of elevation of the top of the top of the tower is found to be 45°. The height (in metres) of the tower is: **(2026)**

(a) $10\sqrt{3}$ (b) $30\sqrt{3}$
(c) 60 (d) 30

Solution: (c)

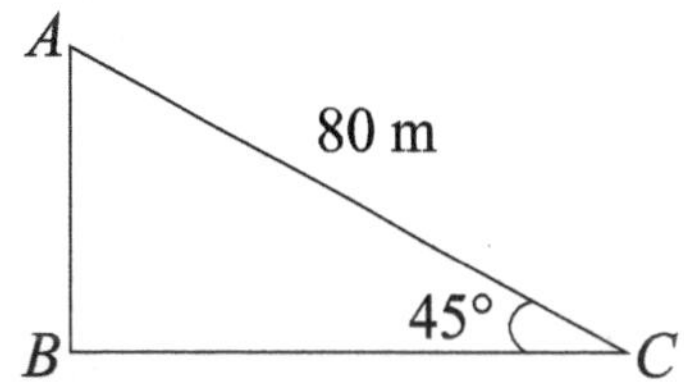

$\frac{AB}{BC} = \tan 45°$

$\Rightarrow \frac{AB}{60} = 1$

$\therefore AB = 60$ m

223. A wire is attached from a point A on the ground to the top of a pole BC making an angle of elevation as 60°. If $AB = 5\sqrt{3}$ m, then length of the wire is: **(2026)**

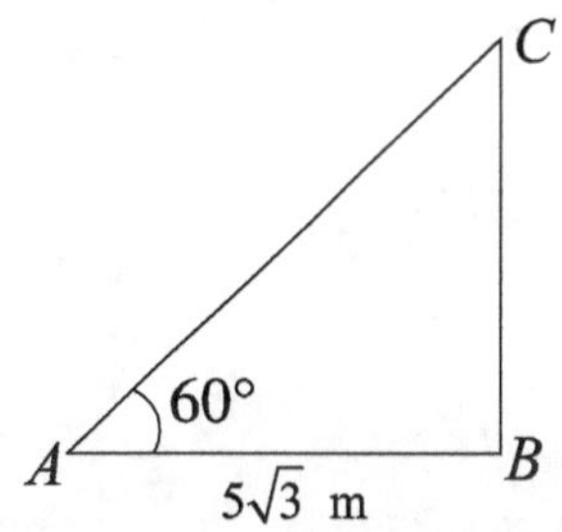

(a) 10 m (b) $10\sqrt{3}$ m
(c) 15 m (d) $\frac{5}{2}\sqrt{3}$ m

Solution: (b)

$\frac{AB}{AC} = \cos 60°$

$\Rightarrow \frac{5\sqrt{3}}{AC} = \frac{1}{2}$

$\therefore AC = 10\sqrt{3}$ m

224. A string of a flying kite is tied to a point on the ground. The length of the string between the kite and the point on the ground is 80 m. The string makes an angle of 30° with the ground. The height of the kite above the ground is: **(2026)**

(a) $20\sqrt{3}$ m (b) 40 m

(c) $40\sqrt{3}$ m (d) $80\sqrt{3}$ m

Solution: (b)

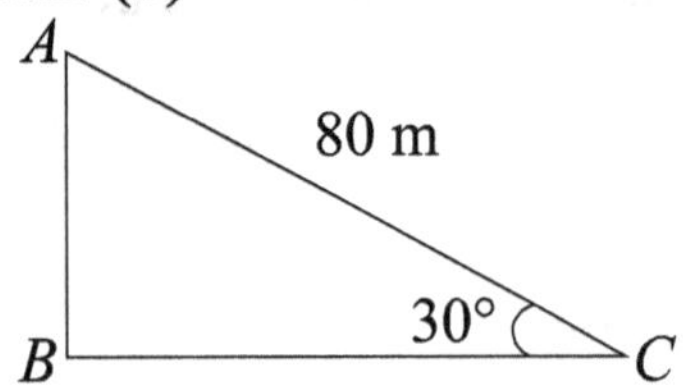

$$\frac{AB}{AC} = \sin 30°$$

$$\Rightarrow \frac{AB}{80} = \frac{1}{2}$$

$\therefore AB = 40$ m

Exercise 8.82

(i) In ΔABC, right angled at C, if $\tan A = \frac{11}{7}$, then the value of $\cot B$ is **(2023)**

(a) $\frac{11}{7}$ (b) $\frac{7}{11}$

(c) $\frac{5}{\sqrt{119}}$ (d) $\frac{3}{\sqrt{117}}$

(ii) In ΔABC, right angled at C, if $\tan B = \frac{4}{3}$, then the value of $\cot A$ is

(a) $\frac{4}{3}$ (b) $\frac{3}{5}$

(c) $\frac{3}{4}$ (d) $\frac{3}{2}$

(iii) In ΔABC, right angled at C, if $\cot A = \frac{3}{5}$, then the value of $\tan B$ is **(2023)**

(a) $\frac{4}{5}$ (b) $\frac{5}{4}$

(c) $\frac{3}{4}$ (d) $\frac{3}{5}$

ANSWER KEYS

Exercise – 8.1

(i) $\frac{3}{5}$ (ii) $\frac{13}{12}$ (iii) $\frac{3}{4}$

(iv) $\frac{\sqrt{5}}{2}$ (v) $\frac{5}{4}$ (vi) $\frac{\sqrt{65}}{7}$

(vii) $\frac{25}{24}$ (viii) $\frac{3}{4}$ (ix) $\frac{4}{3}$

(x) $\frac{24}{7}$ (xi) $\frac{24}{7}$ (xii) $\frac{4}{3}$

(xiii) $\frac{\sqrt{5}}{2}$

Exercise – 8.2

(i) $\frac{7}{4}$ (ii) $\frac{26}{15}$ (iii) $\frac{9}{5}$

(iv) (a) $\frac{4}{5}$ (v) (c) $\frac{5}{8}$ (vi) (c) $\frac{3}{7}$

(vii) (b) $\frac{5}{3}$ (viii) (d) $\frac{7}{3}$ (ix) (a) $\frac{5}{4}$

(x) $\frac{9}{16}$ (xi) $\frac{9}{25}$ (xii) $\frac{81}{100}$

(xiii) $\frac{9}{16}$ (xiv) $\frac{9}{25}$ (xv) $\frac{4}{9}$

Exercise – 8.3

(i) 1 (ii) ±1 (iii) 1 (iv) 1

Exercise – 8.4

(i) 0 (ii) $\frac{a^2-b^2}{a^2+b^2}$ (iii) $\frac{1}{3}$

(iv) $\frac{1}{7}$

Exercise – 8.5

(i) $\frac{5}{6}$ (ii) 0.62 (iii) $\frac{25}{32}$

Exercise – 8.6

(i) 1 (ii) 2

Exercise – 8.7

(i) 1 (ii) –1 (iii) –1

Exercise – 8.8

(i) 2 (ii) 3

Exercise – 8.9

(i) $\cot^2\theta$ (ii) $\tan^4 A$ (iii) $\cos^2 A$

Exercise – 8.10

(i) 1 (ii) $\tan^2 A$ (iii) $\cot^2 A$

Exercise – 8.11

(i) (a) 1 (ii) (c) 1 (iii) (a) –1

Exercise – 8.12

(i) (a) $\sin\theta$ (ii) (c) $\cos\theta$ (iii) (a) $\sin\theta$

Exercise – 8.13

(i) (d) $\operatorname{cosec}^2 A$ (ii) (d) $\operatorname{cosec}^2 A$ (iii) (a) $\sec^2 A$

Exercise – 8.14

(i) (b) –1 (ii) (c) –4 (iii) (b) 3

(iv) (c) –5

Exercise – 8.15

(i) 1

Exercise – 8.16

(i) (b) –2 (ii) (d) –a (iii) (d) $\frac{3}{2}$

Exercise – 8.17

(i) (c) $\frac{\sqrt{1-\cos^2\theta}}{\cos\theta}$ (ii) (b) $\frac{\sqrt{1+\tan^2\theta}}{\tan\theta}$

Exercise – 8.18

(i) (b) 4 (ii) (c) $\frac{3}{2}$ (iii) (d) $\frac{5}{4}$

Exercise – 8.19

(i) (d) $\cot^2\theta - \operatorname{cosec}^2\theta = 1$ (ii) (a) $2\sin^2\theta - 1 = \cos^2\theta - \sin^2\theta$ (iii) (c) $(1-\sin^2\theta)\sec^2\theta = 2$

Exercise – 8.20

(i) $\frac{1}{2}$ **(ii)** $\frac{1}{a^2}$ **(iii)** $\frac{1}{p^2}$

(iv) $\frac{1}{4}$ **(v)** $\frac{1}{3}$

Exercise – 8.21

(i) 36 **(ii)** 4 **(iii)** 144

Exercise – 8.22

(i) (b) a^2 **(ii)** (c) 36 **(iii)** (b) $\frac{1}{a^2}$

Exercise – 8.23

(i) (c) 36 **(ii)** (a) 144 **(iii)** (c) 1

Exercise – 8.24

(i) (c) 1 **(ii)** (a) 1 **(iii)** (c) 1

Exercise – 8.25

(i) –1 **(ii)** $\frac{11}{4}$ **(iii)** –3

Exercise – 8.26

(i) $\frac{1}{2}$ **(ii)** $\frac{1}{6}$ **(iii)** 9

Exercise – 8.27

(i) 4 **(ii)** $2\sqrt{2}$ **(iii)** 4

Exercise – 8.28

(i) 1 **(ii)** 1 **(iii)** 1

Exercise – 8.29

(i) (d) $\frac{\sqrt{3}}{2}$ **(ii)** (b) cos 60° **(iii)** (b) $-\frac{1}{2}$

Exercise – 8.30

(i) (d) 2 **(ii)** (a) $\frac{14}{3}$ **(iii)** (c) 3

Exercise – 8.31

(i) (b) $\frac{\sqrt{2}-1}{2}$ **(ii)** (d) $\frac{\sqrt{2}+1}{2}$ **(iii)** (b) $\frac{3(\sqrt{2}+1)}{2}$

Exercise – 8.32

(i) (d) 7 **(ii)** (c) –1 **(iii)** (d) –2

Exercise – 8.33

(i) (c) –1 **(ii)** (d) $-\frac{1}{2}$ **(iii)** (d) –5

Exercise – 8.34

(i) (c) 0 **(ii)** (c) 2 **(iii)** (b) 2

Exercise – 8.35

(i) 4 **(ii)** 1 **(iii)** 1

Exercise – 8.36

(i) 8 **(ii)** 7 **(iii)** 0

Exercise – 8.37

(i) 2 **(ii)** 3 **(iii)** 1

Exercise – 8.38

(i) (b) –2 **(ii)** (a) –2 **(iii)** (d) 0

Exercise – 8.39

(i) (d) –4 **(ii)** (a) $-\frac{7}{2}$ **(iii)** (a) 4

Exercise – 8.40

(i) 30° **(ii)** 30° **(iii)** 45°

Exercise – 8.41

(i) $\frac{1}{\sqrt{2}}$ **(ii)** $\frac{\sqrt{3}}{2}$ **(iii)** $\frac{1}{2}$

Exercise – 8.42

(i) $\frac{\sqrt{3}}{2}$ **(ii)** $\frac{\sqrt{3}}{2}$ **(iii)** $\frac{3}{2}$

Exercise – 8.43

(i) $A = 52\frac{1}{2}°$ **(ii)** $A = 37\frac{1}{2}°$

Exercise – 8.44

(i) (b) 45° **(ii)** (a) 30° **(iii)** (c) 45°

(iv) (c) 60° **(v)** (b) 45° **(vi)** (d) 45°
(vii) (c) 60° **(viii)** (c) 60°

Exercise – 8.45

(i) (b) $\frac{\sqrt{3}}{2}$ **(ii)** (a) 1 **(iii)** (d) $\frac{1}{2}$

Exercise – 8.46

(i) (c) $\frac{1}{4}$ **(ii)** (d) $\frac{3\sqrt{3}}{2}$ **(iii)** (a) $\sqrt{2}$

Exercise – 8.47

(i) (b) 30° **(ii)** (c) 45° **(iii)** (c) 45°

Exercise – 8.48

(i) (a) 3 **(ii)** (d) $\frac{1}{3}$ **(iii)** (b) 3

Exercise – 8.49

(i) (d) 3 **(ii)** (c) 3 **(iii)** (a) 0

Exercise – 8.50

(i) (c) 60° **(ii)** (b) 30° **(iii)** (d) 90°

Exercise – 8.51

(i) (a) $\frac{1}{\sqrt{2}}$ **(ii)** (c) $\sqrt{3}$ **(iii)** (b) 0

Exercise – 8.52

(i) 0 **(ii)** 0 **(iii)** 0

Exercise – 8.53

(i) 0 **(ii)** 0 **(iii)** 0

Exercise – 8.54

(i) 1 **(ii)** 1 **(iii)** 1

Exercise – 8.55

(i) cos 18° + sin 23° **(ii)** cos 3° + sin 29°
(iii) cos 13° + sin 11°

Exercise – 8.56

(i) 2 **(ii)** 2 **(iii)** 2

Exercise – 8.57

(i) $\frac{3}{2}$ **(ii)** −1 **(iii)** 1
(iv) 0 **(v)** $\frac{1}{2}$ **(vi)** $\frac{5}{3}$

(vii) 1 **(viii)** 1 **(ix)** 1
(x) 1

Exercise – 8.58

(i) 2 **(ii)** 2 **(iii)** 2
(iv) 2 (v) 2

Exercise – 8.59

(i) 2 **(ii)** 2 **(iii)** 0

Exercise – 8.60

(i) 15° **(ii)** 20° **(iii)** 40°

Exercise – 8.61

(i) (b) 30° **(ii)** (d) 50° **(iii)** (a) 25°

Exercise – 8.62

(i) (b) 46° **(ii)** (a) 28° **(iii)** (d) 49°

Exercise – 8.63

(i) 70° **(ii)** 23° **(iii)** 62°

Exercise – 8.64

(i) (b) 45° **(ii)** (a) 45° **(iii)** (c) 45°

Exercise – 8.65

(i) 1 **(ii)** 1 **(iii)** 1 **(iv)** 1
(v) 1

Exercise – 8.66

(i) 1 **(ii)** 1 **(iii)** 1

Exercise – 8.67

(i) 1 **(ii)** 1 **(iii)** 1 **(iv)** 1 **(v)** 1
(vi) 1 **(vii)** 1 **(viii)** 1

Exercise – 8.68

(i) 1 **(ii)** 1 **(iii)** 1

Exercise – 8.69

(i) a **(ii)** a **(iii)** a **(iv)** a **(v)** d

Exercise – 8.70

(i) (d) 44 cm **(ii)** (a) 55 cm **(iii)** (b) $\frac{11}{3}$ cm

Exercise – 8.71

(i) (d) 30° **(ii)** (c) $\left(\frac{25}{2}\right)^{\circ}$ **(iii)** (a) 20°

Exercise – 8.72

(i) (a) $10\sqrt{3}$ **(ii)** (c) $60\sqrt{3}$ **(iii)** (d) 100

Exercise – 8.73

(i) (c) 60° **(ii)** (a) 45°

Exercise – 8.74

(i) (b) $50\sqrt{3}$ **(ii)** (c) $60\sqrt{3}$ **(iii)** (b) $10\sqrt{3}$

Exercise – 8.75

(i) 45° **(ii)** 30° **(iii)** 60°

(iv) (a) 30° **(v)** (c) 45° **(vi)** (b) 60°

(vii) (b) 45° **(viii)** (b) 60° **(ix)** (c) 60°

(x) 30°

Exercise – 8.76

(i) 3 : 1 **(ii)** 1 : 1

Exercise – 8.77

(i) 3 m **(ii)** 27 m **(iii)** 27 m

Exercise – 8.78

(i) $3.5\sqrt{3}$ m **(ii)** $4.5\sqrt{3}$ m **(iii)** 4 m

Exercise – 8.79

(i) $\frac{10\sqrt{3}}{3}$ m **(ii)** 4.84 m **(iii)** 8 m

Exercise – 8.80

(i) 30 m **(ii)** 42.1 m **(iii)** 18.82m

Exercise – 8.81

(i) 15 m **(ii)** 20 m **(iii)** 30m

Exercise – 8.82

(i) (a) $\frac{11}{7}$ **(ii)** (a) $\frac{4}{3}$ **(iii)** (d) $\frac{3}{5}$

9 Mensuration

Key Point Perimeter of a semicircle of radius r unit $= (\pi + 2)r$ unit

1. If the diameter of a semicircular protractor is 14 cm, then find its perimeter **(2009)**

Solution:

Perimeter of a semicircular protractor

$= (\pi + 2)\, r$

$= \left(\frac{22}{7} + 2\right) \times 7$

$= \frac{36}{7} \times 7$

$= 36$ cm.

2. If the radius of a semi-circular protractor is 7 cm, then its perimeter is: **(2023)**

(a) 11 cm (b) 14 cm
(c) 22 cm (d) 36 cm

Solution: (d) 36 cm

Exercise 9.1

(i) If the diameter of a semicircular protractor is 21 cm, then find its perimeter.

(a) 62 cm (b) 52 cm
(c) 54 cm (d) 82 cm

(ii) If the radius of a semi-circular protractor is 14 cm, then its perimeter is:

(a) 14 cm (b) 28 cm
(c) 72 cm (d) 108 cm

(iii) If the perimeter of a semi-circular protractor is 108 cm, then its radius is

(a) 7 cm (b) 14 cm
(c) 21 cm (d) 28 cm

Key Point Circumference of a circle of radius r is $2\pi r$

3. If the difference between the Circumference and the radius of a circle in 37 cm, then using $\pi = \frac{22}{7}$, the circumference (in cm) of the circle is: **(2013)**

(a) 154 (b) 44
(c) 14 (d) 7

Solution: (b)

$2\pi r - r = 37$

$\Rightarrow (2\pi - 1)\, r = 37$

$\Rightarrow \left(2 \times \frac{22}{7} - 1\right) r = 37$

$\Rightarrow \frac{37}{7} r = 37$

$\therefore r = 7$ cm

$\therefore$ Circumference of the circle is $2\pi r = 2 \times \frac{22}{7} \times 7$
$= 44$ cm.

4. A parallelogram having one of its sides 5 cm circumscribes a circle. The perimeter of parallelogram is: **(2025)**

(a) 20 cm
(b) Less than 20 cm
(c) More than 20 cm but less than 40 cm
(d) 40 cm

Solution:

Since the parallelogram circumscribes a circle , then it must be a rhombus.

$\therefore$ Perimeter of the rhombus $= 4 \times 5$

$= 20$ cm

5. If a regular hexagon is inscribed in a circle of radius 3 cm, then its perimeter is **(2025)**

(a) 9 cm (b) 18 cm

(c) 27 cm (d) 30 cm

Solution: (b)

Perimeter of the hexagon = 6 × radius of the circle

$= 6 \times 3$

$= 18$ cm

Exercise 9.2

(i) If the difference between the circumference and the radius of a circle in 74 cm, then using $\pi = \frac{22}{7}$, the circumference (in cm) of the circle is

(a) 44 cm (b) 88 cm

(c) 60 cm (d) 14 cm

(ii) If the difference between the circumference and the diameter of a circle in 30 cm, then using $\pi = \frac{22}{7}$, the circumference (in cm) of the circle is

(a) 14 cm (b) 22 cm

(c) 44 cm (d) 30 cm

(iii) If the sum of circumference and radius of a circle is 51 cm, then using $\pi = \frac{22}{7}$, the diameter (in cm) of the circle is

(a) 7 cm (b) 14 cm

(c) 21 cm (d) 35 cm

Key Point **(i)** The ratio of circumferences of two circles is equal to ratio of their radii

(ii) The ratio of perimeters of two semicircles is equal to ratio of their radii.

6. The circumferences of two circles are in the ratio 4:5. What is the ratio of their radii? **(2023)**

(a) 16 : 25 (b) 25 : 16

(c) $2 : \sqrt{5}$ (d) 4 : 5

Solution: (d)

$2\pi r_1 : 2\pi r_2 = 4 : 5$

$\Rightarrow r_1 : r_2 = 4 : 5$

Exercise 9.3

(i) The circumference of two circles are in the ratio 2 : 3. What is the ratio of their radii?

(a) 1 : 2 (b) 2 : 3

(c) 3 : 2 (d) 1 : 1

(ii) The circumference of two circles are in the ratio 8:15. What is the ratio of their radii?

(a) 8 : 15 (b) 4 : 5

(c) 3 : 5 (d) 16 : 35

(iii) The perimeter of two semicircles are in the ratio 8:7.What is the ratio of their radii?

(a) 4 : 3 (b) 16 : 11

(c) 3 : 4 (d) 8 : 7

7. The circumference of a circle is 39.6 cm. Find its area. **(2020)**

Solution:

$2\pi r = 39.6$

$\Rightarrow 2 \times \frac{22}{7} \times r = 39.6$

$\Rightarrow r = 39.6 \times \frac{7}{2 \times 22}$

$\therefore r = 6.3$ cm

$\therefore$ Area of the circle $= \pi r^2 = \frac{22}{7} \times (6.3)^2 = 124.74$ cm^2

8. If the perimeter and the area of a circle are numerically equal, then the radius of the circle is: **(2023)**

(a) 2 units (b) π units

(c) 4 units (d) 2π units

Solution: (a)

$\pi r^2 = 2\pi r$

$\Rightarrow r = 2$ unit.

9. Area of a quadrant of a circle of radius 7 cm is **(2023)**

(a) 154 cm^2 (b) 77 cm^2

(c) $\frac{77}{2} cm^2$ (d) $\frac{77}{4} cm^2$

Solution:

Required area $= \frac{\pi r^2}{4}$

$= \frac{1}{4} \times \frac{22}{7} \times 7^2 = \frac{77}{2} cm^2$

10. The area of a quadrant of a circle whose circumference is 22 cm is: **(2024)**

(a) $\frac{11}{4} cm^2$ (b) $\frac{11}{2} cm^2$

(c) $\frac{33}{4} cm^2$ (d) $\frac{77}{8} cm^2$

Solution: (d)

$2\pi r = 22$

$\Rightarrow 2 \times \frac{22}{7} \times r = 22$

$\Rightarrow r = \frac{7}{2}$

$\therefore$ required area $= \frac{1}{4} \pi r^2$

$= \frac{1}{4} \times \frac{22}{7} \times \left(\frac{7}{2}\right)^2 = \frac{77}{8} cm^2$

Exercise 9.4

(i) The circumference of a circle is 88 cm, then its area is:

(a) 506 cm^2 (b) 606 cm^2

(c) 616 cm^2 (d) 724 cm^2

(ii) If the perimeter and the area of a circle are numerically equal, then the area of the circle is:

(a) π sq. unit (b) 2π sq. unit

(c) 3π sq unit (d) 4π sq unit

(iii) Area of a quadrant of a circle of radius 3.5 cm is

(a) 9.625 cm^2 (b) 40 cm^2

(c) 42.5 cm^2 (d) 52 cm^2

(iv) Area of a quadrant of a circle whose circumference is 3.5 cm

(a) 86.50 cm^2 (b) 80.625 cm^2

(c) 86.625 cm^2 (d) 24.36 cm^2

Key Point The radii of two circles are r_1 and r_2 unit respectively. The radius of the circle having area equal to the sum of the areas of the two circles is $\sqrt{r_1^2 + r_2^2}$

11. The radii of two circles are 4 cm and 3 cm respectively. The diameter of the circle having area equal to the sum of the areas of the two circles (in cm) is: **(2011)**

(a) 5 (b) 7

(c) 10 (d) 14

Solution: (c)

$\pi r^2 = \pi.4^2 + \pi.3^2$

$\Rightarrow \pi r^2 = 16\pi + 9\pi$

$\Rightarrow \pi r^2 = 25\pi$

$\Rightarrow r^2 = 25$

$\therefore r = 5$

$\therefore$ diameter = 10 cm.

12. If the area of a circle is equal to sum of the areas of two circles of diameters 10 cm and 24 cm, then the diameter of the larger circle (in cm) is **(2012)**

(a) 34 (b) 26

(c) 17 (d) 14

Solution:(b)

Radius of the circles are 5 cm and 12 cm

Now, $\pi r^2 = \pi.5^2 + \pi.12^2$

$\Rightarrow \pi r^2 = 25\pi + 144\pi$

$\Rightarrow \pi r^2 = 169\pi$

$\Rightarrow r^2 = 169$

$\therefore r = 13$ cm

$\therefore$ Diameter of the larger circle is 26 cm.

Key Point If areas of two circles are in the ratio $a : b$, then the ratio of their radius/circumference is $\sqrt{a} : \sqrt{b}$

13. The areas of two circles are in the ratio 9 : 4, then what is the ratio of their circumference? **(2020)**

Solution:

$\pi r_1^2 : \pi r_2^2 = 9:4$

$\Rightarrow \frac{r_1^2}{r_2^2} = \frac{9}{4}$

$\Rightarrow \left(\frac{r_1}{r_2}\right)^2 = \left(\frac{3}{2}\right)^2$

$\therefore \frac{r_1}{r_2} = \frac{3}{2}$

$\therefore$ The ratio of their circumference

$= 2\pi r_1 : 2\pi r_2$

$= r_1 : r_2$

$= 3 : 2$

Exercise 9.5

(i) The radii of two circles are 24 cm and 7 cm respectively. The diameter of the circle having area equal to the sum of the ares of the two circles is:

(a) 24 cm (b) 31 cm
(c) 25 cm (d) 50 cm

(ii) If the area of a circle is equal to sum of the areas of two circles of diameters 8 cm and 6 cm, then circumference of the larger circle (in cm) is:

(a) 10π (b) 12π
(c) 8π (d) 6π

(iii) The areas of two circles are in the ratio 121 : 144, then what is the ratio of their circumference.

(a) 121 : 144 (b) 11 : 12
(c) 10 : 8 (d) 4 : 5

Key Point The area of a semi- circle of radius r is $\frac{\pi r^2}{2}$

14. What is the area of a semi-circle of diameter 'd'? **(2023)**

(a) $\frac{1}{18}\pi d^2$ (b) $\frac{1}{4}\pi d^2$
(c) $\frac{1}{8}\pi d^2$ (d) $\frac{1}{2}\pi d^2$

Solution: (c)

The required area is $\frac{1}{2}\pi\left(\frac{d}{2}\right)^2$

$= \frac{1}{2}\pi.\frac{d^2}{4}$

$= \frac{1}{8}\pi d^2$

15. The area of a semicricle of diameter 'd' is: **(2026)**

(a) $\frac{\pi d^2}{16}$ (b) $\frac{\pi d^2}{4}$
(c) $\frac{\pi d^2}{8}$ (d) $\frac{\pi d^2}{2}$

Solution: (b)

The requied area = πr^2

$= \pi\left(\frac{d}{2}\right)^2$

$= \frac{\pi d^2}{4}$

Exercise 9.6

(i) What is the area of a semi-circle of diameter 14 cm?

(a) 7 cm^2 (b) 11 cm^2
(c) 77 cm^2 (d) 121 cm^2

(ii) What is the area of a semi-circle of radius $7\sqrt{2}$ cm?

(a) 144 cm^2 (b) 154 cm^2
(c) 100 cm^2 (d) $7\sqrt{2}$ cm^2

(iii) What is the area of a semi-circle of perimeter 36 cm?

(a) 36 cm^2 (b) 72 cm^2
(c) 77 cm^2 (d) 100 cm^2

Key Point The radius of the circle inscribed in the triangle is $\frac{2\,\text{area of the triangle}}{\text{Perimeter of the triangle}}$

16. In a right-triangle ABC, right-angled at B, $BC = 12$ cm and $AB = 5$ cm. The radius of the circle inscribed in the triangle (in cm) is **(2014)**

(a) 4 (b) 3

(c) 2 (d) 1

Solution:

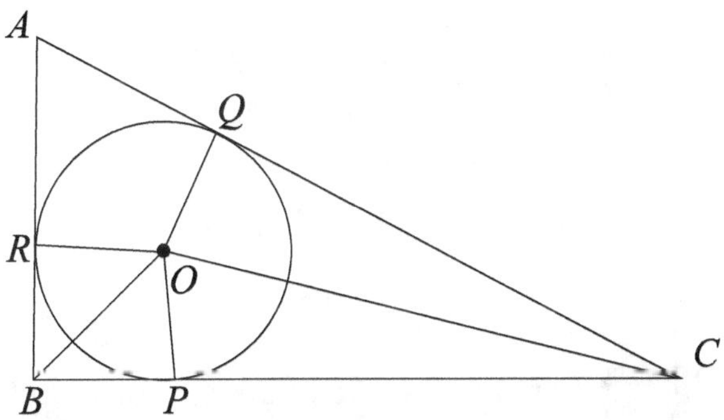

$\text{ar}(\Delta ABC) = \text{ar}\,\Delta BOC + \text{ar}\,\Delta AOC + \text{ar}\,\Delta AOB$

$\Rightarrow \frac{1}{2} \times AB \times BC$

$= \frac{1}{2} \times OP \times BC + \frac{1}{2} OQ \times AC + \frac{1}{2} OR \times AB$

$\Rightarrow \frac{1}{2} \times 5 \times 12$

$= \frac{1}{2} \times r \times 12 + \frac{1}{2} \times r \times 13 + \frac{1}{2} \times r \times 5$

$\Rightarrow 30 = \frac{30r}{2}$

$\therefore r = 2$ cm

Exercise 9.7

(i) In a right-triangle ABC, right-angled at B, $BC = 24$ cm, and $AB = 7$ cm. The radius of the circle inscribed in the triangle (in cm) is:

(a) 2 cm (b) 3 cm

(c) 1.5 cm (d) 2.5 cm

(ii) In a right-triangle ABC, right-angled at B, $BC = 8$ cm, and $AB = 6$ cm. The radius of the circle inscribed in the triangle (in cm) is:

(a) 2 cm (b) 3 cm

(c) 1.5 cm (d) 5 cm

(iii) The radius of the circle inscribed in the triangle whose sides are 8 cm, 11 cm, and 13 cm respectively.

(a) $\sqrt{30}\text{ cm}^2$ (b) $\frac{\sqrt{30}}{2}$ cm

(c) 32 cm^2 (d) $10\sqrt{3}\text{ cm}^2$

Key Point The area of the square inscribed in a circle of radius r unit is $2r^2$ sq. unit.

17. The area of the square inscribed in a circle of radius $5\sqrt{2}$ cm is **(2024)**

(a) 50 cm^2 (b) 100 cm^2

(c) 25 cm^2 (d) 200 cm^2

Solution: (b)

Diagonal of the square

= Diameter of the circle

$\Rightarrow \sqrt{2}\,\text{side} = 2 \times 5\sqrt{2}$

$\Rightarrow$ side = 10 cm

$\therefore$ Area of the square = $10^2 = 100$ cm^2

Exercise 9.8

(i) The area of the square inscribed in a circle of radius $3\sqrt{2}$ cm is:

(a) 18 cm^2 (b) 27 cm^2

(c) 30 cm^2 (d) 36 cm^2

(ii) The area of the square inscribed in a circle of diameter $2\sqrt{2}$ cm is:

(a) 2 cm^2 (b) 4 cm^2

(c) 8 cm^2 (d) $2\sqrt{2}\text{ cm}^2$

(iii) The area of the square inscribed in a circle of perimeter 44 cm is:

(a) 98 cm^2 (b) 44 cm^2

(c) $78\sqrt{2}$ cm^2 (d) 80 cm^2

Key Point If the chord of a circle of radius a cm subtends a right angle at its centre, then the length of the chord will be $\sqrt{2}$ a unit

18. The chord of a circle of radius 10 cm subtends a right angle at its centre. The length of the chord (in cm) is **(2024)**

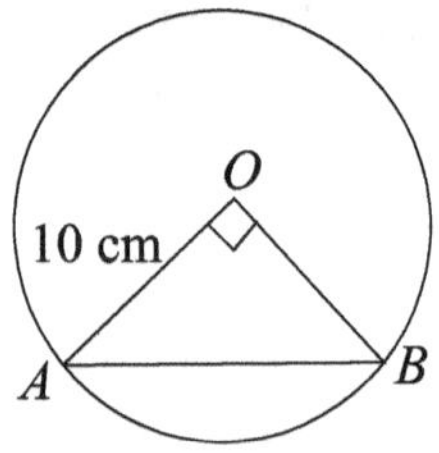

(a) $5\sqrt{2}$ (b) $10\sqrt{2}$

(c) $\frac{5}{\sqrt{2}}$ (d) 5

Solution: (b)

$AB^2 = OA^2 + OB^2$

$\Rightarrow AB^2 = 10^2 + 10^2$

$\Rightarrow AB^2 = 200$

$\therefore AB = 10\sqrt{2}$

Exercise 9.9

(i) the chord of a circle of radius $5\sqrt{2}$ cm subtends a right angle at its centre. The length of the chord (in cm) is:

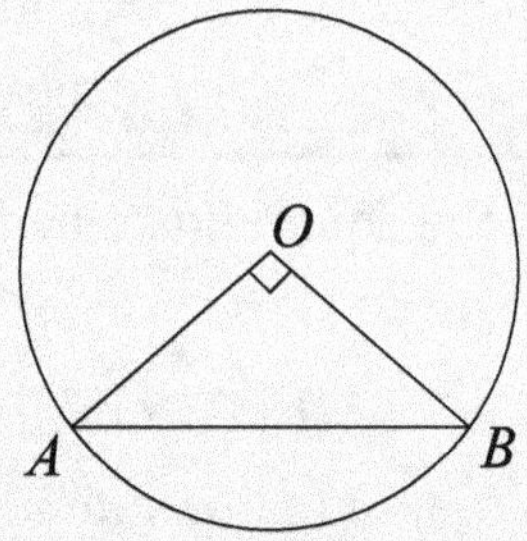

(a) 5 cm (b) $5\sqrt{2}$ cm

(c) 10 cm (d) 12 cm

(ii) the chord of a circle of diameter $8\sqrt{2}$ cm subtends a right angle at its centre. The length of the chord (in cm) is:

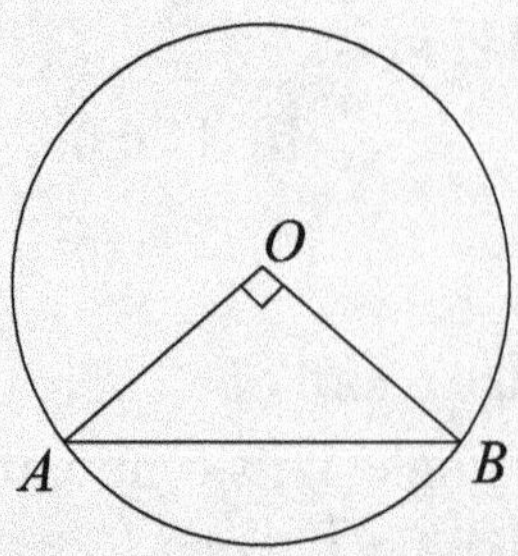

(a) $8\sqrt{2}$ cm (b) $4\sqrt{2}$ cm

(c) 8 cm (d) 12 cm

(iii) The chord of circle of a circumference $44\sqrt{2}$ cm subtends a right angle at its centre. The length of the chord (in cm) is:

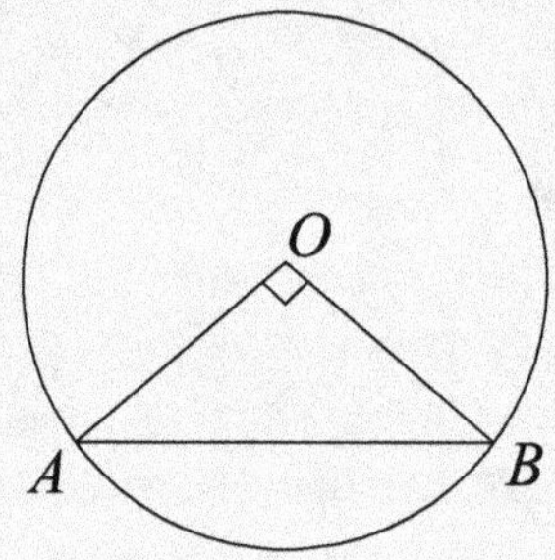

(a) 44 cm (b) $44\sqrt{2}$ cm

(c) 7 cm (d) 14 cm

Key Point Length of an arc of a sector of a circle with radius r and angle with degree measure θ is $\frac{\theta}{360} \times 2\pi r$

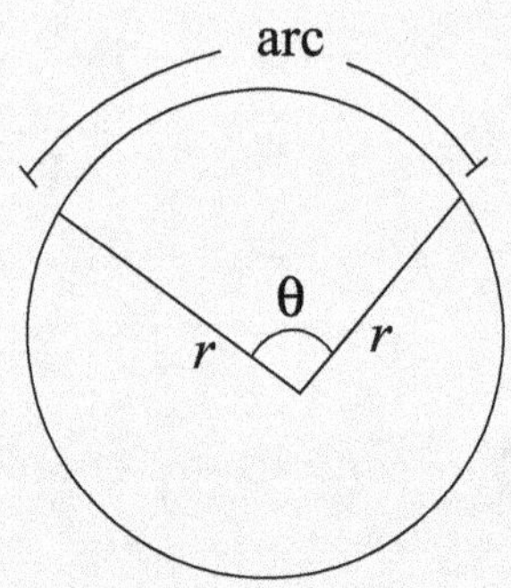

19. In Fig. 4 is a sector of circle of radius 10.5 cm. Find the perimeter of the sector $\left(\text{Take } \pi = \frac{22}{7}\right)$ **(2020)**

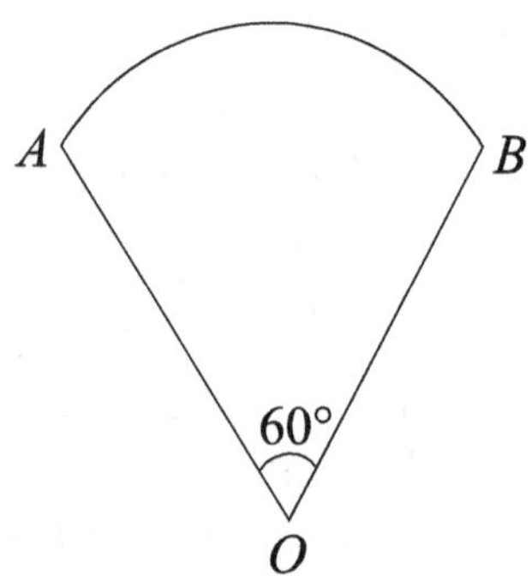

Solution:

Length of arc $AB = \frac{\theta}{360} \times 2\pi r$

$= \frac{60}{360} \times 2 \times \frac{22}{7} \times 10.5 = 11\,\text{cm}$

$\therefore$ Perimeter of the sector

$= 2r + \text{arc } AB$

$= 2 \times 10.5 + 11$

$= 32$ cm

20. The length of the arc of a circle of radius 14 cm. Which subtends an angle of 60° at the centre of the circle is: **(2023)**

(a) $\frac{44}{3}$ cm (b) $\frac{88}{3}$ cm

(c) $\frac{308}{3}$ cm (d) $\frac{616}{3}$ cm

Solution: (a)

Length of arc $= \frac{\theta}{360} \times 2\pi r$

$= \frac{60}{360} \times 2 \times \frac{22}{7} \times 14 = \frac{44}{3}$ cm

21. The perimeter of the sector of a circle of radius 21 cm which subtends an angle of 60° at the centre of circle is: **(2024)**

(a) 22 cm (b) 43 cm

(c) 64 cm (d) 462 cm

Solution: (c)

Length of the arc $= \frac{\theta}{360} \times 2\pi r$

$= \frac{60}{360} \times 2 \times \frac{22}{7} \times 21$

$= 22$ cm

$\therefore$ Perimeter of the sector $= 2r +$ Length of arc

$= 2 \times 21 + 22 = 64$ cm

22. In a circle of radius 21 cm, if an arc subtends an angle of 60° at the centre of the circle, then the length of the arc is: **(2024)**

(a) 11 cm (b) 44 cm

(c) $\frac{22}{7}$ cm (d) 22 cm

Solution: (d)

Length of arc $= \frac{\theta}{360} \times 2\pi r$

$= \frac{60}{360} \times 2 \times \frac{22}{7} \times 21 = 22\,\text{cm}$

23. If an arc subtend an angle of 90° at the centre of a circle, then the ratio of its length to the circumference of the circle is: **(2024)**

(a) 2 : 3 (b) 1 : 4

(c) 4 : 1 (d) 1 : 3

Solution: (b)

Length of the arc $= \frac{2\pi r\theta}{360} = \frac{2\pi r.90}{360} = \frac{2\pi r}{4}$

$\therefore$ The ratio of the length of the arc to the

Circumference of the circle $= \frac{2\pi r}{4} : 2\pi r = 1 : 4$

24. What is the length of the arc of the sector of a circle with radius 14 cm and of central angle 90°? **(2023)**

(a) 22 cm (b) 44 cm

(c) 88 cm (d) 11 cm

Solution: (a)

Length of the sector $\frac{\theta}{360} \times 2\pi r$

$= \frac{90}{360} \times 2 \times \frac{22}{7} \times 14 = 22\,\text{cm}$

25. The perimeter of the shaded region is the given figure is **(2025)**

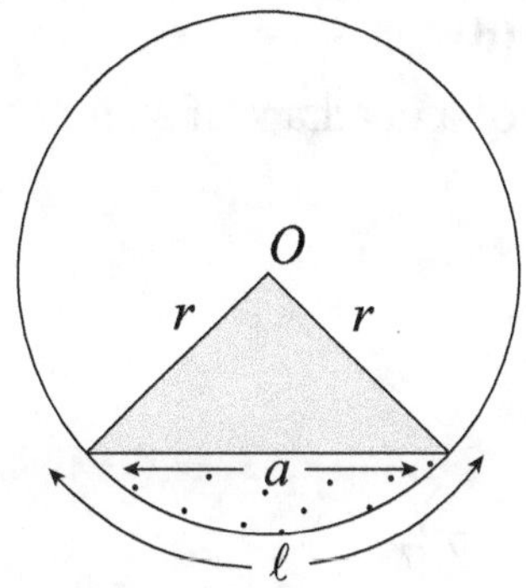

(a) ℓ (b) $\ell + a$
(c) $\ell + 2r$ (d) $\ell + 2r + a$

Solution: (c)

Required perimeter $= r + r + \ell$

$= (\ell + 2r)$

26. The perimeter of a quadrant of a circle of radius r is: **(2025)**

(a) $\frac{1}{4}\pi r^2$ (b) $\frac{1}{4}\pi r^2 + 2r$

(c) $\frac{\pi r}{2}$ (d) $\frac{\pi r}{2} + 2r$

Solution: (d)

Required perimeter $= r + r + \frac{90}{360} \times 2\pi r$

$= 2r + \frac{\pi r}{2}$

27. The length of arc subtending an angle of $210°$ at the center of the circle, is $\frac{44}{3}$ cm. The radius of the circle is: **(2025)**

(a) $2\sqrt{2}$ cm (b) 4 cm

(c) 8 cm (d) $\frac{1}{4}$ cm

Solution:

$\ell = \frac{\theta}{360} \times 2\pi r$

$\Rightarrow \frac{44}{3} = \frac{210}{360} \times 2 \times \frac{22}{7} \times r$

$\Rightarrow r = 4$ cm

28. The perimeter of a quadrant of a circle of radius 7 cm, is: **(2025)**

(a) 18 cm (b) 11 cm
(c) 22 cm (d) 25 cm

Solution: (d)

Perimeter of a quadrant of a circle

$= \frac{\pi r}{2} + 2r$

$= \left(\frac{\pi}{2} + 2\right) r$

$= \left(\frac{22}{7} \times \frac{1}{2} + 2\right) r$

$= \frac{50r}{14}$

$\therefore$ The required perimeter $= \frac{50}{14} \times 7$

$= 25$ cm

29. An arc of length 'ℓ' subtend an angle of $15°$ at the centre of a circle of radius 8.4 cm. The value of ℓ is: **(2025)**

(a) 22 cm (b) 2.2 cm
(c) 9.24 cm (d) 4.2 cm

Solution: (b)

$\ell = \frac{\theta}{360} \times 2\pi r$

$\Rightarrow \ell = \frac{15}{360} \times 2 \times \frac{22}{7} \times 8.4$

$\Rightarrow \ell = 2.2$ cm

30. If an arc of a circle of diameter 10 cm subtends an angle of $144°$ at the center of the circle, then the length of the arc is: **(2025)**

(a) 2π cm (b) 4π cm
(c) 5π cm (d) 6π cm

Solution:

$r = \frac{10}{2} = 5$ cm

$\ell = \frac{\theta}{360} \times 2\pi r$

$\Rightarrow \ell = \frac{144}{360} \times 2 \times \pi \times 5$

$\therefore \ell = 4\pi$ cm

31. In the given figures, O is the centre of circle. XYZ is an arc of the circle subtending an angle of $45°$ at the centre. If the radius of the circle is 32 cm then the length of the arc XYZ is: **(2026)**

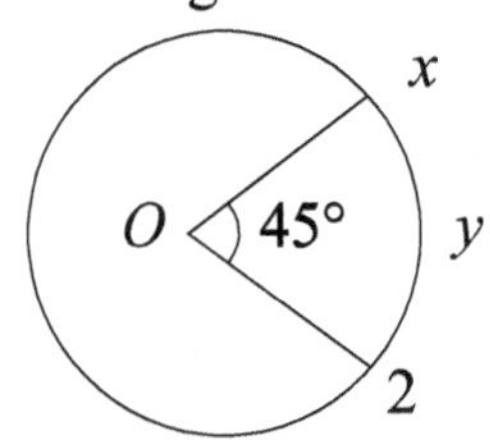

(a) 4π cm (b) 8π cm
(c) 64π cm (d) 128π cm

Solution: (b)

Length of XYZ $= \frac{\theta}{360} \times 2\pi\, r$

$= \frac{45}{360} \times 2\pi \times 32$

$= 8\pi\, cm$

32. The length of the arc of the sector of a cicle with radius 21 cm and of central angle 60°, is: **(2026)**

(a) 22 cm (b) 44 cm

(c) 88 cm (d) 11 cm

Solution: (a)

$\ell = \frac{\theta}{360} \times 2\pi r$

$= \ell = \frac{60}{360} \times 2 \times \frac{22}{7} \times 21$

$= \ell = 22$ cm

Exercise 9.10

(i) In a circle of radius 14 cm, an arc subtends an angle of 30° at the centre, the length of the arc is

(a) 44 cm (b) 28 cm

(c) 11 cm (d) $\frac{22}{3}$ cm

(ii) The length of an arc of a circle of radius 12 cm, subtended by an angle of 45° at the centre is:

(a) 3π cm (b) 6π cm

(c) 9π cm (d) 12π cm

(iii) The perimeter of a sector of a circle of radius 9 cm, subtended by an angle of 60° at the centre is:

(a) $(3\pi + 18)$ cm (b) $(9\pi + 18)$ cm

(c) 12π cm (d) 18π cm

(iv) A pendulum swings through an angle of 30° and describes an arc 8.8 cm in length. Find the length of the pendulum.

(a) 16 cm (b) 16.5 cm

(c) 16.8 cm (d) 17 cm

(v) The perimeter of a sector of a circle having radius r and angle 60° is:

(a) $r\left(\frac{\pi}{6}+1\right)$ (b) $r\left(\frac{\pi}{6}+1\right)$

(c) $\left(\frac{\pi}{6}+1\right)$ (d) $2r\left(\frac{\pi}{6}+1\right)$

Key Point The length of an arc of a circle with radius a cm is ℓ. The angle subtended by the arc at the centre of the circle is $\frac{360l}{2\pi r}$

33. The length of an arc of a circle with radius 12 cm in 10π. The angle subtended by the arc at the centre of the circle, is **(2024)**

(a) 120° (b) 6°

(c) 75° (d) 150°

Solution: (d)

Required angle $= \frac{360l}{2\pi r} = \frac{360 \times 10\pi}{2\pi \times 12} = 150°$

34. A piece of wire 20 cm long is bent into the form of an arc a circle of radius $\frac{60}{\pi}$ cm. The angle subtended by the arc at the centre of the circle is: **(2025)**

(a) 30° (b) 60°

(c) 90° (d) 50°

Solution: (b)

$\frac{\theta}{360} \times 2\pi r = 20$

$\Rightarrow \frac{\theta}{360} \times 2\pi \times \frac{60}{\pi} = 20$

$\Rightarrow \theta = 60$

∴ The required angle is 60°

35. If a large circluar pizza is divided into 5 equal sectors, then the central angle of each sector will be: **(2025)**

(a) 60° (b) 90°

(c) 45° (d) 72°

Solution:

$\ell = \frac{2\pi r}{5}$

Now, $\ell = \frac{\theta}{360} \times 2\pi r$

$\Rightarrow \frac{2\pi r}{5} = \frac{\theta}{360} \times 2\pi r$

$\Rightarrow \theta = 72$

$\therefore$ The required angle is 72°

36. An arc of length 22 cm subtends an angle of $x°$ at the centre of the circle. If radius of circle is 36 cm, the value of x is: **(2025)**

(a) 35 (b) 40
(c) 60 (d) 30

Solution: (a)

$\ell = \frac{\theta}{360} \times 2\pi r$

$\Rightarrow 22 = \frac{x}{360} \times 2 \times \frac{22}{7} \times 36$

$\Rightarrow x = 35$

37. An are of length 2.2 cm subtends an angle θ at the centre of the circle with radius 2.8 cm The value of θ is: **(2026)**

(a) 50° (b) 60°
(c) 45° (d) 30°

Solution:

$s = \frac{\theta}{360} \times 2\pi r$

$\Rightarrow \theta = \frac{360 s}{2\pi r}$

$\Rightarrow \theta = \frac{360 \times 2.2}{2 \times \frac{22}{7} \times 2.8}$

$\Rightarrow \theta = 45°$

38. The length of a pendulum is 70 cm and it describes an arc of length 88 cm when swings. The angle subtended by the arc at the centre is: **(2026)**

(a) 36° (b) 70°
(c) 72° (d) 80°

Solution:

$\ell = \frac{\theta}{360} \times 2\pi r$

$\Rightarrow 88 = \frac{\theta}{360} \times 2 \times \frac{22}{7} \times 70$

$\Rightarrow \theta = \frac{88 \times 360 \times 7}{2 \times 22 \times 70}$

$\Rightarrow \theta = 72°$

39. Arc PQ subtends an angle θ at the centre of the cricle with radius 6.3 cm If $\overparen{PQ} = 11$ cm, then the value of θ is: **(2026)**

(a) 10° (b) 60°
(c) 45° (d) 100°

Solution: (d)

$\overparen{PQ} = \frac{\theta}{360} \times 2\pi r$

$\Rightarrow 11 = \frac{\theta}{360} \times 2 \times \frac{22}{7} \times 6.3$

$\Rightarrow \theta = \frac{360 \times 7 \times 11}{2 \times 22 \times 6.3}$

$\Rightarrow \theta = 100°$

40. The haer hand of *a* clock is 7 cm long the angle swept by it between 7.00 am and 8:10 am is: **(2026)**

(a) $\left(\frac{35}{4}\right)^0$ (b) $\left(\frac{35}{2}\right)^0$

(c) 35° (d) 70°

Solution: (c)

Duration = 1hr 10 min = 70 min

$\therefore$ Required angle $= \left(70 \times \frac{1}{2}\right)^0 = 35°$

Exercise 9.11

(i) The length of an are of a circle with radius 18 cm in 12π. The angle subtended by the are at the centre of the circle, is

(a) 150° (b) 15°
(c) 120° (d) 108°

(ii) The length of an are of a circle with circumference 30π cm is 8π. The angle subtended by the are at the centre of the circle, is

(a) 90° (b) 92°
(c) 88° (d) 96°

(iii) The length of an are of a circle diameter 20 cm is 6π. The angle subtended by the arc at the centre of the circle is

(a) 100° (b) 102°
(c) 106° (d) 108°

Key Point Area of a sector of a circle with radius r and angle with degrees measure θ is $\frac{\theta}{360}\times\pi r^2$

41. Find the area of the sector of a circle of radius 6 cm whose central angle is 30° (Take $\pi = 3.14$) **(2020)**

Solution:

Required area $=\frac{\theta}{360}\times\pi r^2$

$\frac{30}{360}\times3.14\times6^2$

$= 9.42$ sq cm

42. The area of a sector of angle α (in degree) of a circle with radius R is **(2020)**

(a) $\frac{\alpha}{180}\times2\pi R$ (b) $\frac{\alpha}{360}\times2\pi R$

(c) $\frac{\alpha}{180}\times\pi R^2$ (d) $\frac{\alpha}{360}\times\pi R^2$

Solution: (d)

$\frac{\alpha}{360}\times\pi R^2$

43. The minute hand of a clock is 21 cm long the area swept by it in 10 minutes is: **(2024)**

(a) 121 cm² (b) 131 cm²
(c) 231 cm² (d) 172.5 cm²

Solution: (c)

The angle swept by minute hand of a clock in 10 minutes = 60°

$\therefore$ Required area $=\frac{\theta}{360}\times\pi r^2$

$=\frac{60}{360}\times\frac{22}{7}\times(21)^2$

$= 231\text{ cm}^2$

44. Area of a sector of angle θ (in degree) of a circle with radius r is **(2024)**

(a) $\frac{\theta}{180}\times2\pi r$ (b) $\frac{\theta}{180}\times\pi r^2$

(c) $\frac{\theta}{360}\times2\pi r$ (d) $\frac{\theta}{720}\times2\pi r^2$

Solution: (d)

$\frac{\theta}{720}\times2\pi r^2$

45. The area of the sector of a circle with radius 6 cm which subtends an angle of 60° at the centre of the circle is: **(2024)**

(a) $\frac{142}{7}\text{cm}^2$ (b) $\frac{152}{7}\text{cm}^2$

(c) $\frac{132}{7}\text{cm}^2$ (d) $\frac{122}{7}\text{cm}^2$

Solution: (c)

Required area $=\frac{\theta}{360}\times\pi r^2$

$=\frac{60}{360}\times\frac{22}{7}\times6^2$

$=\frac{132}{7}\text{cm}^2$

46. The perimeter of a sector of a circle of radius 14 cm is 68 cm. Find the area of the sector **(2020)**

Solution:

$\frac{2\pi r\theta}{360}+2r=68$

$\Rightarrow 2\times\frac{22}{7}\times14\times\frac{\theta}{360}+2\times14=68$

$\Rightarrow \frac{88\theta}{360}=40$

$\Rightarrow \frac{\theta}{360}=\frac{40}{88}$

$\therefore$ Area of the sector $=\frac{\theta}{360}\times\pi r^2$

$=\frac{40}{88}\times\frac{22}{7}\times(14)^2$

$= 280\text{ cm}^2$

47. The difference of the areas of a minor sector of angle 120° and its corresponding major sector of a circle of radius 21 cm is **(2024)**

(a) 231 cm² (b) 462 cm²
(c) 364.5 cm² (d) 693 cm²

Solution: (b)

Required difference $= \frac{\pi r^2(\theta_2 - \theta_1)}{360}$

$= \frac{22}{7} \times (21)^2 \times \frac{(240-120)}{360}$

$= 462 \text{ cm}^2$

48. In figure-2 $\overparen{PQ}$ and $\overparen{AB}$ are two arcs of concentric circle of radii 7 cm and 3.5 cm respectively with centre O. If $\angle POQ = 30°$, then find area of shaded region **(2020)**

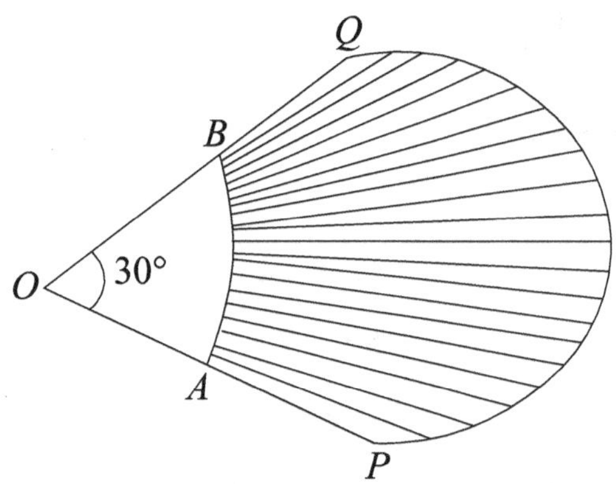

Figure 2

Solution:

Area of shaded region

$= \frac{\theta}{360} \times \pi(R^2 - r^2)$

$= \frac{30}{360} \times \frac{22}{7}\{7^2 - (3.5)^2\}$

$= 9.625 \text{ cm}^2$

49. If a sector of a circle has an area of 40π sq. units and a central angle of 72°, the radius of the circle is: **(2025)**

(a) 200 units (b) 100 units
(c) 20 units (d) $10\sqrt{2}$ units

Solution: (d)

$\frac{\theta}{360} \times \pi r^2 = 40\pi$

$\Rightarrow \frac{72}{360} \times \pi r^2 = 40\pi$

$\Rightarrow r^2 = 200$

$\Rightarrow r = \sqrt{200} = 10\sqrt{2}$ units

50. A sector is cut from a circluar sheet of radius 50 cm, the central angle of the sector being 90°. If another circle of the same area as the sector is formed, the radius of the new circle is **(2025)**

(a) 25 cm (b) 50 cm
(c) 12.5 cm (d) 20 cm

Solution: (a)

Length of the sector $(\ell) = \frac{\theta}{360} \times \pi r^2$

$= \frac{90}{360} \times \pi \times (50)^2$

$= 625\pi \text{ cm}^2$

$\therefore \pi r_1^2 = 625\pi$

$\Rightarrow r_1^2 = 625$

$\therefore r_1 = 25$ cm

51. The ratio of the area of a quadrant of a circle to the area of the some circle is: **(2025)**

(a) 1 : 2 (b) 2 : 1
(c) 1 : 4 (d) 4 : 1

Solution:

Area of the quadrant of a circle $= \frac{90}{360} \times \pi r^2$

$= \frac{\pi r^2}{4}$

$\therefore$ Required ratio $= \frac{\pi r^2}{4} : \pi r^2$

$= 1 : 4$

52. The area of a quadrant of a circle of radius '$2r$' is: **(2025)**

(a) $\frac{1}{4}\pi r^2$ (b) $\frac{1}{2}\pi r^2$
(c) πr^2 (d) $2\pi r^2$

Solution: (c)

Required area $= \frac{90}{360} \times \pi.(2r)^2$

$= \frac{1}{4} \times \pi \times 4r^2$

$= \pi r^2$

53. An arc of a circle is of length 5π cm and the sector it bounds has an area of 20π cm². Its radius is **(2025)**

(a) 10 cm (b) 1 cm

(c) 5 cm (d) 8 cm

Solution: (d)

Area of the sector $= \frac{1}{2} \times$ length of arc $\times$ radius

$\Rightarrow 20\pi = \frac{1}{2} \times 5\pi \times r$

$\Rightarrow r = 8$ cm

54. In a circle of radius 14 cm, the area of the sector made by an arc of length 11 cm with the center, is **(2025)**

(a) 154 cm^2 (b) 102.67 cm^2

(c) 205.33 cm^2 (d) 77 cm^2

Solution:

Area of the sector $= \frac{1}{2} \times$ length of arc $\times$ radius

$= \frac{1}{2} \times 11 \times 14$

$= 77$ cm^2

55. Area of a segment of a circle of redius 'r' and central angle 60° is: **(2026)**

(a) $\frac{\pi r^2}{2} - \frac{1}{2}r^2$ (b) $\frac{2\pi r}{4} - \frac{\sqrt{3}}{4}r^2$

(c) $\frac{\pi r^2}{6} - \frac{\sqrt{3}}{4}r^2$ (d) $\frac{2\pi r}{4} - r^2 \sin 60°$

Solution: (c)

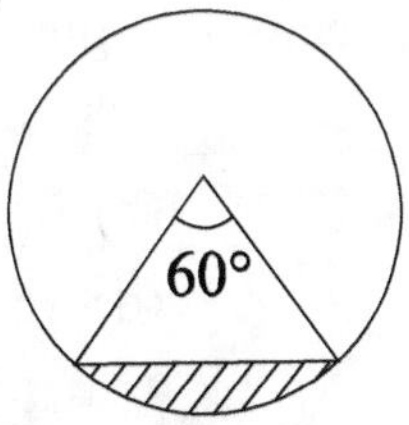

Area of the segment

= Area of sectar – area of tanrgle

$= \frac{60}{360}\pi r^2 - \frac{\sqrt{3}}{4}r^2$

$= \frac{\pi r^2}{6} - \frac{\sqrt{3}}{4}r^2$

56. Shwn in the given figure is a circle with centre O .The are of the minor sector is 7 cm^2. Area of the circle is: **(2026)**

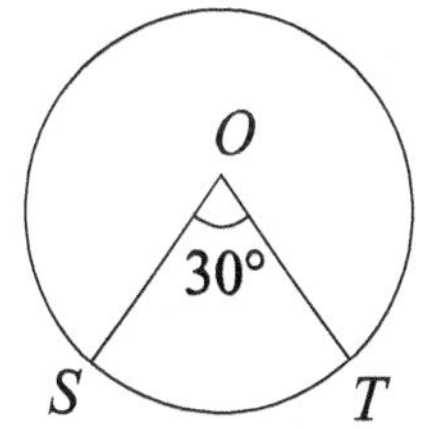

(a) 84π cm^2 (b) $\frac{84}{11}$ cm^2

(c) 84 cm^2 (d) $\frac{\sqrt{89}}{\sqrt{\pi}}$ cm^2

Solution: (c)

Area of sector $= \frac{\text{central angle}}{360} \times$ Area of circle

$= 7 = \frac{30}{360} \times$ Area of circle

= Area of circle = 84 cm^2

Exercise 9.12

(i) In a circle of radius 21 cm, an are subtends an angle of 60° at centre The area of the sector formed by the arc is:

(a) 102 cm^2 (b) 212 cm^2

(c) 240 cm^2 (d) 231 cm^2

(ii) The area of the sector of a circle of radius 5 cm, subtends by an angle of 60° at the centre is:

(a) $\frac{25\pi}{6}$ cm^2 (b) $\frac{3\pi}{2}$ cm^2

(c) $\frac{10\pi}{3}$ cm^2 (d) 2π cm^2

(iii) The area of a quadrat of a circle of radius 4 cm is:

(a) π cm^2 (b) 2π cm^2

(c) 3π cm^2 (d) 4π cm^2

(iv) A cow is tied with a rope of length 14 *m* at the corner of a rectangular find of dimensions 28 m × 22 m, then the area of the field in which are cow can graze is:

(a) 154 m^2 (b) 144 m^2

(c) 100 m^2 (d) 322 m^2

(v) A sector is cut from a circular sheet of radius 3 cm, the angle of the sector being 240°. If another circle of the area some as the sector is formed, then radius of the new circle is

(a) 3 m (b) $2\sqrt{3}$ m

(c) $\sqrt{6}$ m (d) 5 m

(vi) In the given figure, three sectors of a circle of radius 7 cm, making angles of 60°,80° and 40° at the centre are shaded. The area of the shaded region (in cm²) is $\left[\text{use } \pi = \frac{22}{7}\right]$

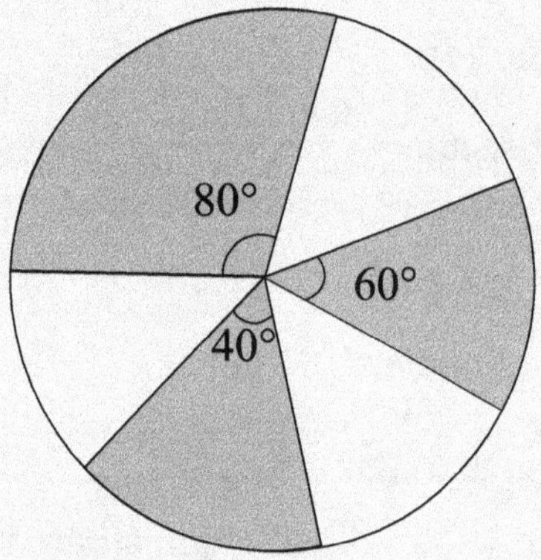

(a) 46 (b) 54

(c) 62 (d) 77

(vii) The length of the minute hand of a clock is 14 cm. the area swept by the minute hand in 5 minute is

(a) 44.2 cm² (b) 51.3 cm²

(c) 102.26 cm² (d) 144 cm²

(viii) The perimeter of a sector of a circle of radius 5.2 cm, is 16.4 cm. The area of the sector is:

(a) 15.6 cm² (b) 30.5 cm²

(c) 16.2 cm² (d) 10 cm²

(ix) A car has two wipers which do not overlap. Each wiper has a blade of length 25 cm sweeping through an angle of 115°. Find the total area cleaned at each sweep of the blades.

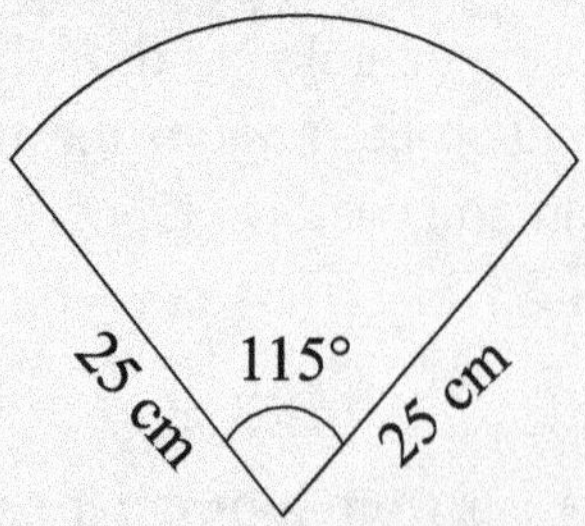

(a) 1200 cm² (b) 1205 cm²

(c) 1210 cm² (d) 1215 cm²

(x) An arc of a circle is of length 5π cm and the sector if bounds has an area of 20π cm², then the radius of the circle is:

(a) 7 cm (b) 8 cm

(c) 10.2 cm (d) 11.5 cm

(xi) In the given figure, O is the centre of a circle. If the area of the sector *OAPB* is $\frac{5}{18}$ times the area of the circle, then the value of x is

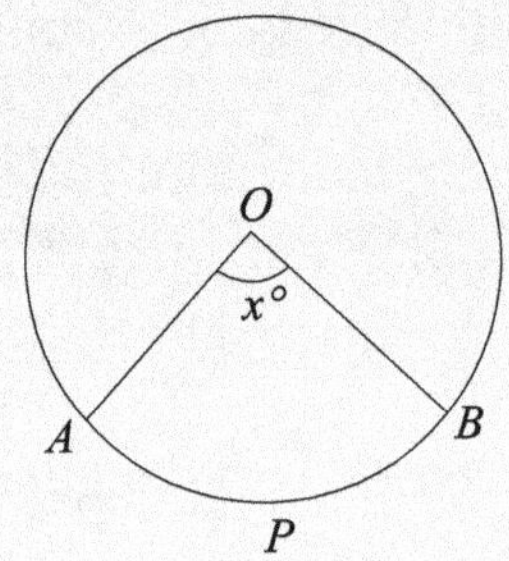

(a) 50° (b) 88°

(c) 92° (d) 100°

(xii) The length of the minute hand of a clock is 6 cm. Find the area swept by it moves from 7:05 p.m to 7:40 p.m.

(a) 60 cm² (b) 66 cm²

(c) 72 cm² (d) 100 cm²

57. The area of the sector of a circle of radius 12 cm is 60π cm². The central angle of the sector is: **(2024)**

(a) 120° (b) 6°

(c) 75° (d) 150°

Solution: (d)

$$\frac{\pi r^2\theta}{360} = 60\pi$$

$$\Rightarrow \frac{\pi}{360} \times 12^2 \times \theta = 60\pi$$

$$\Rightarrow \theta = 150°$$

58. If the area of a sector is one-twelfth if the that of a complete circle, then the angle of the sector is: **(2024)**

(a) 36° (b) 30°

(c) 60° (d) 45°

Solution:

$$\frac{\pi r^2\theta}{360}=\frac{1}{12}\times\pi r^2$$

$\Rightarrow \theta = 30°$

59. If the area of a sector of a circle is $\frac{1}{8}$ of the area of the circle, then the central angle of the sector is: **(2024)**

(a) 30° (b) 45°

(c) 60° (d) 90°

Solution: (b)

$$\frac{\pi r^2\theta}{360}=\frac{1}{8}\pi r^2$$

$\Rightarrow \theta = 45°$

60. The region between a chord and either of the two arcs of a circle in called . **(2024)**

(a) an arc (b) a sector

(c) a segment (d) a semi circle

Solution: (b) a sector

61. If the length of a chord of a circle is equal to its radius, then the angle subtended by chord at the centre is: **(2025)**

(a) 60° (b) 30°

(c) 120° (d) 90°

Solution: (a)

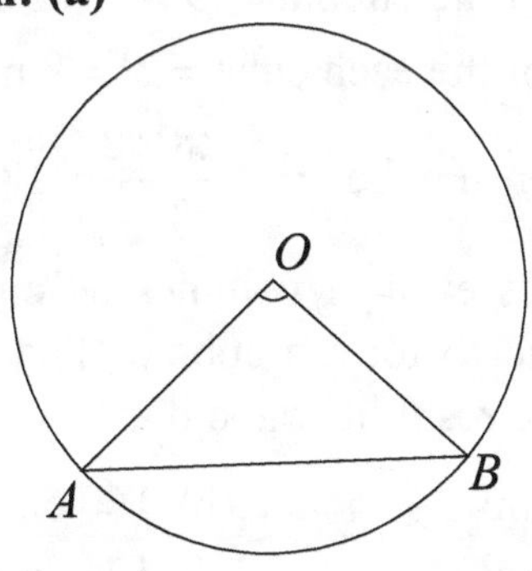

In ΔAOB

$OA = OB = AB$

$\therefore$ ABC is an equilateral triangle

$\therefore$ $\angle AOB = 60°$

62. In the given figure, the shaded region represents: **(2025)**

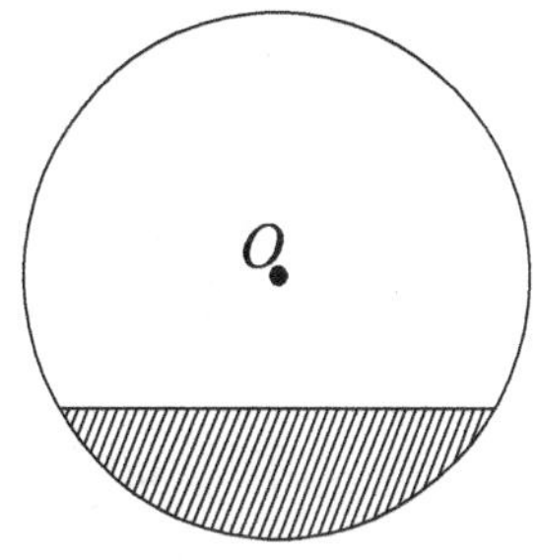

(a) Minor sector (b) Major sector

(c) Minor segment (d) Major segment

Solution: (c)

Minor segment

63. Area of sector of a circle with radius 18 cm is 198 cm² The measure of central angle is: **(2026)**

(a) 70° (b) 14°

(c) 140° (d) 210°

Solution: (a)

$$A=\frac{\theta}{360}\times\pi r^2$$

$$\Rightarrow 198=\frac{\theta}{360}\times\frac{22}{7}\times 18^2$$

$$\Rightarrow \theta=\frac{198\times 360\times 7}{22\times 324}$$

$\Rightarrow \theta = 70°$

Key Point The area of a sector of a circle of radius r unit cut off by an arc, which is ℓ unit long is $\frac{1}{2}rl$

64. The area of a sector of a circle of radius 16 cm cut off by an arc, which is 18.5 cm long is: **(2024)**

(a) 168 cm² (b) 148 cm²

(c) 154 cm² (d) 176 cm²

Solution: (b)

$$A=\frac{1}{2}rl$$

$$=\frac{1}{2}\times 16\times 18.5=148\,\text{cm}^2$$

Exercise 9.13

(i) The area of a sector of a circle of radius 14 cm cut off by an arc, which is 20 cm long is:

(a) 280 cm^2 (b) 288 cm^2
(c) 182 cm^2 (d) 140 cm^2

(ii) The area of a sector of a circle of circumference 44 cm cut off by an arc, which is 12 cm long is:

(a) 42 cm^2 (b) 68 cm^2
(c) 84 cm^2 (d) 102 cm^2

(iii) The area of a sector of a circle of diameter 12 cm cut off by an arc, which is 15 cm long is:

(a) 90 cm^2 (b) 80 cm^2
(c) 112 cm^2 (d) 180 cm^2

Key Point Let ℓ, b and h denotes the length, breadth and height, respectively of a cuboid, then, its total surface area = $2(\ell b + bh + h\ell)$

(ii) Lateral surface area = $2(\ell + b)h$

65. The identical solid cutes of side 'a' are joined end-to-end. The total surface area of the resulting cuboid is: **(2024)**

(a) $6a^2$ (b) $10a^2$
(c) $5a^2$ (d) $4a^2$

Solution: (b)

The length, breadth and height of the cuboid are $2a$, a and a unit respectively.

$\therefore$ TSA $= 2(\ell b + bh + h\ell)$
$= 2(2a.a + a.a + a.2a)$
$= 2(2a^2 + a^2 + 2a^2)$
$= 10a^2$

Exercise 9.14

(i) If two identical solid cube of side 7 cm are joined end to end. Then the total surface area of the resulting cuboid is

(a) 400 cm^2 (b) 450 cm^2
(c) 490 cm^2 (d) 51 cm^2

(ii) If the length, breadth and height of a cuboid are doubled, its surface area becomes

(a) 2 times (b) 4 times
(c) 6 times (d) 8 times

(iii) The cost of painting a cuboidal box of dimensions 2 m × 3 m × 4 m at ₹10 per m^2, is

(a) ₹. 200 (b) ₹ 520
(c) ₹ 610 (d) ₹ 800

(iv) The difference between the total surface area and the lateral surface area of a cuboid of dimentions ℓ, m and n is

(a) 2 ℓb (b) 2 hb
(c) 2 $h\ell$ (d) 2 ℓbh

(v) If the height of a cuboid is doubled while keeping length and breadth the same, its lateral surface area will:

(a) Romain the same
(b) Double
(c) Halve
(d) Become four times

Key Point Let ℓ, b and h denotes the length, breadth and height, respectively, of a cuboid. Then its volume is ℓbh cubic unit.

66. A solid metalic cuboid of dimensions 9 m × 8 m × 2 m is melted and recast into solid cubes of edges 2 m. Find the number of cubes so formed. **(2017)**

Solutions:

Volume of the cuboid = $(9 \times 8 \times 2)m^3$

Volume of the each cube = $2^3 = 8\ m^3$

$\therefore$ number of cubes $= \dfrac{9\times8\times2}{8} = 18$

67. Two cubes each of volumes 64 cm^3 are joined end to end to form a cuboid. The total surface area of the resulting cuboid is: **(2024)**

(a) 192 cm^2 (b) 160 cm^2
(c) 128 cm^2 (d) 176 cm^2

Solution: (b)

$a^3 = 64$

$\Rightarrow a = 4$ cm

$\therefore$ Length, breadth and height of the resulting cuboid are 8 cm, 4 cm, and 4 cm

∴ The total surface area of the resulting cuboid

$= 2(\ell b + bh + h\ell)$

$= 2(8 \times 4 + 4 \times 4 + 4 \times 8)$

$= 2(32 + 16 + 32)$

$= 160 \text{ cm}^2$

68. Water in a river which is 3 m deep and 40 m wide is flowing at the rate of 2 km/h. How much water will fall into the sea in 2 minutes? **(2023)**

(a) 800 m^3 (b) 4000 m^3

(c) 8000 m^3 (d) 2000 m^3

Solution: (c)

Flowing rate of water = 2 km/h

$= \frac{2000}{60} \text{ m/min} = \frac{100}{3} \text{ m/min}$

Length of the water flowing in 2 minute, $l = \frac{200}{3}$ m.

The volume of the water that falls into the sea in

2 minute $= lbh = \frac{200}{3} \times 40 \times 3 = 8000 \text{ m}^3$

Exercise 9.15

(i) A mason constructs a wall of dimension 270 cm × 300 cm × 350 cm with the bricks each of side 22.5 cm × 11.25 cm × 8.75 cm and it is assumed that $\frac{1}{8}$ space is covered by the mortar. Then the number of bricks used to construct the wall is:

(a) 11100 (b) 11200

(c) 11000 (d) 11300

(ii) The breadth of a room is twice its height and is half of its length. The volume of room is 512 dm^3. Its dimensions are

(a) 16 dm, 8 dm, 4 dm

(b) 12 dm, 8 dm, 2 dm

(c) 8 dm, 4 dm, 2 dm

(d) 10 dm, 15 dm, 20 dm

(iii) A river 3 m deep and 40 m wide in flowing at the rate of 2 km per hour. How much water will fall into the sea in a minute?

(a) 4000 m^3 (b) 40 m^3

(c) 400 m^3 (d) 40000 m^3

Key Point If each side of a cube is a unit, then its

(i) Lateral surface area is $4a^2$ sq. unit.

(ii) Total surface area is a^2 sq. unit

(iii) Volume is a^3 cu. unit

69. The total surface area of a cube of side 20 cm is **(2024)**

(a) 240 cm^2 (b) 160 cm^2

(c) 2400 cm^2 (d) 1600 cm^2

Solution: (c)

Total surface area $= 6a^2$

$= 6 \times (20)^2$

$= 6 \times 400 = 2400 \text{ cm}^2$

Exercise 9.16

(i) The lateral surface area of a cube with side length 5 cm is

(a) 100 cm^2 (b) 125 cm^2

(c) 150 cm^2 (d) 200 cm^2

(ii) If the side of a cube in doubled, its volume becomes:

(a) 2 times (b) 4 times

(c) 6 times (d) 8 times

(iii) If the surface area of a cube is 150 cm^2, its length is:

(a) 8 cm (b) 7 cm

(c) 6 cm (d) 5 cm

(iv) If the volume of a cube is 64 cm^3, its lateral surface area is:

(a) 16 cm^2 (b) 64 cm^2

(c) 96 cm^2 (d) 100 cm^2

Key Point If radius of the base and height of a right circular cylinder are r and h respectively, then

(i) Its lateral surface area $2\pi rh$ sq. unit

(ii) Total surface area $2\pi r(h + r)$ sq. unit

(iii) Volume $= \pi r^2 h$ sq. unit.

70. Curved surface area of a cylinder of height 5 cm is 94.2 cm². Radius of the cylinder is (Take $\pi = 3.14$) **(2023)**

(a) 2 cm (b) 3 cm
(c) 2.9 cm (d) 6 cm

Solution: (b)

$2\pi rh = 94.2$

$\Rightarrow 2 \times 3.14 \times r \times 5 = 94.2$

$\Rightarrow r = 3$

71. The area of metal sheet required to make a closed hollow cylinder of height 2.4 m and base radius 0.7 m, is **(2023)**

(a) 10.56 m² (b) 13.52 m²
(c) 13.64 m² (d) 14.08 m²

Solution: (c)

Required area $= 2\pi r(r + h)$

$$= 2 \times \frac{22}{7} \times 0.7(0.7 + 2.4)$$

$$= 2 \times 22 \times 0.1 \times 3.1 = 13.64 \text{ cm}^2$$

72. The curved surface area of a right circular cylinder of height 14 cm is 88 cm². The diameter of its circular base is: **(2023)**

(a) 2 cm (b) 1 cm
(c) 4 cm (d) 7 cm

Solution: (a)

$2\pi rh = 88$

$$\Rightarrow 2 \times \frac{22}{7} \times r \times 14 = 88$$

$\Rightarrow 2r = 2$

73. If the radius of the base of a right circular cylinder is halved, keeping the height the same, then the ratio of the volume of the cylinder thus obtained to the volume of original cylinder is **(2012)**

(a) 1 : 2 (b) 2 : 1
(c) 1 : 4 (d) 4 : 1

Solution: (c)

Volume of the cylinder $= \pi r^2 h$

Volume of the new cylinder $= \pi\left(\frac{r}{2}\right)^2 h = \frac{\pi r^2 h}{4}$

$\therefore$ the ratio of volumes of new cylinder to the old cylinder $= \frac{\pi r^2 h}{4} : \pi r^2 h$

$$= \frac{1}{4} : 1 = 1 : 4$$

74. Two right circular cylinder of equal volumes have their height in the ratio 1 : 2. The ratio of their radii is: **(2025)**

(a) $\sqrt{2} : 1$ (b) 1 : 2
(c) 1 : 4 (d) $1 : \sqrt{2}$

Solution: (a)

$\pi r_1^2 h = \pi r_2^2 .2h$

$$\Rightarrow \frac{r_1^2}{r_2^2} = 2$$

$$\Rightarrow \left(\frac{r_1}{r_2}\right)^2 = 2$$

$$\Rightarrow \frac{r_1}{r_2} = \sqrt{2}$$

$\therefore r_1 : r_2 = \sqrt{2} : 1$

Exercise 9.17

(i) The total surface area of a cylinder with radius 3.5 cm and height 10 cm is

(a) 221 cm² (b) 297 cm²
(c) 306 cm² (d) 440 cm²

(ii) The height of a cylinder whose radius is 7cm and total surface area is 968 cm² is:

(a) 8 cm (b) 12 cm
(c) 14 cm (d) 15 cm

(iii) The volume of a right circular cylinder with radius 7cm and height 10 cm is:

(a) 1540 cm^3 (b) 1440 cm^3
(d) 600 cm^3 (d) 800 cm^3

(iv) The ratio of the radii of two cylinders is 2 : 3, and the ratio of their heights is 5 : 3. The ratio of their volumes is.

(a) 2 : 1 (b) 3 : 4
(c) 20 : 27 (d) 27 : 125

(v) If the radius of a cylinder is doubled and its height is halved, the new volume will be:

(a) Same an before
(b) Twice the original
(c) Half of the original
(d) Four times the original

Key Point If r, h, ℓ are the radius of the base, height and slant height of a cone, then

(i) $l^2 = r^2 + h^2$
(ii) lateral surface area of the cone is πrl
(iii) Total surface area of the cone is $\pi r(l + r)$
(iv) Volume is $\frac{1}{3}\pi r^2 h$

75. Find the height of a cone of radius 5 cm and slant height 13 cm. **(2020)**

Solution:

$\ell^2 = r^2 + h^2$

$\Rightarrow 13^2 = 5^2 + h^2$

$\Rightarrow h^2 = 144$

$\therefore h = 12$ cm

76. The height and radius of a right circular cone are 24 cm and 7 cm respectively. The slant hight of the cone is : **(2024)**

(a) 24 cm (b) 31 cm
(c) 26 cm (d) 25 cm

Solution: (d)

$\ell^2 = r^2 + h^2$

$\Rightarrow \ell^2 = 7^2 + 24^2$

$\Rightarrow \ell^2 = 625$

$\therefore \ell = 25$ cm

77. The curved surface area of a cone having height 24 cm and radius 7 cm is: **(2023)**

(a) 528 cm^2 (b) 1056 cm^2
(c) 550 cm^2 (d) 500 cm^2

Solution: (c)

curve surface area $= \pi r\ell$

$= \pi r\sqrt{r^2 + h^2}$

$= \frac{22}{7} \times 7 \times \sqrt{7^2 + (24)^2}$

$= 22 \times 25$

550 cm^2

78. The curved surface area of a cone radius 7 cm is 550 cm^2. Its slant height is: **(2023)**

(a) 24 cm (b) 25 cm
(c) 14 cm (d) 20 cm

Solution: (b)

$\pi r\ell = 550$

$\Rightarrow \frac{22}{7} \times 7 \times \ell = 550$

$\Rightarrow \ell = 25$ cm

79. If the area of the base of a cone is 51 cm^2 and its volume is 85 cm^3, then the vertical height of the cone is given is **(2023)**

(a) $\frac{5}{6}$ cm (b) $\frac{5}{3}$ cm
(c) $\frac{5}{2}$ cm (d) 5 cm

Solution: (d)

$\frac{1}{3}\pi r^2 h = 85$

$\Rightarrow \frac{1}{3} \times 51 \times h = 85$ $\quad [\because \pi r^2 = 51]$

$\Rightarrow h = 5$ cm

80. The volume of a right circular cone whose area of the base is 156 cm^2 and the vertical height is 8 cm, is **(2023)**

(a) 2496 cm^3 (b) 1248 cm^3
(c) 1664 cm^3 (d) 416 cm^3

Solution: (d)

Required volume $= \frac{1}{3}\pi r^2 h$

$= \frac{1}{3}\times 156\times 8$ $[\because \pi r^2 = 156]$

$= 416\text{ cm}^3$

81. The volume of a cone of radius 'r' and height '$3r$' is **(2023)**

(a) $\frac{1}{3}\pi r^3$ (b) $3\pi r^3$
(c) $9\pi r^3$ (d) πr^3

Solution: (d)

Required volume $= \frac{1}{3}\pi r^2 .3r = \pi r^3$

82. Two cones have their heights in the ratio 1 : 3 and raddii in the ratio 3 : 1. What is the ratio of their volumes? **(2020)**

Solution:

Required ratio $= \frac{1}{3}\pi(3r)^2 h : \frac{1}{3}\pi r^2 .3h$

$= 9 : 3$

$= 3 : 1$

83. The volume of the tangent right circular cone that can be curved out from a solid cube of edge 2 cm is **(2024)**

(a) $\frac{4\pi}{3}$ cu.cm (b) $\frac{5\pi}{3}$ cu.cm

(c) $\frac{8\pi}{3}$ cu.cm (d) $\frac{2\pi}{3}$ cu.cm

Solution: (d)

Radius of the cone $= \frac{2}{2} = 1\text{cm}$

Height of the cone = 2 cm

$\therefore$ Volume of the cone $= \frac{1}{3}\pi r^2 h$

$= \frac{1}{3}\times\pi\times 1^2\times 2$

$= \frac{2\pi}{3}$ cu.cm

84. A cone of height 12 cm and slant height 13 cm is surmountes on a hemisphere having radius equal to that of cone. The entire height of the solid is: **(2025)**

(a) 17 cm (b) 18 cm
(c) 22 cm (d) 23 cm

Solution: (a)

$\ell^2 = h^2 + r^2$

$\Rightarrow 13^2 = 12^2 + r^2$

$\Rightarrow r^2 = 25$

$\therefore r = 5$ cm

Total height $= h + r$

$= 12 + 5$

$= 17$ cm

85. If a cone of greatest possible volume is hollowed out from a solid wooden cylinder, then the ratio of the volume of remaining wood to the volume of cone hollowed out is **(2025)**

(a) 1 : 1 (b) 1 : 3
(c) 2 : 1 (d) 3 : 1

Solution:

Volume of the cylinder $= \pi r^2 h$

Volume of the cone $= \frac{1}{3}\pi r^2 h$

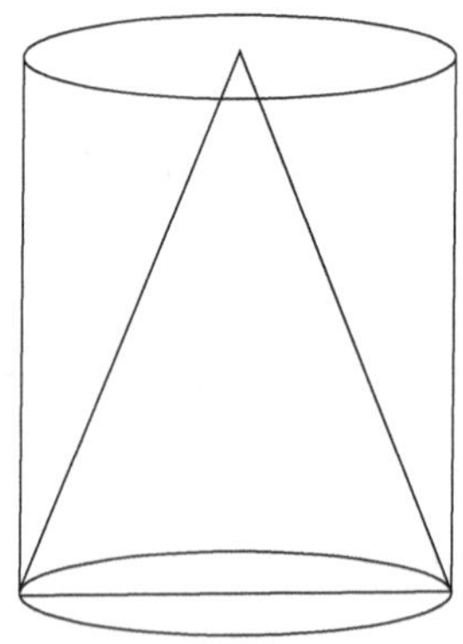

$\therefore$ Volume of the remaining part $= \pi r^2 h - \frac{1}{3}\pi r^2 h$

$= \frac{2}{3}\pi r^2 h$

$\therefore$ The required ratio $= \frac{2}{3}\pi r^2 h : \frac{1}{3}\pi r^2 h$

$= 2 : 1$

86. A cone and cylinder have the same base radius and volume. The height of cone height of cylinder is: **(2025)**

(a) 1 : 1 (b) 3 : 1
(c) 1 : 3 (d) 3 : 2

Solution: (b)

$\frac{1}{3}\pi r^2 h_1 = \pi r^2 h_2$

$\Rightarrow \frac{h_1}{h_2} = 3$

$\therefore h_1 : h_2 = 3 : 1$

87. A cone of height 'h' and radius 'r' is surmounted on a solid cylinder of same dimentions. The total surface area of the entire solid will be: **(2025)**

(a) $2\pi rh + \pi r\sqrt{h^2 + r^2}$

(b) $2\pi rh + \pi r^2 + \pi r\sqrt{h^2 + r^2}$

(c) $2\pi rh + 2\pi r^2 + \pi r\sqrt{h^2 + r^2}$

(d) $2\pi rh + \pi r\sqrt{h^2 + r^2} - \pi r^2$

Solution: (b)

$2\pi rh + \pi r^2 + \pi r\sqrt{h^2 + r^2}$

88. The curved surface area of a cone with base radius 7 cm, is 550 cm^2. The slant height of the cone is: **(2025)**

(a) 25 cm (b) 14 cm
(c) 20 cm (d) 24 cm

Solution: (a)

$\pi r\ell = 550$

$\Rightarrow \frac{22}{7} \times 7 \times \ell = 550$

$\Rightarrow \ell = 25$ cm

89. The largest possible cone is just fitted inside a hollow cube of edge 25 cm. The radius of the base of the cone is: **(2025)**

(a) 5 cm (b) 12.5 cm
(c) 25 cm (d) 10 cm

Solution: (b)

Required radius of the cone is $\frac{25}{2} = 12.5\,\text{cm}$

90. A cone of maximum size is carved out from a solid cube of edge length ℓ. The volume of the cone is: **(2026)**

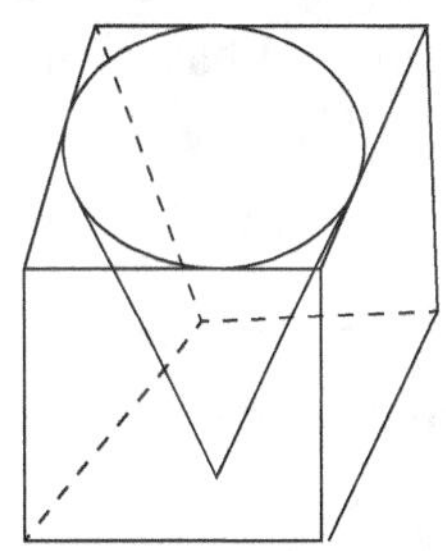

(a) $\frac{\pi\ell^3}{12}$ (b) $\frac{\pi\ell^3}{3}$

(c) $\ell^3\left(1 - \frac{\pi}{3}\right)$ (d) $\frac{\pi\ell^3}{8}$

Solution: (a)

Radius of the cone $= \frac{\ell}{2}$

Height of the cone $= \ell$

$\therefore$ The requied volume $= \frac{1}{3} \times \pi . \left(\frac{\ell}{2}\right)^2 . \ell$

$= \frac{1}{3} \times \pi . \frac{\ell^2}{4} . \ell$

$= \frac{\pi\ell^3}{12}$

91. The total surface area of a soild cone of radius 7 cm and slant height 25 cm is: **(2026)**

(a) 724 cm^2 (b) 709 cm^2
(c) 550 cm^2 (d) 616 cm^2

Solution: (b)

The required surface area is $= \pi r \ell + \pi r^2$

$= \frac{22}{7} \times 7 \times 25 + \frac{22}{7} \times 7^2$

$= 704\,\text{cm}^2$

Exercise 9.18

(i) If the slant height of a cone is 10 cm and the radius is 6 cm, what is its height?

(a) 6 cm (b) 8 cm
(c) 10 cm (d) 15 cm

(ii) The volume of a cone with radius 7 cm and height 12 cm is:

(a) 616 cm^3 (b) 402 cm^3
(c) 800 cm^3 (d) 196 cm^3

(iii) A cone has a base radius of 5 cm and a slant height of 13 cm. What is its curved surface area?

(a) 65π cm^2 (b) 52π cm^2
(c) 36π cm^2 (d) 32π cm^2

(iv) If the height of a cone is tripled while the radius remains the same, its volume becomes:

(a) 3 times (b) 2 times
(c) same (d) 4 times

(v) If the radius of a cone is doubled and the height remains the same, its volume becomes:

(a) same (b) 2 times
(c) 3 times (d) 4 times

(vi) Two cones have the same base radius. If the ratio of their heights is 3:1, what is the ratio of their volumes?

(a) 1 : 3 (b) 3 : 1
(c) 1 : 8 (d) 2 : 3

Key Point If the radius of a sphere is r unit, then its

(i) Surface area = $4\pi r^2$ sq. unit

(ii) Volume $= \frac{4}{3}\pi r^3$ cu.unit

92. The surface area of a sphere 616 cm^2. Find its radius. **(2008)**

Solution:

$4\pi r^2 = 616$

$\Rightarrow 4 \times \frac{22}{7} \times r^2 = 616$

$\Rightarrow r^2 = 49$

$\therefore r = 7$ cm

93. Find the radius of the sphere whose surface area is 36π cm^2. **(2020)**

Solution:

$4\pi r^2 = 36\pi$

$\Rightarrow r^2 = 9$

$\therefore r = 3$ cm

94. The radius of the sphere is $\frac{7}{2}$ cm .The volume of the sphere is: **(2024)**

(a) $\frac{231}{3}$ cu.cm (b) $\frac{539}{12}$ cu.cm

(c) $\frac{539}{3}$ cu.cm (d) 154 cu. cm

Solution:

Required Volume $= \frac{4}{3}\pi r^3$

$= \frac{4}{3} \times \frac{22}{7} \times \left(\frac{7}{2}\right)^3$

$= \frac{4}{3} \times \frac{22}{7} \times \frac{7}{2} \times \frac{7}{2} \times \frac{7}{2}$

$= \frac{539}{3}$ cu.cm

95. The radius of a sphere (in cm) whose volume is 12π cm^2, is **(2020)**

(a) 3 (b) $3\sqrt{3}$

(c) $3^{2/3}$ (d) $3^{1/3}$

Solution: (c)

$\frac{4}{3}\pi r^3 = 12\pi$

$\Rightarrow r^3 = 3^2$

$\Rightarrow r = 3^{2/3}$

96. If the radii of two sphere are in the ratio 2 : 3, then the ratio their respective, volumes is_______. **(2020)**

Solution: Required ratio $= \frac{4}{3}\pi.(2r)^3 : \frac{4}{3}\pi.(3r)^3$

$= 8r^3 : 27r^3 = 8 : 27$

97. If the volumes of two spheres are in the ratio 125 : 64, then the ratio of their surface areas is: **(2024)**

(a) 5 : 4 (b) 25 : 16
(c) 125 : 64 (d) 16 : 5

Solution: (b)

$$\frac{4}{3}\pi r^3 : \frac{4}{3}\pi r_2^3 = 125:64$$

$$\Rightarrow \frac{r_1^3}{r_2^3} = \frac{125}{64}$$

$$\Rightarrow \left(\frac{r_1}{r_2}\right)^3 = \left(\frac{5}{4}\right)^3$$

$$\therefore \frac{r_1}{r_2} = \frac{5}{4}$$

Now, $4\pi r_1^2 : 4\pi r_2^2$

$$= \frac{4\pi r_1^2}{4\pi r_2^2} = \frac{r_1^2}{r_2^2}$$

$$= \left(\frac{r_1}{r_2}\right)^2$$

$$= \left(\frac{5}{4}\right)^2 = \frac{25}{16} = 25:16$$

98. For which of the following solids is the lateral / curved surface area and total surface area the same ? **(2025)**

(a) Cube (b) Cuboid
(c) Hemisphere (d) Sphere

Solution: (d)

Sphere

99. Three tennis balls are just packed in a cyclindrical jar. If radius of each ball is r, volume of air inside the jar is: **(2026)**

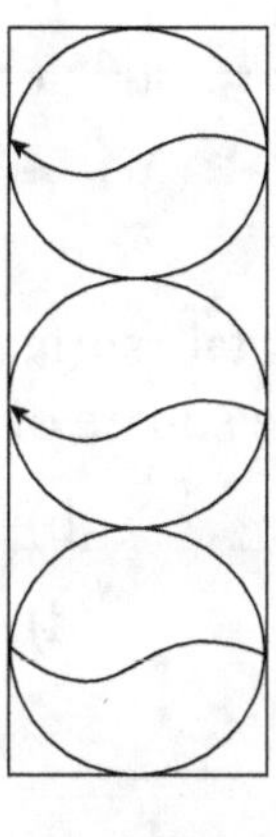

(a) $2\pi r^3$ (b) $3\pi r^3$
(c) $5\pi r^3$ (d) $4\pi r^3$

Solution: (a)

Radius of cylinder = Radius of ball = r

Height of cylinder (h) = $3 \times 2r = 6r$

Volume of 3 balls = $3 \times \frac{4}{3}\pi r^3 = 4\pi r^3$

Volume of air = volume of cylinder-volume of 3 balls

$$= \pi r^2.6r - 4\pi r^3$$

$$= 6\pi r^3 - 4\pi r^3$$

$$= 2\pi r^3$$

100. An ice – cream cone of radius r and height h is completely filled by two spherical scoopes of ice – cream If radius of each spherical scoop in $\frac{r}{2}$, then h:2r equals: **(2026)**

(a) 1 : 8 (b) 1 : 2
(c) 1 : 1 (d) 2 : 1

Solution: (b)

$$\frac{1}{3}\pi r^2 h = 2\times\left(\frac{4}{3}\pi\times\left(\frac{r}{2}\right)^3\right)$$

$$\Rightarrow \frac{1}{3}\pi r^2 h = 2\times\frac{4\pi}{3}\times\frac{r^3}{8}$$

$$\Rightarrow r = h$$

Now h:2r

$= h : 2h$

$= 1 : 2$

101. The radius of a sphere (in cm) whose volume is $36\ \pi$ cm³ is: **(2026)**

(a) 3 (b) $3\sqrt{3}$
(c) $3^{\frac{2}{3}}$ (d) $3^{\frac{1}{3}}$

Solution: (a)

$$V = \frac{4}{3}\pi r^3$$

$$\Rightarrow 36\ \pi = \frac{4}{3}\pi r^3$$

$$\Rightarrow r^3 = 27$$

$$\therefore r = 3 \text{ cm}$$

102. The radius of the largest sphere that can be carved out from a soild cube of side 14 cm is: **(2026)**

(a) 7 cm (b) 14 cm
(c) $14\sqrt{2}$ cm (d) $7\sqrt{2}$ cm

Solution: (a)

Radius of the sphere $= \frac{14}{2}$

$= 7$ cm

Exercise 9.19

(i) If the diameter of a sphere is 14 cm, its surface area is:

(a) 608 cm^2 (b) 616 cm^2
(c) 408 cm^2 (d) 308 cm^2

(ii) The ratio of the surface areas of two spheres with radii $2r$ and r is:

(a) 1 : 1 (b) 2 : 1
(c) 1 : 2 (d) 4 : 1

(iii) If a spheres volume is $\frac{32\pi}{3}$, its surface area is:

(a) 2π (b) 8π
(c) 12π (d) 16π

(iv) If two spheres have volumes in the ratio 64 : 27, their radii are in the ratio:

(a) 4 : 3 (b) 64 : 27
(c) 8 : 9 (d) 2 : 3

(v) If the volume of a sphere is 288π cm^3, its radius is:

(a) 6 cm (b) 8 cm
(c) 12 cm (d) 18 cm

Key Point If the radius of a hemisphere is r unit, then its(i) Surface area $= 2\pi r^2$ unit^2

(ii) Total surface area $= 3\pi r^2$ unit^2

(iii) Total surface area $= \frac{2}{3}\pi r^3$ unit^3

103. Total surface area of a solid hemisphere is: **(2020)**

(a) $3\pi r^2$ (b) $2\pi r^2$
(c) $4\pi r^2$ (d) $\frac{2}{3}\pi r^2$

Solution:

Total surface area of a solid hemisphere

$= 2\pi r^2 + \pi r^2$

$= 3\pi r^2$

104. What is the total surface area of a solid hemisphere of diameters 'd'? **(2023)**

(a) $3\pi d^2$ (b) $2\pi d^2$
(c) $\frac{1}{2}\pi d^2$ (d) $\frac{3}{4}\pi d^2$

Solution: (d)

Total surface area of a hemisphere

$= 2\pi r^2$

$= 3\pi\left(\frac{d}{2}\right)^2$

$= \frac{3}{4}\pi d^2$

105. A solid sphere is cut into two hemisphere. The ratio of the surface areas of sphere to that of two hemisphere taken together is **(2024)**

(a) 1 : 1 (b) 1 : 4
(c) 2 : 3 (d) 3 : 2

Solution: (c)

The surface area of the sphere $= 4\pi r^2$

Total surface areas of two hemispheres

$= 2 \times 3\pi r^2$

$= 6\pi r^2$

$\therefore$ Required ratio $= 2\pi r^2 : 6\pi r^2$

$= 4 : 6$

$= 2 : 3$

106. The ratio of total surface area of a solid hemisphere to the square of its radius is **(2024)**

(a) $2\pi : 1$ (b) $4\pi : 1$
(c) $3\pi : 1$ (d) $1 : 4\pi$

Solution: (c)

Total surface area of a solid hemisphere of radius r is $3\pi r^2$

$\therefore$ Required ratio = $3\pi r^2 : r^2 = 3\pi : 1$

107. The total surface area of a solid hemisphere of radius 7cm is **(2024)**

(a) 98π cm^2 (b) 147π cm^2

(c) 96π cm^2 (d) $228\frac{2}{3}\pi\,\text{cm}^2$

Solution: (b)

Total surface area of the hemisphere = $3\pi r^2$

$= 3 \times \pi \times 7^2$

$= 147\pi$ cm^2

108. Volume and surface area of a solid hemisphere are numerically equal. What is the diameter of hemisphere? **(2017)**

Solution:

$\frac{2}{3}\pi r^3 = 2\pi r^2$

$\Rightarrow r = 3$

109. A cone of height 'h' and radius 'r' is surmounted on a hemisphere of same radius. The total surface area of the entire solid will be: **(2025)**

(a) $\pi rh + \pi r^2$

(b) $\pi r\sqrt{h^2 + r^2} + \pi r^2$

(c) $\pi r\sqrt{h^2 + r^2} + 2\pi r^2$

(d) $\pi r\sqrt{h^2 + r^2} + 3\pi r^2$

Solution: (c)

$\pi r\sqrt{h^2 + r^2} + 2\pi r^2$

110. The total surface area of a solid hemisphere of radius 7 cm is **(2025)**

(a) 98 π cm^2 (b) 196 π cm^2

(c) 147 π cm^2 (d) 174 π cm^2

Solution:

Required area = $3\pi r^2$

$= 3\pi \times 7^2$

$= 147\pi$ cm^2

111. One the top face of the wooden cube of side 7 cm, homispherical depressions of radius 0.35 cm are to be formed by taking out the wood. The maximum number of depressions that can be formed is **(2025)**

(a) 400 (b) 100

(c) 20 (d) 10

Solution: (b)

Diameter of each homisphere = 2×0.35

$= 0.7$ cm

Number of depressions alond one side = $\frac{7}{0.7}$

$= 10$

Maximum number of depressions

= (Number of depressions along one side)^2

$= (10)^2$

$= 100$

112. A hemispherical bowl is made of steel of thickness 1 cm. The outer radius of the bowl is 6 cm. The volume of steel used $\left(\text{in cm}^3\right)$ is: **(2026)**

(a) 182 π (b) $\frac{182}{3}\pi$

(c) $\frac{682}{3}\pi$ (d) $\frac{364}{3}\pi$

Solution: (d)

The required volume of the steel is

$\frac{4}{3}\pi 6^3 - \frac{4}{3}\pi 5^3$

$= \frac{4}{3}\pi\left(6^3 - 5^3\right)$

$= \frac{4}{3}\pi \times 91$

$= \frac{364}{3}\pi\text{cm}^3$

113. A conical cavity of maximum volume is carved out from a wooden soild hemi sphere of radiuas 10 cm. Curved surface area of the cavity carved out is:$\left(\text{use}\,\pi = 3.14\right)$ **(2026)**

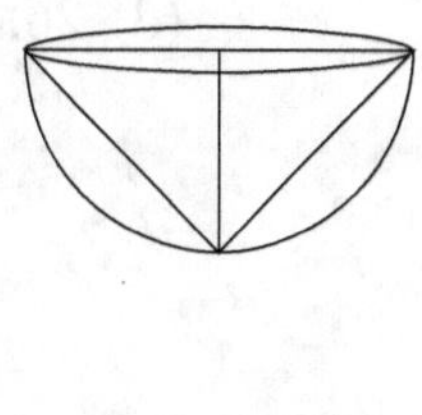

(a) $314\sqrt{2}$ cm² (b) 314 cm²

(c) $\frac{3140}{3}$cm² (d) $3140\sqrt{2}$ cm²

Solution: (a)

Slant height of the cone

$=\sqrt{r^2+h^2}$

$=\sqrt{10^2+10^2}$

$=10\sqrt{2}$ cm²

$\therefore$ $CSA = \pi r \ell$

$=3.14\times10\times10\sqrt{2}$

$=314\sqrt{2}\ cm^2$

114. A camping tent in hemispherical shape of radius 1.4 m, has a door opening of area 0.50 m². Outer surface area of the tent is: **(2026)**

(a) 11.78m^2 (b) 12.32m^2

(c) 11.82m^2 (d) 12.86m^2

Solution: (c)

Requied area $= 2\pi\times(1.4)^2-0.50$

$=2\times\frac{22}{7}\times1.4\times1.4-0.50$

$=2\times22\times0.2\times1.4-0.50$

$=44\times0.28-0.50$

= 12. 32. – 0.50

= 11.82 m²

115. A cylinder of radius r is surmounted on a hemisphere of same radius. If total height of the object is 13 cm then its inner surface area is: **(2026)**

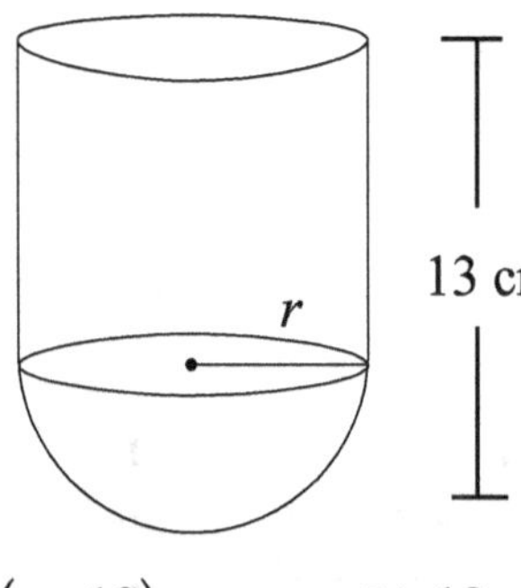

(a) $2\pi r(r+13)$ (b) $13\pi r$

(c) $2\pi(13+r)^2$ (d) $26\pi r$

Solution: (d)

Inner surface area

$=2\pi rh_1+2\pi r^2$

$=2\pi r(h_1+r)$

$=2\pi r\times13$

$=26\pi r$

116. In the adjoining figures, the slant height of the conical part is: **(2026)**

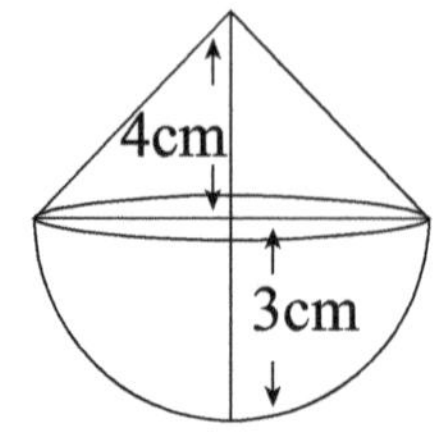

(a) 4 cm (b) 7 cm

(c) 5 cm (d) 25 cm

Solution: (c)

$h = 4$ cm

$r = 3$ cm

$\therefore\ \ell=\sqrt{4^2+3^2}$

$=\sqrt{25}$

= 5 cm.

117. The total surface area of a soild hemisphere of diameter '2d' is: **(2026)**

(a) $3\pi d^2$ (b) $2\pi d^2$

(c) $\frac{1}{2}\pi d^2$ (d) $\frac{3}{4}\pi d^2$

Solution: (a)

$TSA = 2\pi r^2+\pi r^2$

$=2\pi d^2+\pi d^2$

$=3\pi d^2$

Exercise 9.20

(i) If the diameter of a hemisphere is 14 cm, its curved surface area is:

(a) 308 cm² (b) 406 cm²

(c) 322 cm² (d) 516 cm²

(ii) The ratio of the total surface area of a hemisphere to its curved surface area is:

(a) 1 : 2 (b) 2 : 3

(c) 3 : 2 (d) 4 : 9

(iii) If the volume of a hemisphere is 144π cm³, its radius is:

(a) 6 cm (b) 8 cm

(c) 12 cm (d) 4 cm

(iv) The total surface area of a solid hemisphere is numerically equal to its volume when the radius is:

(a) $\frac{3}{2}$ unit (b) $\frac{9}{2}$ unit

(c) 6 unit (d) $\frac{4}{3}$ unit

(iii) A metallic hemisphere is melted and recast into two identical smaller hemispheres. The ratio of the radius of the original hemisphere to that of each new hemisphere is:

(a) 1 : 1 (b) $\sqrt[3]{2}:1$

(c) $1:\sqrt[3]{2}$ (d) 2 : 3

Key Point If a metallic object convented into a new object, then volume of the original object is equal to volume of the new object.

Key Point When a solid item is displaced on water, then the solid displaces a volume of water equal to its own volume

118. A spherical metal ball of radius 8 cm is melted to make 8 smaller identical balls. The radius of the each new ball is ______ cm. **(2020)**

Solution:

volume of the spherical metal ball

$= \frac{4}{3}\pi R^3$

$= \frac{4}{3}\pi 8^3$

volume of each small ball $= \frac{4}{3}\pi r^3$

A.T.Q

$8 \times \frac{4}{3}\pi r^3 = \frac{4}{3}\pi 8^3$

$\Rightarrow r^3 = 64$

$\therefore r = 4$ cm

119. A sphere of diameter 18 cm is dropped into a cylindrical vessel of water of diameter 36 cm,partly filled with water. If the sphere in completely submerged then the water level rises (in cm) by **(2011)**

(a) 3 (b) 4

(c) 5 (d) 6

Solution: (a)

$\frac{4}{3}\pi r^3 = \pi R^2 h$

$\Rightarrow \frac{4}{3} \times 9^3 = 18^2 \times h$

$\therefore h = 3$ cm

Exercise 9.21

(i) If a metallic hemisphere is melted and recast into a sphere, the radius of the new sphere will be:

(a) $\frac{r}{\sqrt[3]{2}}$ (b) r

(c) $\sqrt{r}$ (d) $\sqrt[3]{2}\,r$

(ii) If a metallic hemisphere of radius 8 cm is melted and recast into a new hemisphere of radius 4 cm, how many new hemispheres can be formed?

(a) 1 (b) 2

(c) 4 (d) 8

Exercise 9.22

(i) A sphere of diameter 6 cm is dropped into a cylindrical vessel of water of diameter 12 cm, partly filled with water. If the sphere is completely submerged then the water level rises (in cm) by

(a) 1 cm (b) 2 cm

(c) 3 cm (d) 2.5 cm

(ii) A sphere of 30 cm radius is dropped into a cylindrical vessel of 80 cm diameter, which is partly filled with water, then its level rises by x cm. Find x:

(a) 22 cm (b) 22.5 cm

(c) 18 cm (d) 12 cm

120. A solid is of the form of a cone radius 'r' surmounted on a hemisphere of the same radius. If the height of the cone is the same as the diameter of its base, then the volume of the solid is: **(2023)**

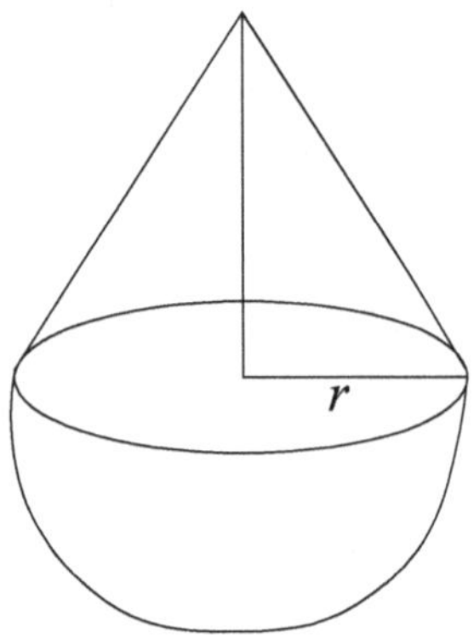

(a) πr^3 (b) $\frac{4}{3}\pi r^3$

(c) $3\pi r^3$ (d) $\frac{2}{3}\pi r^3$

Solution: (b)

Volume of the cone $= \frac{1}{3}\pi r^2 .2r = \frac{2}{3}\pi r^3$

Volume of the hemisphere $= \frac{2}{3}\pi r^3$

$\therefore$ Volume of the solid $= \frac{2}{3}\pi r^3 + \frac{2}{3}\pi r^3$

$= \frac{4}{3}\pi r^3$

Exercise 9.23

(i) A toy is in the form of a cone mounted on a hemisphere of common base radius 7 cm. The total height of the toy is 31 cm. Find the total surface area of the toy.

(a) 465 (b) 1025

(c) 802 (d) 858

(ii) If two solid hemispheres of same base radius 'x' unit are joined together along their bases, then the curve surface area of the new solid formed is (in sq. unit)

(a) πr^2 (b) $2\pi r^2$

(c) $3\pi r^2$ (d) $4\pi r^2$

(iii) A toy is in the form of a cone mounted on a hemisphere and a cylinder. The radius and height of the cone are 3 cm and 4 cm. Find the volume of the given solid?

(a) 198.23 m^3 (b) 206.32 m^3

(c) 207.30 m^3 (d) 198 m^3

(iv) What is the volume of an article which is made by digging out a hemisphere from each end of a solid cylinder?

(a) $\pi r^2 h + 4\pi r^3$ (b) $2\pi rh - \pi r^2$

(c) $\pi r^2 h - \frac{4}{3}\pi r^3$ (d) $\pi r^2 h$.

Question number **121 to 129** are Assertion and Reason based question. Two statements are given, one labelled as Assertion (A) and the other labelled as Reason (R).Select the correct answer to the questions from the codes (A), (B), (C) and D as given bellow.

(a) Both Assertion (A) and Reason (R) are true and Reason (R) is the correct explanation of Assertion (A).

(b) Both Assertion (A) and Reason (R) are true, but Reason (R) is not the correct explanation of Assertion (A).

(c) Assertion (A) is true, but Reason (R) is false.

(d) Assertion (A) is false, but Reason (R) is true.

121. Assertion (A): If the circumference of a circle is 176 cm, then its radius is 28 cm.

Reason(R): circumference = $2\pi \times$ radius of the circle. **(2024)**

Solution: (a)

$2\pi r = 176$

$\Rightarrow 2 \times \frac{22}{7} \times r = 176$

$\Rightarrow r = 28$ cm

122. Assertion (A): The surface area of largent sphere that can be inscribed in a hollow cube side 'a' cm is πa^2 cm^2

Reason(R): The surface area of a sphere of radius 'r' in $\frac{4}{3}\pi r^3$ **(2023)**

Solution: (c)

The surface area of the sphere that can be inscribed in a hollow cube of side 'a' in $4.\pi\left(\frac{a}{2}\right)^2$

$= 4\pi\frac{a^2}{4} = \pi a^2$

123. Assertion (A): Two cubes each of edge length 10 cm are joined together. The total surface area of newly formed cuboid is 1200 cm^2.

Reason(R): Area of each surface of a cube of side 10 cm is 100 cm^2. **(2024)**

Solution: (d)

The length, breadth and height of the newly formed cuboid are 20 cm, 10 cm and 10 cm.

$\therefore$ Total surface area $=2(\ell b + bh + h\ell)$

$= 2(20 \times 10 + 10 \times 10 + 10 \times 20)$

$= 2 \times 500$

$= 1000$ cm^2

Area of each surface of a cube of side 10 cm is $10^2 = 100$ cm^2

124. Assertion (A): The area of canvas cloth required to just cover a heap of rice in the form of a cone of diameter 14 m and height 24 m is 175π sq.m

Reason(R): The curved surface area of a cone of radius r and slant height l is πrl. **(2024)**

Solution: (a)

$\ell^2 = r^2 + h^2$

$\Rightarrow \ell^2 = 7^2 + (24)^2$

$\Rightarrow \ell^3 = 625$

$\therefore \ell = 25$ cm

$\therefore$ Required area $= \pi r\ell$

$= \pi \times 7 \times 25$

$= 175\pi$ sq.cm

125. Assertion (A): The volume of a cone of radius 70 m and height 12 cm is 616 cm^3

Reason(R): The volume of a cone is $\frac{1}{3}\pi r^2 h$ (where r is radius and h is the height) **(2024)**

Solution: (A)

Volume of the cone $= \frac{1}{3}\pi r^2 h$

$= \frac{1}{3} \times \frac{22}{7} \times 7^2 \times 12$

$= 616$ cm^3

126. Assertion(A): If we join two hemispheres of same radius along their bases, then we get a sphere.

Reason(R): The surface area of a sphere of radius r is $3\pi r^2$

Solution: (c)

Assertion(A) is true, but Reason(R) is fasle.

127. Assertion(A): If the total surface area of a solid hemisphere is 462 cm^2, then its radius is 7 cm.

Reason(R): The total surface area of a solid hemisphere of radius r is $3\pi r^2$

Solution: (a)

$3\pi r^2 = 462$

$\Rightarrow 3 \times \frac{22}{7} \times r^2 = 462$

$\Rightarrow r^2 = 49$

$\therefore r = 7$ cm

128. Assertion(A): When a hemisphere of same radius (r) is curved out from one side of a solid wooden cylinder, the total surface area of remaining soild is increased by $2\pi r^2$.

Reason(R): Curved surface area of hemisphere is $2\pi r^2$. **(2025)**

Solution: (a)

Both Assertion (A) and Reason (R) are true and Reason (R) is the correct explanation of Assertion (A).

129. Assertion (A): The surface area of the cuboid formed by joining two cubes of sides 4 cm each, end to end, is 160 cm^2

Reason (R): The surface area of cuboid of dimensions $\ell \times b \times h$ is $(\ell b + bh + h\ell)$ **(2026)**

Solution: (c)

The surface area of the cuboid fromed by joining two cubes of sides 4 cm each end to end/is

$2(8 \times 4 + 4 \times 4 + 4 \times 8)$

$= 2(32 + 16 + 32)$

$= 2 \times 80$

$= 160$ cm^2

Question number **i to x** are Assertion and Reason based question. Two statements are given, one labelled on Assertion (A) and the other labelled an Reason (R). Select the correct answer to these questions from the codes (A), (B), (C) and (D) as given below.

(a) Both Assertion (A) and Reason (R) are true and Reason (R) is the correct explanation of Assertion (A).

(b) Both Assertion (A) and Reason (R) are true, but Reason (R) is not the correct explanation of Assertion (A).

(c) Assertion (A) is true, but reason (R) is false.

(d) Assertion (A) is false, but reason (R) is true.

Exercise 9.24

(i) Assertion (A): If the circumference of a circle is 176 cm, then its radius is 28 cm

Reason (R): circumference = $2\pi \times$ radius

(ii) Assertion (A): If a wire of length 22 cm is bent in the shape of a circle, then area of the circle so formed is 40 cm^2.

Reason (R): circumference of the circle = length of the wire.

(iii) Assertion (A): If the circumference of two circles are in the ratio 3 : 4, then ratio of their areas is 9 : 16.

Reason (R): The circumference of a circle of radius r is $2\pi r$.

(iv) Assertion (A): In a circle of radius 60 cm, the angle of a sector 60°. Then the area of the sector is $18\frac{6}{7}$ cm^2.

Reason (R): Area of the circle with radius r is πr^2.

(v) Assertion (A): If two identical solid cube of side 7 cm are joined end to end. Then the total surface area of the resulting cuboid is 490 cm^2.

Reason (R): Total surface area of cuboid is $lb + bh + h\ell$

(vi) Assertion (A): Total surface area of the cylinder having radius of the base 14 cm and height 30 cm is 3872 cm^2

Reason (R): If r be the radius and h be the height of the cylinder, then total surface area = $2\pi rh + 2\pi r^2$

(vii) Assertion (A): If the height of a come is 24 cm and diameter of the base 14 cm, then the slant height of the come is 15 cm

Reason (R): If r be the radius and h the height of the cone, then slant height $= \sqrt{h^2 + r^2}$

(viii) Assertion (A): If the radius of a come is halved and volume is not changed, then height remains some.

Reason (R): If the radius of a cone in is halved and volume is not changed, then height must be come four times of the original height

(ix) Assertion (A): The radii of two cones are in the ratio 2 : 3 and their volume in the ratio 1 : 3. Then the ratio of their heights is 3 : 2

Reason (R): Volume of the cone $= \frac{1}{3}\pi r^2 h$

(x) Assertion (A): A horse is tied to a peg at are corner of a square shaped grass field of side

15 m by means of a 7 m long rope. The area of that part of the field in which the horse can graze is $\frac{77}{2}$ sqm

Reason (R): Area of the quadrant $= \frac{90}{360} \times \pi r^2$

(xi) Assertion (A): If the perimeter of a circle is equal to that of a square, then the ratio of their areas is 14 : 11.

Reason (R): If the perimeter of a circle is equal to that of a square, then their areas are equal

(xii) Assertion (A): If the volume of two spheres are in the ratio 27 : 8 then their surface area in ratio 9 : 4

Reason (R): Volume of sphere $= \frac{4}{3}\pi r^3$ and surface area $= 4\pi r^2$

ANSWER KEYS

Exercise – 9.1

(i) (c) (ii) (c) (iii) (c)

Exercise – 9.2

(i) (b) (ii) (c) (iii) (b)

Exercise – 9.3

(i) (b) (ii) (a) (iii) (d)

Exercise – 9.4

(i) (c) (ii) (d) (iii) (a) (iv) (d)

Exercise – 9.5

(i) (d) (ii) (a) (iii) (b)

Exercise – 9.6

(i) (c) (ii) (b) (iii) (c)

Exercise – 9.7

(i) (cw) (ii) (a) (iii) (b)

Exercise – 9.8

(i) (d) (ii) (b) (iii) (a)

Exercise – 9.9

(i) (c) (ii) (c) (iii) (d)

Exercise – 9.10

(i) (d) (ii) (b) (iii) (a) (iv) (c) (v) (d)

Exercise – 9.11

(i) (c) (ii) (d) (iii) (d)

Exercise – 9.12

(i) (d) (ii) (a) (iii) (d) (iv) (a) (v) (c) (vi) (d) (vii) (b) (viii) (a) (ix) (d) (x) (b) (xi) (d) (xii) (b)

Exercise – 9.13

(i) (d) (ii) (c) (iii) (a)

Exercise – 9.14

(i) (c) (ii) (b) (iii) (b) (iv) (a) (v) (b)

Exercise – 9.15

(i) (b) (ii) (a) (iii) (d)

Exercise – 9.16

(i) (a) (ii) (d) (iii) (d) (iv) (b)

Exercise – 9.17

(i) (b) (ii) (d) (iii) (a) (iv) (c) (v) (b)

Exercise – 9.18

(i) (c) (ii) (d) (iii) (a) (iv) (a) (v) (d) (vi) (b)

Exercise – 9.19

(i) (b) (ii) (d) (iii) (d) (iv) (a) (v) (a)

Exercise – 9.20

(i) (a) (ii) (c) (iii) (a) (iv) (b)

Exercise – 9.21

(i) (a) (ii) (d) (iii) (b)

Exercise – 9.22

(i) (a) (ii) (b)

Exercise – 9.23

(i) (d) (ii) (d) (iii) (c) (iv) (c)

Exercise – 9.24

(i) (a) (ii) (d) (iii) (b) (iv) (d) (v) (c) (vi) (a) (vii) (d) (viii) (d) (ix) (d) (x) (a) (xi) (c) (xii) (a)

10 Probability

Key Point When a coin is tossed once, all possible outcomes are H and T.

when two coins are tossed simultaneously possible outcomes are HH, HT, TH, TT

When three coins are tossed simultaneously, all possible outcomes are HHH, HHT, HTH, HTT, THH, THT, TTH, TTT.

1. Two coins are tossed simultaneously. Find the probability of getting exactly one head. **(2009)**

Solution:

When two coins are tossed simultaneously all possible outcomes are HH, HT, TH, TT

Total number of possible outcomes = 4

Let E be the event of getting exactly one head.

Then, the favourable outcomes are HT, TH,

Number of favourable outcomes = 2

$\therefore$ P(getting exactly one head) = P(E) = $\frac{2}{4} = \frac{1}{2}$

2. Two coins are tossed simultaneously. The probability of getting at most one head is **(2020)**

(a) $\frac{1}{4}$ (b) $\frac{1}{2}$

(c) $\frac{2}{3}$ (d) $\frac{3}{4}$

Solution:(d)

The favourable outcome are TT, HT, TH

Total number of possible outcome = 4

$\therefore$ P(getting at most one head) = $\frac{3}{4}$

3. A coin is tossed twice. Find the probability of getting head both the times. **(2020)**

Solution:(c)

The possible outcomes are HH, HT, TH, TT

Favourable outcome is HH

$\therefore$ P(getting head both the times) = $\frac{1}{4}$

4. Two coins are tossed together. The probability of getting at least one tail is: **(2023)**

(a) $\frac{1}{4}$ (b) $\frac{1}{2}$

(c) $\frac{3}{4}$ (d) 1

Solution: (c)

The possible outcomes are HH, HT, TH, TT

Favourable outcomes are HT, TH, TT

$\therefore$ P(getting at least one tail) = $\frac{3}{4}$

5. If three coins are tossed simultaneously, what is the probability of getting at most one tail? **(2023)**

(a) $\frac{3}{8}$ (b) $\frac{4}{8}$

(c) $\frac{5}{8}$ (d) $\frac{7}{8}$

Solution: (b)

The possible outcomes are HHH, HHT, HTH, HTT, THH, THT, TTH, TTT

The favourable outcomes are HHH, HHT, HTH, THH

Number of favourable outcomes = 4

$\therefore$ P(getting at most one tail) = $\frac{4}{8}$

6. Two coins are tossed together. The probability of getting almost two heads, is **(2023)**

(a) $\frac{1}{2}$ (b) $\frac{1}{4}$

(c) $\frac{3}{4}$ (d) 1

Solution:(d)

The possible outcomes are HH, HT, TH, TT

The favourable outcomes are HH, HT, TH, TT

$\therefore$Number of favourable outcomes = 4

$\therefore$ P(getting almost two heads) $= \frac{4}{4} = 1$

7. Three coins are tossed together. The probability of getting exactly one tail, is **(2024)**

(a) $\frac{1}{8}$ (b) $\frac{1}{4}$

(c) $\frac{7}{8}$ (d) $\frac{3}{8}$

Solution: (d)

The possible outcomes are

HHH, HHT, HTH, HTT, THH, THT, TTH, TTT

The favourable comes are HHT, HTH, THH.

Number of favourable outcomes = 3

$\therefore$ P(getting exactly one tail) $= \frac{3}{8}$

8. Two coins are tossed simultaneously. The probability of getting at least one head is **(2025)**

(a) $\frac{1}{4}$ (b) $\frac{1}{2}$

(c) $\frac{3}{4}$ (d) 1

9. Two different coins are tossed simultaneously. What is the probability of getting at least one head? **(2020)**

Solution: (c)

The possible outcomes are HH, HT, TH, TT

The favouralbe outcomes are HH, HT, TH

$\therefore$ Number of favourable outcomes = 3

$\therefore$ P(getting at least are head) $= \frac{3}{4}$

10. Three coins are tossed together. The probability that only one coin shows tail, is **(2025)**

(a) $\frac{1}{2}$ (b) $\frac{3}{8}$

(c) $\frac{7}{8}$ (d) 4

Solution: (b)

The possible outcomes are HHH, HHT, HTH, HTT, THH, THT, TTH, TTT.

The favourable outcomes are HHT, HTH, THH.

$\therefore$ Number of favourable outcomes = 3

$\therefore$ P(only one coin shows tail) $= \frac{3}{8}$

11. Three coins are tossed together The porabability of getting exactly one head, is: **(2026)**

(a) $\frac{1}{8}$ (b) $\frac{3}{4}$

(c) $\frac{1}{2}$ (d) $\frac{3}{8}$

Solution:(d)

Number of possible out comen = 8

Favorable out comen are HTT, THT, TTH

Number of favourable out comen = 3

$\therefore$ p(getting exactly me head) $= \frac{3}{8}$

Exercise 10.1

(i) When you toss a coin what is the probability of not getting a head?

(a) $\frac{1}{2}$ (b) 2

(c) 0 (d) 1

(ii) Two unbiased coins are tossed. What are the total number of outcomes.

(a) 1 (b) 2

(c) 4 (d) 8

(iii) Which of the following options is incorrect when one coin is tossed?

(a) $P(H)=\frac{1}{2}$ (b) $P(T)=\frac{1}{2}$

(c) $P(H)+P(T)=1$(d) $P(H)\times P(T)=1$

(iv) Two unbiased coins are tossed. What is the probability of getting no less than 1 head after taking all the possibilities into consideration?

(a) 0 (b) 1

(c) $\frac{1}{2}$ (d) $\frac{3}{4}$

(v) If a coin is tossed 3 times, what is the probability of getting 2 heads and 1 tail?

(a) $\frac{1}{8}$ (b) $\frac{3}{8}$

(c) $\frac{1}{4}$ (d) $\frac{7}{8}$

(vi) If a coin is tossed 4 times, what is the probability of getting exactly 2 heads?

(a) $\frac{1}{2}$ (b) $\frac{5}{8}$

(c) $\frac{1}{4}$ (d) $\frac{3}{8}$

Key Point **(i)** A deck of playing cards has in all 52 cards

(ii) It has 13 cards each of four suits, namely spades, clubs, hearts and diamonds.

(iii) Cards of spades and clubs are black cards.

(iv) Total number of black cards is 26.

(v) Cards of hearts and diamonds are red cards.

(vi) Total number of red cards is 26.

(vii) The cards in each suit are ace, king, queen, Jack, 10, 9, 8, 7, 6, 5, 4, 3, 2, 1.

(viii) King, queen and Jack cards are called face cards.

Thus, there are in all 12 face cards.

12. A card is drawn at random from a well shuffled pack of 52 playing cards. Find the probability of getting a red face card: **(2010)**

13. A card is selected at random a deck of 52 playing cards. The probability of it being a red face card is: **(2025)**

(a) $\frac{3}{13}$ (b) $\frac{2}{13}$

(c) $\frac{1}{2}$ (d) $\frac{3}{26}$

Solution: (d)

Number of possible outcomes = 52

Number of favourable outcomes (red face cards) = 6

$\therefore$ P(getting a red face card) $= \frac{6}{52} = \frac{3}{26}$

14. A card is drawn at random from a well shuffled playing cards. Find the probability of getting neither a red card nor a queen. **(2016)**

Solution:

Number of possible outcomes = 52

Number of favourable outcomes = 24

P(getting neither a red card nor a queen) $= \frac{24}{52}$ $= \frac{6}{13}$

15. A card is drawn at random from a well shuffled deck of 52 playing cards. What is the probability of getting a black king? **(2020)**

16. One card is drawn at random from a well shuffled deck of 52 playing cards. What is the probability of getting a black king? **(2023)**

(a) $\frac{1}{26}$ (b) $\frac{1}{13}$

(c) $\frac{1}{52}$ (d) $\frac{1}{2}$

Solution: (a)

Number of possible outcomes = 52

Number of favourable outcomes = 2

P(getting a black king) $= \frac{2}{52} = \frac{1}{26}$

17. A card is drawn at random from a well shuffled pack of 52 cards. Find the probability of getting a red king? **(2020)**

18. A card is drawn at random from a well shuffled deck of 52 playing cards. What is the probability that it is a red king is: **(2023)**

(a) $\frac{1}{13}$ (b) $\frac{1}{52}$

(c) $\frac{1}{26}$ (d) $\frac{2}{13}$

19. One card is drawn at random from a well shuffled deck of 52 playing cards. The probability that it is a red king is: **(2023)**

(a) $\frac{1}{52}$ (b) $\frac{1}{26}$

(c) $\frac{2}{26}$ (d) $\frac{2}{13}$

Solution:

Number of possible outcomes = 52

Number of favourable outcomes = 2

$\therefore$ P(getting a red king) $= \frac{2}{52} = \frac{1}{26}$

20. From a well shuffled deck of 52 playing cards, a card is drawn at random. What is the probability of getting a red queen? **(2023)**

21. A card is drawn at random from a well shuffled deck of 52 playing cards. The probability that drawn card is a red queen, is **(2024)**

(a) $\frac{1}{13}$ (b) $\frac{2}{13}$

(c) $\frac{1}{52}$ (d) $\frac{1}{26}$

Solution: (d)

Number of possible outcomes = 52

Number of favourable outcomes = 2

$\therefore$ P(getting a red queen) $= \frac{2}{52} = \frac{1}{26}$

22. One card is drawn at random from a well shuffled deck of 52 cards. What is the probability of getting a jack? **(2020)**

(a) $\frac{3}{26}$ (b) $\frac{1}{52}$

(c) $\frac{1}{13}$ (d) $\frac{3}{52}$

Solution: (c)

Number of possible outcomes = 52

Number of favourable outcomes = 4

$\therefore$ P(getting a jack) $= \frac{4}{52} = \frac{1}{13}$

23. A card is drawn at random from a well shuffled deck of 52 cards. The probability of getting a red card is **(2023)**

(a) $\frac{1}{26}$ (b) $\frac{1}{13}$

(c) $\frac{1}{4}$ (d) $\frac{1}{2}$

Solution: (d)

Number of possible outcomes = 52

Number of favourable outcomes = 26

$\therefore$ P(getting a red card) $= \frac{26}{52} = \frac{1}{2}$

24. A card is drawn at random from a well-shuffled deck of 52 playing cards. The probability of getting an ace of spade is: **(2023)**

(a) $\frac{1}{13}$ (b) $\frac{3}{52}$

(c) $\frac{1}{26}$ (d) $\frac{1}{52}$

Solution: (d)

Number of possible outcomes = 52

Number of favourable outcome = 1

$\therefore$ *P*(getting an ace of spade) $= \frac{1}{52}$

25. From a well-shuffled deck of 52 cards, a card is drawn at random. What is the probability of getting king of hearts? **(2023)**

(a) $\frac{1}{52}$ (b) $\frac{1}{26}$

(c) $\frac{1}{13}$ (d) $\frac{12}{13}$

Solution: (a)

Total number of possible outcomes = 52

Number of favourable outcome = 1

$\therefore$ P(getting king of hearts) $= \frac{1}{52}$

26. One card is drawn at random from a well shuffled deck of 52 playing cards. What is the probability of getting '4' of hearts? **(2023)**

(a) $\frac{1}{52}$ (b) $\frac{1}{13}$

(c) $\frac{1}{26}$ (d) $\frac{1}{6}$

Solution: (a)

Total number of possible outcomes = 52

Number of favourable outcome = 1

$\therefore$ P(getting 4 of hearts) $= \frac{1}{52}$

27. A card is drawn at random from a well-shuffled pack of 52 cards. The probability that the card drawn is not an ace is **(2023)**

(a) $\frac{1}{13}$ (b) $\frac{9}{13}$

(c) $\frac{4}{13}$ (d) $\frac{12}{13}$

Solution: (d)

Total number of possible outcomes = 52

Number of favourable outcome = 52 – 4 = 48

$\therefore$ P(not an ace) $= \frac{48}{52} = \frac{12}{13}$

28. One card is drawn at random from a well shuffled pack of 52 playing cards. The probability that the drawn card is a queen is: **(2023)**

(a) $\frac{4}{3}$ (b) $\frac{4}{52}$

(c) $\frac{2}{13}$ (d) $\frac{1}{26}$

Solution: (b)

Total number of possible outcomes = 52

Number of favourable outcomes = 4

$\therefore$ P(drawn card is a queen) $= \frac{4}{52}$

29. A card is drawn at random from a well shuffled deck of 52 playing cards. The probability of getting a face card is: **(2023)**

(a) $\frac{1}{2}$ (b) $\frac{3}{13}$

(c) $\frac{4}{13}$ (d) $\frac{1}{13}$

Solution: (b)

Total number of possible outcomes = 52

Number of favourable outcomes = 12

$\therefore$ *P*(getting a face card) $= \frac{12}{52} = \frac{3}{13}$

30. A card is drawn at random from a deck of 52 playing cards. The probability that the drawn card is not a red face card is **(2023)**

(a) $\frac{3}{26}$ (b) $\frac{23}{26}$

(c) $\frac{7}{52}$ (d) $\frac{23}{52}$

Solution: (b)

Total number of possible outcomes = 52

Number of favourable outcomes = 52 – 6 = 46

$\therefore$ P(drawn card is not a red face card) $= \frac{46}{52} = \frac{23}{26}$

31. A card is drawn at random from a well shuffled deck of 52 playing cards. The probability that drawn card shows number '9' is: **(2025)**

(a) $\frac{1}{26}$ (b) $\frac{4}{13}$

(c) $\frac{1}{52}$ (d) $\frac{1}{13}$

Solution: (d)

Total number of possible outcomes = 52

Number of favourable outcomes = 4

$\therefore$ P(drawn card shows number 9) $= \frac{4}{52} = \frac{1}{13}$

32. A black card is lost from a deck of 52 playing cards. Rest of the cards one shuffled and one card is drawn at random from the available cards. The probability that drawn card is 'King of hearts', is **(2025)**

(a) $\frac{1}{52}$ (b) $\frac{1}{4}$

(c) $\frac{1}{51}$ (d) $\frac{1}{26}$

Solution: (c)

Total number of possible outcomes = 51

Number of favourable outcomes = 1

$\therefore$ P(drawn card is king of hearts) $= \frac{1}{51}$

33. A card is drawn at random from well shuffled deck of 52 playing cards The probability that it is either a ten or *a* king is: **(2026)**

(a) $\frac{1}{26}$ (b) $\frac{2}{13}$

(c) $\frac{1}{13}$ (d) $\frac{8}{26}$

Solution:

The number of possible out comet is 8

$\therefore P(E) = \frac{8}{52} = \frac{2}{13}$

34. A card is drawn from a well shuffled deck of 52 playing cards The probability of getting a queen of spades is: **(2026)**

(a) $\frac{1}{26}$ (b) $\frac{1}{52}$

(c) 0 (d) $\frac{1}{4}$

Solution: (b)

The possible outcome is 1

$\therefore p(E) = \frac{1}{52}$

Exercise 10.2

(i) The probability of drawing a card that is either a heart or a king is:

(a) $\frac{17}{52}$ (b) $\frac{4}{13}$

(c) $\frac{1}{2}$ (d) $\frac{5}{13}$

(ii) If a card is drawn at random, the probability that it is not a diamond, is:

(a) $\frac{1}{4}$ (b) $\frac{3}{4}$

(c) $\frac{1}{2}$ (d) $\frac{13}{52}$

(iii) The probability of drawing a 2 of any suit is:

(a) $\frac{1}{26}$ (b) $\frac{1}{4}$

(c) $\frac{1}{13}$ (d) $\frac{7}{52}$

(iv) A card is drawn from a well shuffled deck of cards. What is the probability that the card drawn in neither a king nor a queen.

(a) $\frac{11}{13}$ (b) $\frac{12}{13}$

(c) $\frac{1}{2}$ (d) $\frac{3}{4}$

(v) A card is selected from a deck of 52 cards. The probability of being not a red face card is.

(a) $\frac{3}{52}$ (b) $\frac{3}{13}$

(c) $\frac{1}{52}$ (d) $\frac{23}{26}$

(vi) A card is drawn at random from a well-shuffled deck of playing cards. Find the probability of drawing a card which is neither a king nor a red card.

(a) $\frac{3}{13}$ (b) $\frac{4}{13}$

(c) $\frac{6}{13}$ (d) $\frac{1}{52}$

(vii) A red face card in lost from a deck of 52 playing cards. Rest of the cards are shuffled and one card is drawn at random from the available cards. The probability that drawn card is a face card is.

(a) $\frac{1}{52}$ (b) $\frac{12}{52}$

(c) $\frac{11}{51}$ (d) $\frac{13}{51}$

(viii) Three face cards of spade are removed from a well shuffled pack of 52 cards and a card is drawn from the remaining pack. The probability of getting a black face card is

(a) $\frac{2}{25}$ (b) $\frac{3}{49}$

(c) $\frac{7}{52}$ (d) $\frac{1}{51}$

Key Point **(1)** If a die is thrown once, then total possible outcomes are 1, 2, 3, 4, 5 and 6. **(2)** If a die is thrown twice or two dice are thrown simultaneously, then total possible outcomes will be $6^2 = 36$ and they are

(1, 1), (1, 2), (1, 3), (1, 4), (1, 5), (1, 6),
(2, 1), (2, 2), (2, 3) , (2, 4), (2, 5), (2, 6),
(3, 1), (3, 2), (3, 3), (3, 4), (3, 5), (3, 6),
(4, 1), (4, 2), (4, 3), (4, 4), (4, 5), (4, 6),

(5, 1), (5, 2), (5, 3), (5, 4), (5, 5), (5, 6)
(6, 1), (6, 2), (6, 3), (6, 4), (6, 5), (6, 6)

(3) If a die is thrown n times or n dice are thrown simultaneously, then total number of possible outcomes is 6^n.

35. Two dice are thrown together, the probability of getting the same number on both dice is **(2012)**

(a) $\frac{1}{2}$ (b) $\frac{1}{3}$

(c) $\frac{1}{6}$ (d) $\frac{1}{12}$

Solution: (c)

Total number of possible outcomes = 36

Favourable outcomes are (1, 1), (2, 2), (3, 3), (4, 4), (5, 5), (6, 6)

∴ Number of favourable outcomes = 6

∴ P (getting same number on both sides) $= \frac{6}{36}$

$= \frac{1}{6}$

36. The probability of getting an even number, when a dice is thrown once, is: **(2013)**

Solution:

Number of possible outcomes = 6

Favourable outcomes are = 2, 4, 6

∴ Number of favourable outcomes = 3

∴ P(getting an even number) $= \frac{3}{6} = \frac{1}{2}$

37. Two different dice are tossed together. Find the probability that the product of the two numbers on the top of the dice is 6: **(2015)**

Solution:

Number of possible outcomes = 36

Favourable outcomes are = (1, 6), (2, 3), (3, 2), (6, 1),

∴ Number of favourable outcomes = 4

∴ P(product of the two numbers on the top of the dice is 6) $= \frac{4}{36} = \frac{1}{9}$

38. If a pair of dice is thrown once, then what is the probability of getting a sum of 8? **(2020)**

39. The probabi lity of getting a sum of 8, when two dice are thrown simultaneously, is **(2024)**

(a) $\frac{1}{2}$ (b) $\frac{1}{9}$

(c) $\frac{1}{6}$ (d) $\frac{5}{36}$

Solution:

Number of possible outcomes = 36

Favourable outcomes are = (2, 6), (3, 5), (4, 4), (5, 3), (6, 2)

∴ Number of favourable outcomes = 5

∴ P(getting a sum of 8) $= \frac{5}{36}$

40. A die is thrown once. The probability of getting an odd number is: **(2020)**

(a) 1 (b) $\frac{1}{2}$

(b) $\frac{4}{6}$ (d) $\frac{2}{6}$

Solution: (b)

The number of possible outcomes = 6

Favourable outcomes are = 1, 3, 5

∴ Number of favourable outcomes = 3

∴ P(getting an odd number) $= \frac{3}{6} = \frac{1}{2}$

41. A dice is thrown once. If getting a six, is a success. Then find the probability of a failure. **(2020)**

Solution:

The number of possible outcomes = 6

Favourable outcomes are 1, 2, 3, 4, 5

∴ Number of favourable outcomes = 5

∴ P(a failure) $= \frac{5}{6}$

42. A die is thrown once. Find the probability of getting a number less than 7. **(2023)**

(a) $\frac{5}{6}$ (b) 1

(c) $\frac{1}{6}$ (d) 0

Solution:

The number of possible outcomes = 6

Favourable outcomes are 1, 2, 3, 4, 5, 6

∴ Number of favourable outcomes = 6

∴ P(getting a number less than 7) $= \frac{6}{6} = 1$

43. A die is thrown once. The probability of getting an odd prime number is **(2023)**

(a) $\frac{1}{2}$ (b) $\frac{1}{6}$

(c) $\frac{1}{3}$ (d) $\frac{2}{3}$

44. An unbiased die is thrown. The probability of getting an odd prime number is: **(2023)**

(a) $\frac{1}{6}$ (b) $\frac{1}{2}$

(c) $\frac{2}{3}$ (d) $\frac{1}{3}$

45. A die is rolled once. What is the probability of getting an odd prime number? **(2024)**

(a) $\frac{1}{6}$ (b) $\frac{1}{3}$

(c) $\frac{2}{3}$ (d) $\frac{3}{4}$

Solution:

The number of possible outcomes = 6

Favourable outcomes are 3, 5

Number of favourable outcomes = 2

∴ P(getting an odd prime number) $= \frac{2}{6} = \frac{1}{3}$

46. A die is rolled once. The probability that a composite number comes up, is: **(2023)**

(a) $\frac{1}{2}$ (b) $\frac{2}{3}$

(c) $\frac{1}{3}$ (d) 0

Solution: (a)

The number of possible outcomes = 6

Favourable outcomes are 4, 6

∴ Number of favourable outcomes = 2

∴ P(getting a composite number) $= \frac{2}{6} = \frac{1}{3}$

47. Two dice are rolled together. The probability that the sum of the numbers that appeared is 9, is: **(2023)**

(a) $\frac{5}{36}$ (b) $\frac{1}{9}$

(c) $\frac{1}{12}$ (d) $\frac{1}{6}$

Solution: (b)

Total number of possible outcomes = 36

Favourable outcomes are = (3, 6), (4, 5), (5, 4), (6, 3)

Total number of favourable outcomes = 4

∴ P(sum of the numbers that appeared is 9) $= \frac{4}{36} = \frac{1}{9}$

48. Two dice are rolled together. The probability of getting a doublet is: **(2024)**

(a) $\frac{2}{36}$ (b) $\frac{1}{36}$

(c) $\frac{1}{6}$ (d) $\frac{5}{6}$

Solution: (c)

Total number of possible outcomes = 36

Favourable outcomes are = (1, 1), (2, 2), (3, 3), (4, 4), (5, 5), (6, 6)

∴ Total number of favourable outcomes = 6

∴ P(getting a doublet) $= \frac{6}{36} = \frac{1}{6}$

49. In an experiment of throwing a pair of dice, the probability of not getting a doublet is: **(2025)**

(a) $\frac{1}{6}$ (b) $\frac{5}{6}$

(c) $\frac{1}{5}$ (d) $\frac{1}{30}$

Solution: (b)

Total number of possible outcomes = 36

Number of favourable outcomes = 36 – 6 = 30

∴ P(not getting a doublet) $= \frac{30}{36} = \frac{5}{6}$

50. On a throw of a die, if getting 6 is considered success then probability of losing the game is: **(2024)**

(a) 0 (b) 1

(c) $\frac{1}{6}$ (d) $\frac{5}{6}$

Solution: (d)

Total number of possible outcomes = 6

Total number of favourable outcomes = 5

$\therefore$ P(losing the game) = $\frac{5}{6}$

51. A fair die is thrown once. The probability of getting a composite number less than 5 is: **(2024)**

(a) $\frac{1}{3}$ (b) $\frac{2}{3}$

(c) 0 (d) $\frac{1}{6}$

Solution: (a)

Total number of possible outcomes = 6

Favourable outcome is 2

$\therefore$ Number of favourable outcome = 1

$\therefore$ *P*(getting a composite number less than 5) = $\frac{1}{6}$

52. When a die is thrown, the probability of getting an odd number greater than 3 is **(2024)**

(a) $\frac{1}{6}$ (b) $\frac{1}{3}$

(c) $\frac{1}{2}$ (d) 0

Solution:

Total number of possible outcomes = 6

Favourable outcome is 5

$\therefore$ P(getting an odd number greater than 3) = $\frac{1}{6}$

53. Two dice are rolled together. The probability of getting sum of numbers on the two dice as 2, 3 or 5 is: **(2024)**

(a) $\frac{7}{36}$ (b) $\frac{11}{36}$

(c) $\frac{5}{36}$ (d) $\frac{4}{9}$

Solution: (a)

Total number of possible outcomes = 36

Favourable outcomes are = (1, 1), (1, 2), (1, 4), (2, 1), (2, 3), (4, 1), (3, 2)

$\therefore$ Number of favourable outcomes = 7

P (getting sum of numbers on the two dice as 2, 3 or 5) = $\frac{7}{36}$

54. Two dice are rolled together. The probability of getting the sum of the two number to be more than 10, is **(2024)**

(a) $\frac{1}{9}$ (b) $\frac{1}{6}$

(c) $\frac{7}{12}$ (d) $\frac{1}{12}$

55. Two dice are rolled together. What is the probability of getting a sum greater than 10? **(2023)**

(a) $\frac{1}{9}$ (b) $\frac{1}{6}$

(c) $\frac{1}{12}$ (d) $\frac{5}{18}$

Solution: (a)

Total number of possible outcomes = 36

Favourable outcomes are = (5, 6), (6, 5), (6, 6),

$\therefore$ Number of favourable outcomes = 3

$\therefore$ P(getting a sum greater than 10) = $\frac{3}{36} = \frac{1}{12}$

56. A die is thrown once . What is the probability of getting a number less than 3? **(2020)**

Solution:

Total number of possible outcomes = 6

Favourable outcomes are 1, 2

$\therefore$ Number of favourable outcomes = 2

$\therefore$ P(getting a number less than 3) = $\frac{2}{6} = \frac{1}{3}$

57. Two dice are tossed simultaneously. The probability of getting odd number on both the dice is: **(2024)**

(a) $\frac{6}{36}$ (b) $\frac{3}{36}$

(c) $\frac{12}{36}$ (d) $\frac{9}{36}$

Solution: (d)

Total number of possible outcomes = 36

Favourable outcomes are = (1, 1), (1, 3), (1, 5), (3, 1), (3, 3), (3, 5), (5, 1), (5, 3) (5, 5)

∴ Number of favourable outcomes = 9

∴ P(getting odd number on both the dice) $= \frac{9}{36}$

58. The probability of throwing a number greater than 2 with a fair die is. **(2024)**

(a) $\frac{2}{3}$ (b) $\frac{1}{3}$

(c) $\frac{1}{2}$ (d) $\frac{5}{6}$

Solution: (a)

Total number of possible outcomes = 6

Favourable outcomes are 3, 4, 5, 6

Number of favourable outcomes = 4

∴ P(getting a number greater than 2) $= \frac{4}{6} = \frac{2}{3}$

59. A die is thrown once . What is the probability of getting a prime number. **(2020)**

Solution:

Total number of possible outcomes = 6

Favourable outcomes are 2, 3, 5

Number of favourable outcomes = 3

∴ P(getting a prime number) $= \frac{3}{6} = \frac{1}{2}$

60 Two dice are thrown simultaneously. What is the probability that the sum of the two numbers appearing on the top is 13? **(2020)**

Solution:

The number of possible outcomes is 36

Number of favourable outcomes = 0

∴ P (sum of the two numbers appearing on the top is 13) $= \frac{0}{36} = 0$

61. A pair of dice is thrown once. The probability that sum of numbers appearing on top faces is atleast 4 is **(2025)**

(a) $\frac{1}{11}$ (b) $\frac{10}{11}$

(c) $\frac{5}{6}$ (d) $\frac{11}{12}$

Solution:

The number of possible outcomes = 36

Favourable outcomes are (1, 3), (1, 4), (1, 5), (1, 6), (2, 2), (2, 3), (2, 4), (2, 5), (2, 6) , (3, 1), (3, 2), (3, 3), (3, 4), (3, 5), (3, 6), (4, 1), (4, 2), (4, 3), (4, 4), (4, 5), (4, 6), (5, 1), (5, 2), (5, 3), (5, 4), (5, 5), (5, 6), (6, 1), (6, 2), (6, 3) (6, 4), (6, 5), (6, 6)

∴ Number favourable outcomes = 33

∴ P (sum of numbers appearing on top faces is atleast 4) $= \frac{33}{36} = \frac{11}{12}$

62. The probability of getting a composite number greater than 3 on throwing a dice is **(2025)**

(a) $\frac{1}{6}$ (b) $\frac{1}{3}$

(c) $\frac{1}{2}$ (d) $\frac{2}{3}$

Solution:

The number of possible outcomes = 6

Favourable outcomes are 4, 6

Number of favourable outcome = 2

∴ P (getting a composite number greater than 3) $= \frac{2}{6} = \frac{1}{3}$

63. The total number of outcomes in the experiment of simultaneous thrown of three dice is **(2025)**

(a) 6 (b) 18

(c) 36 (d) 216

Solution: (d)

$6 \times 6 \times 6 = 216$

64. Two dice are rolled together. The probability that at least one of them shows a six, is **(2025)**

(a) $\frac{12}{36}$ (b) $\frac{5}{36}$

(c) $\frac{11}{36}$ (d) $\frac{6}{36}$

Solution: (c)

Number of possible outcomes = 36

Favourable outcomes are (1, 6), (2, 6), (3, 6), (4, 6), (5, 6), (6, 6), (6, 1), (6, 2), (6, 3), (6, 4), (6, 5)

Number of favourable outcomes = 11

$\therefore P(\text{at least one of than shows a six}) = \frac{11}{36}$

65. Two dice are rolled together. The probability that only one die shows number 4, is **(2025)**

(a) $\frac{11}{36}$ (b) $\frac{1}{3}$

(c) $\frac{5}{18}$ (d) $\frac{1}{4}$

Solution: (c)

The total number of possible outcomes = 36

Favourable outcomes are (1, 4), (2, 4), (3, 4), (5, 4), (6, 4), (4, 1), (4, 2), (4, 3), (4, 5), (4, 6)

$\therefore$ Total number of favourable outcomes = 10

$\therefore$ P(only one die shows number 4) $= \frac{10}{36} = \frac{5}{18}$

66. Two dice are rolled together. The probability that the sum of the numbers obtained is divisible by 6, is: **(2026)**

(a) $\frac{1}{6}$ (b) $\frac{11}{36}$

(c) $\frac{1}{12}$ (d) $\frac{1}{4}$

Solution: (a)

Number of possible outcomes = 36

favouraite outcomes are

(1, 5), (2, 4), (3, 3), (4, 2), (5, 1), (6, 6)

Number of favourable outcomes

= 6

$\therefore P(\text{sum of numbers obtained is divisible by 6})$

$= \frac{6}{36} = \frac{1}{6}$

67. Two dice are rolled together The probability of getting an out come (x, y) where $x > y$, is **(2026)**

(a) $\frac{5}{12}$ (b) $\frac{5}{6}$

(c) 1 (d) 0

Solution: (a)

Possible outcomes are

(2, 1), (3, 1), (3, 2), (4, 1), (4, 2), (4, 3) (5, 1), (5, 2), (5, 3),

(5, 4), (6, 1), (6, 2), (6, 3), (6, 4), (6, 5)

Number of possible outcomes = 15

$\therefore P(E) = \frac{15}{36} = \frac{5}{12}$

68. A die is thrown once. Probability of getting a number other then 3 is: **(2026)**

(a) $\frac{1}{6}$ (b) $\frac{3}{6}$

(c) $\frac{5}{6}$ (d) 1

Solution:

The number of possible outcomes = 5

$\therefore P(E) = \frac{5}{6}$

69. Two differnt dice are rolled together. The probability that both the obtained numbers are less than 4 is: **(2026)**

(a) $\frac{2}{9}$ (b) $\frac{7}{36}$

(c) $\frac{1}{4}$ (d) $\frac{2}{3}$

Solution: (c)

The possible outcomes are (1, 1), (1, 2), (1, 3), (2, 1), (2, 2), (2, 3), (3, 1), (3, 2), (3, 3)

$\therefore$ The number of possible outcomes = 9

$P(E) = \frac{9}{36} = \frac{1}{4}$

70. The probability of getting sum greater than 10,. When two dice are rolled together is: **(2026)**

(a) $\frac{1}{9}$ (b) $\frac{1}{18}$

(c) $\frac{1}{12}$ (d) 1

Solution: (c)

The possible outcomes are (5, 6), (6, 5), (6, 6)

$\therefore$ The number of possible outcomes = 3

$\therefore P(E) = \frac{3}{36} = \frac{1}{12}$

Exercise 10.3

(i) The probability of getting a composite number greater then 2 on throwing a dice is

(a) $\frac{1}{6}$ (b) $\frac{2}{6}$

(c) $\frac{3}{6}$ (d) 1

(ii) Two die are rolled together. The probability that atleast one of them shows a 4, is

(a) $\frac{11}{36}$ (b) $\frac{3}{36}$

(c) $\frac{12}{36}$ (d) $\frac{10}{36}$

(iii) Two dice are rolled together. The probability that only one die shows number 1, is

(a) $\frac{10}{36}$ (b) $\frac{11}{36}$

(c) $\frac{1}{2}$ (d) $\frac{2}{3}$

(iv) The probability of getting 8 in rolling fair die is__________.

(a) 1 (b) 2

(c) 0 (d) 8

(v) A die is rolled. What is the probability that the number appearing on upper face is less than 4.

(a) $\frac{1}{2}$ (b) $\frac{1}{3}$

(c) 1 (d) $\frac{5}{6}$

(vi) Two dice are rolled simultaneously. A is an event that product of numbers on the uppermost face is 12, then $P(A) = ?$

(a) $\frac{1}{6}$ (b) $\frac{1}{9}$

(c) $\frac{2}{9}$ (d) $\frac{1}{3}$

(vii) A die is thrown, the probability of getting a perfect square is__________

(a) $\frac{1}{2}$ (b) 1

(c) $\frac{1}{3}$ (d) $\frac{2}{3}$

(viii) Two dice are thrown simultaneously. E is the event that sum of numbers on the uppermost face is at least 10 than $P(E) =$

(a) $\frac{1}{6}$ (b) $\frac{2}{6}$

(c) $\frac{3}{6}$ (d) $\frac{2}{3}$

(ix) This dice are rolled. What is the probability that the sum is 7?

(a) $\frac{1}{6}$ (b) $\frac{1}{12}$

(c) $\frac{5}{36}$ (d) $\frac{7}{36}$

(x) Two dice are thrown together. The probability that sum of the two numbers will be a multiple of 4, is

(a) $\frac{1}{2}$ (b) $\frac{1}{3}$

(c) $\frac{1}{8}$ (d) $\frac{1}{4}$

Key Point A bag contains a nos red, b nos white and c nos green balls. A ball is drawn at a random from the bag. Then probability of getting

(i)a red ball $= \dfrac{a}{a+b+c}$

(ii)a white ball $= \dfrac{b}{a+b+c}$

(iii)a green ball $= \dfrac{c}{a+b+c}$

(iv)not a red ball $= \dfrac{b+c}{a+b+c}$

(v)not a white ball $= \dfrac{a+c}{a+b+c}$

(vi)not a green ball $= \dfrac{a+b}{a+b+c}$.

71. A bag contains 4 red, 5 white and 6 green balls. A ball is drawn at a random from the bag. Find the probability of getting not a red ball. **(2020)**

Solution:

Total number of balls = 4 + 5 + 6 = 15

Number of favourable outcomes = 5 + 6 = 11

$\therefore$ P(getting not a red ball) = $\frac{11}{15}$

72. A bag contains 5 red, 4 blue and 3 green balls. A ball is drawn at random from the bag. Find the probability of getting a ball not of blue colour. **(2020)**

Solution:

Total number of balls = 5 + 4 + 3 = 12

Number of favourable outcomes = 5 + 3 = 8

$\therefore$ P(getting a ball not of blue colour) = $\frac{8}{12} = \frac{2}{3}$

73. A bag has 4 red balls and 2 yellow balls. A ball is drawn at random from the bag without looking into the bag. What is the probability of getting a yellow ball? **(2024)**

(a) $\frac{1}{6}$ (b) $\frac{2}{3}$

(c) $\frac{1}{3}$ (d) 1

Solution: (c)

Total number of balls = 4 + 2 = 6

P(getting a yellow ball) = $\frac{2}{6} = \frac{1}{3}$

74. A bag contains 5 pinks, 8 blue and 7 yellow balls. One ball is drawn at random from the bag. What is the probability of getting neither a blue nor a pink ball? **(2023)**

(a) $\frac{1}{4}$ (b) $\frac{2}{5}$

(c) $\frac{7}{20}$ (d) $\frac{13}{20}$

Solution: (c)

Total number of balls = 5 + 8 + 7 = 20

P(getting neither a blue nor a pink ball) = $\frac{7}{20}$

75. A bag contains 3 red, 5 black and 7 white balls. A ball is drawn from the bag at random. The probability that the ball drawn is not black is **(2020)**

(a) $\frac{1}{3}$ (b) $\frac{9}{15}$

(c) $\frac{5}{10}$ (d) $\frac{2}{3}$

Solution: (d)

Total number of balls = 3 + 5 + 7 = 15

Number of favourable outcomes = 3 + 7 = 10

$\therefore$ P(getting a nor black ball) = $\frac{10}{15}$

76. A bag contains 3 red, 4 white and 7 green balls. A ball is drawn at random. The probability that the ball drawn is not of red colour is: **(2025)**

(a) $\frac{1}{11}$ (b) $\frac{2}{14}$

(c) $\frac{11}{14}$ (d) $\frac{3}{11}$

Solution: (c)

Total number of balls = 3 + 4 + 7 = 14

Number of favourable outcomes = 4 + 7 = 11

$\therefore$ P(getting a non-red colour ball) = $\frac{11}{14}$

77. The number of red balls in a bag is 10 more than the number of black balls. If the probability of drawing a red ball at random from this bag is $\frac{3}{5}$, then the total number of balls in the bag is: **(2025)**

(a) 50 (b) 60

(c) 80 (d) 40

Solution: (a)

Let the number of black ball is x.

$\therefore$ Number of red ball = $x + 10$

$\therefore$ Total number of balls = $x + (x + 10) = 2x + 10$

Now, P (getting a red ball) = $\frac{3}{5}$

$\Rightarrow \frac{x+10}{2x+10} = \frac{3}{5}$

$\Rightarrow 5x + 50 = 6x + 30$

$\Rightarrow -x = -20$

$\therefore x = 20$

$\therefore$ Total number of balls = $2 \times 20 + 10 = 50$

78. A bag contains red balls and black balls in the ratio 3 : 7. A ball is drawn at random. The probability that ball so drawn in black in colour, is **(2025)**

(a) $\frac{3}{7}$ (b) 0.3

(c) 0.7 (d) $\frac{1}{7}$

Solution: (c)

Let the number of red ball = $3x$

Number of black ball = $7x$

$\therefore$ Total number of balls = $3x + 7x = 10x$

$\therefore$ P(getting a black ball) $= \frac{7x}{10x} = \frac{7}{10} = 0.7$

79. A bag contains red coloured, blue coloured and green coloured balls in the ratio 2 : 3 : 4. A ball is drawn at random from the given bag. The probability that the ball so drawn being not of blue colour is

(a) $\frac{1}{9}$ (b) $\frac{1}{3}$

(c) $\frac{2}{3}$ (d) $\frac{8}{9}$

Solution: (c)

Let the number of red, blue and green colour balls are $2x$, $3x$ and $4x$

$\therefore$ Total number of ball = $2x + 3x + 4x = 9x$

$\therefore$ P(getting non blue colour ball) $= \frac{2x+4x}{9x}$

$= \frac{6x}{9x} = \frac{2}{3}$

80. From a bag containing yellow, red and green balls. The probability of darwing a yellow or red ball is equal to that of drawing a green ball from the bag is: **(2026)**

(a) $\frac{1}{3}$ (b) $\frac{1}{2}$

(c) 1 (d) Not possible of find

Solution: (b)

$P(y \text{ or } r) = P(a)$

and $P(y \text{ or } r) + P(a) = 1$

$\Rightarrow P(a) + P(a) = 1$

$\Rightarrow 2P(a) = 1$

$\therefore P(a) = \frac{1}{2}$

81. A bag contains some red and some white balls. *A* ball is drawn at random from the bag. If the probablity of getting a red ball is $\frac{2}{7}$, then the probability of getting a white ball is: **(2026)**

(a) $\frac{1}{14}$ (b) $\frac{5}{7}$

(c) $\frac{1}{7}$ (d) $\frac{2}{7}$

Solution: (b)

P (red) + P (white) = 1

$\Rightarrow \frac{2}{7} + P(\text{white}) = 1$

$\Rightarrow \therefore P(\text{white}) = \frac{5}{7}$

Exercise 10.4

(i) A bag contains 6 red, 3 white and 6 green balls. A ball is drawn at a random from a bag. Find the probability of getting not a white ball.

(a) $\frac{1}{5}$ (b) $\frac{2}{5}$

(c) $\frac{3}{5}$ (d) $\frac{4}{5}$

(ii) A bag contains 5 red balls and some blue balls. If the probability of drawing a blue ball is double that of a red ball, than the number of blue balls in the bag is

(a) 5 (b) 10

(c) 15 (d) None of these

(iii) A bag contains 7 black, 5 red and 3 white balls. A ball is drawn from the bag at random. Find the probability that the ball drawn is black or white.

(a) $\frac{2}{3}$ (b) $\frac{7}{15}$

(c) $\frac{1}{3}$ (d) $\frac{1}{5}$

(iv) The number of red balls in a bag is 8 more, than the number of black balls. If the probability of drawing a red ball at random from this bag is $\frac{5}{7}$, then the total number of balls in the bag is:

(a) 20 (b) 22
(c) 24 (d) 30

(v) A bag contains red balls and black balls in the ratio 2 : 3. A ball is drawn at random. The probability that ball so drawn in black in colour, is

(a) $\frac{2}{5}$ (b) $\frac{3}{5}$
(c) 1 (d) $\frac{1}{5}$

(vi) A bag contains red coloured, blue coloured and green coloured balls in the ratio 5 : 4 : 3. A ball is drawn at random from the given bag. The probability that the ball so drawn being not of blue colour is

(a) $\frac{2}{3}$ (b) $\frac{4}{5}$
(c) $\frac{8}{11}$ (d) $\frac{1}{3}$

Key Point **(i)** For some event E, $0 \le P(E) \le 1$.
(ii) $P(E) = 0$, when E is an impossible event.
(iii) $P(E) = 1$, when E is a sure event.
(iv) $P(E) + P(\text{not } E) = 1$

82 Which of the following cannot be the probability of an event? **(2011)**

(a) 1.5 (b) $\frac{3}{5}$
(c) 25% (d) 0.3

Solution: (a)

Probability of any event cannot be more than 1.

∴ 1.5 cannot be the probability of any event.

83. Which of the following numbers cannot be the probability of an event? **(2023)**

(a) 0.5 (b) 5%
(c) $\frac{1}{0.5}$ (d) $\frac{0.5}{14}$

Solution: (c)

$\frac{1}{0.5} = 2 > 1$

84. Which of the following cannot be the probability of an event?

(a) 0.1 (b) $\frac{5}{3}$
(c) 3% (d) $\frac{1}{3}$

Solution: (b)

$\frac{5}{3} > 1$

85. Which of the following is not probability of an event?

(a) 0.89 (b) 52%
(c) $\frac{1}{13}\%$ (d) $\frac{1}{0.89}$

Solution: (d)

$\frac{1}{0.89} = \frac{100}{89} > 1$

86. Which of the following numbers cannot be the probability of happening of an event. **(2023)**

(a) 0 (b) $\frac{7}{0.01}$
(c) 0.07 (d) 0.07/3

Solution: (b)

$\frac{7}{0.01} = 700 > 1$

87. The probability of an impossible event is. **(2020)**

(a) 1 (b) $\frac{1}{2}$
(c) not defined (d) 0

Solution: (d) 0

88. Let E be an event such that $P(\text{not } E) = \frac{1}{5}$, then $P(E)$ is equal to **(2023)**

(a) $\frac{1}{5}$ (b) $\frac{2}{5}$
(c) 0 (d) $\frac{4}{5}$

Solution: (d)

$P(E) + P(\text{not } E) = 1$

$\Rightarrow P(E) + \frac{1}{5} = 1$

$\Rightarrow P(E) = \frac{4}{5}$

89. If probability of winning a game is p, then probability of losing the game is **(2024)**

(a) $1+p$ (b) $-p$

(c) $p-1$ (d) $1-p$

Solution: (d)

$P(\text{winning a game}) + P(\text{lossing a game}) = 1$

$\Rightarrow p + P(\text{losing a game}) = 1$

$\therefore P(\text{losing a game}) = 1 - p$

90. If $P(A)$ denotes the probability of an event A, then **(2024)**

(a) $P(A) < 0$ (b) $P(A) > 1$

(c) $0 \le P(A) \le 1$ (d) $-1 \le P(A) \le 1$

Solution: (c)

$0 \le P(A) \le 1$

91. Which of the following relationship is correct **(2024)**

(a) $P(E) = 1 + P(\bar{E})$

(b) $P(\bar{E}) - P(E) = 1$

(c) $P(E) + P(\bar{E}) = 1$

(d) $P(E) = 2P(\bar{E})$

Solution: (c)

$P(E) + P(\bar{E}) = 1$

92. If the probability of an event is 'p', what is the probability of its complementary event? **(2024)**

(a) $1-p$ (b) $p-1$

(c) p (d) $\frac{1}{p}$

Solution: (a)

$1-p$

93. For an event E, if $P(E) + P(\bar{E}) = q$, then the value of $q^2 - 4$ is **(2024)**

(a) –3 (b) 3

(c) 5 (d) –5

Solution: (a)

$q = P(E) + P(\bar{E})$

$\Rightarrow q = 1$

$\therefore q^2 - 4 = 1^2 - 4 = -3$

94. Probability of happening of an event is denoted by 'p' and probability of non-happening of the event is denoted by 'q'. Relation between 'p' and 'q' is **(2023)**

(a) $p + q = 1$ (b) $p = 1, q = 1$

(c) $p = q - 1$ (d) $p + q + 1 = 0$

Solution: (a)

$p + q = 1$

95. If the probability of winning a game is 0.07, what is the probability of lossing it? **(2020)**

Solution:

Probability of lossing the game

= 1 – probability of winning the game

$= 1 - 0.07 = 0.93$

96. The probability that it will rain tomorrow is 0.85. What is the probability that it will not rain tomorrow? **(2020)**

Solution:

Probability that it will not rain tomorrow $= 1 - 0.85 = 0.15$

97. The probability of an event that is not sure to happen is________ **(2020)**

Solution: O $\le$ P $<$ 1

98. If the probability of an event E happening is 0.02, then the $P(\bar{E})$_____ **(2020)**

Solution:

$P(\bar{E}) = 1 - 0.2 = 0.98$

99. If E is an event such that $P(E) = 1\%$, then $P(\bar{E})$ is equal to **(2020)**

(a) 0.09 (b) 0.99

(c) $\frac{1}{99}$ (d) 0.90

Solution: (b)

$P(E)+P(\bar{E})=1$

$\Rightarrow 1\%+P(\bar{E})=1$

$\Rightarrow \frac{1}{100}+P(\bar{E})=1$

$\therefore P(\bar{E})=\frac{99}{100}=0.99$

100. Which of the following can not be the proba bility of an event? **(2026)**

(a) $\frac{39}{100}$ (b) $\frac{0.001}{20}$

(c) $\frac{10}{0.2}$ (d) 10%

Solution: (c)

$\frac{10}{0.2}=50>1$

Exercise 10.5

(i) Which of the following can not be the probability of an event?

(a) 2.5% (b) $\frac{2}{3}$

(c) $\frac{1}{0.2}$ (d) 0.2

(ii) Let E be an event such that $P(\text{not } E)=\frac{2}{3}$, then $P(E)$ is equal to

(a) $\frac{2}{3}$ (b) 1

(c) $\frac{3}{2}$ (d) $\frac{1}{3}$

(iii) The probability that cannot exist among the following:

(a) $\frac{1}{3}$ (b) −1.3

(c) 25% (d) 0.45

(iv) What is the sum of the probability of occurring an event E and the probability of not occurring the event E?

(a) 0 (b) −1

(c) 1 (d) ∞

(v) What is the name of the event for which the probability is zero?

(a) Impossible event

(b) Random event

(c) Exhaustive events

(d) Mutual events

(vi) The probability of occurring an events is 0.28. Find the probability of not occurring the event.

(a) 0.28 (b) 0.82

(c) 0.72 (d) 0.80

101. A box contains 90 discs, numbered from 1 to 90. If one disc in drawn at random from the box, the probability that it bears a prime-number less then 23, is **(2013, 2023)**

(a) $\frac{7}{90}$ (b) $\frac{10}{90}$

(c) $\frac{4}{45}$ (d) $\frac{9}{89}$

Solution:

Total number possible outcomes = 90

Favourable outcomes are 2, 3, 5, 7, 11, 13, 17, 19

∴ Number of favourable outcomes = 8

∴ P(a prime number less than 23) $=\frac{8}{90}=\frac{4}{45}$

102. The probability that a number selected at random from the numbers 1, 2, 3, ..., 15 is a multiple of 4, is **(2014)**

(a) $\frac{4}{15}$ (b) $\frac{2}{15}$

(c) $\frac{1}{5}$ (d) $\frac{1}{3}$

103. What is the probability that a number selected randomly from the number 1, 2, 3, ..., 15 is a multiple of **(2024)**

(a) $\frac{4}{15}$ (b) $\frac{6}{15}$

(c) $\frac{3}{15}$ (d) $\frac{5}{17}$

104. The probability that a number selected at random from the numbers 1, 2, 3, ..., 15 is a multiple of 4, is **(2020)**

(a) $\frac{4}{15}$ (b) $\frac{2}{15}$

(c) $\frac{1}{15}$ (d) $\frac{1}{5}$

Solution:

To number of possible outcomes = 15

Favourable outcomes are 4, 8, 12

$\therefore$ Total number of favourable outcomes = 3

$\therefore P(\text{multiple of } 4) = \frac{3}{15} = \frac{1}{5}$

105. Cards marked with number 3, 4, 5, .., 50 are placed in a box and mixed throughly. A card is drawn at random from the box. Find the probability that the selected card bears a perfect square number. **(2016)**

Solution:

Total number of outcomes = 48

Number of favourable outcomes = 6

$\therefore P(\text{selected card bears a perfect square number})$
$= \frac{6}{48} = \frac{1}{8}$

106. From the data 1, 4, 7, 9, 16, 21, 25, if all the even numbers are removed, then the probability of getting a random a prime number from the remaining is: **(2024)**

(a) $\frac{2}{5}$ (b) $\frac{1}{5}$

(c) $\frac{1}{7}$ (d) $\frac{2}{7}$

Solution: (b)

Total number of possible outcomes = 5

Favourable outcome is 7

Number of favourable outcome = 1

$\therefore$ P(getting a prime number) = $\frac{1}{5}$

107. 20 tickets, on which numbers 1 to 20 are written are mixed throughly and then a ticket is drawn at random out of them. Find the probability that the number on the drawn ticket is a multiple of 3 or 7. **(2016)**

Solution:

Total number of possible outcomes = 20

Favourable outcomes are 3, 6, 7, 9, 12, 14, 15, 18

$\therefore$ P(getting a drawn ticket is a multiple of 3 or 7)
$= \frac{8}{20} = \frac{2}{5}$

108. A number is choosen at random from the numbers −3, −2, −1, 0, 1, 2, 3. What will be the probability that square of this number is less than or equal to 1? **(2017)**

Solution:

Total number of possible outcomes = 7

Favourable outcomes are $(-1)^2$, 0^2, 1^2

$\therefore$ Number of favourable outcomes = 3

$\therefore P(\text{square of the given number is less than or equal to } 1) = \frac{3}{7}$.

109. A game of chance consists of spinning an arrow which comes to rest pointing at one of the numbers 1 , 2, 3, 4, 5, 6, 7, 8 and there are equally likely outcomes. Find the probability that the arrow will point at any factor of 8. **(2015)**

Solution:

Total number of possible outcomes = 8

Number of favourable outcomes = 4

$\therefore P(\text{getting any factor of } 8) = \frac{4}{8} = \frac{1}{2}$

110. A box contains cards numbered 6 to 55. A card is drawn at random from the box. The probability that the drawn card has a number which is a perfect square, is **(2024)**

(a) $\frac{7}{50}$ (b) $\frac{7}{55}$

(c) $\frac{1}{10}$ (d) $\frac{5}{49}$

Solution:

Number of possible outcomes = 50

Favourable outcomes are 9, 16, 25, 36, 49

∴ Number of favourable outcomes = 5

∴ P(getting a perfect square) = $\frac{5}{50} = \frac{1}{10}$.

111. A box contains cards numbered 6 to 50. A card is drawn at random from the box. The probability that the drawn card has a number which is a perfect square, is: **(2024)**

(a) $\frac{5}{44}$ (b) $\frac{1}{9}$

(c) $\frac{1}{11}$ (d) $\frac{7}{45}$

Solution: (b)

Total number of possible outcomes = 45

Favourable outcomes are = 9, 16, 25, 36, 49

Number of favourable outcomes = 5

∴ P(getting a perfect square) = $\frac{5}{45} = \frac{1}{9}$

112. A beg contains 100 cards numbered 1 to 100. A card is drawn at random from the bag. What is the probability that the number on the card is a perfect cube? **(2023)**

(a) $\frac{1}{20}$ (b) $\frac{3}{50}$

(c) $\frac{1}{25}$ (d) $\frac{7}{100}$

Solution:

The number of possible outcomes is 100

Favourable outcomes are 1, 8, 27, 64

Number of favourable outcomes is 4

∴ P(getting a perfect cube) = $\frac{4}{100} = \frac{1}{25}$

113. Cards bearing numbers 3 to 20 are placed in a bag and mixed throughly. A card is taken out of the bag at random. What is the probability that the number on the card taken at is an even number?

(a) $\frac{9}{17}$ (b) $\frac{1}{2}$

(c) $\frac{5}{9}$ (d) $\frac{7}{18}$

Solution: (b)

The number of possible outcomes is 18

Favourable outcomes are 4, 6, 8, 10 12, 14, 16, 18, 20

∴ Number of favourable outcomes = 9

∴ P(getting an even numbers) = $\frac{9}{18} = \frac{1}{2}$

114. One ticket is drawn at random from a bag containing tickets numbered 1 to 40. The probability that the selected ticket has a number which is a multiple of 7 is **(2024)**

(a) $\frac{1}{7}$ (b) $\frac{1}{18}$

(c) $\frac{1}{5}$ (d) $\frac{7}{40}$

Solution: (b)

The number of possible outcomes = 40

Favourable outcomes are 7, 14, 21, 28, 35

∴ Number of favourable outcomes = 5

∴ P(getting a number which is a multiple of 7)

$= \frac{5}{40} = \frac{1}{8}$

115. A letter of English alphabet is chosen at random. Determine the probability that the chosen letter is a consonant **(2015, 2020)**

Solution:

The total number of possible outcomes = 26

Number of favourable outcomes = 21

∴ P(chosen letter is a consonant) = $\frac{21}{26}$

116. A letter is chosen at random from the 'MATHEMATICS'. What is the probability that it will be a vowel? **(2024)**

(a) $\frac{3}{11}$ (b) $\frac{4}{11}$

(c) $\frac{5}{11}$ (d) $\frac{7}{11}$

Solution: (b)

Total number of favourable outcomes = 11

Favourable outcomes are = A, E, A, I

∴ Number of favourable outcomes = 4

∴ P(chosen letter is a vowel) = $\frac{4}{11}$

117. From the letters of the word 'MOBILE', a letter is selected at random. The probability that the selected letter is a vowel is. **(2024)**

(a) $\frac{3}{7}$ (b) $\frac{1}{6}$

(d) $\frac{1}{2}$ (d) $\frac{1}{3}$

Solution:

The number of possible outcomes = 6

Favourable outcomes are O, I, E

$\therefore$ Number of favourable outcomes = 3

$\therefore$ P(selected letter is a vowel) $= \frac{3}{6} = \frac{1}{2}$

118. If a number x is chosen at random from the number −3, −2, −1, 0, 1, 2, 3. Then find the probability of $x^3 < 4$ **(2020)**

Solution:

Total number of possible outcomes = 7

Favourable outcomes are = −3, −2, −1, 0, 1

$\therefore$ Total number of favourable outcomes = 5

$\therefore$ $P(n^3 < 4) = \frac{5}{7}$

119. A girl calculated that the probability of her winning the first prize in a lottery is 0.08. If 6000 tickets are sold, how many tickets has she bought? **(2023)**

(a) 40 (b) 240

(c) 480 (d) 750

Solution: (c)

Let she has sold x tickets

$\therefore \frac{x}{6000} = 0.08$

$\Rightarrow x = 480$

120. In a group of 20 people, 5 can't swim. If one person in selected at random then the probability that he/she can swim is **(2023)**

(a) $\frac{3}{4}$ (b) $\frac{1}{3}$

(c) 1 (d) $\frac{1}{4}$

Solution: (a)

Total number person can swim = 20 − 5 = 15

$\therefore$ P(selected person can swim) $= \frac{15}{20} = \frac{3}{4}$

121. In a survey it is found that every fifth person has a vehicle. The probability of a person Not having a vehicle is **(2023)**

(a) $\frac{1}{5}$ (b) 5%

(c) $\frac{4}{5}$ (d) 95%

Solution: (c)

P(vehicle) $= \frac{1}{5}$

$\therefore$ P(No vehicle) = 1 − P(vehicle)

$= 1 - \frac{1}{5} = \frac{4}{5}$

122. Two friends were born in the year 2000. The probability that they have the some birthday is **(2024)**

(a) $\frac{1}{365}$ (b) $\frac{364}{365}$

(c) $\frac{1}{366}$ (d) $\frac{365}{366}$

Solution: (a)

Since year 2000 in a leap year.

Total number of possible outcomes = 366

Favourable outcomes = 1

$\therefore$ P(same birthday) $= \frac{1}{366}$

123. What is the probability that a randomly taken leap year has 52 Sundays? **(2020)**

Solution:

A Leap year has 366 days i.e 52 weeks and 2 days. The possible pair of the remaining two days are. (Sunday, Monday), (Monday, Tuesday), (Tuesday, Wednesday), (Wednesday, Thursday), (Thursday, Friday), (Friday, Saturday), (Saturday, Sunday).

There are total 7 pairs and out of 7 pairs, Only 2 pairs have Sunday. The remaining 5 pairs do not include Sunday.

$\therefore$ P(randomly taken leap year has exactly 52 Sundays) $= \frac{5}{7}$

124. The probability of drawn an even prime number out of numbers from 1 to 30 is **(2025)**

(a) $\frac{1}{30}$ (b) $\frac{4}{15}$

(c) $\frac{7}{30}$ (d) 0

Solution: (a)

The total number of possible outcomes = 30

Total number of favourable outcomes = 1

$\therefore P(\text{even prime number}) = \frac{1}{30}$

125. In a cricket match; a batsman hits the boundary 7 times out of the 42 balls he plays. The probability of his not hitting a boundary is: **(2025)**

(a) $\frac{1}{7}$ (b) $\frac{2}{7}$

(c) $\frac{5}{6}$ (d) $\frac{1}{6}$

Solution: (c)

Number of balls for which the batsman does not hit a boundary = 42 – 7 = 35

$\therefore P(\text{batsman not hitting a boundary}) = \frac{35}{42} = \frac{5}{6}$

126. Letter A to F are mentioned on six faces of a die such that each face has a different letter. Two such dice are thrown simultaneously. The probability that vowels turn up to both the dice is: **(2025)**

(a) $\frac{1}{4}$ (b) $\frac{1}{3}$

(c) $\frac{1}{9}$ (d) $\frac{1}{36}$

Solution: (c)

Total number of possible outcomes = 36

Favourable outcomes are = AA, AE, EA, EE

$\therefore$ Number of favourable outcomes = 4

$\therefore P(\text{vowels turn up to both the dice}) = \frac{4}{36} = \frac{1}{9}$

127. A plggy work contains ₹ 1 coins and ₹ 2 coins in the ratio 9 : 11 respectively. The plggy bank is accidently dropped and a coin pops out of it. The probability that it is a ₹2 coin is **(2025)**

(a) $\frac{9}{11}$ (b) 0.45

(c) 0.55 (d) $\frac{1}{11}$

Solution: (c)

Let the total number of one rupee coin be $9x$ and two rupee coins be $11x$.

$\therefore P(₹2 \text{ coin}) = \frac{11x}{20x} = \frac{11}{20} = 0.55$

128. The probability that a-2 digit number less then 20, selected at random will be a multiple of 2 and not a multiple of 3, is **(2025)**

(a) $\frac{1}{2}$ (b) $\frac{1}{5}$

(c) $\frac{3}{10}$ (d) $\frac{3}{11}$

Solution: (c)

Total number of possible outcomes = 10

Favourable outcomes are 10, 14, 16,

$\therefore P(E) = \frac{3}{10}$

129. In a random experiment of throwing a die, which of the following is a sure event? **(2025)**

(a) Getting a number between 1 and 6.

(b) Getting an odd number <7

(c) Getting an even number <7

(d) Getting a natural number <7.

Solution: (d)

Getting a natural number <7.

130. In a family of 3 children, the probability of having at least one boy is **(2014)**

(a) $\frac{7}{8}$ (b) $\frac{1}{8}$

(c) $\frac{5}{8}$ (d) $\frac{3}{4}$

Solution: (a)

The possible outcomes are BBB, BBG, BGB, BGG, GGG, GGB, GBB, GBG.

$\therefore$ The total number of outcomes = 8

$\therefore$ Number of favourable outcomes = 7

$\therefore P(\text{at least one boy}) = \frac{7}{8}$

131. The probabilty for a randomly selected number out of 1, 2, 3, 4, …, 25 to be a composite number is: **(2026)**

(a) $\frac{15}{25}$ (b) $\frac{10}{25}$

(c) $\frac{11}{25}$ (d) $\frac{9}{25}$

Solution: (a)

The favourable outcomes are

4, 6, 8, 9, 10, 12, 14, 15, 16, 18,

20, 21, 22, 24, 25

The number of favourable outcomes is 15

$\therefore P(\text{getting a composite number}) = \frac{15}{25}$

132. Meera calculates that the probablityof her winning the first prize in a lottery is 0.08. If total 800 tickets were sold, the number of tickects bought by her is: **(2026)**

(a) 64 (b) 640

(c) 100 (d) 10

Solution: (a)

$$\text{P(winning)} = \frac{\text{Number of tickets bought}}{\text{total number of tickets}}$$

$\Rightarrow 0.08 = \frac{5}{800}$

$\Rightarrow x = 64$

133. A letter is selected from the letters of the word FEBUARY. The probability that it is a vowel is: **(2026)**

(a) $\frac{1}{8}$ (b) $\frac{2}{8}$

(c) $\frac{3}{8}$ (d) $\frac{3}{7}$

Solution: (c)

The number of possible outcomes = 3

$\therefore P(E) = \frac{3}{8}$

Exercise 10.6

(i) A box contains 80 discs, numbered from 1 to 80. If one disc is drawn at random from the box, the probability that it bears a square number less than 60, is

(a) $\frac{1}{2}$ (b) $\frac{7}{80}$

(c) $\frac{5}{16}$ (d) $\frac{3}{16}$

(ii) The probability that a number selected at random from the numbers 1, 2, 3, ..., 15 is a multiple of 6 is

(a) $\frac{1}{3}$ (b) $\frac{5}{6}$

(c) $\frac{6}{15}$ (d) $\frac{2}{15}$

(iii) Cards marked with number 6, 7, 8, .., 100 are placed in a box and mixed throughly. A card is drawn at random from the box. Find the probability that the selected card bears a perfect square number.

(a) $\frac{6}{95}$ (b) $\frac{5}{17}$

(c) $\frac{7}{95}$ (d) $\frac{3}{95}$

(iv) From the data 2, 3, 5, 8, 13, 21, 34, 55. If all the prime numbers are removed , then the probability of getting a random a even number from the remaining is

(a) $\frac{1}{2}$ (b) $\frac{3}{4}$

(c) 1 (d) $\frac{3}{5}$

(v) 30 tickets, on which numbers 1 to 30 are written are mixed throughly and then a ticket in drawn at random out of them. Find the probability that the number on the drawn ticket is a multiple of 4 or 6.

(a) $\frac{2}{5}$ (b) $\frac{10}{30}$

(c) $\frac{7}{8}$ (d) $\frac{1}{2}$

(vi) A number is choosen at random from the numbers –6, –5, –4, –3, –2, –1, 0, 1, 2, 3, 4, 5, 6. What will be the probability that cube of the number is less than or equal to 3.

(a) $\frac{3}{5}$ (b) $\frac{7}{13}$

(c) $\frac{8}{13}$ (d) $\frac{6}{14}$

(vii) A game of chance consists of spinning an arrow which comes to rest pointing at one of the numbers 1 , 2, 3, 4, 5, 6, 7, 8 and these are equally likely outcomes. Find the probability that the arrow will point at any factor of 6.

(a) $\frac{1}{6}$ (b) $\frac{1}{3}$

(c) $\frac{5}{8}$ (d) $\frac{1}{2}$

(viii) A box contains cards numbered 6 to 55. A card is drawn at random from the box. The probability that the drawn card has a number which is a perfect cube, is

(a) $\frac{3}{50}$ (b) $\frac{1}{25}$

(c) $\frac{3}{5}$ (d) $\frac{3}{4}$

(ix) Cards bearing numbers 1 to 30 are placed in a bag and mixed throughly. A card is taken out of the bag at random. What is the probability that the number on the card taken out is an prime number

(a) $\frac{5}{6}$ (b) $\frac{1}{6}$

(c) $\frac{3}{5}$ (d) $\frac{1}{3}$

(x) One ticket is drawn at random from a bag containing tickets numbered 1 to 80. The probability that the selected ticket has a number which is a multiple of 9 is

(a) $\frac{1}{10}$ (b) $\frac{3}{80}$

(c) $\frac{5}{11}$ (d) $\frac{1}{2}$

(xi) A letter is chosen at random from the 'ENVIRONMENT'. What is the probability that it will be a vowel?

(a) $\frac{3}{11}$ (b) $\frac{4}{11}$

(c) $\frac{1}{3}$ (d) $\frac{7}{11}$

(xii) If a number x is chosen at random from the number –5, –4, –3, –2, –1, 0, 1, 2, 3, 4, 5. Then find the probability of $x^3 < 2$.

(a) $\frac{3}{11}$ (b) $\frac{4}{11}$

(c) $\frac{5}{11}$ (d) $\frac{7}{11}$

(xiii) A boy calculated that the probability of her winning the fist prize in a lottery is 0.05. If 2000 tickets are sold, how many tickets has she bought?

(a) 100 (b) 200

(c) 862 (d) 1000

(xiv) In a group of 40 people, 8 can't swim. If one person in selected at random then the probability that he/she can swim is

(a) $\frac{1}{5}$ (b) $\frac{2}{5}$

(c) $\frac{3}{5}$ (d) $\frac{4}{5}$

(xv) Two friends were born in the year 2004. The probability that they have the different birthday is

(a) $\frac{1}{366}$ (b) $\frac{365}{366}$

(c) $\frac{1}{365}$ (d) $\frac{364}{365}$

(xvi) Raju and Ritesh two friends. The probability that both will have the some birthday in a non-leap year is

(a) $\frac{1}{366}$ (b) $\frac{3}{5}$

(c) $\frac{364}{365}$ (d) $\frac{1}{365}$

(xvii) What is the probability that a randomly taken leap year her 53 Sundays?

(a) $\frac{1}{7}$ (b) $\frac{2}{7}$

(c) $\frac{3}{7}$ (d) $\frac{5}{7}$

(xviii) In a cricket match, a batsman hits the boundary 21 times out of the 60 balls he plays. The probability of his not hitting a boundary is

(a) $\frac{11}{20}$ (b) $\frac{13}{20}$

(c) $\frac{3}{5}$ (d) $\frac{1}{60}$

(xix) Letter A to F are mentioned on six faces of a die such that each face has a different letter. Two such dice are thrown simultaneously. The probability that consonants turn up to both the dice is

(a) $\frac{1}{9}$ (b) $\frac{2}{3}$

(c) $\frac{4}{9}$ (d) $\frac{5}{12}$

(xx) The probability that a 2 digit number less then 30, selected at random will be a multiple of 2 and not a multiple of 3

(a) $\frac{1}{2}$ (b) 5%

(c) $\frac{7}{20}$ (d) 0.80

(xxi) In a family of 3 children, the probability of having at least two girls is

(a) $\frac{1}{2}$ (b) $\frac{7}{8}$

(c) $\frac{3}{8}$ (d) $\frac{3}{4}$

134. The probability of selecting a rotter apple randomly from a heap of 900 apples is 0.18. What is the number of rotten apple in the heap? **(2017)**

Solution:

Let the number of rotten apple be x

$\therefore \frac{x}{900} = 0.18$

$\Rightarrow x = 162$

135. If the probability of a player winning a game is 0.79, then the probability of his lossing the same game is: **(2024)**

(a) 1.79 (b) 0.31

(c) 0.21% (d) 0.21

Solution: (d)

Probability of the player lossing the game $= 1 - 0.49$

$= 0.21$

136. The probability for a leap year (selected at random) will contain 53 Tuesdays is 3-2 **(2024)**

(a) $\frac{1}{7}$ (b) $\frac{2}{7}$

(c) $\frac{3}{7}$ (d) $\frac{4}{7}$

Solution: (b)

A Leap year has 52 weeks and 2 days.

Hence, a leap year will have 52 Tuesdays for sure.

Now, the 2 extra days can be {(Monday, Tuesday), (Tuesday, Wednesday), (Wednesday, Thursday), (Thursday, Friday), (Friday, Saturday), (Saturday, Sunday), (Sunday, Monday),}.

$\therefore$ Favourable outcomes are = (Monday, Tuesday), (Tuesday, Wednesday).

$\therefore$ Number favourable outcomes = 2

$\therefore$ P(a leap year will contain 53 Tuesdays) $= \frac{2}{7}$.

Direction: Question number **137 to 148** are Assertion and Reason based question. Two statements are given, one labelled as Assertion (A) and the other labelled as Reason (R).Select the correct answer to the questions from the codes (A), (B), (C) and D as given bellow.

(a) Both Assertion (A) and Reason (R) are true and Reason (R) is the correct explanation of Assertion (A).

(b) Both Assertion (A) and Reason (R) are true, but Reason (R) is not the correct explanation of Assertion (A).

(c) Assertion (A) is true, but Reason (R) is false.

(d) Assertion (A) is false, but Reason (R) is true.

137. Assertion (A): The probability of getting a prime number, when a die is thrown once, is $\frac{2}{3}$

Reason (R): On the faces of a die, prime numbers are 2, 3, 5 **(2023)**

Solution: (d)

$P(\text{getting a prime number}) = \frac{3}{6} = \frac{1}{2}$

138. Assertion (A): When two coins are tossed together the probability of getting not tail is $\frac{1}{4}$

Reason (R): The probability $P(E)$ of an event E satisfies: $0 \leq P(E) \leq I$ **(2023)**

Solution: (b)

When two coins are tossed together, the possible outcome, will be HH, HT, TH, TT.

Favourable outcomes is HH

$\therefore P(\text{getting not tail}) = \frac{1}{4}$

139. Assertion (A): The probability that a leap year has 53 Sunday is $\frac{2}{7}$.

Reason (R): The probability that a non-leap year has 53 Sunday is $\frac{1}{7}$. **(2023)**

Solution: (b)

The leap year has 52 weeks and 2 days

$\therefore$ Required probability $= \frac{2}{7}$

The non leap year has 365 day i.e 52 weeks and 1 day

$\therefore$ Required probability $= \frac{1}{7}$

140. Assertion (A): The probability of getting 8 on rolling a die is zero (0)

Reason (R): The probability of an impossible event is zero (0). **(2024)**

Solution: (a)

Required probability $= \frac{0}{6} = 0$

141. Assertion (A): The probability of getting an even number, when a die is thrown once is $\frac{1}{2}$.

Reason (R): In a die, total number of possible outcomes is 6 and the even numbers are 2, 4 and 6. **(2024)**

Solution: (a)

Required probability $= \frac{3}{6} = \frac{1}{2}$

142. Assertion (A): The probability of selecting a number at random from the number 1 to 20 is 1.

Reason (R): For any event E, if $P(E) = 1$, then E is called a sure event. **(2025)**

Solution: (d)

Required probability $= \frac{1}{20}$

143. In an experiment of throwing a die

Assertion (A): Event E_1: getting a number less then 3 and event E_2: getting a number greater than 3 are complementary events.

Reason (R): If two events E and F are complementary events, then $P(E) + P(F) = 1$ **(2025)**

Solution: (d)

Assertion (A) is false, and Reason (R) is true

144. Assertion (A): If E is an event such that $P(E) = \frac{1}{999}$, then $P(\bar{E}) = 0.001$

Reason (R): $P(E) + P(\bar{E}) = 1$. **(2025)**

Solution: (d)

$P(E) + P(\bar{E}) = 1$

$\Rightarrow \frac{1}{999} + P(\bar{E}) = 1$

$\Rightarrow P(\bar{E}) = 1 - \frac{1}{999} = \frac{998}{999}$

Question number **i to x** are Assertion and Reason based question. Two statements are given, one labelled on Assertion (A) and the other labelled an Reason (R). Select the correct answer to these questions from the codes (A), (B), (C) and (D) as given below.

(a) Both Assertion (A) and Reason (R) are true and Reason (R) is the correct explanation of Assertion (A).

(b) Both Assertion (A) and Reason (R) are true, but Reason (R) is not the correct explanation of Assertion (A).

(c) Assertion (A) is true, but reason (R) is false.

(d) Assertion (A) is false, but reason (R) is true.

145. Assertion (A): The events 'getting 2' and ''not getting 2'' in a single thraw of an unbiased die are not equally likely events.

Reason (R): The probability of getting 2 in a single throw of an unbiased die is $\frac{1}{6}$ **(2026)**

Solution: (d)

Assertion (A) is false, but Reason (R)is true

146. Assertion (A): The probability that a leap year has 53 mondays is $\frac{2}{7}$

Reason (R): The probablity that a nonleap year has 53 monday is $\frac{5}{7}$

Solution: (c)

The leap year has 52 weeks and 2 days

$\therefore$ Required probability $=\frac{2}{7}$

The non leap year has 52 weeks and 1 extra day

$\therefore$ Required probability $=\frac{1}{7}$

147. Assertion (A): From a bag contaianing 5 red balls, 2 white balls and 3 green balls, the probability of drawing a non-white ball is $\frac{4}{5}$

Reason(R): For any event E,

$P(E)+P(\text{not } E)=1$

Solution: (a)

P (drawing white ball)

$=\frac{2}{10}$

$=\frac{1}{5}$

$\therefore P$ (drawing a not white ball)

$=1-\frac{1}{5}$

$=\frac{4}{5}$

148. Assertion (A): If probability of happening of an event in $0.2p$, $p > 0$, then p can't be more than 5

Reason(R): $P(E)=1-p(E)$ for an event E

Solution: (b)

$0 \leq 0.2p \leq 1$

$\Rightarrow 0 \leq \frac{p}{5} \leq 1$

$\Rightarrow 0 \leq p \leq 5$

Exercise 10.7

(i) **Assertion (A):** If $P(E)=0.52$ then $P(\bar{E})=0.48$

Reason (R): $P(\bar{E})=1-P(E)$

(ii) **Assertion (A):** The probability of an event of getting 4 or not 4 on throwing a die are both equal to $\frac{1}{2}$.

Reason (R): The probability of an event (E) is defined as

$$P(E)=\frac{\text{Number of outcomes favourable to E}}{\text{Total Number of possible outcomes}}$$

(iii) **Assertion (A):** The probability of selecting a prime number from 1 to 20 is $\frac{2}{5}$ where as, the probability of selecting an even number is $\frac{1}{2}$.

Reason (R): The probability of getting even number on the throw of a die is $\frac{1}{6}$.

(iv) **Assertion (A):** The probability of getting a vowel on chosen from the letters of the English alphabet is $\frac{5}{26}$.

Reason (R): There are 5 vowels in the English alphabet.

(v) **Assertion (A):** Two players A and B play a cricket match. The probability of A winning the match is 0.48. The probability of B winning the match is to 0.52.

Reason (R): The sum of the probability of complementary events is 1.

(vi) Assertion (A): A bag contains 8 black balls and some white balls. If the probability of drawing a white ball from the bag is five times that of a black ball, then the number of white ball in the bag is 32.

Reason (R): Probability of getting black ball is $\frac{1}{5}$.

(vii) Assertion (A): If the probability of winning a game is $\frac{2}{3}$, then the probability of losing the game is $\frac{1}{3}$

Reason (R): For any event E, we have $P(E) + P(\text{ not } E) = 1$

(viii) Assertion (A): Cards numbered 1, 2, 3,..., 20 are put in a box and mixed throughly. One card is then drawn at random. The probability of drawing an even number is $\frac{1}{3}$.

Reason (R): For any event E, we have $0 \leq P(E) \leq 1$.

(ix) Assertion (A): Six face cards are removed from a deck of cards and the cards are well shuffled. Then the probability of drawing a face card is $\frac{2}{3}$.

Reason (R): King, Queen and Jack are know as face cards. So, there are 12 face cards in total.

(x) Assertion (A): A number is selected from the number 1 to 50. The probability that it will be a prime number is $\frac{7}{50}$.

Reason (R): The exists 8 prime numbers from natural numbers 1 to 20.

ANSWER KEYS

EXERCISE – 10.1

(i) (a) $\frac{1}{2}$ **(ii)** (c) 4 **(iii)** (d) $P(H) \times P(T) = 1$

(iv) (d). $\frac{3}{4}$ **(v)** (b) $\frac{3}{8}$ **(vi)** (d) $\frac{3}{8}$

EXERCISE – 10.2

(i) (b) $\frac{4}{13}$ **(ii)** (b) $\frac{3}{4}$ **(iii)** (c) $\frac{1}{13}$ **(iv)** (a) $\frac{11}{13}$

(v) (d) $\frac{23}{26}$ **(vi)** (c) $\frac{6}{13}$ **(vii)** (c) $\frac{11}{51}$ **(viii)** (b) $\frac{3}{49}$

EXERCISE – 10.3

(i) (b) $\frac{2}{6}$ **(ii)** (a) $\frac{11}{36}$ **(iii)** $\frac{10}{36}$ **(iv)** (c) 0

(v) (a) $\frac{1}{2}$ **(vi)** (b) $\frac{1}{9}$ **(vii)** (c) $\frac{1}{3}$ **(viii)** (a) $\frac{1}{6}$

(ix) (a) $\frac{1}{6}$ **(x)** (d) $\frac{1}{4}$

EXERCISE – 10.4

(i) (d) $\frac{4}{5}$ **(ii)** (b) 10 **(iii)** (a) $\frac{2}{3}$ **(iv)** (c) 24

(v) (b) $\frac{3}{5}$ **(vi)** (a) $\frac{2}{3}$

EXERCISE – 10.5

(i) (c) $\frac{1}{0.2}$ **(ii)** (d) $\frac{1}{3}$ **(iii)** (b) -1.3

(iv) (c) 1 **(v)** (a) impossible event.

(vi) (b) 0.82

EXERCISE – 10.6

(i) (b) $\frac{7}{80}$ **(ii)** (d) $\frac{2}{15}$ **(iii)** (d) $\frac{3}{95}$

(iv) (a) $\frac{1}{2}$ **(v)** (a) $\frac{2}{5}$ **(vi)** (c) $\frac{8}{13}$

(vii) (d) $\frac{1}{2}$ **(viii)** (b) $\frac{1}{25}$

(ix) (d) $\frac{1}{3}$ **(x)** (a) $\frac{1}{10}$ **(xi)** (b) $\frac{4}{11}$

(xii) (d) $\frac{7}{11}$ **(xiii)** (d) 1000 **(xiv)** (d) $\frac{4}{5}$

(xv) (b) $\frac{365}{366}$ **(xvi)** (d) $\frac{1}{365}$ **(xvii)** (b) $\frac{2}{7}$

(xviii)(b) $\frac{13}{20}$ **(xix)** (c) $\frac{4}{9}$ **(xx)** (c) $\frac{7}{20}$

(xxi) (a) $\frac{1}{2}$

EXERCISE – 10.7

(i) (a) **(ii)** (d) **(iii)** (b)
(iv) (a) **(v)** (a) **(vi)** (b)
(vii) (a) **(viii)** (d) **(ix)** (d)
(x) 1

11 Statistics

Key Point Class mark of a class = $\frac{\text{upper limit} + \text{lower limit}}{2}$

1. Find the class marks of classes 10 – 20 and 35 – 55 **(2008)**

Solution:

Class mark of 10 – 20 = $\frac{10+20}{2} = 15$

Class mark of 35 – 55 = $\frac{35+55}{2} = 45$

2. Find the class mark of class 10 – 25 and 35 – 55 **(2008)**

Solution:

Class mark of 10 – 25 = $\frac{10+25}{2} = 17.5$

Class mark of 35 – 55 = $\frac{35+55}{2} = 45$

3. Find the class marks of the classes 10.25 and 35.55 **(2020)**

Solution:

Class mark of the classes 10.25 and

$35.55 = \frac{10.25+35.55}{2}$

$= \frac{45.80}{2}$

$= 22.90$

4. Find the class mark of the classes 15 – 35 and 45 – 60 **(2020)**

Solution:

Class mark of 15 – 35 = $\frac{15+35}{2} = 25$

Class mark of 45 – 60 = $\frac{45+60}{2} = 52.5$

5. Find the class marks of the class 20 – 50 and 35 – 60 **(2020)**

Solution:

Class mark of 20 – 50 = $\frac{20+50}{2}$

$- 35$

Class mark of 35 – 60 = $\frac{35+60}{2}$

$= 47.5$

Exercise 11.1

(i) Find the class mark of the class 100 – 110

(a) 50 (b) 100

(c) 55 (d) 60

(ii) Find the class mark of the classes 17.25 and 20.75

(a) 17 (b) 20

(c) 19 (d) 32

(iii) Find the class mark of the class 95 – 105

(a) 100 (b) 95

(c) 105 (d) 110

Key Point Arithmetic mean: The average of a given set of numbers is called the a arithmetic mean (AM) or mean

(i) mean = $\frac{\text{Sum of observations}}{\text{Total number of observations}}$

(ii) The mean of n observations,

$x_1, x_2, x_3, \ldots, x_n$ is given by

$$\bar{x} = \frac{x_1 + x_2 + x_3 + \ldots + x_n}{n}$$

6. The mean of the following frequency distribution is 62.8. Find the missing frequency x **(2007)**

Class	0 – 20	20 – 40	40 – 60	60 – 80	80 – 100	100 – 120
Frequency	5	8	x	12	7	8

Solution:

Class	Classmark (x_i)	Frequen of (f_i)	$f_i x_i$
0–20	10	5	50
20–40	30	8	240
40–60	50	x	$50x$
60–80	70	12	840
80–100	90	7	630
100–120	110	8	880
		$\Sigma f_i = 40 + x$	$\Sigma f_i x_i = 2640 + 50x$

Now, $\bar{x} = \dfrac{\sum f_i x_i}{\sum f_i}$

$\Rightarrow 62.8 = \dfrac{2640 + 50x}{40 + x}$

$\Rightarrow 2640 + 50x = 2512 + 62.8x$

$\Rightarrow 12.8x = 128$

$\Rightarrow x = 10$

7. If the mean of the first n natural number is 15, find n **(2020)**

Solution:

$$\frac{1+2+3+\ldots+n}{n} = 15$$

$$\Rightarrow \frac{\frac{n(n+1)}{2}}{n} = 15$$

$$\Rightarrow \frac{n+1}{2} = 15$$

$\Rightarrow n + 1 = 30$

$\Rightarrow n = 29$

8. If the mean of 6, 7, x, 8, y, 14 is 9, then **(2023)**

(a) $x + y = 21$ (b) $x + y = 19$

(c) $x - y = 19$ (d) $x - y = 21$

Solution: (b)

$$\frac{6+7+x+8+y+14}{6} = 9$$

$$\Rightarrow \frac{35+x+y}{6} = 9$$

$\Rightarrow 35 + x + y = 54$

$\Rightarrow x + y = 19$

9. The mean of first ten natural numbers is **(2023)**

(a) 5.5 (b) 55

(c) 45 (d) 4.5

Solution:

Sum of first ten natural numbers $= 1+2+3+\ldots+10$

$$= \frac{10(10+1)}{2}$$

$= 55$

$\therefore$ mean of first ten natural numbers $= \dfrac{55}{10} = 5.5$

10. For some data $x_1, x_2, \ldots, x_n$ with respective frequenien $f_1, f_2 \ldots, f_n$ the value of $\sum_1^n f_i(x_i - \bar{x})$ is equal to **(2025)**

(a) $n\bar{x}$ (b) 1

(c) $\sum f_i$ (d) 0

Solution: (d)

$$\bar{x} = \frac{f_1x_1 + f_2x_2 + \ldots + f_nx_n}{f_1 + f_2 + f_3 \ldots + f_n}$$

$$(f_1 + f_2 \ldots . f_n)\bar{x} = f_1x_1 + f_2x_2 + \ldots + f_nx_n$$

Now $\sum_{1}^{n} f_i\left(x_i - \bar{x}\right)$

$= f_1\left(x_1-\bar{x}\right)+f_2\left(x_2-\bar{x}\right)+....+f_n\left(x_n-\bar{x}\right)$

$= \left(f_1x_1+f_2x_2+.....f_nx_n\right)-\left(f_1+f_2+...+f_n\right)\bar{x}$

$= \left(f_1+f_2+.....+f_n\right)\bar{x}-\left(f_1+f_2+.....+f_n\right)\bar{x}$

$= 0$

11. Let 'a' be the assumed mean and 'h' be the class size for a grouped data. Which of the following is not the correct formula to find the mean of the grouped data? **(2026)**

(a) $\bar{x}=\dfrac{\sum f_i x_i}{\sum f_i}$

(b) $\bar{x}=a+\dfrac{\sum f_i d_i}{\sum f_i}$, where $d_i = x_i - a$

(c) $\bar{x}=a+\dfrac{\sum f_i u_i}{\sum f_i}$, where $u_i=\dfrac{x_i-4}{h}$

(d) $\bar{x}=a+\dfrac{\sum f_i y_i}{\sum f_i}\times h$, where $y_i=\dfrac{x_i-a}{h}$

Solution: (c)

$\bar{x}=a+\dfrac{\sum f_i u_i}{\sum f_i}$, where $u_i=\dfrac{x_i-a}{h}$

12. While calculating mean of a grouped frequency distribution, step deviatian method was used $\left(\dfrac{x-a}{h}=u\right)$. It was found that $\bar{x}=64$, $h = 5$ and $a = 62.5$ the value of $\bar{u}$ is **(2026)**

(a) 0.5 (b) 1.5

(c) 0.3 (d) 7.5

Solution: (c)

$\bar{x}=a+\dfrac{\sum f_i u_i}{\sum f_i}\times h\ [\because u=\dfrac{x-a}{h}]$

$\Rightarrow 64 = 62.5+\bar{u}\times 5\ [\because \bar{u}=\dfrac{\sum f_i u_i}{\sum f_i}]$

$\Rightarrow 5\bar{u}=1.5$

$\Rightarrow \bar{u}=0.3$

Exercise 11.2

(i) If the mean of 6, 8, 9, x, 13, is 10, find the value of x.

(a) 10 (b) 12

(c) 16 (d) 14

(ii) If $d_i = x_i - a, \sum_{i=l}^{n} f_i = 25, a = 250, \bar{x} = 250$, then find the value of $\sum_{i=l}^{n} f_i d_i$

(a) 1 (b) – 1

(c) 3 (d) 0

(iii) If the mean of frequency distribution is 7.5 and $\sum f_i x_i = 120+3k$, $\sum f_i = 30$, then k is equal to

(a) 20 (b) 25

(c) 35 (d) 40

Key Point Median class : The class whose cumulative frequency is just greter than $\left(\dfrac{N}{2}\right)$

$N=\sum f_i$ in the median class

For ungrouped frequency distribution

Median = Value of $\left(\dfrac{n+1}{2}\right)$ observation, when n is odd

$=\dfrac{1}{2}\times$[value of $\left(\dfrac{n}{2}\right)^{th}$ observation

+value of $\left(\dfrac{n}{2}+1\right)^{th}$ observation]

when n is even

13. For the following frequency distribution

Class	0–5	5–10	10–15	15–20	20–25
Frequency	8	10	19	25	8

The upper limit of median class is **(2020)**

(a) 15 (b) 10

(c) 20 (d) 25

Solution: (a)

Class	Frequency	Cumulative frequency
0-5	8	8
5-10	10	18
10-15	19	37
15-20	25	62
20-25	8	70

Here $N = 70$ $\therefore \frac{N}{2} = 35$

The cumulative frequency just greater than 35 is 37

∴ The median class is 10 – 15

∴ The upper limit of median class is 15

14. The median class for the data given below is:

Class	20 – 40	40 – 60	60 – 80	80 – 100	100 – 120
Frequency	10	12	14	13	17

(2023)

(a) 80 – 100 (b) 20 – 40
(c) 40 – 60 (d) 60 – 80

Solution: (d)

Class	Frequency	Cumulative frequency
20 – 40	10	10
40 – 60	12	22
60 – 80	14	36
80 – 100	13	49
100 – 120	17	66
	$N = 66$	

Here $N = 66$ $\therefore \frac{N}{2} = 33$

The cumulative frequency to just greater than 33 is 36

∴ the median class is 60 – 80

15. The median group in the following frequency distribution is:

Class	0 – 10	10 – 20	20 – 30	30 – 40	40 – 50	50 – 60
Frequency	5	8	20	15	7	5

(2024)

(a) 10 – 20 (b) 20 – 30
(c) 30 – 40 (d) 40 – 50

Solution: (b)

Class	Frequency	Cumulative frequency
0 – 10	5	5
10 – 20	8	13
20 – 30	20	33
30 – 40	15	48
40 – 50	7	55
50 – 60	5	60
	$N = 60$	

Here $N = 60$ $\therefore \frac{N}{2} = 30$

The cumulative frequency just greater than 30 is 33

∴ The median group is 20 – 30

16. The class mark of the median class of the following data is

Class Interval	10 – 25	25 – 40	40 – 55	55 – 70	70 – 85	85 – 100
Frequency	2	3	7	6	6	6

(2025)

(a) 40 (b) 55
(c) 47.5 (d) 62.5

Solution: (d)

Class	Frequency	Cumulative frequency
10 – 25	2	2
25 – 40	3	5
40 – 55	7	12
55 – 70	6	18
70 – 85	6	24
85 – 100	6	30
	$N = 30$	

Here $N = 30$ $\therefore \frac{N}{2} = 15$

The cumulative frequency just greater than 15 is 18

$\therefore$ The median class is 55 – 70

$\therefore$ Class mark of the median class is

$$\frac{55+70}{2} = \frac{125}{2} = 62.5$$

17. Following data shows the marks obtained by 100 students in a class test:

Mark obtained	20	29	28	33	42	38	43	25
No. of Students	6	28	24	15	2	4	1	20

The median will be the average of which two observations? **(2025)**

(a) 29 and 33 (b) 25 and 38

(c) 28 and 29 (d) 33 and 38

Solution : (c)

Marks	No. of Students	Cumulative frequency
20	6	6
25	20	26
28	24	50
29	28	78
33	15	93
38	4	97
42	2	99
43	1	100
	$N = 100$	

Here $N = 100(\text{even})$ $\therefore \frac{N}{2} = 50$

$$\therefore \text{Median} = \frac{1}{2}\left[\left(\frac{n}{2}\right)^{\text{th}} \text{term} + \left(\frac{n}{2}+1\right)^{\text{th}} \text{term}\right]$$

$$= \frac{1}{2}\left[50^{\text{th}} \text{ term} + 51^{\text{th}} \text{ term}\right]$$

$$= \frac{1}{2}(28+29)$$

18. The median of first seven prime numbers is **(2023)**

(a) 5 (b) 7

(c) 11 (d) 13

Solution: (b)

The first seven prime numbers are 2, 3, 5, 7, 11, 13, 17

Here $n = 7$

$\therefore$ Median at the position $\frac{7+1}{2} = 4^{th}$

position = 7

19.

class	0 – 10	10 – 20	20 – 30	30 – 40	40 – 50
frequemcy	3	5	7	9	11

The upper limit of the median class of the above data is: **(2026)**

(a) 10 (b) 20

(c) 30 (d) 40

Solution:(d)

Class	frequency	cummulative frequency
0-10	3	3
10-20	5	8
20-30	7	15
30-40	9	24
40-50	11	35

Here $N = 35$ $\therefore \frac{N}{2} = 17.5$

The cummulative frequency just greater than 17.5 is 24

$\therefore$ The median class is 30 – 40

$\therefore$ The upper limit of median class is 40

Exercise 11.3

(i) The median of first ten even numbers is

(a) 7 (b) 8

(c) 9 (d) 10

(ii) The median of first nine odd numbers is

(a) 9 (b) 7

(c) 2 (d) 11

(iii) The median of first seven composite numbers is

(a) 7 (b) 9

(c) 10 (d) 8

(iv) The median group in the following frequency distribution is

Class interval	0 – 8	8 – 16	16 – 24	24 – 32	32 – 40	32– 40
Frequency	8	10	16	24	15	7

(a)16 – 24 (b) 24 – 32

(c) 32 – 40 (d) 40 – 48

Key Point The class having maximum frequency called the modal class.

20. The distribution belows gives the marks obtained by 80 students on a test

Marks	less than 10	less than 20	less than 30	less than 40	less than 50	less than 60
No. of Students	3	12	27	57	75	80

The modal class of this distribution is **(2023)**

(a) 10 – 20 (b) 20 – 30

(c) 30 – 40 (d) 50 – 60

21. For the following distribution

Marks Below	10	20	30	40	50	60
No. of Students	3	12	27	57	75	80

The modal class is: **(2023)**

(a) 10 – 20 (b) 20 – 30

(c) 30 – 40 (d) 50 – 60

22.

Marks	Below 10	Below 20	Below 30	Below 40	Below 50
No. of Students	3	12	27	57	75

For the above distribution, the modal class is **(2024)**

(a) 10 – 20 (b) 20 – 30

(c) 30 – 40 (d) 40 – 50

Solution:

Marks	0 – 10	10 – 20	20 – 30	30 – 40	40 – 50	50 – 60
No. of students	3	9	15	30	18	5

∴ The class 30 – 40 has the maximum frequency

∴ The modal class is 30 – 40

23. For the following distribution, the lower limit of modal class is:

Class interval	1 – 10	11 – 20	21 – 30	31 – 40	41 – 50
Frequency	10	12	20	5	3

(2024)

(a) 21 (b) 30.5

(c) 21.5 (d) 20.5

Solution: (d)

Class	0.5 – 10.5	10.5 – 20.5	20.5 – 30.5	30.5 – 40.5	40.5 – 50.5
Frequency	10	12	20	5	3

The class 20.5 – 30.5 has the maximum frequency

∴ The modal class is 20.5 – 30.5

∴ The lower limit at modal class is 20.5

24. The following distribution gives the daily income of 50 workers of a factory

Income (in ₹)	400 – 424	425 – 449	450 – 474	475 – 499	500 – 529
No. of workers	12	14	8	6	10

The lower limit of the modal class is **(2024)**

(a) 425 (b) 449

(c) 424.5 (d) 425.5

Solution: (c)

Class	399.5 – 424.5	424.5 – 449.5	449.5 – 474.5	474.5 – 499.5	499.5 – 524.5
No. of workers	12	14	8	6	10

The class 424.5 – 449.5 has the maximum frequency

∴ The modal class is 424.5 – 449.5

∴ The lower limit of the modal class is 424.5

25. The following distribution shows the number of runs scored by some batsman in test matches

Run sored	3000 – 4000	4000 – 5000	5000 – 6000	6000 – 7000
No. of Batsman	5	10	9	8

The lower limit of the modal class is **(2025)**

(a) 3000 (b) 4000

(c) 5000 (d) 6000

Solution: (b)

The modal class is 4000 – 5000 has the maximum frequency

∴ The modal class is 4000 – 5000

∴ The lower limit of the modal class is 4000

26. For the following distribution:

Class	0 – 10	10 – 20	20 – 30	30 – 40	40 – 50
Frequency	10	12	15	20	9

The sum of lower limits of the median class and modal class is **(2025)**

(a) 30 (b) 50

(c) 40 (d) 60

Solution: (b)

Class	**Frequency**	**Cumulative frequency**
0 – 10	10	10
10 – 20	12	22
20 – 30	15	37
30 – 40	20	57
40 – 50	9	66

Here $N = 66$ $\therefore \frac{N}{2} = 33$

The cumulative frequency just greater than 33 is 37

∴ The median class is 20 – 30

∴ The lower limit of median class in 20

The class 40 – 50 has the maximum frequency

∴ The modal class is 40 – 50

∴ The lower limit of modal class is 40

∴ The sum of lower limits of median class and modal class is $20 + 40 = 60$

27. The annual rainfall record of a city for 66 days in given in the following table:

Rainfall (in cm)	0 – 10	10 – 20	20 – 30	30 – 40	40 – 50	50 – 60
Number of days	22	10	8	15	5	6

The difference of upper limits of modal and median class is: **(2024)**

(a) 10 (b) 15

(c) 20 (d) 30

Solution: (c)

Class	**Frequency**	**Cumulative frequency**
0 – 10	22	22
10 – 20	10	32
20 – 30	8	40
30 – 40	15	55
40 – 50	5	60
50 – 60	6	66

Here $N = 66 \quad \therefore \frac{N}{2} = 33$

The cumulative frequency just greater than 33 is 40

∴ The median class is 20 – 30

∴ The upper limit of median class is 30

The class 0 – 10 has the maximum frequency

∴ The modal class is 0 – 10

∴ The upper limit of modal class is 10

∴ The difference of upper limits of modal class and median class is 30 – 10 = 20

28. For the following distribution:

Class	0 – 5	5 – 10	10 – 15	15 – 20	20 – 25
Frequency	10	15	12	20	9

The sum of lower limits of median class and modal class is: **(2023)**

(a) 15 (b) 25

(c) 30 (d) 35

29. Consider the following distribution:

Class	0 – 5	5 – 10	10 – 15	15 – 20	20 – 25
Frequency	10	15	12	20	9

The sum of lower limits of the median class and the modal class is **(2020)**

(a) 15 (b) 25

(c) 30 (d) 35

Solution:

Class	Frequency	Cumulative frequency
0 – 5	10	10
5 – 10	15	25
10 – 15	12	37
15 – 20	20	57
20 – 25	9	66
	$N = 66$	

$N = 66 \quad \frac{N}{2} = 33$

The cumulative frequency just grater than 33 is 37

∴ The median class is 10 – 15

∴ The lower limit of median class is 10

The class 15 – 20 has the maximum frequency

∴ The modal class is 15 – 20

∴ The lower limit of modal class is 15

∴ The sum of lower limits of the median class and the modal class is 10 + 15 = 25

Exercise 11.4

(i) For the following distribution

Class	0 – 10	10 – 20	20 – 30	30 – 40	40 – 50	50 – 60	60 – 70	70 – 80
Freque-ncy	5	8	7	12	28	20	10	10

The sum of lower limits of median class and modal class is

(a) 24 (b) 36

(c) 80 (d) 40

(ii) For the following distribution

Class	0 – 10	10 – 20	20 – 30	30 – 40	40 – 50
Frequency	8	30	6	5	21

The modal class is

(a) 0 – 10 (b) 10 – 20

(c) 40 – 50 (d) 20 – 30

(iii) For the following distribution

Class	0 – 15	15 – 30	30 – 45	45 – 60	60 – 75
Frequency	15	12	18	16	9

The difference of the upper limit of the median class the lower limit of the modal class is

(a) 10 (b) 12

(c) 14 (d) 15

Key Point Mode = 3(Median) – 2 (Mean)

30. Using the empirical formula, find the mode of a distribution whose mean is 8.32 and the median is 8.05 **(2020)**

Solution:

Mode = 3(Median) – 2(Mean)

$= 3 \times 8.05 - 2 \times 8.32$

$= 24.15 - 16.64$

$= 7.51$

31. The mean and median of a distribution are 14 and 15 respectively. The value of mode is **(2020)**

(a) 16 (b) 17

(c) 18 (d) 13

Solution:

Mode = 3(Median) – 2(Mean)

$= 3 \times 15 - 2 \times 14$

$= 45 - 28$

$= 17$

32. The median and mode respectively of a frequency distribution are 26 and 29. Then its mean is **(2020)**

(a) 27.5 (b) 24.5

(c) 28.4 (d) 25.8

Solution: (b)

Mode = 3(Median) – 2(Mean)

$\Rightarrow 29 = 3 \times 26 - 2$ mean

$\Rightarrow 2$ mean $= 78 - 29$

$\Rightarrow 2$mean $= 49$

$\Rightarrow$ mean $= 24.5$

33. Median and mode of a distribution are 25 and 21 respectively. Median of the data using empirical relationship is **(2023)**

(a) 27 (b) 29

(c) 18 (d) $\frac{29}{3}$

Solution: (b)

Mode = 3(Median) – 2(Mean)

$\Rightarrow 21 = 3 \times 25 - 2$ (Mean)

$\Rightarrow 2$ mean $= 75 - 21$

$\Rightarrow 2$ Mean $= 54$

$\therefore$ Mean $= 27$

34. If the mean and mode of a distribution are 15 and 18 respectively, then the median of the distribution is : **(2023)**

(a) 17 (b) 15

(c) 16 (d) 18

Solution: (c)

Mode = 3(Median) – 2(Mean)

$\Rightarrow 18 = 3$Median $- 2 \times 15$

$\Rightarrow 3$Median $= 48$

$\therefore$ Median $= 16$

35. If the mean and the median of a data are 12 and 15 respectively then its mode is **(2023)**

(a) 13.5 (b) 21

(c) 6 (d) 14

Solution: (b)

Mode = 3(Median) – 2(Mean)

$\Rightarrow$ Mode $= 3 \times 15 - 2 \times 12$

$= 45 - 24$

$= 21$

36. The empirical relation between the mode, median and mean of a distribution is : **(2023)**

(a) Mode = 3 Median – 2 Mean

(b) Mode = 3 Mean – 2 Median

(c) Mode = 2 Median – 3 Mean

(d) Mode = 2 Mean – 3 Median

Solution: (a)

Mode = 3Median – 2Mean.

37. If the mean and mode of a data are 24 and 12 respectively, then is median is **(2025)**

(a) 25 (b) 18

(c) 20 (d) 22

Solution: (c)

Mode = 3Median – 2Mean

$\Rightarrow 12 = 3$Median $- 2 \times 24$

$\Rightarrow 3$Median $=60$

$\therefore$ Median $= 20$

38. If the mode of some observations is 10 and sum of mean and median in 25, then the mean and median respectively are. **(2025)**

(a) 12 and 13 (b) 13 and 12

(c) 10 and 15 (d) 15 and 10

Solution: (a)

mode = 3 median – 2 mean

$\Rightarrow$ 10 = 3 median – 2mean

$\Rightarrow$ 3 median – 2mean = 10 (i)

and median + mean = 25 (ii)

Solving the equation (i) and (ii) we get

median = 12 and mean = 13

39. If mode and median of given set of observations are 13 and 11 respectively, then the value of mean is : **(2025)**

(a) 17 (b) 7

(c) 10 (d) 28

Solution: (c)

Mode = 3 (Median) – 2(Mean)

$\Rightarrow 13 = 3 \times 11 - 2\text{Mean}$

$\Rightarrow$ 2Mean = 20

$\therefore$ Mean = 10

40. If mean and mode of given set of observation are 10 and 13 respectively, then the value of median is **(2025)**

(a) 19 (b) 4

(c) 11 (d) 43

Solution: (c)

Mode = 3 (Median) – 2 (Mean)

$\Rightarrow 13 = 3\text{ Median} - 2 \times 10$

$\Rightarrow$ 3 Median = 33

$\therefore$ Median = 11

41. If mean and median of given set of observations are 10 and 11 respectively, the value of mode is **(2025)**

(a) 10.5 (b) 8

(c) 13 (d) 21

Solution: (c)

Mode = 3 (Median) – 2(Mean)

$= 3 \times 11 - 2 \times 10$

$= 13$

42. What is the mode of a data if median and mean of the same data are 9.6 and 10.5, respectively

(a) 7.8 (b) 12.3

(c) 8.4 (d) 7

Solution: (a)

Mode = 3(Median) – 2(Mean)

$= 3 \times 9.6 - 2 \times 10.5$

$= 28.8 - 21$

$= 7.8$

43. Mode and Mean of a data are $15x$ and $18x$, respectively. Then the median of the data is **(2025)**

(a) x (b) $11x$

(c) $17x$ (d) $34x$

Solution: (c)

Mode = 3(Median) – 2(Mean)

$\Rightarrow 15x = 3\text{ Median} - 2 \times 18x$

$\Rightarrow 3\text{ Median} = 36x + 15x$

$\Rightarrow 3\text{ Median} = 51x$

$\therefore \text{Median} = 17x$

44. The medium and mode of a distribution are 25.2 and 26.1 respectively The mean of the distribution is: **(2026)**

(a) 24.75 (b) 24.25

(c) 24.3 (d) 25.5

Solution: (a)

Mode = 3 median – 2 mean

$\Rightarrow 26.1 = 3 \times 25.2 - 2$ mean

$\Rightarrow 26.1 = 75.6 - 2$ mean

$\Rightarrow$ 2 mean = 49.5

$\Rightarrow$ mean = 24.75

45. The mean and median of a fequency distribution are 43 and 43.4 respectively The mode of the distribution is: **(2026)**

(a) 43.4 (b) 42.4

(c) 44.2 (d) 49.3

Solution: (c)

mode = 3 median – 2 mean

$= 3 \times 43.4 - 2 \times 43$

$= 13.0.2 - 86$

= 44.2

46. If the mean and mode of a data are 12 and 21 respectively, then its median is: **(2026)**

(a) 6 (b) 13.5

(c) 15 (d) 14

Solution: (c)

mode = 3 median – 2 mean

$\Rightarrow 21 = 3 \times \text{median} - 2 \times 12$

$\Rightarrow$ 3 median = 45

$\Rightarrow$ median = 15

47. If for a data, median is 5 and mode is 4, then mean is equal to: **(2026)**

(a) 7 (b) 11

(c) $\frac{11}{2}$ (d) $\frac{14}{3}$

Solution: (c)

mode = 3 median – 2 mean

$\Rightarrow 4 = 3 \times 5 - 2$ mean

$\Rightarrow -2$ mean $= -11$

$\Rightarrow$ mean $= \frac{11}{2}$

48. Mean and median of a frequency distribution are 43 and 40 respectively the value mode is:**(2026)**

Solution: (a)

mode = 3 median – 2 mean

$= 3 \times 40 - 2 \times 43$

= 120 – 86

= 34

Exercise 11.5

(i) The median when it is given that mode and mean are 8 and 9 respectively, is :

(a) 8.33 (b) 8.52

(c) 8.57 (d) 8.67

(ii) If the mean and median of a data are 18 and 16 respectively their mode is

(a) 8 (b) 18

(c) 16 (d) 12

(iii) If mean = (3 median – mode) a, then the value of a is

(a) 1 (b) 2

(c) $\frac{1}{2}$ (d) – 1

(iv) If the mode of some observations is 12 and sum of mean and median is 15, then the mean and median 26, then the mean and median respectively are

(a) 12, 13 (b) 12.32, 13.34

(c) 12.80, 13.20 (d) 12.92, 13.22

49. The cumulative frequency table is useful in determining **(2020)**

(a) mean (b) median

(c) mode (d) all of these

Solution: (b)

Median

50. The time in second, taken by 150 athletes to run a 100 m on hurdle race are tabulated below

Time (sec)	13 – 14	14 – 15	15 – 16	16 – 17	17 – 18	18 – 19
No of Athletes	2	4	5	71	48	20

The number of athletes who completed the race in less than 17 seconds is **(2023)**

(a) 11 (b) 71

(c) 82 (d) 68

Solution: (c)

Time (in sec)	**Cumulative frequency**
less than 14	2
less than 15	6
less than 16	11
less than 17	82
less than 18	130
less than 19	150

∴ The number of athletes who completed the race in less then 17 second is 82

51. If a certain variable x divides a statistical data arranged in order into two equal parts, then the value of x is called the **(2024)**

(a) mean (b) median

(c) mode (d) range

Solution: (b)

Median

52. After an examination, a teacher wants to know the makes obtained by maximum number of the students in her class. She requires to calculate__________ of marks. **(2024)**

(a) median (b) mode

(c) mean (d) range

Solution:(b)

Mode

53. The middle-most observation of every data arranged in order in called **(2024)**

(a) mode (b) median

(c) mean (d) deviation

Solution: (b) Median

54. Which of the following depends on all observations of a given data ? **(2025)**

(a) median (b) mean

(c) Range (d) mode

Solution(b) mean

Exercise 11.6

(i) For a frequency distribution, the mode is the value that has the:

(a) Highest frequency

(b) Lowest frequency

(c) Middle frequency

(d) None of these

(ii) The median class in a grouped data is the class interval with:

(a) Highest frequency

(b) Lowest frequency

(c) Cumulative frequency just greater than $\frac{N}{2}$

(d) Non of these

(iii) Which of the following is not a measure of central tendency

(a) mean (b) mode

(c) median (d) Standrad deviation

(iv) Median =

(a) $\ell+\left\{h\times\frac{\left(\frac{N}{2}-c.f\right)}{f}\right\}$

(b) $\ell+\left\{h\times\frac{\left(c.f-\frac{N}{2}\right)}{f}\right\}$

(c) $\ell-\left\{h\times\frac{\left(\frac{N}{2}-cf\right)}{f}\right\}$

(d) Non of these

(v) Mode =

(a) $x_k+h.\left\{\frac{(f_{k-1}-f_k)}{(2f_k-f_{k-1}-f_{k+1})}\right\}$

(b) $x_k+h\left\{\frac{(f_k-f_{k-1})}{(2f_k-f_{k-1}-f_{k+1})}\right\}$

(c) $x_k+h.\left\{\frac{(f_k-f_{k-1})}{(f_k-2f_{k-1}-f_{k+1})}\right\}$

(d) $x_k+h\left\{\frac{(f_k-f_{k-1})}{(f_k-f_{k-1}-2f_{k+1})}\right\}$

55. Assertion (A): median marks of student in a class test in 16 It means half of the class got marks less than 16.

Reason (R): median divides the distribution in two equal parts **(2025)**

Solution: (a)

Both Assertion(A) and Reason (R) are true and Reason (R) in the correct explanation of Assertion(A).

56. Assertion: If the difference of the mode and median of a data is 24, then the difference of the median and mean is 12.

Reason (R): mode = 3 mean – 2 median **(2025)**

Solution: (d)

mode – median = 24 (Given)

$\Rightarrow$ mode = 24 + median

$\Rightarrow$ mode = 3 median – 2 mean

$\Rightarrow$ 24 + median = 3 median – 2 mean

$\Rightarrow$ 2 median – 2 mean = 24

$\Rightarrow$ median – mean = 12

57. Assertion (A): The mean of first 'n' natural numbers is $\frac{n-1}{2}$

Reason (R): The sum of first 'n' natural numbers is $\frac{n(n+1)}{2}$

Solution: (d) Assertion (A) is false but Reason (R) is true

58. Assertion (A): median of a data is the value of $\frac{N}{2}$, where N represents sum of all frequenices.

Reason (R): median divides the whole distribution in two equal parts **(2026)**

Solution: (d)

Assertion (A) is false but Reason (R) is true.

Exercise 11.7

(i) **Assertion (A):** If the value of mode and mean is 16 and 28 respectively. The median is 24

Reason (R): Median = (mode + 2 mean) / 2

(ii) **Assertion (A):** consider the following frequency distribution

Class	0 – 10	10 – 20	20 – 30	30 – 40	40 – 50	50 – 60
Freque-ncy	8	12	16	21	2	7

The modal class is 30 – 40

Reason(R): The class having the maximum frequency is called the modal class,

(iii) **Assertion (A):** consider the following frequency distribution

Class	11 – 15	16 – 20	21 – 25	26 – 30	31 – 35	36 – 40	41 – 45	46 – 50
Frequ-ency	2	3	6	7	14	12	4	2

The median class is 30.5 – 35.5

Reason(R): If $N = \sum f_i$, then the class when cumulative frequency is just greater than $\left(\frac{N}{2}\right)$ is the median class

(iv) **Assertion (A):** The mean of the data set 2, 3, 5, 7, 11, is 5.6

Reason(R): The mean is calculated by dividing the sum of all observations by the number of observations.

(v) **Assertion (A):** The median of the data 2, 3, 4, 5, 6, is 4.

Reason(R): The median is the middle most value when data is arranged in ascending or descending order

ANSWER KEYS

Exercise – 11.1

(i) (c) 55 **(ii)** (c) 19 **(iii)** (a) 100

Exercise – 11.2

(i) (d) 14 **(ii)** (d) 0 **(iii)** (c) 35

Exercise – 11.3

(i) (c) 9 **(ii)** (a) 9 **(iii)** (b) 9

(iv) (b)24 – 32

Exercise – 11.4

(i) (c) 80 **(ii)** (b) 10 – 20 **(iii)** (d) 15

Exercise – 11.5

(i) (d) 8.67 **(ii)** (d) 12 **(iii)** (c) $\frac{1}{2}$

(iv) (c)12.80, 13.20

Exercise – 11.6

(i) a **(ii)** c **(iii)** d

(iv) a **(v)** b

Exercise – 11.7

(i) c **(ii)** a **(iii)** a

(iv) a **(v)** a

www.ingramcontent.com/pod-product-compliance
Lightning Source LLC
LaVergne TN
LVHW080021210926
848481LV00011B/3073